AF678956

Glioma Cell Biology

Aleksi Sedo • Rolf Mentlein
Editors

Glioma Cell Biology

Springer

Editors
Aleksi Sedo
Biochemistry and Experimental Oncology
Charles University Prague 1st
Faculty of Medicine
Prague
Czech Republic

Rolf Mentlein
Anatomisches Institut
Christian-Albrechts-Universität zu Kiel
Kiel
Germany

ISBN 978-3-7091-1430-8 ISBN 978-3-7091-1431-5 (eBook)
DOI 10.1007/978-3-7091-1431-5
Springer Wien Heidelberg New York Dordrecht London

Library of Congress Control Number: 2014955657

© Springer-Verlag Wien 2014
This work is subject to copyright. All rights are reserved by the Publisher, whether the whole or part of the material is concerned, specifically the rights of translation, reprinting, reuse of illustrations, recitation, broadcasting, reproduction on microfilms or in any other physical way, and transmission or information storage and retrieval, electronic adaptation, computer software, or by similar or dissimilar methodology now known or hereafter developed. Exempted from this legal reservation are brief excerpts in connection with reviews or scholarly analysis or material supplied specifically for the purpose of being entered and executed on a computer system, for exclusive use by the purchaser of the work. Duplication of this publication or parts thereof is permitted only under the provisions of the Copyright Law of the Publisher's location, in its current version, and permission for use must always be obtained from Springer. Permissions for use may be obtained through RightsLink at the Copyright Clearance Center. Violations are liable to prosecution under the respective Copyright Law.
The use of general descriptive names, registered names, trademarks, service marks, etc. in this publication does not imply, even in the absence of a specific statement, that such names are exempt from the relevant protective laws and regulations and therefore free for general use.
While the advice and information in this book are believed to be true and accurate at the date of publication, neither the authors nor the editors nor the publisher can accept any legal responsibility for any errors or omissions that may be made. The publisher makes no warranty, express or implied, with respect to the material contained herein.

Printed on acid-free paper

Springer is part of Springer Science+Business Media (www.springer.com)

Preface

Gliomas are fatal malignant diseases, but also represent excellent models for tumor research with the aim to eventually discover new and appropriate therapeutics against this and other types of cancer. Gliomas are characterized by unregulated growth, apoptosis-resistance, diffuse invasion, strongly increased angiogenesis, and immunosuppression—all hallmarks of other tumor entities, but here focused in a particularly pronounced manner. However, there are also characteristics unique to gliomas, in particular the exceptional brain microenvironment with specialized cells and extracellular matrix. All this results in biological properties of this tumor.

Recently published books dedicated to the glioma are either partially or entirely dedicated to its diagnostic and/or clinical aspects. This book focusses on various aspects of glioma cell biology. They are systematically covered in chapters written by experts in their field, first looking on the "inner space"—biology of the glioma cells themselves, then on their "joint venture"—bidirectional interactions among the microenvironment of the brain tumor, and finally on the experimental models available for glioma research.

The book starts with an overview of the cancer stem cell hypothesis and its implications for gliomas. As gliomas appear to be extremely heterogeneous tumors (the term *glioblastoma multiforme*—now often only glioblastoma—for the most common and malignant form implies this), they have become a paradigm for the tumor stem cell hypothesis (generation of tumor cells from the stem-like cells) versus classical opinions of a clonal origin of tumors (mutations in differentiated cells). The molecular mechanisms driving the malignant phenotype are exemplified in further chapters. Epigenetic changes and the role of microRNAs are summarized, followed by chapters on altered signal transduction mechanisms and the role of apoptosis and autophagy in glioma cells.

The following part is more centered around the complex interactions of glioma cells with the individual cellular partner populations and with the whole extracellular microenvironment of the brain. First, the various types of growth factors mediating autocrine as well as paracrine interactions between tumor and tumor stroma cells are reviewed. Then, the particular interactions between glioma and endothelial cells are highlighted (angiogenesis factors) followed by a chapter on the communications of glioma cells with the immune system. Thus, the three abovementioned hallmarks of gliomas—dysregulated growth, strong vascularization, and immunosuppression—are illuminated on the molecular level. Next to communication factors, the particular cellular and extracellular components of the glioma and brain microenvironment are reviewed. Besides few other cell types, microglia cells/macrophages constitute the dominant stroma cells of this brain tumor. Therefore, the origin, special properties, and the bilateral communication of these cells with glioma cells are highlighted. Often neglected, the unique brain extracellular matrix and the adhesion molecules mediating its interaction with tumor cells are reviewed in two chapters. These are — together with the glioma-associated or produced proteases and motility factors — of exceptional importance for the understanding of the highly invasive character of gliomas. Indeed, the invasion of the brain tissue without clear margins is the reason for the still relatively poor outcome of surgical tumor resection.

Finally in the third part, the preclinical models are introduced, in which newly identified targets can be tested. Two chapters highlight the general methods and special constraints to investigate these fatal brain tumors in animal models.

Thus, the composition of this book follows the general concept: analysis of molecular alterations of malignant cells, viewing this under different hypotheses, explaining the special phenotype of a tumor in its cellular/extracellular milieu of the host organ, extracting putative therapeutic targets (that will be described in different chapters) from these perceptions, and applying these to the preclinical models to cure the patient.

We are aware that we could not review all aspects of glioma cell biology. Besides the fact that our space is limited, all of experts invited hadn't had the time to contribute. Our view on gliomas might of course be influenced by our own research concepts and topics. Additionally, as many of the authors declared, not all relevant works could be cited due to the space (or knowledge) limitations. We therefore apologize for this to all who do not see them adequately mentioned as well as for our mistakes (which surely do occur!).

Nonetheless, we hope that this book will be helpful and encouraging for researchers and physicians in understanding the various aspects of tumor biology, particularly concerning the brain, and this concise information will be another a step in the combat against these diseases.

Prague, Czech Republic
Kiel, Germany
May 2014

Aleksi Sedo
Rolf Mentlein

Contents

List of Contributors

Irene Appolloni IRCCS AOU San Martino – IST, Genoa, Italy

Andrej Besse Masaryk Memorial Cancer Institute, Central European Institute of Technology, Brno, Czech Republic

Petr Busek First Faculty of Medicine, Institute of Biochemistry and Experimental Oncology, Charles University in Prague, Prague, Czech Republic

Filippo Calzolari Institute for Stem Cell Research, National Research Center for Environment and Health, Neuherberg, Germany

Mujeeburahim Cheerathodi Department of Neurosurgery, Unit 1004 S5.8136c, University of Texas M.D. Anderson Cancer Center, Houston, TX. USA

Selby Chen Department of Neurological Surgery, Mayo Clinic, Rochester, MN, USA

Iwona Anna Ciechomska Laboratory of Molecular Neurobiology, Nencki Institute of Experimental Biology, Warsaw, Poland

Moneeb Ehtesham Department of Neurosurgery, Vanderbilt University Medical Center, Nashville, TN, USA

Aleksandra Ellert-Miklaszewska Laboratory of Molecular Neurobiology, Nencki Institute of Experimental Biology, Warsaw, Poland

Takuya Furuta Department of Neurosurgery, Graduate School of Medical Science, Kanazawa University, Kanazawa, Japan

Jun-Ichiro Hamada Department of Neurosurgery, Graduate School of Medical Science, Kanazawa University, Kanazawa, Japan

Petra Hamerlik Brain Tumor Biology, Danish Cancer Society Research Center, Copenhagen, Denmark

Kirsten Hattermann Department of Anatomy, University of Kiel, Kiel, Germany

Yutaka Hayashi Department of Neurosurgery, Graduate School of Medical Science, Kanazawa University, Kanazawa, Japan

Bozena Kaminska Laboratory of Molecular Neurobiology, Nencki Institute of Experimental Biology, Warsaw, Poland

Imad Saeed Khan Department of Neurosurgery, Vanderbilt University Medical Center, Nashville, TN, USA

Daisuke Kita Department of Neurosurgery, Graduate School of Medical Science, Kanazawa University, Kanazawa, Japan

Evzen Krepela First Faculty of Medicine, Institute of Biochemistry and Experimental Oncology, Charles University in Prague, Prague, Czech Republic

Paolo Malatesta Department of Experimental Medicine (DIMES), University of Genoa, Genoa, Italy

Christian Mawrin Research Group for Perisynaptic Extracellular Matrix, Leibniz Institute for Neurobiology, Magdeburg, Germany

Joseph H. McCarty Department of Neurosurgery, Unit 1004 S5.8136c, University of Texas M.D. Anderson Cancer Center, Houston, TX, USA

Rolf Mentlein Department of Anatomy, University of Kiel, Kiel, Germany

Mitsutoshi Nakada Department of Neurosurgery, Graduate School of Medical Science, Kanazawa University, Kanazawa, Japan

Vivaldo Moura Neto Instituto de Ciências Biomédicas, Universidade Federal do Rio de Janeiro, Rio de Janeiro, RJ, Brazil

Ian F. Parney Department of Neurological Surgery, Mayo Clinic, Rochester, MN, USA

Martin Prevorovsky Faculty of Sciences, Department of Cell Biology, Charles University in Prague, Prague, Czech Republic

Vinay K. Puduvalli Division of Neuro-oncology, Arthur G. James Cancer Hospital and The Richard J. Solove Research Institute, The Ohio State University Wexner Medical Center, 320 W 10th Avenue, Starling Loving Hall Suite M410, Columbus OH 43210

Jiri Sana Masaryk Memorial Cancer Institute, Central European Institute of Technology, Brno, Czech Republic

Aleksi Sedo First Faculty of Medicine, Institute of Biochemistry and Experimental Oncology, Charles University in Prague, Prague, Czech Republic

Constanze Seidenbecher Research Group for Perisynaptic Extracellular Matrix, Leibniz Institute for Neurobiology, Magdeburg, Germany

Ondrej Slaby Masaryk Memorial Cancer Institute, Central European Institute of Technology, Brno, Czech Republic

Takuya Watanabe Department of Neurosurgery, Graduate School of Medical Science, Kanazawa University, Kanazawa, Japan

Part I

The Inner Space: Molecular Mechanisms Driving the Malignant Phenotype of Glioma Cells

Cancer Stem Cells and Glioblastoma

1

Petra Hamerlik

Abstract

Gliomas are tumors of astroglial origin and the World health Organization (WHO) classifies them based on histological criteria into four grades of ascending malignancy. *Glioblastoma multiforme* (GBM, WHO grade IV) is among the most lethal of human cancers with conventional therapy offering only palliation. GBM accounts for the most frequent type of primary brain tumors in Europe and the USA, comprising more than a half of all gliomas, with a 5-year survival of patients of no more than 5 %. Despite concerted efforts and advances in currently available therapies, the expected survival of GBM patients remains dismal. Highly infiltrative character renders complete surgical resection impossible and together with notoriously known radio- and chemoresistance accounts for high recurrence rates and mortality of nearly 95 %. Traditionally, the approach to cancer treatment has been to eradicate all of the cancerous cells to achieve "cure" and was based on the idea that the vast majority of cells have tumorigenic potential. One reason for the lack of clinical advances is the lack of understanding of the GBM biology in general and the cellular origin of this disease in particular. The cancer stem cell hypothesis postulates that cancers contain a subset of highly aggressive cells that propagate and maintain the tumors through unlimited self-renewal and potent tumorigenicity. Within GBM, a distinct population of $CD133^+$ cells has been documented to display stem cell properties in vitro, in particular self-renewal, unlimited proliferative potential, capacity for multi-lineage differentiation, and recapitulation of patient's phenotype upon orthotopic implantation in immunocompromized host. The investigation and study of cancer stem cells received enormous

P. Hamerlik (✉)
Brain Tumor Biology, Danish Cancer Society Research Center, Strandboulevarden 49, 2100-DK Copenhagen, Denmark

Laboratory of Genome Integrity, Institute of Molecular and Translational Medicine, Palacky University, Hnevotinska 3, Olomouc 775 15, Czech Republic
e-mail: pkn@cancer.dk

A. Sedo and R. Mentlein (eds.), *Glioma Cell Biology*,
DOI 10.1007/978-3-7091-1431-5_1, © Springer-Verlag Wien 2014

attention over the past decade, yet it's relevance to therapeutic resistance remains controversial. Although the cancer stem cell hypothesis may have multiple implications in therapeutic management of glioblastoma, as well as other brain tumor malignancies, caution must be exercised as targeting a rare population of tumorigenic cells without consideration of the largely heterogeneous tumor bulk comprised of proliferative cells may not change overall patient survival.

Keywords
Heterogeneity • Self-renewal • Therapeutic resistance • GBM • Microenvironment

Abbreviations

ABC	ATP-binding cassette
bFGF	Basic fibroblast growth factor
Bmi1	B lymphoma Mo-MLV insertion region 1 homolog
CSC	Cancer stem cell
DDR	DNA damage response
EGF	Epidermal growth factor
GBM	*Glioblastoma multiforme*
IR	Ionizing radiation
NSC	Neural stem cell
PTC1	Patched-1
RTK	Receptor tyrosine kinase
Shh	Sonic hedgehog
Smo	Smoothened
TCGA	The cancer genome atlas
TIC	Tissue-initiating cell
TMZ	Temozolomide
VEGF	Vascular endothelial cell growth factor
WHO	World Health Organization

1.1 Cancer Stem Cell Hypothesis

The adult human brain has been for many years thought to be a static, fully differentiated organ. Today, it is generally accepted that both neural stem cells (NCS) and glial progenitor cells in multiple regions of the adult brain persist throughout life. The self-renewing and multipotent neural stem cells (NSCs) have been isolated from subventricular zone, the lining of the lateral ventricles, the dentate gyrus and the hippocampus, as wells as the subcortical white matter (Doetsch et al. 1997; Fukuda et al. 2003; Gage 2000; Kim and Morshead 2003).

While NSCs comprise a relatively quiescent cell population, these cells have the potential to proliferate and migrate extensively, characterizing the adult brain as dynamic system with surprisingly high plasticity (Altman and Chorover 1963; Altman and Das 1965; Doetsch et al. 1999). NSCs have been associated with tissue repair after stroke and severe injuries, and have been suggested as tools for treatment of neurological disorders, such as Alzheimer's disease. In the light of these facts, cancer can be considered organ system with aberrant activation of developmental and wound response pathways (Rich 2008; Rich and Eyler 2008). Recent evidence suggests that within the heterogeneous tumor mass, there is a cell subpopulation with the unique capacity for sustained self-renewal and tumor propagation in vivo.

Historically, the approach to cancer treatment has been to eradicate all cancerous cells, where individual cells are equal in respect to their potential to proliferate, self-renew, and drive tumor growth. This notion, known as the stochastic or clonal evolution model (Fig. 1.1a) of tumorigenesis, proposes that a transformed single cell gains unlimited proliferative capacity (Chen et al. 2010; Li et al. 2007a; Shackleton et al. 2009). During early stages of tumorigenesis, a single or very few cells transform, where "pro-survival" mutations allow for clonal expansion of the "fittest" cells, resulting in a symbiotic coexistence of various subpopulations within the heterogeneous tumor mass. Importantly, during the lifetime of the tumor, any of the cancer cells can participate in tumor progression or develop resistance resulting in disease recurrence. This model has been challenged by the recently revived hierarchical model or the cancer stem cell hypothesis (Reya et al. 2001; Rich 2008; Sanai et al. 2005). The cancer stem cell hypothesis (Fig. 1.1b) postulates that there is a rare subpopulation of cancer cells with stem-like cell properties, including the ability to self-renew, that gives rise to multi-differentiated progeny and sustained proliferation. In contrast to stochastic model, the multipotent nature of these cells results in heterogeneity within tumor as a result of aberrant differentiation and epigenetic modification of the progeny, whereas the vast majority of progeny does not contribute to tumor growth and recurrence after therapeutic intervention.

This concept is not new, as already in 1855, Rudolph Virchow followed by Julius Cohnheim proposed (Rahman et al. 2011) that cancer develops from activation of dormant embryonic-tissue remnants. Their observations were based on the histological similarities (proliferation index and degree of differentiation) between fetal tissues and cancer. In the 1960s, ethically questionable experiments performed by Brunschwig, Southam, and Levin demonstrated a low frequency of tumor initiation when tumor cells harvested from patients with malignancy were injected subcutaneously into the same or different patients (Brunschwig et al. 1965). According to their results, tumors were formed only when more than 10^6 cells were injected. This and further reports showing (Bruce and Van Der Gaag 1963; Brunschwig et al. 1965) the clonogenic potential of lymphoma cells in vivo lead to hypothesis that tumor growth may be initiated and maintained by a minority of cancer cells, not the entire population. In 1994, John Dick and colleagues published their seminal findings that human acute myeloid leukemia is organized as a hierarchy that originates from a primitive hematopoietic cell and this report turned into a

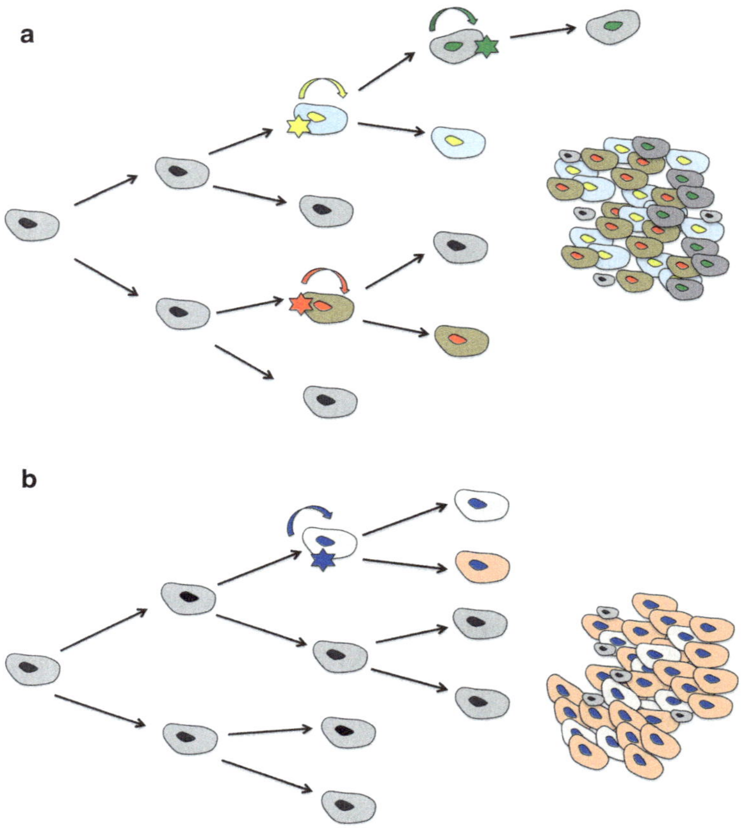

Fig. 1.1 Different models of tumorigenesis: stochastic or clonal evolution model (**a**) versus hierarchical model or the cancer stem cell hypothesis (**b**)

paradigm for later studies, which proposed the existence of a similar model for solid tumors (Lapidot et al. 1994). The cancer stem cells (CSCs) hypothesis and clonal evolution model do not contradict each other, instead highlight the importance of abnormal differentiation program in tumorigenesis, thereby suggesting a key role of cellular hierarchy in tumor evolution.

1.2 Evidence of CSCs in Gliomas

Traditionally, gliomas have been thought to originate from a parenchymal differentiated glial cell, which undergoes a series of genetic alternations accompanied by a dedifferentiation process (Jellinger 1978). The persistence of proliferative pool of NSCs and progenitor cells in the adult brain encouraged multiple investigators to evaluate these as putative cells of origin (Ligon et al. 2007; Noble and Dietrich 2002; Pardal et al. 2003). Park et al have already suggested the existence of CSCs in 1971 (Stopschinski et al. 2013). The concept of CSCs was first extended to brain tumors by Ignatova et al. (2002), who isolated clonogenic, neurosphere-forming stem-like cells from human GBM (Ignatova et al. 2002). GSCs isolated using neurosphere culture were subsequently shown to differentiate into multi-lineage progeny and formed tumors transplantation (Rahman et al. 2011). In 2003, studies by Singh et al. showed that cancer-initiating cells are enriched in the $CD133^+$ population and injection of only as many as 100 $CD133^+$ cells initiated tumor whereas $CD133^-$ could not, even when 10,000 cells were injected (Singh et al. 2003, 2004). These reports were followed by outburst of similar studies in GBM as well as other solid tumors (Rich and Eyler 2008; Strauss et al. 2012; Yan et al. 2013). GBM stem cells share many characteristics with NSCs, such as self-renewal, neurosphere formation, marker expression, multilineage differentiation, high motility, and localization in highly specialized "stem cell" niche (Gilbertson and Rich 2007; Kim and Morshead 2003; Reya et al. 2001). However, the term "stem cell" in gliomas refers to their function, not their origin. The true "cell of origin" has not yet been identified. The proper terminology remains unsettled, which limits our ability to effectively communicate the precise meaning of these labels and inform literature searches (Rich 2008), with most groups using the term cancer stem cells (CSCs), tumor-initiating/propagating cells (TICs), or cancer stem-like cells. Although controversial, the concept of cancer stem cell hypothesis recognizes the intra-tumoral heterogeneity and provides a novel framework to study tumor biology as wells as resistance of aggressive and genetically unstable cancer cells to current treatment in GBM.

1.3 Cancer Stem Cells: Identification and Quantitation

1.3.1 CSC in the Context Inter-/Intra-tumoral Heterogeneity

The defining features of CSCs in GBM are evolving and prospective functional enrichment approaches poses challenges. There are ongoing disputes across the field due to technical variance and lack of universal markers, suggesting that there is not a single marker to identify CSCs, probably due to inter- and intra-patient heterogeneity or lack of absolute fingerprints. The Cancer Genome Atlas initiative (TCGA) and GBM genome sequencing efforts resulted in integration of multidimensional genomic data and molecular classification of glioblastoma into proneural, classical, and mesenchymal subtypes that have potential implications for patient prognosis and therapeutic management (Phillips et al. 2006; Verhaak et al. 2010; Verhaak and Valk 2010). The evidence of inter-patient heterogeneity fueled numerous studies, some of them investigating the differences of CSCs pools in individual subtypes (Bhat et al. 2013; Mao et al. 2013).

The intra-tumoral heterogeneity is not only caused by the macroscopic localization of the tumor (i.e., brain stem versus brain hemispheres) (Schonberg et al. 2013; Stopschinski et al. 2013) but also by the microenvironmental factors (hypoxia, acidosis) within the tumor as such. The multi-lineage differentiation capacity of CSCs into committed progenitors or terminally differentiated cells further enhances the degree of intra-tumoral heterogeneity, which may in turn contribute to the inter-patient variability and support the notion of molecular subclassification of glioblastoma (Huse et al. 2011; Lottaz et al. 2010; Mao et al. 2013; Verhaak et al. 2010). $CD133^+$ CSCs isolated from proneural (Joo et al. 2008; Lottaz et al. 2010) have shown to share similarities with $CD133^-$ cells in mesenchymal GBM subtype. Recently, Ichiro et al. reported the differential properties of CSCs isolated from proneural and mesenchymal subtypes of GBM in respect to their metabolic demands and resistance to ionizing radiation. Albeit these findings suggest that the heterogeneity of CSCs corresponds to the heterogeneity of GBM, where unfortunately a functional and experimental proof is still lacking (Huse et al. 2011).

1.3.2 Technical Aspects of CSC Identification and Isolation

The most popular way of identification and isolation of CSCs in GBM uses specific cell surface markers, such as CD133, CD15, CD44, and others or their combinations (Cheng et al. 2010; Hambardzumyan et al. 2008; Persano et al. 2013). Other approaches take advantage of the known overexpression of multidrug resistance genes (encoding pumps like ABCG2) responsible for exclusion of Hoechst 33342 dye and marking so-called side population which has been postulated to be enriched in CSCs. Novel and not as much utilized technique is the use of Aldefluor assay (based on the measurement of alcohol dehydrogenase 1 activity; ALDH1). Both "side population" and Aldefluor assay are imperfect, as they require substantial in vitro culture time (Rich 2008).

The first and most critical technical problem may occur already at the stage of tumor dissociation. The time post-surgery and the method of enzymatic tissue digestion dictate the recovery rates of most CSC-specific surface antigens and account for viability rates later affecting the engraftment efficiencies in immunocompromized hosts. Despite the progresses in defining proper cell culture media allowing for CSC maintenance, extensive passaging in vitro influences the metabolic and expression prolife, as well as genomic stability of cultured cells (Eyler and Rich 2008; Lee et al. 2006). The growth of derived CSCs in the mouse environment might be more complex and contribute to technical difficulties associated with their maintenance. Upon biopsy dissociation and derivation of a primary sphere culture, one practically strips off any cells and components of the supportive microenvironment, which can't be properly substituted and so only cells with minimal dependence on their "niche" survive and will be propagated in vitro or in vivo (Rahman et al. 2011). The frequency of CSCs is often quantitated based on the number of cells, which are tumorigenic, when transplanted into immunocompromized mice. Although this method has been considered as the most reliable, it seems that one may greatly underestimate the frequency of CSCs depending on the animal model used (nude mice vs. SCID mice or SCID with no residual immunity). Using NOD/SCID interleukin-2 receptor gamma chain null mice, Quintana et al. (2008) found that 27 % of melanoma cells could form a tumor with a single cell transplant, suggesting that these cells are much more common, at least in some human tumors than anticipated (Quintana et al. 2008).

1.3.3 CSC Markers

1.3.3.1 CD133

CD133 is a penta-membrane glycoprotein (also known as Prominin-1) and was first discovered as a cell surface marker for hematopoietic stem cells (Miraglia 1997). Uchida et al. (2000) have described its expression in human fetal brain as a marker for neural stem cells (Uchida et al. 2000). Very little is known about the cellular function of CD133 (Fargeas et al. 2003). CD133 knockout mice manifest with a progressive photoreceptor degeneration resulting in vision loss (Zacchigna et al. 2009). Several groups have reported that CD133 is a marker of poor survival in astrocytomas (Beier et al. 2008; Hermansen et al. 2011; Joo et al. 2008; Mak et al. 2011; Zeppernick et al. 2008), with CD133+ cells localized to clusters near vascularized regions or as solitary cells invading non-neoplastic brain parenchyma. It has been demonstrated that the expression of CD133 is cell cycle-dependent and may be upregulated by hypoxia and acidosis (Beier et al. 2007; Jaksch et al. 2008). The biological function of CD133 in CSCs biology remains elusive; however, a recent report by Wei et al. (2013) implies its pro-survival role upstream from phosphotidylinositol 3-kinase (PI3K)/Akt kinase signaling (Wei et al. 2013).

1.3.3.2 CD15

CD15 is (also known as SSEA-1 and Lewis-X Antigen) a carbohydrate adhesion molecule associated with glycolipids and glycoproteins. First reports have shown its expression in NCS derived from human embryonic stem cells and embryonic neural stem cells (Barraud et al. 2007; Pruszak et al. 2007). This marker has been used as an alternative to CD133 in identifying GBM-derived CSCs and the frequency of expression varied from 2.4 to 70 % (Son et al. 2009). Later reports indicate that CD15 labels actively proliferate progenitor rather than cancer stem cells (Cheng et al. 2010).

1.3.3.3 CD44

The CD44 proteins form a ubiquitously expressed family of cell surface adhesion molecules involved in cell–cell and cell–matrix interactions (Ishimoto et al. 2010; Jin et al. 2006). CD44 is expressed in multiple tumors types as well as normal tissues where it functions in the regulation of cell proliferation, cell migration, transmission of survival signals, and other cell–cell and cell–matrix interactions. GBM tumor initiation was reported to be attenuated by targeting TGF-β and its receptor CD44 in GBM-derived CSCs localized in vascular niche in vivo (Anido et al. 2010). Another reports by Jijiwa et al. (2011) and Mao et al. (2013) propose CD44 as a complementary marker to CD133 for CSCs identification and isolation in mesenchymal GBM subtype (Jijiwa et al. 2011; Mao et al. 2013).

1.3.3.4 CSC Transcription Factors

Central to regulation of survival, maintenance, self-renewal, and transduction of extracellular signals from cellular microenvironment into CSCs are these transcription factors: Sox2, Oct4, Nanog, c-Myc, Olig2, and Bmi1 (Schonberg et al. 2013; Stopschinski et al. 2013; Yan et al. 2013; Zhou et al. 2009). Increased Oct4 expression correlates with the degree of malignancy in gliomas (Hambardzumyan et al. 2008; Schonberg et al. 2013) and its inhibition leads to decreased sphere formation and differentiation of CSCs (Ikushima et al. 2011). Sox2 and Nanog interact with Oct4 and so contribute to CSC tumor-initiating capacities. c-Myc has been for decades known as oncogene with high frequency of genomic as well as regulatory alternations contributing to cancer progression (Sheiness et al. 1978; Vennstrom et al. 1982). shRNA-mediated knockdown of c-Myc in CSCs lead to abrogation of tumor initiation in orthotopic GBM model, demonstrating the importance of c-Myc for CSCs tumorigenecity and maintenance (Wang et al. 2008). Olig2 has been long known as a basic helix-loop-helix (bHLH) transcription factor in CNS (Dimou et al. 2008) with functions in the oligodendroctye lineage as well as multipotential neuron/glia progenitor maintenance (Zhu et al. 2012). Of the $CD133^+$ subpopulation of GBM cells, nearly 98 % are positive for Olig2, which is crucial for their proliferation and cell cycle progression (Ligon et al. 2007). Bmi1 is a polycomb group protein (component of the Polycomb Repressive Complex 1) belonging to epigenetic silencers with crucial function during embryonic development (Acquati et al. 2013). Bmi1 has been found enriched in CSCs and is required for their self-renewal (Facchino et al. 2010) and its inhibition leads to radiosensitization of CSCs.

1.4 Pathways Regulating CSCs

1.4.1 Notch Signaling

Notch proteins include four transmembrane receptors, which mediate cell–cell communication as well as cellular proliferation, differentiation, and apoptosis (Schonberg et al. 2013; Wang et al. 2012). There are five ligands that bind Notch receptors: Delta-like 1, 3, 4 and Jagged-1, -2 (Ohishi et al. 2002; Schonberg et al. 2013). The activation of Notch requires sequential proteolytic cleavages by the γ-secretase complex to release its intracellular domain (NICD) and translocate it from membrane to nucleus (Cheng et al. 2010). Notch signaling promotes the proliferation of normal neural stem cells and is indispensable for maintenance of neural progenitors both in vitro and in vivo. Inhibiting Notch by a γ-secretase inhibitor (GSI-18) induces CSC differentiation and apoptosis (Fan et al. 2006) and sensitizes CSCs to radiation (Guo et al. 2009; Purow et al. 2005; Radtke and Raj 2003; Ronchini and Capobianco 2001). Additionally, in a K-Ras-induced murine gliomas model, Notch activates intermediate filament protein and stem cell marker, nestin, further supporting Notch role in maintaining the stem cell phenotype of GBM-derived CSCs (Shih and Holland 2006). Other Notch regulators like Delta/Notch-like epidermal growth factor-related receptor (DNER) and the Notch ligand Delta-like 4 (DLL4) have also been reported to regulate GBM growth and progression (Li et al. 2007b; Sun et al. 2009).

1.4.2 Wnt/β-Catenin Signaling

The canonical Wnt signaling cascade is one of the key regulators in embryonic and adult stem cells. Wnt proteins bind to cell surface receptors of the Frizzled family and their activation leads to nuclear accumulation of nuclear β-catenin, which promotes transcription of multiple target genes including c-Myc and cyclin D1 (Pu et al. 2009; Tanaka et al. 2013). In brain, the Wnt pathway regulates development, in particular the proliferation and self-renewal of NSCs and progenitors cells in fetal ventricular zone, the postnatal subventricular zone, and hippocampus (Kalani et al. 2008; Nusse et al. 2008). Primarily, Wnt signaling and its alternations were linked to medulloblastoma, but recent reports indicate that Wnt/β-catenin pathway may be associated with GSCs maintenance and resistance to ionizing radiation (Chen et al. 2007; Woodward et al. 2007).

1.4.3 Sonic Hedgehog Signaling

Sonic hedgehog (Shh) is a key regulator of cell fate determination and proliferation of adult stem cell including neural stem cells. Upon binding of Shh to its associated transmembrane receptor Patched-1 (PTC1), membrane protein Smoothened (Smo) gets released and activates Gli transcription factors. Once Gli is activated, it translocates into nucleus, where it induces or represses the transcription of downstream genes, such as Wnt, IGF, and PDGFR-α, c-Myc, and cyclin D1 (Dietrich et al. 2008, 2010). Shh pathway abrogation has been reported to deplete CSCs in GBM and increase their radio-resistance (Clement et al. 2007).

1.4.4 Phosphotidylinositol 3-Kinase/Akt Signaling

Receptor tyrosine kinases (RTKs) transduce oncogenic signaling from growth factors, among others Epidermal Growth Factor (EGF) and basic Fibroblast Growth Factor (bFGF), two mitogens commonly used to propagate CSCs in vitro (Lee et al. 2006). Among the most studied and frequently mutated in GBM is the EGFR-mediated growth signaling through PI3K/Akt kinase (Rich 2008; Tanaka et al. 2013). Malignant gliomas, GBM in particular, frequently display EGFR amplifications and/or constitutive activation of EGFRvIII variant that result in elevated PI3K/Akt signaling. Transducing primary astrocytes with c-myc and Akt induces tumorigenicity and increases expression of several stem cell markers (Schonberg et al. 2013). Interestingly, CD133 was found to directly interact with p85 regulatory subunit of PI3K, where knockdown of CD133 resulted in decreased

PI3K/Akt signaling and ultimately reduced CSCs self-renewal and tumorigenicity (Wei et al. 2013). This finding is in concordance with previous studies, where inhibition of Akt disrupted CSC invasion potential, proliferation, and maintenance in vitro and in vivo, primarily by increasing the rates of apoptosis (Eyler et al. 2008; Gallia et al. 2009).

1.5 Radio- and Chemo-resistance of CSCs

Standard of care treatment in GBM currently involves the use of both ionizing radiation (IR) and DNA-alkylating agent temozolomide (TMZ). In GBM, DNA damage responses (DDR) were shown preferentially activated in the pool of $CD133^+$ CSCs when compared to their negative counterparts, possibly contributing to lower rates of apoptosis after IR. Moreover, IR treatment of mice bearing orthotopic GBM tumors resulted in enrichment of CSCs. CSC were shown to have higher metabolic activity (measured by ATP production) resulting in higher reactive oxygen species levels (ROS) and consequently higher level of oxidative damage to DNA (Venere et al. 2014). Intriguingly, reports on actual DNA repair efficiency of CSCs versus non-CSCs are discrepant, most probably due to technical issues accompanying CSC isolation and maintenance in vitro (McCord et al. 2009b; Ropolo et al. 2009). The cell surface adhesion protein and GSC marker, L1Cam (CD171), further enhances the DDR activation via direct regulation of NBS1 and ATM/Chk1/Chk2 pathway in response to IR-induced double strand DNA breaks (Cheng et al. 2011). The polycomb group protein, Bmi1, represents additional levels of DDR response regulation. Bmi1 contributes to radioresistance of CSCs by remodeling the chromatin structure, which leads to impairment of repair factor recruitment to damaged DNA (Facchino et al. 2010).

Chemotherapeutic management of GBM has undergone considerable changes in the last two decades. Since the late 1970s, alkylating substances such as nimustine (ACNU), carmustine (BCNU), and lomutine (CCNU) were the main choices (Beier et al. 2011). Introduction of TMZ as standard treatment in addition to radiotherapy and surgical resection improved both the overall survival and progression-free survival in patients with newly diagnosed GBM (Stupp et al. 2005). Compared to non-CSCs, CSCs exhibit significantly higher expression of O6-methylguanine-DNA-methyltransferase (MGMT), which makes them more resistant to TMZ treatment (Binello and Germano 2011). In addition to TMZ, GBM-derived CSCs are more resistant to several other chemotherapeutic agents, including carboplatin, paclitaxel, and etoposide (Capper et al. 2009; Liu et al. 2006).

Notably, methylations and other epigenetic modifications may also impair the effect of chemotherapeutics; for examples, CSCs have a hypermethylated caspase-8 promoter that renders them resistant to therapies utilizing the tumor necrosis factor-related apoptosis-inducing ligand (TRAIL) pathway (Capper et al. 2009). The most straightforward mechanism that may be actively contributing to CSCs' chemoresistance is the overexpression of ABC (ATP-binding cassette) transporters (Bleau et al. 2009a, b). Moreover, their overexpression correlates with the levels of several stem-cell markers such as CD133, nestin (Yamamoto et al. 2009), or Notch-

1 and Nanog (Bourguignon et al. 2008; Patrawala et al. 2005). Ectopic expression of CD133 results in an upregulation of ABC transporter upon treatment of CSCs with anti-cancer drugs, camptothecin and doxorubicin (Angelastro and Lame 2010). Interestingly, a report by Venere et al. (2014) reported preferential sensitivity of CSCs (compared to non-CSCs) to olaparib—a potent small molecule inhibitor of the nuclear enzyme poly(ADP-ribose) polymerase (PARP) (Venere et al. 2014). It acts by binding to PARP, inhibiting PARP-mediated repair of single strand DNA breaks. Furthermore, inhibition of PARP sensitized CSCs to IR, albeit opening a new window for therapeutic intervention in glioblastoma.

1.6 Microenvironmental Regulation of CSC

Over 30 years ago, Schofield (1978) proposed the existence of unique spatially defined regions within each tumor, which were suspected to provide factors necessary to the survival and development of cells capable to regenerate tissues in adult organisms (Schofield 1978). A number of studies have clearly demonstrated that the so-called niche directs proliferation, differentiation of cells, and as such constitutes a key regulator of stem-cell fate (Blanpain and Fuchs 2006; Calvi et al. 2003; Fuchs et al. 2004). The key role played by the local microenvironment in the initiation and progression of tumors is becoming increasingly clear.

A series of recently published studies have shown that aberrant vascular stem cell niches, reminiscent of those observed in normal brain, exist in glioblastoma and other types of brain tumors and support CSCs (Gilbertson and Rich 2007). Calabrese et al. (2007) provided convincing data that brain tumors orchestrate vascular niches that maintain CSC pool and disruption of these ablates the fraction of self-renewing tumor cells ultimately leading the tumor growth arrest (Calabrese et al. 2007). Bao et al. showed that CSCs secrete high levels of pro-angiogenic cytokines, among the most abundant being Vascular Endothelial Growth Factor (VEGF)—a key factor in tumor angiogenesis (Bao et al. 2006). They have demonstrated that freshly resected CSCs, but not non-CSCs, human glioblastoma cells readily form highly vascular and hemorrhagic tumors in vivo (Bao et al. 2006). Furthermore, cultures enriched for CSCs induced higher levels of endothelial progenitor cell proliferation, recruitment, and mobilization compared to non-CSC cultures. When VEGF or stromal-derived factor 1 (SDF-1, CXCL12) is inhibited in CSCs, all aspects of angiogenesis were dramatically reduced (Bao et al. 2006; Folkins et al. 2009). Hamerlik et al. (2012) have reported autocrine VEGF/VEGFR2 signaling as an evasion tool of CSCs to anti-angiogenic therapy (Hamerlik et al. 2012; Knizetova et al. 2008). Knocking-down VEGFR2 not only decreased VEGF secretion, but it significantly decreased CSCs' viability, self-renewal potential, and tumorigenicity. Moreover, abrogation of VEGFR2 signaling sensitized GBM cells to IR (Hamerlik et al. 2012; Knizetova et al. 2008). Wang et al. (2010) and Ricci-Vitiani et al. (2010) proposed the neoplastic origin of tumor endothelium and CSC as a possible source to endothelial progenitors (Ricci-Vitiani et al. 2010; Wang et al. 2010). In contrast, Cheng et al. (2013) have shown CSCs to

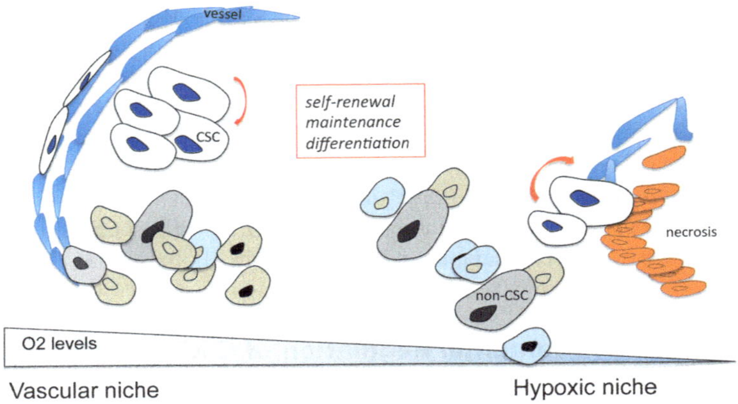

Fig. 1.2 The vascular and hypoxic niches in gliomas as drivers for cancer stem cell self-renewal, maintenance, and differentiation, see text

differentiate into pericytes supporting vessel function and so tumor growth (Cheng et al. 2013). Collectively, this data add yet another complexity to CSC–vascular niche interaction.

Hypoxia has been reported to play a crucial role in the maintenance and regulation of NSC, leading to a recognition of a "hypoxic nice" (Mazumdar et al. 2010; Roitbak et al. 2008). Evans et al. analyzed oxygenation of normal brain and gliomas (Evans et al. 2004, 2008). Their measurements showed that physiological oxygen levels in healthy brain range between 12.5 and 2.5 %, while GBM tumors showed mild to moderate/severe hypoxia with oxygen tensions ranging between 2.5 and 0.1 %. Several reports have established key regulatory functions of the hypoxic niche in CSC maintenance and survival (Bar 2011; Heddleston et al. 2009; Li et al. 2009a). Expression of several of stem cell markers (e.g., CD133, A2B5, Nestin, Oct-4, and Sox2) is upregulated (Li et al. 2009a, b; McCord et al. 2009a; Seidel et al. 2010), whereas the expression of differentiation markers (GFAP) is downregulated under hypoxia. Treatments that would efficiently disrupt aberrant tumor niche(s) would therefore prove active against glioblastoma (Gilbertson and Rich 2007).

Conclusion and Future Perspectives

The emergence of cancer stem cells and recent advances in our understanding of signaling pathway crucial to their self-renewal, survival, and tumorigenic potential have led to the development of novel targeted therapies, currently being evaluated in clinical trials (Sathornsumetee and Rich 2008; Schonberg et al. 2013; Tanaka et al. 2013). Phase I and phase II clinical trials using γ-secretase inhibitors to inhibit Notch signaling or an oral hedgehog antagonist (vismodegib) are ongoing (Tanaka et al. 2013). The current challenge is to develop experimental models encompassing the complexity of this highly heterogeneous malignancy (Fig. 1.2), including three-dimensional cyto-

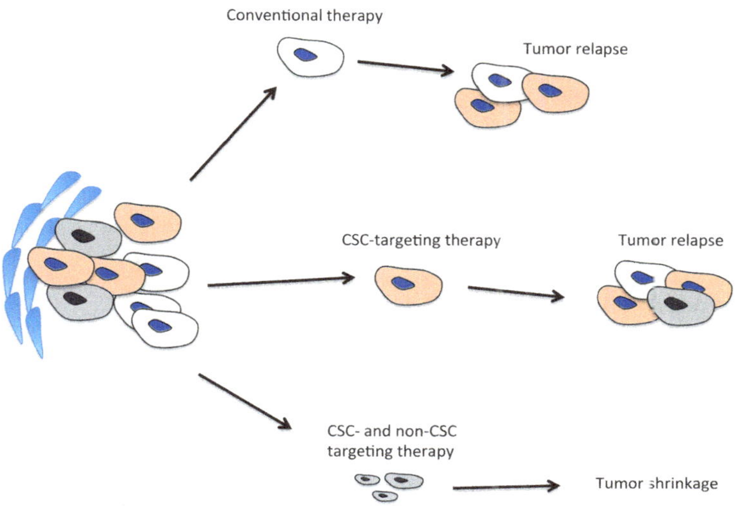

Fig. 1.3 Therapeutic implication of cancer stem cell hypothesis. Conventional therapies fail to target CSCs and often result in tumor relapse. CSC-targeting therapies delay significantly the tumor growth, but due to high plasticity of remaining tumor cells (hypoxia/acidosis-induced stemness), recurrence is inevitable. Treatment modalities combining the CSC-targeting and conventional approaches may be beneficial for successful therapeutic intervention in gliomas

architecture, vascular and hypoxic niches, and stromal component contribution, as targeting every single component may represent exciting new approach for cancer treatment.

Despite great advances in glioblastoma treatment modalities, the effects on patient survival are dismal. Although many compounds demonstrated strong efficacy in preclinical studies, very few of them showed similar effect in clinical trials, due to negligible antitumoral activity and/or severe side effects, which might be reflecting the intra- as well as inter-tumoral heterogeneity common to GBM. For this reason, a better understanding of CSC biology within the complex tumor microenvironment and its interplay with resistance to currently available therapies must be improved. Scientists advocating CSC hypothesis suggested CSC-directed therapy to be the most likely successful cancer treatment. Because of the high GBM cell plasticity, the capacity of non-CSCs to redifferentiate into CSC, only targeting cancer stem cells would ultimately lead to disease recurrence (Eyler and Rich 2008; Persano et al. 2013; Rich 2008). Therefore, a combined treatment targeting both CSCs and their differentiated progeny is more likely to be efficient approach in glioblastoma management (Fig. 1.3).

Acknowledgments The Danish Council for Independent Research/Medical Sciences ID4765/11-105457; Czech Ministry of Health (NT11065-5); Danish Cancer Research Society; NovoNordisk Foundation; Lundbeckfonden.

References

Acquati S, Greco A, Licastro D, Bhagat H, Ceric D, Rossini Z, Grieve J, Shaked-Rabi M, Henriquez NV, Brandner S et al (2013) Epigenetic regulation of survivin by Bmi1 is cell type specific during corticogenesis and in gliomas. Stem Cells 31:190–202

Altman J, Chorover SL (1963) Autoradiographic investigation of the distribution and utilization of intraventricularly injected adenine-3h, uracil-3h and thymidine-3h in the brains of rats. J Physiol 169:770–779

Altman J, Das GD (1965) Autoradiographic and histological evidence of postnatal hippocampal neurogenesis in rats. J Comp Neurol 124:319–335

Angelastro JM, Lame MW (2010) Overexpression of CD133 promotes drug resistance in C6 glioma cells. Mol Cancer Res 8:1105–1115

Anido J, Saez-Borderias A, Gonzalez-Junca A, Rodon L, Folch G, Carmona MA, Prieto-Sanchez RM, Barba I, Martinez-Saez E, Prudkin L et al (2010) TGF-beta receptor inhibitors target the CD44(high)/Id1(high) glioma-initiating cell population in human glioblastoma. Cancer Cell 18:655–668

Bao S, Wu Q, Sathornsumetee S, Hao Y, Li Z, Hjelmeland AB, Shi Q, McLendon RE, Bigner DD, Rich JN (2006) Stem cell-like glioma cells promote tumor angiogenesis through vascular endothelial growth factor. Cancer Res 66:7843–7848

Bar EE (2011) Glioblastoma, cancer stem cells and hypoxia. Brain Pathol 21:119–129

Barraud P, Stott S, Mollgard K, Parmar M, Bjorklund A (2007) In vitro characterization of a human neural progenitor cell coexpressing SSEA4 and CD133. J Neurosci Res 85:250–259

Beier D, Hau P, Proescholdt M, Lohmeier A, Wischhusen J, Oefner PJ, Aigner L, Brawanski A, Bogdahn U, Beier CP (2007) CD133(+) and CD133(−) glioblastoma-derived cancer stem cells show differential growth characteristics and molecular profiles. Cancer Res 67:4010–4015

Beier D, Schulz JB, Beier CP (2011) Chemoresistance of glioblastoma cancer stem cells—much more complex than expected. Mol Cancer 10:128

Beier D, Wischhusen J, Dietmaier W, Hau P, Proescholdt M, Brawanski A, Bogdahn U, Beier CP (2008) CD133 expression and cancer stem cells predict prognosis in high-grade oligodendroglial tumors. Brain Pathol 18:370–377

Bhat KP, Balasubramaniyan V, Vaillant B, Ezhilarasan R, Hummelink K, Hollingsworth F, Wani K, Heathcock L, James JD, Goodman LD et al (2013) Mesenchymal differentiation mediated by NF-kappaB promotes radiation resistance in glioblastoma. Cancer Cell 24:331–346

Binello E, Germano IM (2011) Targeting glioma stem cells: a novel framework for brain tumors. Cancer Sci 102:1958–1966

Blanpain C, Fuchs E (2006) Epidermal stem cells of the skin. Annu Rev Cell Dev Biol 22:339–373

Bleau AM, Hambardzumyan D, Ozawa T, Fomchenko EI, Huse JT, Brennan CW, Holland EC (2009a) PTEN/PI3K/Akt pathway regulates the side population phenotype and ABCG2 activity in glioma tumor stem-like cells. Cell Stem Cell 4:226–235

Bleau AM, Huse JT, Holland EC (2009b) The ABCG2 resistance network of glioblastoma. Cell Cycle 8:2936–2944

Bourguignon LY, Peyrollier K, Xia W, Gilad E (2008) Hyaluronan-CD44 interaction activates stem cell marker Nanog, Stat-3-mediated MDR1 gene expression, and ankyrin-regulated multidrug efflux in breast and ovarian tumor cells. J Biol Chem 283:17635–17651

Bruce WR, Van Der Gaag H (1963) A quantitative assay for the number of murine lymphoma cells capable of proliferation in vivo. Nature 199:79–80

Brunschwig A, Southam CM, Levin AG (1965) Host resistance to cancer. Clinical experiments by homotransplants, autotransplants and admixture of autologous leucocytes. Ann Surgery 162:416–425

Calabrese C, Poppleton H, Kocak M, Hogg TL, Fuller C, Hamner B, Oh EY, Gaber MW, Finklestein D, Allen M et al (2007) A perivascular niche for brain tumor stem cells. Cancer Cell 11:69–82

Calvi LM, Adams GB, Weibrecht KW, Weber JM, Olson DP, Knight MC, Martin RP, Schipani E, Divieti P, Bringhurst FR et al (2003) Osteoblastic cells regulate the haematopoietic stem cell niche. Nature 425:841–846

Capper D, Gaiser T, Hartmann C, Habel A, Mueller W, Herold-Mende C, von Deimling A, Siegelin MD (2009) Stem-cell-like glioma cells are resistant to TRAIL/Apo2L and exhibit down-regulation of caspase-8 by promoter methylation. Acta Neuropathol 117:445–456

Chen MS, Woodward WA, Behbod F, Peddibhotla S, Alfaro MP, Buchholz TA, Rosen JM (2007) Wnt/beta-catenin mediates radiation resistance of Sca1+ progenitors in an immortalized mammary gland cell line. J Cell Sci 120:468–477

Chen R, Nishimura MC, Bumbaca SM, Kharbanda S, Forrest WF, Kasman IM, Greve JM, Soriano RH, Gilmour LL, Rivers CS et al (2010) A hierarchy of self-renewing tumor-initiating cell types in glioblastoma. Cancer Cell 17:362–375

Cheng L, Bao S, Rich JN (2010) Potential therapeutic implications of cancer stem cells in glioblastoma. Biochem Pharmacol 80:654–665

Cheng L, Huang Z, Zhou W, Wu Q, Donnola S, Liu JK, Fang X, Sloan AE, Mao Y, Lathia JD et al (2013) Glioblastoma stem cells generate vascular pericytes to support vessel function and tumor growth. Cell 153:139–152

Cheng L, Wu Q, Huang Z, Guryanova OA, Huang Q, Shou W, Rich JN, Bao S (2011) L1CAM regulates DNA damage checkpoint response of glioblastoma stem cells through NBS1. EMBO J 30:800–813

Clement V, Sanchez P, de Tribolet N, Radovanovic I, Ruiz i Altaba A (2007) HEDGEHOG-GLI1 signaling regulates human glioma growth, cancer stem cell self-renewal, and tumorigenicity. Curr Biol 17:165–172

Dietrich J, Diamond EL, Kesari S (2010) Glioma stem cell signaling: therapeutic opportunities and challenges. Expert Rev Anticancer Ther 10:709–722

Dietrich J, Imitola J, Kesari S (2008) Mechanisms of disease: the role of stem cells in the biology and treatment of gliomas. Nat Clin Pract Oncol 5:393–404

Dimou L, Simon C, Kirchhoff F, Takebayashi H, Gotz M (2008) Progeny of Olig2-expressing progenitors in the gray and white matter of the adult mouse cerebral cortex. J Neurosci 28:10434–10442

Doetsch F, Garcia-Verdugo JM, Alvarez-Buylla A (1997) Cellular composition and three-dimensional organization of the subventricular germinal zone in the adult mammalian brain. J Neurosci 17:5046–5061

Doetsch F, Garcia-Verdugo JM, Alvarez-Buylla A (1999) Regeneration of a germinal layer in the adult mammalian brain. Proc Natl Acad Sci U S A 96:11619–11624

Evans SM, Jenkins KW, Jenkins WT, Dilling T, Judy KD, Schrlau A, Judkins A, Hahn SM, Koch CJ (2008) Imaging and analytical methods as applied to the evaluation of vasculature and hypoxia in human brain tumors. Radiat Res 170:677–690

Evans SM, Judy KD, Dunphy I, Jenkins WT, Nelson PT, Collins R, Wileyto EP, Jenkins K, Hahn SM, Stevens CW et al (2004) Comparative measurements of hypoxia in human brain tumors using needle electrodes and EF5 binding. Cancer Res 64:1886–1892

Eyler CE, Foo WC, LaFiura KM, McLendon RE, Hjelmeland AB, Rich JN (2008) Brain cancer stem cells display preferential sensitivity to Akt inhibition. Stem Cells 26:3027–3036

Eyler CE, Rich JN (2008) Survival of the fittest: cancer stem cells in therapeutic resistance and angiogenesis. J Clin Oncol 26:2839–2845

Facchino S, Abdouh M, Chatoo W, Bernier G (2010) BMI1 confers radioresistance to normal and cancerous neural stem cells through recruitment of the DNA damage response machinery. J Neurosci 30:10096–10111

Fan X, Matsui W, Khaki L, Stearns D, Chun J, Li YM, Eberhart CG (2006) Notch pathway inhibition depletes stem-like cells and blocks engraftment in embryonal brain tumors. Cancer Res 66:7445–7452

Fargeas CA, Corbeil D, Huttner WB (2003) AC133 antigen, CD133, prominin-1, prominin-2, etc.: prominin family gene products in need of a rational nomenclature. Stem Cells 21:506–508

Folkins C, Shaked Y, Man S, Tang T, Lee CR, Zhu Z, Hoffman RM, Kerbel RS (2009) Glioma tumor stem-like cells promote tumor angiogenesis and vasculogenesis via vascular endothelial growth factor and stromal-derived factor 1. Cancer Res 69:7243–7251

Fuchs E, Tumbar T, Guasch G (2004) Socializing with the neighbors: stem cells and their niche. Cell 116:769–778

Fukuda S, Kato F, Tozuka Y, Yamaguchi M, Miyamoto Y, Hisatsune T (2003) Two distinct subpopulations of nestin-positive cells in adult mouse dentate gyrus. J Neurosci 23:9357–9366

Gage FH (2000) Mammalian neural stem cells. Science 287:1433–1438

Gallia GL, Tyler BM, Hann CL, Siu IM, Giranda VL, Vescovi AL, Brem H, Riggins GJ (2009) Inhibition of Akt inhibits growth of glioblastoma and glioblastoma stem-like cells. Mol Cancer Ther 8:386–393

Gilbertson RJ, Rich JN (2007) Making a tumour's bed: glioblastoma stem cells and the vascular niche. Nat Rev Cancer 7:733–736

Guo D, Ye J, Dai J, Li L, Chen F, Ma D, Ji C (2009) Notch-1 regulates Akt signaling pathway and the expression of cell cycle regulatory proteins cyclin D1, CDK2 and p21 in T-ALL cell lines. Leuk Res 33:678–685

Hambardzumyan D, Becher OJ, Holland EC (2008) Cancer stem cells and survival pathways. Cell Cycle 7:1371–1378

Hamerlik P, Lathia JD, Rasmussen R, Wu Q, Bartkova J, Lee M, Moudry P, Bartek J Jr, Fischer W, Lukas J et al (2012) Autocrine VEGF-VEGFR2-Neuropilin-1 signaling promotes glioma stem-like cell viability and tumor growth. J Exp Med 209:507–520

Heddleston JM, Li Z, McLendon RE, Hjelmeland AB, Rich JN (2009) The hypoxic microenvironment maintains glioblastoma stem cells and promotes reprogramming towards a cancer stem cell phenotype. Cell Cycle 8:3274–3284

Hermansen SK, Christensen KG, Jensen SS, Kristensen BW (2011) Inconsistent immunohistochemical expression patterns of four different CD133 antibody clones in glioblastoma. J Histochem Cytochem 59:391–407

Huse JT, Phillips HS, Brennan CW (2011) Molecular subclassification of diffuse gliomas: seeing order in the chaos. Glia 59:1190–1199

Ignatova TN, Kukekov VG, Laywell ED, Suslov ON, Vrionis FD, Steindler DA (2002) Human cortical glial tumors contain neural stem-like cells expressing astroglial and neuronal markers in vitro. Glia 39:193–206

Ikushima H, Todo T, Ino Y, Takahashi M, Saito N, Miyazawa K, Miyazono K (2011) Glioma-initiating cells retain their tumorigenicity through integration of the Sox axis and Oct4 protein. J Biol Chem 286:41434–41441

Ishimoto T, Oshima H, Oshima M, Kai K, Torii R, Masuko T, Baba H, Saya H, Nagano O (2010) CD44+ slow-cycling tumor cell expansion is triggered by cooperative actions of Wnt and prostaglandin E2 in gastric tumorigenesis. Cancer Sci 101:673–678

Jaksch M, Munera J, Bajpai R, Terskikh A, Oshima RG (2008) Cell cycle-dependent variation of a CD133 epitope in human embryonic stem cell, colon cancer, and melanoma cell lines. Cancer Res 68:7882–7886

Jellinger K (1978) Glioblastoma multiforme: morphology and biology. Acta Neurochir 42:5–32

Jijiwa M, Demir H, Gupta S, Leung C, Joshi K, Orozco N, Huang T, Yildiz VO, Shibahara I, de Jesus JA et al (2011) CD44v6 regulates growth of brain tumor stem cells partially through the AKT-mediated pathway. PLoS One 6:e24217

Jin L, Hope KJ, Zhai Q, Smadja-Joffe F, Dick JE (2006) Targeting of CD44 eradicates human acute myeloid leukemic stem cells. Nat Med 12:1167–1174

Joo KM, Kim SY, Jin X, Song SY, Kong DS, Lee JI, Jeon JW, Kim MH, Kang BG, Jung Y et al (2008) Clinical and biological implications of CD133-positive and CD133-negative cells in glioblastomas. Lab Invest 88:808–815

Kalani MY, Cheshier SH, Cord BJ, Bababeygy SR, Vogel H, Weissman IL, Palmer TD, Nusse R (2008) Wnt-mediated self-renewal of neural stem/progenitor cells. Proc Natl Acad Sci U S A 105:16970–16975

Kim M, Morshead CM (2003) Distinct populations of forebrain neural stem and progenitor cells can be isolated using side-population analysis. J Neurosci 23:10703–10709

Knizetova P, Ehrmann J, Hlobilkova A, Vancova I, Kalita O, Kolar Z, Bartek J (2008) Autocrine regulation of glioblastoma cell cycle progression, viability and radioresistance through the VEGF-VEGFR2 (KDR) interplay. Cell Cycle 7:2553–2561

Lapidot T, Sirard C, Vormoor J, Murdoch B, Hoang T, Caceres-Cortes J, Minden M, Paterson B, Caligiuri MA, Dick JE (1994) A cell initiating human acute myeloid leukaemia after transplantation into SCID mice. Nature 367:645–648

Lee J, Kotliarova S, Kotliarov Y, Li A, Su Q, Donin NM, Pastorino S, Purow BW, Christopher N, Zhang W et al (2006) Tumor stem cells derived from glioblastomas cultured in bFGF and EGF more closely mirror the phenotype and genotype of primary tumors than do serum-cultured cell lines. Cancer Cell 9:391–403

Li F, Tiede B, Massague J, Kang Y (2007a) Beyond tumorigenesis: cancer stem cells in metastasis. Cell Res 17:3–14

Li JL, Sainson RC, Shi W, Leek R, Harrington LS, Preusser M, Biswas S, Turley H, Heikamp E, Hainfellner JA, Harris AL (2007b) Delta-like 4 Notch ligand regulates tumor angiogenesis, improves tumor vascular function, and promotes tumor growth in vivo. Cancer Res 67:11244–11253

Li Z, Bao S, Wu Q, Wang H, Eyler C, Sathornsumetee S, Shi Q, Cao Y, Lathia J, McLendon RE et al (2009a) Hypoxia-inducible factors regulate tumorigenic capacity of glioma stem cells. Cancer Cell 15:501–513

Li Z, Wang H, Eyler CE, Hjelmeland AB, Rich JN (2009b) Turning cancer stem cells inside out: an exploration of glioma stem cell signaling pathways. J Biol Chem 284:16705–16709

Ligon KL, Huillard E, Mehta S, Kesari S, Liu H, Alberta JA, Bachoo RM, Kane M, Louis DN, Depinho RA et al (2007) Olig2-regulated lineage-restricted pathway controls replication competence in neural stem cells and malignant glioma. Neuron 53:503–517

Liu G, Yuan X, Zeng Z, Tunici P, Ng H, Abdulkadir IR, Lu L, Irvin D, Black KL, Yu JS (2006) Analysis of gene expression and chemoresistance of CD133+ cancer stem cells in glioblastoma. Mol Cancer 5:67

Lottaz C, Beier D, Meyer K, Kumar P, Hermann A, Schwarz J, Junker M, Oefner PJ, Bogdahn U, Wischhusen J et al (2010) Transcriptional profiles of CD133+ and CD133– glioblastoma-derived cancer stem cell lines suggest different cells of origin. Cancer Res 70:2030–2040

Mak AB, Blakely KM, Williams RA, Penttila PA, Shukalyuk AI, Osman KT, Kasimer D, Ketela T, Moffat J (2011) CD133 protein N-glycosylation processing contributes to cell surface recognition of the primitive cell marker AC133 epitope. J Biol Chem 286:41046–41056

Mao P, Joshi K, Li J, Kim SH, Li P, Santana-Santos L, Luthra S, Chandran UR, Benos PV, Smith L et al (2013) Mesenchymal glioma stem cells are maintained by activated glycolytic metabolism involving aldehyde dehydrogenase 1A3. Proc Natl Acad Sci U S A 110:8644–8649

Mazumdar J, O'Brien WT, Johnson RS, LaManna JC, Chavez JC, Klein PS, Simon MC (2010) O2 regulates stem cells through Wnt/beta-catenin signalling. Nat Cell Biol 12:1007–1013

McCord AM, Jamal M, Shankavaram UT, Lang FF, Camphausen K, Tofilon PJ (2009a) Physiologic oxygen concentration enhances the stem-like properties of CD133+ human glioblastoma cells in vitro. Mol Cancer Res 7:489–497

McCord AM, Jamal M, Williams ES, Camphausen K, Tofilon PJ (2009b) CD133+ glioblastoma stem-like cells are radiosensitive with a defective DNA damage response compared with established cell lines. Clin Cancer Res 15:5145–5153

Miraglia S, Godfrey W, Yin AH, Atkins K, Warnke R, Holden JT, Bray RA Waller EK, Buck DW (1997) Novel five-transmembrane hematopoietic stem cell antigen: isolation characterization, and molecular cloning. Blood 90:5013–21

Noble M, Dietrich J (2002) Intersections between neurobiology and oncology: tumor origin, treatment and repair of treatment-associated damage. Trends Neurosci 25:103–107

Nusse R, Fuerer C, Ching W, Harnish K, Logan C, Zeng A, ten Berge D, Kalani Y (2008) Wnt signaling and stem cell control. Cold Spring Harb Symp Quant Biol 73:59–66

Ohishi K, Varnum-Finney B, Bernstein ID (2002) The notch pathway: modulation of cell fate decisions in hematopoiesis. Int J Hematol 75:449–459

Pardal R, Clarke MF, Morrison SJ (2003) Applying the principles of stem-cell biology to cancer. Nat Rev Cancer 3:895–902

Patrawala L, Calhoun T, Schneider-Broussard R, Zhou J, Claypool K, Tang DG (2005) Side population is enriched in tumorigenic, stem-like cancer cells, whereas ABCG2+ and ABCG2− cancer cells are similarly tumorigenic. Cancer Res 65:6207–6219

Persano L, Rampazzo E, Basso G, Viola G (2013) Glioblastoma cancer stem cells: role of the microenvironment and therapeutic targeting. Biochem Pharmacol 85:612–622

Phillips HS, Kharbanda S, Chen R, Forrest WF, Soriano RH, Wu TD, Misra A, Nigro JM, Colman H, Soroceanu L et al (2006) Molecular subclasses of high-grade glioma predict prognosis, delineate a pattern of disease progression, and resemble stages in neurogenesis. Cancer Cell 9:157–173

Pruszak J, Sonntag KC, Aung MH, Sanchez-Pernaute R, Isacson O (2007) Markers and methods for cell sorting of human embryonic stem cell-derived neural cell populations. Stem Cells 25:2257–2268

Pu P, Zhang Z, Kang C, Jiang R, Jia Z, Wang G, Jiang H (2009) Downregulation of Wnt2 and beta-catenin by siRNA suppresses malignant glioma cell growth. Cancer Gene Ther 16:351–361

Purow BW, Haque RM, Noel MW, Su Q, Burdick MJ, Lee J, Sundaresan T, Pastorino S, Park JK, Mikolaenko I et al (2005) Expression of Notch-1 and its ligands, Delta-like-1 and Jagged-1, is critical for glioma cell survival and proliferation. Cancer Res 65:2353–2363

Quintana E, Shackleton M, Sabel MS, Fullen DR, Johnson TM, Morrison SJ (2008) Efficient tumour formation by single human melanoma cells. Nature 456:593–598

Radtke F, Raj K (2003) The role of Notch in tumorigenesis: oncogene or tumour suppressor? Nat Rev Cancer 3:756–767

Rahman M, Deleyrolle L, Vedam-Mai V, Azari H, Abd-El-Barr M, Reynolds BA (2011) The cancer stem cell hypothesis: failures and pitfalls. Neurosurgery 68:531–545, discussion 545

Reya T, Morrison SJ, Clarke MF, Weissman IL (2001) Stem cells, cancer, and cancer stem cells. Nature 414:105–111

Ricci-Vitiani L, Pallini R, Biffoni M, Todaro M, Invernici G, Cenci T, Maira G, Parati EA, Stassi G, Larocca LM, De Maria R (2010) Tumour vascularization via endothelial differentiation of glioblastoma stem-like cells. Nature 468:824–828

Rich JN (2008) The implications of the cancer stem cell hypothesis for neuro-oncology and neurology. Future Neurol 3:265–273

Rich JN, Eyler CE (2008) Cancer stem cells in brain tumor biology. Cold Spring Harb Symp Quant Biol 73:411–420

Roitbak T, Li L, Cunningham LA (2008) Neural stem/progenitor cells promote endothelial cell morphogenesis and protect endothelial cells against ischemia via HIF-1alpha-regulated VEGF signaling. J Cereb Blood Flow Metab 28:1530–1542

Ronchini C, Capobianco AJ (2001) Induction of cyclin D1 transcription and CDK2 activity by Notch(ic): implication for cell cycle disruption in transformation by Notch(ic). Mol Cell Biol 21:5925–5934

Ropolo M, Daga A, Griffero F, Foresta M, Casartelli G, Zunino A, Poggi A, Cappelli E, Zona G, Spaziante R et al (2009) Comparative analysis of DNA repair in stem and nonstem glioma cell cultures. Mol Cancer Res 7:383–392

Sanai N, Alvarez-Buylla A, Berger MS (2005) Neural stem cells and the origin of gliomas. New Engl J Med 353:811–822

Sathornsumetee S, Rich JN (2008) Designer therapies for glioblastoma multiforme. Ann NY Acad Sci 1142:108–132

Schofield R (1978) The relationship between the spleen colony-forming cell and the haemopoietic stem cell. Blood Cells 4:7–25

Schonberg DL, Lubelski D, Miller TE, Rich JN (2013) Brain tumor stem cells: molecular characteristics and their impact on therapy. Mol Aspects Med. doi:10.1016/j.mam.2013.06.004

Seidel S, Garvalov BK, Wirta V, von Stechow L, Schanzer A, Meletis K, Wolter M, Sommerlad D, Henze AT, Nister M et al (2010) A hypoxic niche regulates glioblastoma stem cells through hypoxia inducible factor 2 alpha. Brain 133:983–995

Shackleton M, Quintana E, Fearon ER, Morrison SJ (2009) Heterogeneity in cancer: cancer stem cells versus clonal evolution. Cell 138:822–829

Sheiness D, Fanshier L, Bishop JM (1978) Identification of nucleotide sequences which may encode the oncogenic capacity of avian retrovirus MC29. J Virol 28:600–610

Shih AH, Holland EC (2006) Notch signaling enhances nestin expression in gliomas. Neoplasia 8:1072–1082

Singh SK, Clarke ID, Terasaki M, Bonn VE, Hawkins C, Squire J, Dirks PB (2003) Identification of a cancer stem cell in human brain tumors. Cancer Res 63:5821–5828

Singh SK, Hawkins C, Clarke ID, Squire JA, Bayani J, Hide T, Henkelman RM, Cusimano MD, Dirks PB (2004) Identification of human brain tumour initiating cells. Nature 432:396–401

Son MJ, Woolard K, Nam DH, Lee J, Fine HA (2009) SSEA-1 is an enrichment marker for tumor-initiating cells in human glioblastoma. Cell Stem Cell 4:440–452

Stopschinski BE, Beier CP, Beier D (2013) Glioblastoma cancer stem cells–from concept to clinical application. Cancer Lett 338:32–40

Strauss R, Hamerlik P, Lieber A, Bartek J (2012) Regulation of stem cell plasticity: mechanisms and relevance to tissue biology and cancer. Mol Ther 20:887–897

Stupp R, Mason WP, van den Bent MJ, Weller M, Fisher B, Taphoorn MJ, Belanger K, Brandes AA, Marosi C, Bogdahn U et al (2005) Radiotherapy plus concomitant and adjuvant temozolomide for glioblastoma. New Engl J Med 352:987–996

Sun P, Xia S, Lal B, Eberhart CG, Quinones-Hinojosa A, Maciaczyk J, Matsui W, Dimeco F, Piccirillo SM, Vescovi AL, Laterra J (2009) DNER, an epigenetically modulated gene, regulates glioblastoma-derived neurosphere cell differentiation and tumor propagation. Stem Cells 27:1473–1486

Tanaka S, Louis DN, Curry WT, Batchelor TT, Dietrich J (2013) Diagnostic and therapeutic avenues for glioblastoma: no longer a dead end? Nat Rev Clin Oncol 10:14–26

Uchida N, Buck DW, He D, Reitsma MJ, Masek M, Phan TV, Tsukamoto AS, Gage FH, Weissman IL (2000) Direct isolation of human central nervous system stem cells. Proc Natl Acad Sci U S A 97:14720–14725

Venere M, Hamerlik P, Wu Q, Rasmussen RD, Song LA, Vasanji A, Tenley N, Flavahan WA, Hjelmeland AB, Bartek J, Rich JN (2013) Therapeutic targeting of constitutive PARP activation compromises stem cell phenotype and survival of glioblastoma-initiating cells. Cell Death Differ 21(2):258–269

Vennstrom B, Sheiness D, Zabielski J, Bishop JM (1982) Isolation and characterization of c-myc, a cellular homolog of the oncogene (v-myc) of avian myelocytomatosis virus strain 29. J Virol 42:773–779

Verhaak RG, Hoadley KA, Purdom E, Wang V, Qi Y, Wilkerson MD, Miller CR, Ding L, Golub T, Mesirov JP et al (2010) Integrated genomic analysis identifies clinically relevant subtypes of glioblastoma characterized by abnormalities in PDGFRA, IDH1, EGFR, and NF1. Cancer Cell 17:98–110

Verhaak RG, Valk PJ (2010) Genes predictive of outcome and novel molecular classification schemes in adult acute myeloid leukemia. Cancer Treat Res 145:67–83

Wang J, Sullenger BA, Rich JN (2012) Notch signaling in cancer stem cells. Adv Exp Med Biol 727:174–185

Wang J, Wang H, Li Z, Wu Q, Lathia JD, McLendon RE, Hjelmeland AB, Rich JN (2008) c-Myc is required for maintenance of glioma cancer stem cells. PLoS One 3:e3769

Wang R, Chadalavada K, Wilshire J, Kowalik U, Hovinga KE, Geber A, Fligelman B, Leversha M, Brennan C, Tabar V (2010) Glioblastoma stem-like cells give rise to tumour endothelium. Nature 468:829–833

Wei Y, Jiang Y, Zou F, Liu Y, Wang S, Xu N, Xu W, Cui C, Xing Y, Liu Y et al (2013) Activation of PI3K/Akt pathway by CD133-p85 interaction promotes tumorigenic capacity of glioma stem cells. Proc Natl Acad Sci U S A 110:6829–6834

Woodward WA, Chen MS, Behbod F, Alfaro MP, Buchholz TA, Rosen JM (2007) WNT/beta-catenin mediates radiation resistance of mouse mammary progenitor cells. Proc Natl Acad Sci U S A 104:618–623

Yamamoto A, Shofuda T, Islam MO, Nakamura Y, Yamasaki M, Okano H, Kanemura Y (2009) ABCB1 is predominantly expressed in human fetal neural stem/progenitor cells at an early development stage. J Neurosci Res 87:2615–2623

Yan K, Yang K, Rich JN (2013) The evolving landscape of glioblastoma stem cells. Curr Opin Neurol 26(6):701–707

Zacchigna S, Oh H, Wilsch-Brauninger M, Missol-Kolka E, Jaszai J, Jansen S, Tanimoto N, Tonagel F, Seeliger M, Huttner WB et al (2009) Loss of the cholesterol-binding protein prominin-1/CD133 causes disk dysmorphogenesis and photoreceptor degeneration. J Neurosci 29:2297–2308

Zeppernick F, Ahmadi R, Campos B, Dictus C, Helmke BM, Becker N, Lichter P, Unterberg A, Radlwimmer B, Herold-Mende CC (2008) Stem cell marker CD133 affects clinical outcome in glioma patients. Clin Cancer Res 14:123–129

Zhou BB, Zhang H, Damelin M, Geles KG, Grindley JC, Dirks PB (2009) Tumour-initiating cells: challenges and opportunities for anticancer drug discovery. Nat Rev Drug Discov 8:806–823

Zhu X, Zuo H, Maher BJ, Serwanski DR, LoTurco JJ, Lu QR, Nishiyama A (2012) Olig2-dependent developmental fate switch of NG2 cells. Development 139:2299–2307

Epigenetic Changes in Gliomas

2

Vinay K. Puduvalli

Abstract

Epigenetic alterations have been fundamentally implicated in the pathobiology of several malignancies including gliomas. Such changes involve tumor specific promoter methylation, histone modifications, and changes in microRNA. Several such recently reported alterations have also been seen to highly correlate with outcome highlighting their clinical significance, including the following: Promoter methylation of *MGMT* correlates with improved survival in glioblastoma patients who receive chemoradiation therapy and in anaplastic glioma patients treated with radiation alone as frontline therapy. A glioma CpG island methylator phenotype (G-CIMP) confers better survival to patients with the proneural subtype of glioblastoma. Mutations of *IDH1* are commonly seen as early events in low grade and anaplastic gliomas and are also associated with a hypermethylator phenotype. Mutations of *H3F3A* which encodes the histone variant H3.3 have been reported in pediatric high grade gliomas which in turn have been linked to reduction in the histone H3 lysine trimethylation mark (H3K27me3) and possibly DNA hypomethylation as mechanisms for activated gene expression. Lastly, altered expression of several microRNA has been implicated in glioma biology. These findings point to a rapidly expanding recognition of the critical role of epigenetic changes in gliomagenesis and pathobiology. This in turn has given rise to clinical trials that either utilize epigenetic alterations as prognostic markers for patient stratification or in attempts at therapeutic targeting of epigenetic states of gliomas. However, much remains to be learned about the complex interplay between various epigenetic factors in glioma cells and their influence on genetic alterations; such insights will provide a foundation for therapeutic strategies aimed at modulating epigenetic factors to target high and low grade gliomas.

V.K. Puduvalli (✉)
Division of Neuro-oncology, Ohio State University Comprehensive Cancer Center, Columbus, OH 43220, USA
e-mail: Vinay.Puduvalli@osumc.edu

A. Sedo and R. Mentlein (eds.), *Glioma Cell Biology*,
DOI 10.1007/978-3-7091-1431-5_2, © Springer-Verlag Wien 2014

Keywords
Epigenetics • Gliomas • Histone modification • Methylation • Prognosis

Abbreviations

CSC	Cancer stem cell
GBM	Glioblastoma
G-CIMP	Glioma CpG island methylator phenotype
H3K	Histone H3 lysine
HAT	Histone acetyl transferases
HDAC	Histone deacetylase
IDH	Isocitrate dehydrogenase
MECP	Methyl-CpG-binding protein
MGMT	O6-methyl-guanine DNA methyl transferase
OS	Overall survival
PCV	Procarbazine, CCNU, Vincristine regimen
PFS	Progression-free survival
PRC	Polycomb repressor complex
RT	Radiation therapy
TCGA	The cancer genome atlas
TMZ	Temozolomide
VEGFR	Vascular endothelial cell growth factor receptor
WHO	World Health Organization

2.1 Epigenetics in Normal Cells and Development

The term "epigenetics" encompasses the spectrum of modifications of the human genome which alter gene expression without affecting the DNA sequence. These include DNA methylation, changes in chromatin structure through posttranslational modifications of nucleosomal histones, and the effects of several classes of non-coding RNA. The organization of the central nervous system during development in humans involves a set of rapid changes that sets the foundation for long term functional programs of the adult nervous system. Key to such changes is the cell's ability to systematically turn on programs that decide its developmental fate and to turn off those that had maintained it in a more undifferentiated state. The selection of such programs is critically driven by epigenetic processes which guide a cell through a series of steps that channel it to its final differentiated and mature state. Aberrations in such processes can hence result in abnormal maturation and functional states that can divert the cells fate into several disease states. A deeper understanding of the normal role of epigenetic processes can provide critical insights into the pathogenic mechanisms of such diseases including malignancies.

From a developmental perspective, there are several well characterized epigenetic processes that play a major role in cell and organism fate. For instance, the silencing of one of the X chromosomes in females needed to equalize the dosage of X-linked genes between the two sexes is a result of a combination of dense hypermethylation, silencing histone marks and chromatin compaction which are initiated by specific noncoding RNA. A second process is genomic imprinting which involves epigenetic processes that govern monoallelic expression from only one parental allele and the silencing of the other through a combination of noncoding RNA, repressive histone marks, and methylation through as yet poorly characterized epigenetic mechanisms. Thirdly, lineage restriction that allows embryonic cells to follow specific differentiation pathways to form specialized differentiated cells that can assume the functional hierarchies inherent to various organs including the brain is another process regulated by epigenetic mechanisms. Key to this function is the role of proteins such as Repressor element 1 (RE-1) silencing transcription factor/neuron restrictive silencing factor (REST/NRSF) and its co-repressor, co-REST, which cooperate to recruit epigenetic mechanisms laying down repressive histone marks and causing chromatin compaction. Lastly, epigenetic mechanisms are critical in terminal differentiation of cells and maintenance of their mature phenotype which is especially important in post-mitotic states in neurons but also play a significant role in glial cells.

In normal cells, epigenetic modifications involve the DNA itself or various histone and nonhistone proteins that are closely associated with DNA that can sculpt the three-dimensional structure of chromatin, control access of various regulatory elements to their respective recognition sites in gene promoters and regulators, and respond to internal and environmental signals to modify gene expression. These controls can provide a means for the cell to fine-tune its responses to internal and external stimuli and allow adaptations to counter the threats to the cell and to fulfil the various demands related to its function.

Methylation of the promoter sequences of genes is a well-recognized covalent DNA modification that introduces a methyl group at the cytosine residue typically in CpG sites within regions of DNA (where a cytosine occurs next to a guanine separated by a phosphate). High frequency of CpG sites occur in regions of the DNA called CpG islands which are found in the promoters of nearly half the known human genes; these regions are considered essential for gene regulation and methylated promoters usually result in silencing of gene expression. Promoter methylation can facilitate binding of methyl-binding proteins and subsequently modify chromatin structure; this in turn can prevent transcription factors from accessing the promoter and induce silencing of gene expression. Although gene bodies are on the whole CpG-poor, extensive methylation of these sites is noticed but such methylation is not associated with gene repression despite their association with repressive marks such as H3K9me3 and being bound by methyl-CpG-binding protein 2 (MECP2), unlike that seen in promoters and transcription start sites; gene body methylation sites are related to cancer-causing mutations in somatic cells. In general, normal cells show genome-wide hypermethylation and promoter specific hypomethylation to allow expression of essential genes whereas in cancer cells the converse is often seen. Such aberrations in methylation are seen frequently in cancer and may be pathogenic in this setting.

Histones are the main protein components of chromatin and form the supporting framework around which the DNA strand is wound allowing efficient compaction of the long DNA strand and additionally allowing a remarkable mechanism for control of gene expression. The evolutionally conserved core histones, H2A, H2B, H3, and H4, form an octameric chromatin unit around which the DNA strand is wound forming the nucleosome. Histones H1 and H5 are linker histones that support the ends of the DNA strand between nucleosomes. In addition to their role in chromatin structure, histones are subject to posttranslational modifications which provide a dynamic and flexible way to control gene expression; such modifications include lysine (K) methylation, arginine methylation, lysine acetylation, and phosphorylation of serine, threonine, or tyrosine residues which can be combined in various ways to control gene expression that constitute a "histone code." These histone "marks" can represent repression or activation states of gene expression; for instance, histone H3 lysine (K) methylation (-Me) can occur as mono- (H3K4Me) di- (H3K4Me2) or tri- (H3K4Me3) methylation which has different "meanings" related to gene regulation. Activation marks include H3K4me3 occurring at gene promoters and H3K36Me3 in the gene body; conversely, H3K27Me3 and H3K9Me2/3 are repressive marks. Acetylation (Ac) of the lysine tail of histone H3 is another modification that regulates chromatin structure; lysine acetylation is governed by the dynamic equilibrium between histone acetyl transferases (HAT) and histone deacetylases (HDAC) with acetylation resulting in relaxation of the chromatin structure and deacetylation resulting in chromatin compaction. Such changes in chromatin structure can regulate access of transcription factors to gene regulatory elements and control gene expression.

Polycomb-group proteins are another component of the epigenetic machinery in cells that alter chromatin structure and control gene expression (Jones 2012). PRC2 (Polycomb repressor complex 2) acts as methyltransferase generating H3K27Me2/3 which in turn influences diverse processes such as stem cell maintenance and differentiation. The methyltransferase activity of PRC2 resides in its catalytic subunit, EZH2, which methylates the lysine residue of H3K27 (McCabe et al. 2012). PRC1 functions to ubiquitylate histone H2A and participates in chromatin compaction including in X-inactivation.

The untranslated portion of the genome, long considered a region of 'junk' DNA consisting of nonfunctional repetitive elements and viral sequences left over from evolutionary processes, has been recognized to contain several noncoding sequences that are transcribed. Noncoding RNAs constitute an expanding family of molecules that appear to have critical regulatory roles in normal and pathological states. Their role was initially identified with the discovery of micro-RNA as key regulators of gene function and this has opened a new vista to the complexity of gene regulation in normal and diseased tissue. Although not considered strictly epigenetic by conventional definition, their unique role in gene regulation best fits in this category of biology and exciting insights into their function are being gained through intensive research into their regulation and impact on gene expression.

2.2 Epigenetics in Cancer

Epigenomic processes, being complex and having crucial roles in decoding genomic data, are often involved in aberrations of the various components of the processes that subsequently drive development and progression of cancer. Such aberrations can involve changes both in DNA methylation patterns and in the genes that encode regulators of epigenetic processes. The role of epigenetics in cancer and other disease states has expanded its definition to include processes that influence chromatin structure and function and are relevant to the malignant process but do not involve a change in DNA sequence. There has been intense debate about whether epigenetic changes set the stage for subsequent genetic changes such as mutations, amplifications, deletions, or translocations leading to malignancies; however, emerging data suggests that there is a complex and often iterative interaction between these processes with reciprocal contributions to each other that continue throughout the course of a cancer.

One of the earliest of epigenetic changes identified in cancer cells was global hypomethylation of the genome with selective hypermethylation of promoters of specific genes unlike normal cells which have extensive intergenic regions of hypermethylation along with hypomethylation of gene promoters (Fig. 2.1). Approximately 5–10 % of normally unmethylated CpG promoter islands are abnormally methylated in cancer involving not only coding regions that affect protein production but various noncoding RNAs that can influence tumorigenesis (Baylin and Jones 2011). Recurrent heterozygous functionally relevant mutations have also been seen in *DNMT3A*, a methyltransferase that methylate hemimethylated DNA and function as a de novo methyltransferase to effect DNA methylation, in up to 25 % of patients with acute myeloid leukemia (AML) (Ley et al. 2010). Similarly, the TET family of DNA hydroxylases, which are responsible for converting 5-methylcytosine to 5-hydroxymethylcytosine and subsequently to 5-carboxylcytosine, are sometimes subject to mutations and when these result in loss of function, can cause genome-wide hypermethylation which appear to be associated with the malignant process (Cimmino et al. 2011).

Histone acetyltransferases (HAT) and histone deacetylases (HDAC) regulate nucleosomal configuration and chromatin structure. Involvement of HATs was among the earliest identified epigenetic changes in the malignant process involving chromatin; they are bound by several viral oncoproteins during tumorigenesis (Bannister and Kouzarides 1996). In addition, genetic alterations that disrupt their function are closely associated with malignancies (Iyer et al. 2004; Wang et al. 2005). HDACs are also implicated in malignancies although somatic mutations in the genes encoding these proteins are uncommon; for instance, HDACs can be recruited by chimeric fusion proteins to effect aberrant gene silencing. In addition, inhibitors of HDAC reverse several aspects of the malignant process including inducing apoptosis, growth arrest, decreased invasion, and angiogenesis, suggesting that higher levels of HDACs or specific interactions with other tumor related factors may be critical to malignancies and that these are reversible by HDAC inhibition. Given that HDACs can also deacetylate nonhistone proteins,

Fig. 2.1 Differences in DNA methylation patterns between normal and tumor cells

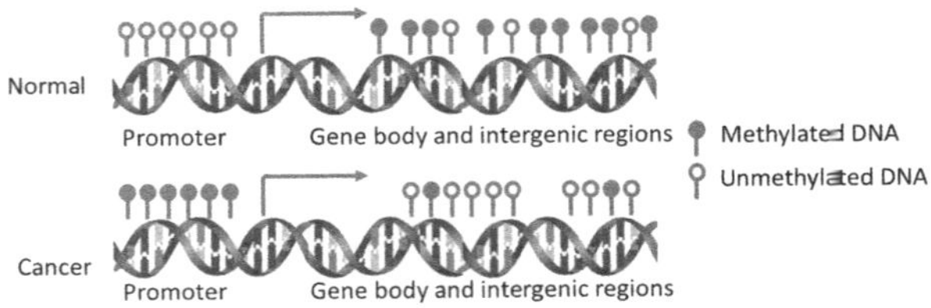

it is possible that some of their antitumor effects are related to such non-chromatin effects.

Readers of histone acetylation such as the BET family of bromodomain proteins (such as BRD2, BRD3, and BRD4) regulate transcriptional elongation and influence transition through the cell cycle. Recent studies have shown that the highly aggressive NUT-midline carcinomas are associated with recurrent translocations involving *BRD3/4* (Filippakopoulos et al. 2010). BET inhibitors have recently entered clinical trials and have been highly effective in NUT-midline carcinomas. BET family of proteins is likely to have a role in several other tumors and inhibitors of these proteins are likely to be increasingly useful in such malignancies (Dawson et al. 2011). Other epigenetic targets include histone demethylases such as the Jumonji class of demethylases which can potentially be inhibited as a class by high levels of 2-hydroxuglutarate (2-HG) aberrantly formed when mutations in the *IDH1/2* genes result in altered substrate specificity for their protein products which now convert α-ketoglurate into 2-HG (see Sect. 2.3.3). Mutations of histone genes have recently been implicated in pediatric brain tumors (see Sect. 2.3.5). Lastly noncoding RNA typified by microRNA (miRNA) are strongly implicated in the malignant process by altering the expression of various genes by modulating transcriptional and post-transcription process potentially involving multiple targets. A miRNA signature with high and low levels of expression of a set of miRNA was identified in chronic lymphocytic leukemia that was associated with disease progression and clinical outcome (Calin et al. 2005). Germ-line and somatic mutations were also identified in several genes encoding such miRNA. Similarly, a long noncoding RNA called HOTAIR which helps the coordinated activity of the PRC2 complex (involved in methylating H3K27) and a transcriptional co-repressor complex is aberrantly overexpressed in advanced breast and colorectal cancer (Kogo et al. 2011). These findings have provided the basis for a plethora of epigenetic inhibitors and modulators that are rapidly being translated into the clinical arena.

2.3 Epigenetic Changes in Gliomas

2.3.1 MGMT Promoter Methylation

Naturally occurring alkylation of DNA occurs in human cells at the N^7 and O^6 positions of guanine. Alkylation at the O^6 residue can be mutagenic by inducing mismatch errors during DNA replication; these adducts are removed by O^6 methyl-guanine methyl-transferase (MGMT) which is ubiquitously expressed in human cells (Gerson et al. 1986). MGMT is considered a "suicide" enzyme because it is depleted after accepting an alkyl group and requires de novo synthesis for replenishing its levels. Given that alkylator therapy has been the mainstay of cytotoxic approaches against gliomas for several decades, there has been an intense interest in the regulation of *MGMT* and the role of differences in MGMT expression in the response to alkylator therapy.

In preclinical studies, cells which do not express MGMT (initially termed as Mer−) were shown to have increased sensitivity to alkylators such as nitrosoureas (Zlotogorski and Erickson 1983). Further studies into the differential expression of MGMT in cells led to the initial identification of a complex correlation between methylation of the gene and MGMT expression (Pieper et al. 1990). Cells that lack MGMT expression were seen to silence the *MGMT* gene by methylation of its promoter (Costello et al. 1994a; Harris et al. 1994; Qian and Brent 1997; von Wronski et al. 1992). In contrast, heavy methylation of the regions in the body of *MGMT* containing exon sequences was seen in MGMT expressing cells and correlated with increased mRNA expression (Costello et al. 1994b). There have been conflicting reports about the correlation of MGMT activity or MGMT protein levels and alkylator sensitivity (Baer et al. 1993; Bobola et al. 1996) but several studies suggested that low MGMT levels were correlated with improved outcome (Belanich et al. 1996; Chen et al. 1999; Friedman et al. 1998; Jaeckle et al. 1998; Lee et al. 1996; Silber et al. 1999). Esteller et al. first reported the correlation between *MGMT* promoter methylation and clinical outcome in patients with high grade gliomas treated with alkylators using a methylation specific polymerase chain reaction (PCR) assay (Herman et al. 1996), and suggested that *MGMT* promoter methylation could be used as a method of patient selection for treatment with alkylators (Esteller et al. 2000). This concept was tested in an international multi-center phase III trial by Stupp and colleagues using the alkylating agent, temozolomide, in combination with radiation therapy and subsequently as adjuvant treatment in patients with newly diagnosed WHO grade IV gliomas which established this regimen as the new standard of care for patients with newly diagnosed glioblastoma (Stupp et al. 2005). A companion study by Hegi et al. in a subset of the patients in this trial strongly suggested that *MGMT* promoter methylation as determined by methylation specific PCR is an independent prognostic factor for benefit from temozolomide (Hegi et al. 2005). Confirmation of this finding was reported by Gilbert et al. in a subsequent international phase III study of chemoradiation using dose-dense versus standard dose temozolomide in which

MGMT promoter methylation was associated with improved survival and progression (Gilbert et al. 2013).

Prognostic markers help identify patients with differences in outcomes related to intrinsic differences in their tumors regardless of treatments used. *Predictive* markers on the other hand are those that predict responses to specific therapies. In the case of MGMT promoter methylation and gliomas, controversy exists about whether this finding is a prognostic marker or predictive of outcome to temozolomide therapy. In the setting of chemoradiation therapy using temozolomide in patients with newly diagnosed glioblastoma, MGMT promoter methylation has been regarded as a predictive marker as patients who had this finding and received temozolomide had a clearly improved outcome compared with those who did not. However, in the study by Stupp et al., in the RT only arm, patients who had MGMT promoter methylation appeared to have a better outcome (although statistical significance for this comparison was not provided) compared to those without this finding (median OS 15.3 months versus 11.8 months for unmethylated and median PFS 5.9 months versus 4.4 months for unmethylated). Further, the 2-year survival rate for the RT alone arm was 22.7 % for the methylated group compared to <2 % for the unmethylated one. In addition, a "tail" of longer OS and PFS was seen in patients with unmethylated *MGMT* promoter treated with temozolomide and RT compared to those who received RT alone also reflected in the 2 year survival rate of 13.8 % versus <2 % (Hegi et al. 2005). These results raise the possibility that MGMT promoter methylation may be more a prognostic marker not solely related to temozolomide therapy although the possibility of the effect of efficacy of adjuvant chemotherapy may have contributed to these differences cannot be excluded. Thus, while a predictive role for MGMT promoter methylation related to temozolomide therapy for patients with newly diagnosed glioblastoma remains possible, available data does not rule out the possibility that this may be predominantly prognostic as well. Indeed, several subsequent multicenter phase III trials (RTOG-0525/EORTC26052-22053, RTOG-0825, and AVAGlio) confirmed the prognostic value of MGMT promoter methylation but could not answer the question of predictive value of this marker given that patients in both arms of these trials received temozolomide (Chinot et al. 2014; Gilbert et al. 2013, 2014). A smaller study by Brandes et al. suggested that the MGMT promoter methylation was of prognostic value for temozolomide therapy for patients with glioblastoma only at initial diagnosis and not for recurrent disease; the study also suggested that the methylation status of the promoter could change during the course of the disease (Brandes et al. 2010). Given that the EORTC study had an upper limit for age of 70 years for enrollment, a study to assess the role of MGMT promoter methylation in the elderly patient with glioblastoma was conducted with patients who were 70 years of age or older who received standard chemoRT and adjuvant temozolomide. MGMT promoter methylation status was seen to be a strong independent prognostic factor for survival and progression free survival upon multifactorial analysis (Minniti et al. 2011). Large scale data in the pediatric high grade glioma population is currently lacking; reports of the frequency of MGMT promoter methylation in this population shows significant disparity between studies

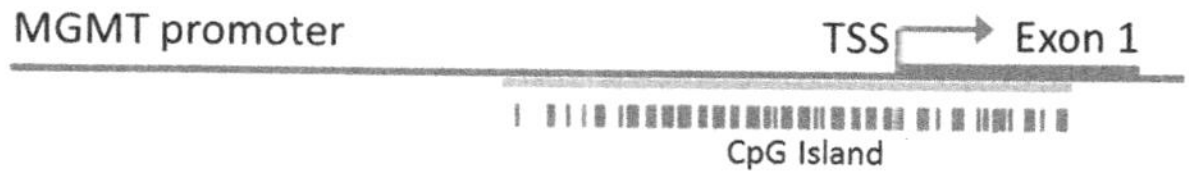

Fig. 2.2 MGMT promoter and CpG Island

presumably because of the small numbers of patients assessed. For instance, a small study of 10 patients showed *MGMT* promoter methylation in 40 % of cases and correlation with longer overall survival (Donson et al. 2007). Another small retrospective study ($N=24$) reported that 50 % of pediatric glioblastoma patients had MGMT promoter methylation but did not provide correlation with clinical outcome (Srivastava et al. 2010). One retrospective study of a small set of pediatric patients ($N=24$) who were more uniformly treated by enrollment in the German HIT-GBM trial showed that *MGMT* promoter methylation was seen in 77 % of patients and was associated with longer event-free survival compared with the unmethylated cohort whereas MGMT protein levels did not exhibit this correlation (Schlosser et al. 2010). However, other retrospective studies have reported 0–30 % cases of high grade gliomas with *MGMT* promoter methylation (Buttarelli et al. 2010; Fassan et al. 2011; Lee et al. 2011) emphasizing the need for analysis of a larger patient population to accurately establish the true proportion of pediatric gliomas with this finding.

Assessment of the role of *MGMT* promoter methylation in patients with anaplastic oligodendroglioma showed different results from those in glioblastoma. In the EORTC Study 26,951 patients with anaplastic oligodendroglioma were randomized to RT only or RT followed by adjuvant PCV chemotherapy; a subset of patients from this study had assessment of *MGMT* promoter methylation. The majority of these patients (80 %) had promoter methylation which strongly correlated with 1p/19q co-deletion and both were independent prognostic factors for survival. This prognostic value was seen both in the RT only and the RT/PCV arm for clinical outcome suggesting that MGMT promoter methylation identifies a tumor subset with better outcome to treatment regardless of the use of alkylators (van den Bent et al. 2009). This also brings to focus a bigger question—can the same molecular marker related to alkylator therapy be predictive in one brain tumor type and prognostic in another with similar treatment paradigms? While possible, this seems unlikely; the alternative explanation that methylation of the *MGMT* promoter may be a part of a methylated state of several other gene promoters that may be tightly correlated with the clinical outcome in response to RT and alkylator therapy remains to be explored (Fig. 2.2).

The *MGMT* gene, located on chromosome 10q, has a CpG island in its promoter region which consists of 762bp and has 97 CpG sites which can potentially be methylated and lead to gene silencing; however, it is still unclear as to methylation of which of the CpG sites is most associated with gene silencing. Deborah et al. identified two regions DMR1 and DMR2 with the latter including the region most commonly analyzed by MSP assays (Malley et al. 2011). DMR2 contained several CpG sites (83, 86, 87, and 89) which were critical for MGMT expression. Shah et al. analyzed all CpG sites in the MGMT promoter to identify the sites and methylation patterns that best predicted outcomes and proposed a methylation specific multiplex ligation-dependent probe amplification (MS-MLPA) assay using data from across the entire promoter as an alternative to the conventional gel-based MSP assay and identified four groups with various degrees of promoter CpG site methylation (Shah et al. 2011). The current commercially available test (OncoMethylome Sciences, licensed to LabCorp), developed as a refinement of the standard gel-based MS-PCR, is a high-throughput real-time MSP assay which provides a more valid result from FFPE samples (Vlassenbroeck et al. 2008). Measurements using this method showed a bimodal distribution with mean ratio values of 0.7 and 68.1 that define unmethylated and methylated promoter respectively with a cutoff ratio value of 8. The threshold for 95 % probability for unmethylated promoter was reported to be 4 whereas that for methylated promoter was 16.

Methods such as these for detecting of *MGMT* promoter methylation have come under scrutiny given that various trials have utilized different methods to identify this marker using formalin fixed paraffin embedded tissue specimens (Weller et al. 2010). In the landmark trial that established the role of temozolomide in the treatment of glioblastoma, Hegi et al. used a methylation specific PCR assay which is a nonquantitative method initially established by Esteller et al. which uses primers that distinguish between methylated and unmethylated sites (Esteller et al. 2000; Hegi et al. 2005). This assay interrogated the sites contained in the DMR2 region of the CpG island. In contrast, RTOG-0525, the trial of dose-dense versus standard temozolomide utilized the validated high throughput real-time MSP assay and confirmed the prognostic value of this marker in patients with glioblastoma (Gilbert et al. 2013). There continues to be debate about whether all patients with newly diagnosed glioblastoma should have testing for *MGMT* promoter methylation and whether this should be used to select patients for treatment with temozolomide and RT in routine clinical practice.

2.3.2 Glioma CpG Island Methylator Phenotype

Promoter methylation of specific genes can impact the expression of those genes and consequently affect the levels of their protein products and their cellular function. However, tumor cells exhibit widespread epigenetic changes across the genome with characteristic global hypomethylation in the intergenic regions and hypermethylation of promoter regions of multiple genes (Jones and Baylin 2007). The Cancer Genome Atlas (TCGA) project was set up in the USA to identify genome-wide changes of the DNA sequence, copy number, DNA methylation, gene expression, and associate these with clinical information for patients with malignancies; one of the first subprojects undertaken under this project was to molecularly characterize glioblastomas both in terms of genetic and epigenetic changes. Using data related to genome-wide methylation patterns obtained using the Illumina GoldenGate platform and subsequently with the Infinium platform in a set of 272 specimens of newly diagnosed glioblastoma from TCGA, Noushmehr et al. reported the identification of a glioma-CpG island methylator phenotype (gCIMP) which describes hypermethylation of a large number of gene loci in a subset of specimens (Noushmehr et al. 2010) and was similar to the CIMP pattern seen in colorectal cancer. The gCIMP cluster was highly enriched for the proneural genetic subtype of glioblastoma (Verhaak et al. 2010) with 21/24 (88 %) of gCIMP tumors being proneural in subtype; conversely, gCIMP tumors represented a distinct subtype of proneural glioblastoma comprising about 30 % of that tumor subtype. Two other methylation clusters which were moderately enriched for the classical and mesenchymal genetic subtypes of glioblastoma were also described.

From a clinical perspective, patients with G-CIMP glioblastoma were significantly younger (median age 36 years) compared to those with non-G-CIMP tumors (median age 59 years). Of note, patients with the G-CIMP proneural subtype had a significantly better survival (median 150 weeks) compared with those who have non-G-CIMP proneural glioblastoma (median 42 weeks). Overall, patients with G-CIMP proneural glioblastoma had the best survival compared to all other genetic subtypes. Additionally, G-CIMP tumors were highly associated with IDH-1 mutations with 78 % of G-CIMP tumor having IDH-1 mutations. The study also showed the G-CIMP IDH-1 mutant tumors were mostly enriched in recurrent or secondary glioblastoma subtypes.

2.3.3 IDH1 Mutation and Relationship to Epigenetic Changes in Gliomas

An unbiased genomic analysis of glioblastoma by Parsons et al. revealed previously unidentified heterozygous somatic mutations in isocitrate dehydrogenase 1 (*IDH1*) gene predominantly in young patients and in those with secondary glioblastoma which were associated with longer survival (median overall survival of 3.8 years as compared to 1.1 years for patients with wild-type *IDH1*). These mutations affect a highly conserved histidine residue at amino acid position 132 of IDH1 which is located in the binding pocket of isocitrate to the enzyme which functions in the cytosol catalyzing the conversion of isocitrate to α-ketoglutarate. The mutations occur most frequently in transcript position 395 (guanine changed to adenine) which replace the arginine at amino acid residue 132 of the protein with histidine (R132H).

A more detailed follow-up analysis by the same group reported by Yan et al. reported additional mutations changing the conserved arginine residue to cysteine (R132C), leucine (R132L), serine (R132S), and glycine (R132G). In addition, mutations were noted in IDH2 in position 172 replacing arginine at this site to glycine (R172G), methionine (R172M), or lysine (R172K) again affecting the binding site of isocitrate to the enzyme analogous to the R132 site in IDH1 (Yan et al. 2009). Although these mutations were reminiscent of activating mutations seen in other genes, assessment of enzymatic activity of the mutants in a paradigm of exogenous overexpression demonstrated that mutant proteins had a lower level of enzymatic activity. Also, in paired samples from low grade tumors and from their subsequent counterparts after progression to higher grade glioma, IDH mutations were seen to be among the earliest of gene alterations occurring along with *TP53* mutations but not with other well-known mutations most often seen in primary glioblastoma (Ichimura et al. 2009; Watanabe et al. 2009).

Patients with *IDH* mutations had an improved overall survival compared to those with wild-type *IDH* in both glioblastoma and anaplastic astrocytoma. Of particular relevance, patients with glioblastoma with *IDH* mutations appeared to have a median overall survival (31 months) that was better than those with anaplastic astrocytoma with wild-type *IDH* (20 months) which suggests a potential for this prognostic marker to be superior to histologic subtype in terms of predicting survival outcomes, a retrospective finding that will need to be prospectively validated (Yan et al. 2009).

Although mutations of *IDH1* results in a loss of ability to convert isocitrate to α-ketoglutarate, the heterozygous nature of the mutation suggested that this was not simply a loss of function alteration; in an elegant study analyzing the changes in metabolites in cells overexpressing mutant IDH1 compared with those that overexpressed wild-type IDH1, Dang and colleagues showed that mutant IDH1 has an altered substrate specificity and unlike wild-type IDH1, catalyzes a NADPH-dependent reduction of α-ketoglutarate to 2-hydroxyglutarate (2-HG) which constituted a gain of function mutation (Dang et al. 2009). This was confirmed in glioma specimens with mutant IDH1 which showed increased 2-HG levels. These functions suggested a role for 2-HG as an oncometabolite with a role in early stages of glioma formation although the mechanism by which increased 2-HG influenced glioma formation remained uncertain.

Insights into the possible mechanism emerged when it was demonstrated that 2-HG could function as a competitive inhibitor of α-ketoglutarate-dependent dioxygenases including AlkB family of oxidative demethylases, the histone lysine demethylases, and the TET family of methylcytosine hydroxylases (Chowdhury et al. 2011; Figueroa et al. 2010; Xu et al. 2011b). 2-HG expression in cells with IDH mutations was shown to cause global hypermethylation which was similar to that induced by *TET2* loss of function mutations in AML cells suggesting a possible interaction between 2-HG expression and TET enzyme function. The TET family of enzymes are important regulators of epigenetic controls of gene expression via demethylation of DNA. They catalyze the conversion of 5-methyl cytosine (5mC) initially to 5-hydroxymethylcytosine (5hmC) and subsequently to 5-formylcytosine (5fC) and 5-carboxylcytosine (5caC) (He et al. 2011; Ito et al. 2010); these residues are susceptible to guanine-cytosine base-pair mismatches and subsequent base-excision repair that result in unmethylated cytosine. This results in widespread demethylation in specific loci of DNA with variable global demethylation. Elevated levels of 2-HG that can result from *IDH* mutations can inhibit TET2 function and prevent demethylation of DNA causing genome-wide areas of hypermethylation In some cases of lower grade gliomas without *IDH* mutations, TET2 promoter methylation that may cause gene silencing have been reported (Kim et al. 2011). These findings suggested an intimate association between IDH mutations, lowered α-ketoglutarate levels, elevated 2-HG levels, and the epigenetic regulator mechanisms of glioma cells (Losman and Kaelin 2013). Of note, unlike acute myelogenous leukemia in which *IDH* mutations are associated with a worse prognosis, glioma patients with mutant IDH have better survival than their counterparts with wild-type proteins. This may seem apparently contradictory to the statement that *IDH* mutations are among the earliest oncogenic events and that 2-HG is an oncometabolite that drives tumor progression. One possible explanation for this is that the role of IDH may be different in these two tumor types. Additionally, the genes that are hypermethylated and presumable silenced may have different implication to tumor biology and hence may yield different outcomes. The frequent co-occurrence of *TP53* mutations in gliomas in association with the *IDH1*^{R132H} mutations is not seen in AML suggesting differences in the role of these mutations in gliomas. Accumulation of 2-HG in IDH1 mutant cells appeared to inhibit proliferation rate and has an adverse effect on glioma cells. Glioma cell lines stably expressing *IDH1*^{R132H} have lower proliferation and migration rates suggesting that this mutation and consequent increase in 2-HG are inhibitory to glioma growth (Bralten et al. 2011). Corresponding to this, IDH mutant gliomas have shown a better outcome and increased sensitivity to chemotherapy (Houillier et al. 2010; SongTao et al. 2012).

Expression of mutant IDH1 can be detected by routine immunohistochemistry using a R132H mutation specific IDH1 antibody (Capper et al. 2009). This antibody has become very useful for identifying such mutant protein without the need for sequencing the gene. The antibody has also demonstrated utility in demonstrating the extent of IDH1 mutant tumor by highlighting the infiltrative mutant IDH1 positive cells against the wild-type background of normal brain cells (Capper et al. 2010); further, the antibody helps discriminate tumor cells from reactive astrocytes in the . However, approximately 20 % of immunohistochemistry negative tumors may harbor IDH1 mutations that can only be detected by sequencing. This may be important in patients with low grade or anaplastic tumors or those with secondary glioblastoma in whom the likelihood of IDH1 mutation is highest. Such mutations have not been seen in pilocytic astrocytoma and ependymoma patients; thus in cases where there is diagnostic uncertainty related to these tumors, the presence of mutant IDH1 may suggest alternative diagnosis. The ability to detection mutant IDH1 in routine clinical practice using immunohistochemistry provides an easy way to identify tumors with a better overall prognosis and as a surrogate for the epigenetic changes induced by this mutation.

2.3.4 IDH1 Mutation and Relationship to G-CIMP

IDH mutations are seen in young patients with grade 2 or 3 gliomas or secondary glioblastoma and are associated with improved survival. *IDH* mutations are also associated with increased 2-HG which in turn results in hypermethylation of DNA. G-CIMP is predominantly seen in a subset of proneural subtype of glioblastoma or in lower grade tumors in young people and is also associated with improved survival. The association of G-CIMP with IDH mutations raised the possibility that the epigenetic effects of IDH mutations may be the mechanism for the hypermethylator phenotype seen with G-CIMP. In a study examining whether expression of mutant IDH1 directly induces a hypermethylator phenotype similar to G-CIMP, exogenous expression of mutant IDH1 resulted in hypermethylation of a set of genes that accurately classified a set of low grade gliomas with endogenous mutant IDH1 into G-CIMP and non-G-CIMP tumors. Upon analysis of TCGA data using the gene set associated with IDH mutation, the gene signature was able to separate only the proneural glioblastoma into two subgroups with different prognoses, but not the other subtypes, similar to G-CIMP status (Turcan et al. 2012). These results strongly suggested that IDH-1 mutations drive the methylation that characterizes G-CIMP status. This could potentially mean that IDH1 status that can be detected by immunohistochemistry could be used as a surrogate for the G-CIMP status in patients with proneural glioblastoma and in lower grade gliomas.

2.3.5 Histone H3 and Chromatin Remodeling Gene Mutations in Pediatric Gliomas

Brain tumors are the most frequent solid tumors in childhood and are a significant cause for morbidity and mortality in this age group. However, these tumors have not been as well characterized as adult brain tumors until recently. Pediatric gliomas constitute a distinct set of tumors with significant differences from their adult counterparts in clinical characteristics and in responses to therapy. With the advent of genome-wide assessment of alterations in malignancies, genomic studies in pediatric tumors have also been recently undertaken. Glioblastoma and diffuse intrinsic pontine glioma (DIPG) are aggressive tumors in pediatric patients (Broniscer and Gajjar 2004). Given the limited tissue-based studies that have been conducted on these tumors, their molecular nature has been poorly characterized.

The initial glimpse into the genomic landscape of these tumors came from whole-exome sequencing study of pediatric glioblastoma which revealed for the first time that mutations in histones can be associated with malignancies. Nearly half of these tumors show somatic mutations in the genes encoding histone H3.3 and components of a chromatin remodeling complex including ATRX and DAXX, which chaperones histone H3 for replication independent chromatin assembly at telomeres; 31 % showed somatic mutations within *H3F3A* gene consisting of an adenine-to-thymine or guanine-to-adenine transversion resulting respectively in a substitution of the lysine at position 27 to methionine (K27M) and that of glycine at position 34 to arginine or valine (G34R/G34V) in the tail of histone H3.3 which is involved in key regulatory posttranslational modifications. In addition, mutations in *ATRX* and *DAXX* were seen in 31 % of all samples overall and in all tumors that had G34R/G34V H3.3 mutations (Schwartzentruber et al. 2012). A subsequent whole genome DNA sequencing study of diffuse infiltrative pontine glioma (DIPG) and matched normal tissue showed that this tumor type also had somatic mutations in

H3F3A and *HIST1H3B* (encoding a related isoform histone H3.1), with p.K27M in their N-terminal tail. Additional studies showed that this finding was seen in 78 % of DIPG and 22 % of non-brain stem high grade gliomas. Further, somatic mutation resulting in a p.G34R alteration in *H3F3A* was seen in 15 % of non-brain stem high grade gliomas but not in the DIPGs. These mutations were not seen in pediatric low grade gliomas, ependymoma, medulloblastoma, or non-CNS pediatric tumors. In a study of thalamic gliomas in young adults, *H3F3A* K27M mutations were seen in >90 % of patients younger than 50 years. Interestingly, the specimens from young adults which had a *H3F3A* K27M mutation also showed a high frequency (nearly 80 %) of unmethylated *MGMT* promoter. Mechanistic studies to understand the functional relevance of these mutations have shown that the mutant histone H3 is aberrantly recruited to and inhibits the activity of polycomb repressor complex 2 (PRC2, which methylates H3K27) by interacting with its EZH2 subunit and consequently reduces the overall amounts of H3K27 trimethylation (H3K27me3) which in turn resulted in activation of genes that are known to be involved in gliomagenesis; simultaneously, a set of genes showed an increased level of H3K27me3 occupancy in their promoters and consequent reduction in expression—these included genes whose protein products are known to have immunity and cell cycle regulatory functions (Bender et al. 2013).

Mutations in other genes involved in epigenetic regulation have also been noted; a study assessing pediatric high grade gliomas identified truncating loss-of-function mutations in *SETD2*, which encodes a H3K36 trimethyltransferase, in 15 % of the patients (Fontebasso et al. 2013). These tumors occurred in the cerebral hemispheres and showed a significant decrease in H3K36me3 levels. The mutations were also mutually exclusive of *H3F3A* mutations in these tumors.

2.4 Noncoding RNA and Brain Tumors

More than 98 % of the genome consists predominantly of long stretches of repeat sequences and fragments of viral sequences which are thought to be remnants of evolutionary processes and have long been considered "junk" DNA. However, in recent years, it has become increasingly evident that these regions contain numerous inter- and intragenic sequences that are transcribed into special types of RNA but are not translated into proteins (Alexander et al. 2010). Such noncoding RNA have been of specific interest given their increasingly evident role in health and disease. The first of these RNA to be discovered were microRNA (miRNA) which are small noncoding RNA containing about 22 base pairs and which are conserved across species. They function as transcriptional and posttranscriptional regulators of gene expression by base-paring with 3′ or 5′ untranslated area of mRNA, preventing their translation and facilitating their degradation. A miRNA can target multiple RNA and conversely, multiple miRNA can target a single mRNA; this has made it challenging to delineate the regulatory networks of miRNA (Pasquinelli 2012). Aberrant expression or regulation of miRNA is closely associated with malignancies including brain tumors. Over a thousand miRNAs have been identified and in a comprehensive review of the expression of these miRNA in brain tumors, it was reported that 253 miRNAs were significantly upregulated and 95 significantly downregulated with 17 others that were inconsistently reported to have altered levels (Moller et al. 2013).

One of the first reports of a role for miRNA in brain tumors was the identification of significant overexpression of miR-21 in glioblastoma and glioma cell lines (Chan et al. 2005). The higher levels of this miRNA appears to have a functional relevance in tumor cell survival; knockdown of miR-21 in glioma cells led to apoptosis via caspases activation suggesting a protective role for this miRNA in high grade gliomas. Subsequent studies showed that miR-21 promotes glioma invasion by targeting matrix metalloproteinase regulators (Gabriely et al. 2008). A more global microarray-based analysis of miRNA expression in glioblastoma and glioma cells lines reported overexpression of miR-221and miR-10b and reduced expression of miR-128 and miR-181 family but did not identify miR-21 overexpression (Ciafre et al. 2005). miR10b expression was related to genes encoding products related to invasion and migration including metalloproteinases, uPAR (urokinase receptor), and RhoC (Ras homolog gene family member C) (Sasayama et al. 2009; Sun et al. 2011). A critical insight into the relationship between miRNA expression and known regulators of glioma emerged when it was shown that miR-26a was a regulator of PTEN expression and was frequently amplified at the human glioma in association with monoallelic *PTEN* loss. A direct role for miR-26a was seen in a mouse model of glioma in which overexpression of miR26a enhanced tumor formation providing evidence for a novel miRNA-based mechanism for PTEN regulation in glioma(Huse et al. 2009). Conversely, another miRNA, miR-7, was reported to inhibit EGFR and Akt pathway and its downregulation in glioblastoma allowed a more aggressive phenotype (Kefas et al. 2008).

In another functional paradigm for miRNA, miR-25 and -32 were shown to be involved in an autoregulatory loop in glioma cells in which p53 mediated their repression by downregulating E2F1 and MYC which are transcriptional regulators of these miRNA. miR-25 and -32 in turn downregulate Mdm2 and TSC1, the negative regulators of p53 and the mTOR, respectively allowing p53 accumulation and causing decreased cell proliferation in vitro and in vivo through cell cycle arrest. However, in the setting of defective or nonfunctional p53 as frequently seen in tumors, MYC-driven expression of miR-25 and miR-32 can drive proliferation and survival supporting tumorigenesis (Suh et al. 2012).

Several other types of noncoding RNA have been identified and appear to play a role in malignancies; these include transcribed ultraconserved regions (T-UCRs), PIWI-interacting RNAs (piRNAs), large intergenic noncoding RNAs (lincRNAs) small nucleolar RNAs (snoRNAs), and long noncoding RNAs (lncRNAs) (Esteller 2011). The roles for these in brain tumors are only beginning to be understood. It is also clear that there is a complex interplay between known growth factors and signal transduction molecules and their regulation through noncoding RNA function (positive or negative) than will be critical in gaining insights into new targets for therapy.

2.5 Clinical Trials of Epigenetic Modulators in Gliomas

Several preclinical studies have shown that histone deacetylase (HDAC) inhibitors can induce apoptosis, reduce migration, cause cell cycle arrest, and inhibit angiogenesis in glioma models in vitro and in vivo (Sawa et al. 2001; 2002; Xu et al. 2011a). A phase II trial of vorinostat (suberoyl-anilide hydroxamic acid, SAHA), an orally bioavailable HDAC inhibitor in patients with recurrent glioblastoma, met its efficacy endpoint with a 6-month progression-free survival rate (PFS-6) of 15.2 % which compared favorably with the historical PFS-6 of ~10 % for prior single agent trials of signal transduction agents (Galanis et al. 2009). Based on preclinical studies suggesting a synergy between HDAC inhibition and proteasome inhibition (Asklund et al. 2012), a phase II study of vorinostat in combination with the proteasome inhibitor, bortezomib, was conducted in patients with recurrent glioblastoma; however, the study was closed early after interim analysis showed a PFS-6 of 0 % (Friday et al. 2012). A phase I trial of vorinostat with temozolomide showed that the combination was well tolerated and led to an intergroup (Alliance and adult brain tumor consortium) phase II study of vorinostat in combination with chemoradiation therapy for patients with newly diagnosed glioblastoma which is ongoing. In the setting of recurrent disease, an adaptive randomized multicenter phase II trial comparing bevacizumab alone versus vorinostat plus bevacizumab in patients with recurrent glioblastoma using a novel adaptive randomized Bayesian design is currently accruing patients in the USA through the Brain Tumor Trials Collaborative (BTTC). Another BTTC adaptive randomized trial of vorinostat with or without isotretinoin and dose dense temozolomide in patients with recurrent glioblastoma is also open to accrual.

Valproic acid (VPA), which has long been used as an anticonvulsant in the treatment of patients with gliomas, is also known to be an HDAC inhibitor. In a retrospective analysis of a EORTC/NCIC trial comparing RT with RT plus TMZ, patients receiving VPA alone as an anticonvulsant had an improved survival with TMZ/RT than patients receiving another anticonvulsant alone or patients not receiving any anticonvulsant, an effect that may potentially be related to its HDAC inhibitory effects (Weller et al. 2011). An ongoing study against pediatric high grade gliomas of VPA combined with radiation followed by adjuvant VPA and bevacizumab seeks to examine the efficacy of VPA mediated HDAC inhibition in sensitizing these tumors to radiation therapy and bevacizumab. A similar study of VPA combined with temozolomide and RT in adults with high grade gliomas is currently ongoing.

2.6 Summary and Future Directions

The influence of epigenetics on the malignant process, initially considered as a "passenger" event in the evolution of tumors has recently gained center stage due to its rapidly expanding role in all aspects of tumorigenesis. Of particular note, nearly all major biomarkers associated with prognosis in gliomas are epigenetic in nature, including *MGMT* promoter methylation, IDH1 mutations and their influence on global methylation, and G-CIMP and histone H3 mutations. Thus epigenetic alterations have gained center stage in the pathobiology and clinical status of these tumors. Various studies indicate that epigenetic processes by virtue of their ability to influence a large number of genes and consequently, various biological functions regulated by their protein products, may be critical in establishing the complex aberrant signaling networks established by tumors and contribute to their heterogeneity, adaptive treatment resistance, and clinical course. Conversely, a better understanding of these changes will very likely yield new opportunities for therapeutic interventions against brain tumors.

References

Alexander RP, Fang G, Rozowsky J, Snyder M, Gerstein MB (2010) Annotating non-coding regions of the genome. Nat Rev Genet 11:559–571

Asklund T, Kvarnbrink S, Holmlund C, Wibom C, Bergenheim T, Henriksson R, Hedman H (2012) Synergistic killing of glioblastoma stem-like cells by bortezomib and HDAC inhibitors. Anticancer Res 32:2407–2413

Baer JC, Freeman AA, Newlands ES, Watson AJ, Rafferty JA, Margison GP (1993) Depletion of O6-alkylguanine-DNA alkyltransferase correlates with potentiation of temozolomide and CCNU toxicity in human tumour cells. Br J Cancer 67:1299–1302

Bannister AJ, Kouzarides T (1996) The CBP co-activator is a histone acetyltransferase. Nature 384:641–643

Baylin SB, Jones PA (2011) A decade of exploring the cancer epigenome – biological and translational implications. Nat Rev Cancer 11:726–734

Belanich M, Randall T, Pastor MA, Kibitel JT, Alas LG, Dolan ME, Schold SC Jr, Gander M, Lejeune FJ, Li BF et al (1996) Intracellular Localization and intercellular heterogeneity of the human DNA repair protein O(6)-methylguanine-DNA methyltransferase. Cancer Chemother Pharmacol 37:547–555

Bender S, Tang Y, Lindroth AM, Hovestadt V, Jones DT, Kool M, Zapatka M, Northcott PA, Sturm D, Wang W et al (2013) Reduced H3K27me3 and DNA hypomethylation are major drivers of gene expression in K27M mutant pediatric high-grade gliomas. Cancer Cell 24:660–672

Bobola MS, Tseng SH, Blank A, Berger MS, Silber JR (1996) Role of O6-methylguanine-DNA methyltransferase in resistance of human brain tumor cell lines to the clinically relevant methylating agents temozolomide and streptozotocin. Clin Cancer Res 2:735–741

Bralten LB, Kloosterhof NK, Balvers R, Sacchetti A, Lapre L, Lamfers M, Leenstra S, de Jonge H, Kros JM, Jansen EE et al (2011) IDH1 R132H decreases proliferation of glioma cell lines in vitro and in vivo. Ann Neurol 69:455–463

Brandes AA, Franceschi E, Tosoni A, Bartolini S, Bacci A, Agati R, Ghimenton C, Turazzi S, Talacchi A, Skrap M et al (2010) O(6)-methylguanine DNA-methyltransferase methylation status can change between first surgery for newly diagnosed glioblastoma and second surgery for recurrence: clinical implications. Neuro Oncol 12:283–288

Broniscer A, Gajjar A (2004) Supratentorial high-grade astrocytoma and diffuse brainstem glioma: two challenges for the pediatric oncologist. Oncologist 9:197–206

Buttarelli FR, Massimino M, Antonelli M, Lauriola L, Nozza P, Donofrio V, Arcella A, Oliva MA, Di Rocco C, Giangaspero F (2010) Evaluation status and prognostic significance of O6-methylguanine-DNA methyltransferase (MGMT) promoter methylation in pediatric high grade gliomas. Childs Nerv Syst 26:1051–1056

Calin GA, Ferracin M, Cimmino A, Di Leva G, Shimizu M, Wojcik SE, Iorio MV, Visone R, Sever NI, Fabbri M et al (2005) A MicroRNA signature associated with prognosis and progression in chronic lymphocytic leukemia. N Engl J Med 353:1793–1801

Capper D, Weissert S, Balss J, Habel A, Meyer J, Jager D, Ackermann U, Tessmer C, Korshunov A, Zentgraf H et al (2010) Characterization of R132H mutation-specific IDH1 antibody binding in brain tumors. Brain Pathol 20:245–254

Capper D, Zentgraf H, Balss J, Hartmann C, von Deimling A (2009) Monoclonal antibody specific for IDH1 R132H mutation. Acta Neuropathol 118:599–601

Chan JA, Krichevsky AM, Kosik KS (2005) MicroRNA-21 is an antiapoptotic factor in human glioblastoma cells. Cancer Res 65:6029–6033

Chen ZP, Yarosh D, Garcia Y, Tampieri D, Mohr G, Malapetsa A, Langleben A, Panasci LC (1999) Relationship between O6-methylguanine-DNA methyltransferase levels and clinical response induced by chloroethylnitrosourea therapy in glioma patients. Can J Neurol Sci 26:104–109

Chinot OL, Wick W, Mason W, Henriksson R, Saran F, Nishikawa R, Carpentier AF, Hoang-Xuan K, Kavan P, Cernea D et al (2014) Bevacizumab plus radiotherapy-temozolomide for newly diagnosed glioblastoma. N Engl J Med 370:709–722

Chowdhury R, Yeoh KK, Tian YM, Hillringhaus L, Bagg EA, Rose NR, Leung IK, Li XS, Woon EC, Yang M et al (2011) The oncometabolite 2-hydroxyglutarate inhibits histone lysine demethylases. EMBO Rep 12:463–469

Ciafre SA, Galardi S, Mangiola A, Ferracin M, Liu CG, Sabatino G, Negrini M, Maira G, Croce CM, Farace MG (2005) Extensive modulation of a set of microRNAs in primary glioblastoma. Biochem Biophys Res Commun 334:1351–1358

Cimmino L, Abdel-Wahab O, Levine RL, Aifantis I (2011) TET family proteins and their role in stem cell differentiation and transformation. Cell Stem Cell 9:193–204

Costello JF, Futscher BW, Kroes RA, Pieper RO (1994a) Methylation-related chromatin structure is associated with exclusion of transcription factors from and suppressed expression of the O-6-methylguanine DNA methyltransferase gene in human glioma cell lines. Mol Cell Biol 14:6515–6521

Costello JF, Futscher BW, Tano K, Graunke DM, Pieper RO (1994b) Graded methylation in the promoter and body of the O6-methylguanine DNA methyltransferase (MGMT) gene correlates with MGMT expression in human glioma cells. J Biol Chem 269:17228–17237

Dang L, White DW, Gross S, Bennett BD, Bittinger MA, Driggers EM, Fantin VR, Jang HG, Jin S, Keenan MC et al (2009) Cancer-associated IDH1 mutations produce 2-hydroxyglutarate. Nature 462:739–744

Dawson MA, Prinjha RK, Dittmann A, Giotopoulos G, Bantscheff M, Chan WI, Robson SC, Chung CW, Hopf C, Savitski MM et al (2011) Inhibition of BET recruitment to chromatin as an effective treatment for MLL-fusion leukaemia. Nature 478:529–533

Donson AM, Addo-Yobo SO, Handler MH, Gore L, Foreman NK (2007) MGMT promoter methylation correlates with survival benefit and sensitivity to temozolomide in pediatric glioblastoma. Pediatr Blood Cancer 48:403–407

Esteller M (2011) Non-coding RNAs in human disease. Nat Rev Genet 12:861–874

Esteller M, Garcia-Foncillas J, Andion E, Goodman SN, Hidalgo OF, Vanaclocha V, Baylin SB, Herman JG (2000) Inactivation of the DNA-repair gene MGMT and the clinical response of gliomas to alkylating agents. N Engl J Med 343:1350–1354

Fassan M, Tassone E, Onisto M, Perilongo G, D'Avella D, Gardiman MP (2011) MGMT promoter methylation in pediatric high-grade gliomas. Childs Nerv Syst 27:7–8

Figueroa ME, Abdel-Wahab O, Lu C, Ward PS, Patel J, Shih A, Li Y, Bhagwat N, Vasanthakumar A, Fernandez HF et al (2010) Leukemic IDH1 and IDH2 mutations result in a hypermethylation phenotype, disrupt TET2 function, and impair hematopoietic differentiation. Cancer Cell 18:553–567

Filippakopoulos P, Qi J, Picaud S, Shen Y, Smith WB, Fedorov O, Morse EM, Keates T, Hickman TT, Felletar I et al (2010) Selective inhibition of BET bromodomains. Nature 468:1067–1073

Fontebasso AM, Schwartzentruber J, Khuong-Quang DA, Liu XY, Sturm D, Korshunov A, Jones DT, Witt H, Kool M, Albrecht S et al (2013) Mutations in SETD2 and genes affecting histone H3K36 methylation target hemispheric high-grade gliomas. Acta Neuropathol 125:659–669

Friday BB, Anderson SK, Buckner J, Yu C, Giannini C, Geoffroy F, Schwerkoske J, Mazurczak M, Gross H, Pajon E et al (2012) Phase II trial of vorinostat in combination with bortezomib in recurrent glioblastoma: a north central cancer treatment group study. Neuro Oncol 14:215–221

Friedman HS, McLendon RE, Kerby T, Dugan M, Bigner SH, Henry AJ, Ashley DM, Krischer J, Lovell S, Rasheed K et al (1998) DNA mismatch repair and O6-alkylguanine-DNA alkyltransferase analysis and response to Temodal in newly diagnosed malignant glioma. J Clin Oncol 16:3851–3857

Gabriely G, Wurdinger T, Kesari S, Esau CC, Burchard J, Linsley PS, Krichevsky AM (2008) MicroRNA 21 promotes glioma invasion by targeting matrix metalloproteinase regulators. Mol Cell Biol 28:5369–5380

Galanis E, Jaeckle KA, Maurer MJ, Reid JM, Ames MM, Hardwick JS, Reilly JF, Loboda A, Nebozhyn M, Fantin VR et al (2009) Phase II trial of vorinostat in recurrent glioblastoma multiforme: a north central cancer treatment group study. J Clin Oncol 27:2052–2058

Gerson SL, Trey JE, Miller K, Berger NA (1986) Comparison of O6-alkylguanine-DNA alkyltransferase activity based on cellular DNA content in human, rat and mouse tissues. Carcinogenesis 7:745–749

Gilbert MR, Dignam JJ, Armstrong TS, Wefel JS, Blumenthal DT, Vogelbaum MA, Colman H, Chakravarti A, Pugh S, Won M et al (2014) A randomized trial of bevacizumab for newly diagnosed glioblastoma. N Engl J Med 370:699–708

Gilbert MR, Wang M, Aldape KD, Stupp R, Hegi ME, Jaeckle KA, Armstrong TS, Wefel JS, Won M, Blumenthal DT et al (2013) Dose-dense temozolomide for newly diagnosed glioblastoma: a randomized phase III clinical trial. J Clin Oncol 31:4085–4091

Harris LC, Remack JS, Brent TP (1994) In vitro methylation of the human O6-methylguanine-DNA methyltransferase promoter reduces transcription. Biochim Biophys Acta 1217:141–146

He YF, Li BZ, Li Z, Liu P, Wang Y, Tang Q, Ding J, Jia Y, Chen Z, Li L et al (2011) Tet-mediated formation of 5-carboxylcytosine and its excision by TDG in mammalian DNA. Science 333:1303–1307

Hegi ME, Diserens AC, Gorlia T, Hamou MF, de Tribolet N, Weller M, Kros JM, Hainfellner JA, Mason W, Mariani L et al (2005) MGMT gene silencing and benefit from temozolomide in glioblastoma. N Engl J Med 352:997–1003

Herman JG, Graff JR, Myohanen S, Nelkin BD, Baylin SB (1996) Methylation-specific PCR: a novel PCR assay for methylation status of CpG islands. Proc Natl Acad Sci U S A 93:9821–9826

Houillier C, Wang X, Kaloshi G, Mokhtari K, Guillevin R, Laffaire J, Paris S, Boisselier B, Idbaih A, Laigle-Donadey F et al (2010) IDH1 or IDH2 mutations predict longer survival and response to temozolomide in low-grade gliomas. Neurology 75:1560–1566

Huse JT, Brennan C, Hambardzumyan D, Wee B, Pena J, Rouhanifard SH, Sohn-Lee C, le Sage C, Agami R, Tuschl T, Holland EC (2009) The PTEN-regulating microRNA miR-26a is amplified in high-grade glioma and facilitates gliomagenesis in vivo. Genes Dev 23:1327–1337

Ichimura K, Pearson DM, Kocialkowski S, Backlund LM, Chan R, Jones DT, Collins VP (2009) IDH1 mutations are present in the majority of common adult gliomas but rare in primary glioblastomas. Neuro Oncol 11:341–347

Ito S, D'Alessio AC, Taranova OV, Hong K, Sowers LC, Zhang Y (2010) Role of Tet proteins in 5mC to 5hmC conversion, ES-cell self-renewal and inner cell mass specification. Nature 466:1129–1133

Iyer NG, Ozdag H, Caldas C (2004) p300/CBP and cancer. Oncogene 23:4225–4231

Jaeckle KA, Eyre HJ, Townsend JJ, Schulman S, Knudson HM, Belanich M, Yarosh DB, Bearman SI, Giroux DJ, Schold SC (1998) Correlation of tumor O6 methylguanine-DNA methyltransferase levels with survival of malignant astrocytoma patients treated with bis-chloroethylnitrosourea: a Southwest Oncology Group study. J Clin Oncol 16:3310–3315

Jones PA (2012) Functions of DNA methylation: islands, start sites, gene bodies and beyond. Nat Rev Genet 13:484–492

Jones PA, Baylin SB (2007) The epigenomics of cancer. Cell 128:683–692

Kefas B, Godlewski J, Comeau L, Li Y, Abounader R, Hawkinson M, Lee J, Fine H, Chiocca EA, Lawler S, Purow B (2008) microRNA-7 inhibits the epidermal growth factor receptor and the Akt pathway and is down-regulated in glioblastoma. Cancer Res 68:3566–3572

Kim YH, Pierscianek D, Mittelbronn M, Vital A, Mariani L, Hasselblatt M, Ohgaki H (2011) TET2 promoter methylation in low-grade diffuse gliomas lacking IDH1/2 mutations. J Clin Pathol 64:850–852

Kogo R, Shimamura T, Mimori K, Kawahara K, Imoto S, Sudo T, Tanaka F, Shibata K, Suzuki A, Komune S et al (2011) Long noncoding RNA HOTAIR regulates polycomb-dependent chromatin modification and is associated with poor prognosis in colorectal cancers. Cancer Res 71:6320–6326

Lee JY, Park CK, Park SH, Wang KC, Cho BK, Kim SK (2011) MGMT promoter gene methylation in pediatric glioblastoma: analysis using MS-MLPA. Childs Nerv Syst 27:1877–1883

Lee SM, Reid H, Elder RH, Thatcher N, Margison GP (1996) Inter- and intracellular heterogeneity of O6-alkylguanine-DNA alkyltransferase expression in human brain tumors: possible significance in nitrosourea therapy. Carcinogenesis 17:637–641

Ley TJ, Ding L, Walter MJ, McLellan MD, Lamprecht T, Larson DE, Kandoth C, Payton JE, Baty J, Welch J et al (2010) DNMT3A mutations in acute myeloid leukemia. N Engl J Med 363:2424–2433

Losman JA, Kaelin WG Jr (2013) What a difference a hydroxyl makes: mutant IDH, (R)-2-hydroxyglutarate, and cancer. Genes Dev 27:836–852

Malley DS, Hamoudi RA, Kocialkowski S, Pearson DM, Collins VP, Ichimura K (2011) A distinct region of the MGMT CpG island critical for transcriptional regulation is preferentially methylated in glioblastoma cells and xenografts. Acta Neuropathol 121:651–661

McCabe MT, Ott HM, Ganji G, Korenchuk S, Thompson C, Van Aller GS, Liu Y, Graves AP, Della Pietra A III, Diaz E et al (2012) EZH2 inhibition as a therapeutic strategy for lymphoma with EZH2-activating mutations. Nature 492:108–112

Minniti G, Salvati M, Arcella A, Buttarelli F, D'Elia A, Lanzetta G, Esposito V, Scarpino S, Maurizi Enrici R, Giangaspero F (2011) Correlation between O6-methylguanine-DNA methyltransferase and survival in elderly patients with glioblastoma treated with radiotherapy plus concomitant and adjuvant temozolomide. J Neurooncol 102:311–316

Møller HG, Rasmussen AP, Andersen HH, Johnsen KB, Henriksen M, Duroux M (2013) A systematic review of microRNA in glioblastoma multiforme: micro-modulators in the mesenchymal mode of migration and invasion. Mol Neurobiol 47:131–144

Noushmehr H, Weisenberger DJ, Diefes K, Phillips HS, Pujara K, Berman BP, Pan F, Pelloski CE, Sulman EP, Bhat KP et al (2010) Identification of a CpG island methylator phenotype that defines a distinct subgroup of glioma. Cancer Cell 17:510–522

Pasquinelli AE (2012) MicroRNAs and their targets: recognition, regulation and an emerging reciprocal relationship. Nat Rev Genet 13:271–282

Pieper RO, Futscher BW, Dong Q, Ellis TM, Erickson LC (1990) Comparison of O-6-methylguanine DNA methyltransferase (MGMT) mRNA levels in Mer+ and Mer- human tumor cell lines containing the MGMT gene by the polymerase chain reaction technique. Cancer Commun 2:13–20

Qian XC, Brent TP (1997) Methylation hot spots in the 5′ flanking region denote silencing of the O6-methylguanine-DNA methyltransferase gene. Cancer Res 57:3672–3677

Sasayama T, Nishihara M, Kondoh T, Hosoda K, Kohmura E (2009) MicroRNA-10b is overexpressed in malignant glioma and associated with tumor invasive factors, uPAR and RhoC. Int J Cancer 125:1407–1413

Sawa H, Murakami H, Ohshima Y, Murakami M, Yamazaki I, Tamura Y, Mima T, Satone A, Ide W, Hashimoto I, Kamada H (2002) Histone deacetylase inhibitors such as sodium butyrate and trichostatin A inhibit vascular endothelial growth factor (VEGF) secretion from human glioblastoma cells. Brain Tumor Pathol 19:77–81

Sawa H, Murakami H, Ohshima Y, Sugino T, Nakajyo T, Kisanuki T, Tamura Y, Satone A, Ide W, Hashimoto I, Kamada H (2001) Histone deacetylase inhibitors such as sodium butyrate and trichostatin A induce apoptosis through an increase of the bcl-2-related protein Bad. Brain Tumor Pathol 18:109–114

Schlosser S, Wagner S, Muhlisch J, Hasselblatt M, Gerss J, Wolff JE, Fruhwald MC (2010) MGMT as a potential stratification marker in relapsed high-grade glioma of children: the HIT-GBM experience. Pediatr Blood Cancer 54:228–237

Schwartzentruber J, Korshunov A, Liu XY, Jones DT, Pfaff E, Jacob K, Sturm D, Fontebasso AM, Quang DA, Tonjes M et al (2012) Driver mutations in histone H3.3 and chromatin remodelling genes in paediatric glioblastoma. Nature 482:226–231

Shah N, Lin B, Sibenaller Z, Ryken T, Lee H, Yoon JG, Rostad S, Foltz G (2011) Comprehensive analysis of MGMT promoter methylation: correlation with MGMT expression and clinical response in GBM. PLoS One 6:e16146

Silber JR, Blank A, Bobola MS, Ghatan S, Kolstoe DD, Berger MS (1999) O6-methylguanine-DNA methyltransferase-deficient phenotype in human gliomas: frequency and time to tumor progression after alkylating agent-based chemotherapy. Clin Cancer Res 5:807–814

SongTao Q, Lei Y, Si G, YanQing D, HuiXia H, XueLin Z, LanXiao W, Fei Y (2012) IDH mutations predict longer survival and response to temozolomide in secondary glioblastoma. Cancer Sci 103:269–273

Srivastava A, Jain A, Jha P, Suri V, Sharma MC, Mallick S, Puri T, Gupta DK, Gupta A, Sarkar C (2010) MGMT gene promoter methylation in pediatric glioblastomas. Childs Nerv Syst 26:1613–1618

Stupp R, Mason WP, van den Bent MJ, Weller M, Fisher B, Taphoorn MJ, Belanger K, Brandes AA, Marosi C, Bogdahn U et al (2005) Radiotherapy plus concomitant and adjuvant temozolomide for glioblastoma. N Engl J Med 352:987–996

Suh SS, Yoo JY, Nuovo GJ, Jeon YJ, Kim S, Lee TJ, Kim T, Bakacs A, Alder H, Kaur B et al (2012) MicroRNAs/TP53 feedback circuitry in glioblastoma multiforme. Proc Natl Acad Sci U S A 109:5316–5321

Sun L, Yan W, Wang Y, Sun G, Luo H, Zhang J, Wang X, You Y, Yang Z, Liu N (2011) MicroRNA-10b induces glioma cell invasion by modulating MMP-14 and uPAR expression via HOXD10. Brain Res 1389:9–18

Turcan S, Rohle D, Goenka A, Walsh LA, Fang F, Yilmaz E, Campos C, Fabius AW, Lu C, Ward PS et al (2012) IDH1 mutation is sufficient to establish the glioma hypermethylator phenotype. Nature 483:479–483

van den Bent MJ, Dubbink HJ, Sanson M, van der Lee-Haarloo CR, Hegi M, Jeuken JW, Ibdaih A, Brandes AA, Taphoorn MJ, Frenay M et al (2009) MGMT promoter methylation is prognostic but not predictive for outcome to adjuvant PCV chemotherapy in anaplastic oligodendroglial tumors: a report from EORTC Brain Tumor Group Study 26951. J Clin Oncol 27:5881–5886

Verhaak RG, Hoadley KA, Purdom E, Wang V, Qi Y, Wilkerson MD, Miller CR, Ding L, Golub T, Mesirov JP et al (2010) Integrated genomic analysis identifies clinically relevant subtypes of glioblastoma characterized by abnormalities in PDGFRA, IDH1, EGFR, and NF1. Cancer Cell 17:98–110

Vlassenbroeck I, Califice S, Diserens AC, Migliavacca E, Straub J, Di Stefano I, Moreau F, Hamou MF, Renard I, Delorenzi M et al (2008) Validation of real-time methylation-specific PCR to determine O6-methylguanine-DNA methyltransferase gene promoter methylation in glioma. J Mol Diagn 10:332–337

von Wronski MA, Harris LC, Tano K, Mitra S, Bigner DD, Brent TP (1992) Cytosine methylation and suppression of O6-methylguanine-DNA methyltransferase expression in human rhabdomyosarcoma cell lines and xenografts. Oncol Res 4:167–174

Wang J, Iwasaki H, Krivtsov A, Febbo PG, Thorner AR, Ernst P, Anastasiadou E, Kutok JL, Kogan SC, Zinkel SS et al (2005) Conditional MLL-CBP targets GMP and models therapy-related myeloproliferative disease. EMBO J 24:368–381

Watanabe T, Nobusawa S, Kleihues P, Ohgaki H (2009) IDH1 mutations are early events in the development of astrocytomas and oligodendrogliomas. Am J Pathol 174:1149–1153

Weller M, Gorlia T, Cairncross JG, van den Bent MJ, Mason W, Belanger K, Brandes AA, Bogdahn U, Macdonald DR, Forsyth P et al (2011) Prolonged survival with valproic acid use in the EORTC/NCIC temozolomide trial for glioblastoma. Neurology 77:1156–1164

Weller M, Stupp R, Reifenberger G, Brandes AA, van den Bent MJ, Wick W, Hegi ME (2010) MGMT promoter methylation in malignant gliomas: ready for personalized medicine? Nat Rev Neurol 6:39–51

Xu J, Sampath D, Lang FF, Prabhu S, Rao G, Fuller GN, Liu Y, Puduvalli VK (2011a) Vorinostat modulates cell cycle regulatory proteins in glioma cells and human glioma slice cultures. J Neurooncol 105:241–251

Xu W, Yang H, Liu Y, Yang Y, Wang P, Kim SH, Ito S, Yang C, Wang P, Xiao MT et al (2011b) Oncometabolite 2-hydroxyglutarate is a competitive inhibitor of alpha-ketoglutarate-dependent dioxygenases. Cancer Cell 19:17–30

Yan H, Parsons DW, Jin G, McLendon R, Rasheed BA, Yuan W, Kos I, Batinic-Haberle I, Jones S, Riggins GJ et al (2009) IDH1 and IDH2 mutations in gliomas. N Engl J Med 360:765–773

Zlotogorski C, Erickson LC (1983) Pretreatment of normal human fibroblasts and human colon carcinoma cells with MNNG allows chloroethylnitrosourea to produce DNA interstrand crosslinks not observed in cells treated with chloroethylnitrosourea alone. Carcinogenesis 4:759–763

Signaling Cascades Driving the Malignant Phenotype of Glioma Cells

3

Mitsutoshi Nakada, Daisuke Kita, Takuya Furuta, Takuya Watanabe, Yutaka Hayashi, and Jun-Ichiro Hamada

Abstract

The most malignant brain tumor *glioblastoma multiforme* (GBM) is characterized by exponential growth and diffuse invasiveness. These characteristics are attributed to the varieties of common genetic lesions in genes encoding signaling proteins. These DNA damages lead to either activating mutations (Ras, PI3K, and Akt) or loss of function of tumor suppressor proteins (TP53, Rb, and PTEN). The signaling pathways induced by the altered signaling proteins play a role in the maintenance of the GBM malignant phenotype. Recent studies have elucidated various genetic alterations and critical signaling pathways involved in GBM. This has improved our understanding of GBM cell proliferation, migration, and invasion. Additionally, new exciting areas of research, such as stem cell biology and "-omics" analyses, have been recently employed for GBM studies. Many efforts are now directed to identification of signaling molecules involved in gliomagenesis and glioma stem cell maintenance. Signaling pathway analysis has resulted in novel GBM subclassifications that could integrate with conventional histopathological features providing information regarding the molecular mechanism of pathogenicity. The identification of aberrant signal transduction pathways and their role in glioma development and progression may contribute to the development of novel therapeutic targets. This chapter provides an overview of the current understanding of GBM oncogenomics, signaling pathways, and glioma stem cell signaling as well as the molecular circuitry regulating several key cellular processes.

Keywords

Glioma • Glioblastoma • Signaling • Tyrosine kinase receptor

M. Nakada (✉) • D. Kita • T. Furuta • T. Watanabe • Y. Hayashi • J.-I. Hamada
Department of Neurosurgery, Graduate School of Medical Science, Kanazawa University, 13-1 Takara-machi, Kanazawa 920-8641, Japan
e-mail: mnakada@med.kanazawa-u.ac.jp

A. Sedo and R. Mentlein (eds.), *Glioma Cell Biology*,
DOI 10.1007/978-3-7091-1431-5_3, © Springer-Verlag Wien 2014

Abbreviations

ALK	Activin receptor-like kinase
bFGF	Basic fibroblast growth factor
BMP	Bone marrow protein
ECM	Extracellular matrix
EGFR	Epidermal growth factor receptor
ERK	Extracellular signal-regulated kinase
Frz	Frizzled receptor
GBM	*Glioblastoma multiforme*
Gli	Gliotactin
GSC	Glioma stem cells
HB-EGF	Heparin-binding EGF-like growth factor
IDH1	Isocitrate dehydrogenase-1
IL-2	Interleukin-2
iPS	Induced pluripotent
JNK	c-Jun N-terminal kinase
LIF	Leukemia inhibitory factor
LIFR	LIF receptor
MAPK	Mitogen-activated protein kinase (ERK1 and ERK2)
MAPKK	MAPK kinase
MMP	Matrix metalloproteinase
NICD	Intracellular domain of the Notch receptor
NSC	Neural stem cell
PDGFRA	Platelet-derived growth factor receptor A
PDK1	Phosphoinositide-dependent kinase-1
PI3K	Phosphatidylinositide-3-kinase
PIP3	Phosphatidylinositol 3,4,5-trisphosphate
Ptch	Patched homolog
PTEN	Phosphatase and tensin homolog
RTK	Receptor tyrosine kinase
Shh	Sonic hedgehog
Smo	Smoothened homolog
Sox2	Sry-related HMG-Box factor 2
TCGA	The Cancer Genome Atlas
TGF	Transforming growth factor
TMZ	Temozolomide
TNF-α	Tumor necrosis factor-α
VEGFR	Vascular endothelial cell growth factor receptor

3.1 Introduction

Glioblastoma (GBM) is the most common adult primary tumor of the central nervous system. GBMs are characterized by rapid growth and diffuse invasiveness into the adjacent brain parenchyma (Nakada et al. 2007, 2013). The abnormal behavior of glioma cells is due to various genetic abnormalities as well as aberrant cell signaling. Hyperactivation of these signaling pathways drives tumorigenesis and supports tumor growth. To date, we do not fully understand the mechanisms underlying malignant phenotype of GBM. However, cytogenetic and molecular genetic studies over the past two decades have revealed a number of recurrent chromosomal abnormalities and genetic alterations in GBM (Goodenberger and Jenkins 2012; Gupta and Salunke 2012; Marumoto and Saya 2012; Nakada et al. 2011; Tanaka et al. 2013; Van Meir et al. 2010). Importantly, advances in molecular technologies, such as high-density microarray and genome sequencing, have enabled evaluation of the genetic and epigenetic changes in these tumors at the genome-wide level. This work has revealed a number of key genetic abnormalities as well as some critical signaling pathways responsible for GBM initiation, proliferation, migration, and invasion.

Among many key signaling pathways in GBM, receptor tyrosine kinase (RTK)/phosphatidylinositide-3-kinase (PI3K)/Akt, TP53/MDM2, Rb/cyclin-dependent kinase 2A (CDKN2A), Ras/extracellular signal-regulated kinase (ERK), and transforming growth factor (TGF) have been thoroughly characterized (Nakada et al. 2011; Van Meir et al. 2010). These signaling pathways represent a common therapeutic target for most GBM owing to its central position in the signaling cascade, affecting proliferation, apoptosis, and invasion. However, almost all clinical trials using small-molecule inhibitors that target the individual signaling pathways for GBM have, to date, failed. A major barrier in the development of a therapeutic strategy is that each GBM exhibits specific genetic characteristics. To address these differences, recent work has established a molecular subclassification for GBM based on its dominant signaling pathway. The ultimate goal of this characterization is to promote individualized therapy based on the molecular GBM subtype.

It is known that cancer cells and stem cells share common defining features of an undifferentiated state and an unlimited capacity for self-renewal; this concept has led to the cancer stem cell hypothesis. Recent reports have described that glioma-initiating stem-like cells or glioma stem cells (GSCs) may be responsible for the development and recurrence of GBM (Natsume et al. 2011; Sampetrean and Saya 2013; Tamase et al. 2009). Moreover, this hypothesis has led to studies on the role of signaling pathways in GSCs.

This chapter discusses the growing understanding of the complex signaling pathways that underlie GBM formation and progression.

3.2 GBM Subclassification and Its Signaling

In a comprehensive study carried out by The Cancer Genome Atlas (TCGA) project, 601 cancer-related candidate genes were sequenced in more than 200 human GBM samples (Network 2008). The project also analyzed genome-wide DNA copy number changes, DNA methylation status, and protein-coding and noncoding RNA expression. In a similar but complementary study, 20,661 protein-encoding genes were sequenced from 22 GBM samples and integrated the genetic alteration information with DNA copy number (Parsons et al. 2008). These integrative genomic studies provided a comprehensive view of the complicated genetic landscape of GBM and revealed a set of core signaling pathways commonly activated in GBM. These pathways include the RTK pathway, the TP53 pathway, and the RB pathway. The majority of GBM exhibits genetic alterations in all three pathways that, in turn, promote accelerated cell proliferation and enhanced cell survival while allowing the tumor cell to escape cell cycle checkpoints, senescence, and apoptosis.

Overall, genome-wide studies have revealed that tumor histology correlates with distinct gene expression signatures. Additionally, molecular profiles can be used to identify tumor subclasses that would otherwise be indistinguishable by standard morphological characterization. Unsupervised clustering analysis of the TCGA data set enabled subdivision of GBM into four subtypes: termed the proneural, neural, mesenchymal, and classical. This subdivision is based on the similarity of genomic features (Verhaak et al. 2010). Each group shows a different aberration and gene expression, which may predict therapy efficacy (Fig. 3.1).

The proneural type was associated with younger age, platelet-derived growth factor receptor A (*PDGFRA*) abnormalities, isocitrate dehydrogenase-1 (*IDH1*) mutation, and *TP53* mutation, all of which are associated with secondary GBM. The prognostic value of the molecular subclassification has also been evaluated, suggesting that the proneural subtype generally correlates with a marginally improved survival (Vitucci et al. 2011). However, this subtype is resistant to temozolomide (TMZ) and radiation therapy.

The neural subtype exhibits gene expression signatures that are similar to those found in normal brain tissue, with activation of neuronal markers such as Fbxo3, Syt1, and Gabra, suggesting that this subtype consists of cells of a differentiated phenotype (Network 2008). Significant response to radiochemotherapy was not observed in this subtype (Verhaak et al. 2010).

The classical subtype exhibits epidermal growth factor receptor (EGFR) abnormalities without *IDH1* and *TP53* mutations and is most sensitive to radiochemotherapy.

The mesenchymal subtype, characterized by high expression of CHI3L1 and MET, and NF1 mutation/deletion, usually shows a poor outcome (median overall survival, 1.2 year for proneural versus 1.0 years for mesenchymal) (Verhaak et al. 2010; Vitucci et al. 2011). However, several studies have confirmed that this subtype is sensitive to combinational radiation and chemotherapy, which could

	Proneural	Neural	Classical	Mesenchymal
Genetic alterations	PDGFRA P53 IDH1 NKX2-2 PI3K	Fbxo3 Syt1 Gabra	EGFR PTEN CDKN2A	CHI3L1 MET NF1 CD44 PTEN
Signaling				NF-κB TGF-β
Age	Young			
Prognosis	Better			Poor
Radiochemosensitivity	Poor		Best	Better

Fig. 3.1 Subclassification of glioblastoma

reflect its high proliferative index or higher levels of microvascular endothelial proliferation (Gravendeel et al. 2009; Verhaak et al. 2010).

A large number of these classification studies have now been carried out, and they provide interesting insights into the molecular nature of GBM (Verhaak et al. 2010). Furthermore, clinical trials designed on the basis of these molecular subtypes are ongoing.

3.3 Glioma Stem Cell Signaling

Gliomas contain a heterogeneous cell population, and only the GSC subset has the ability to extensively proliferate and form new tumors (Ebben et al. 2010; Natsume et al. 2011; Sampetrean and Saya 2013; Singh et al. 2004). The dysregulation and/or abnormally functioning key signaling pathways in GSCs potentially lead to gliomagenesis. The origin of GSC is still controversial. Accumulating evidence suggests that GSCs might originate from the transformation of neural stem cells (NSCs) and their progenitors (Fig. 3.2).

GSCs share several characteristics with NSCs: the expression of NSC antigens, such as nestin, CD133, Musashi-1, and Bmi-1 (Galli et al. 2004; Hemmati et al. 2003; Ignatova et al. 2002), and the ability to grow as nonadherent spheres when cultured in the presence of EGF and basic fibroblast growth factor (bFGF) under serum-free conditions (Vescovi et al. 2006). Some investigators suggest that differentiated cells could revert to stem cell or their progenitor cell state (Fig. 3.2) (Bachoo et al. 2002; Dai et al. 2001; Uhrbom et al. 2002). Experimental depletion of tumor suppressors and activation of RTKs and their ligands are likely to induce

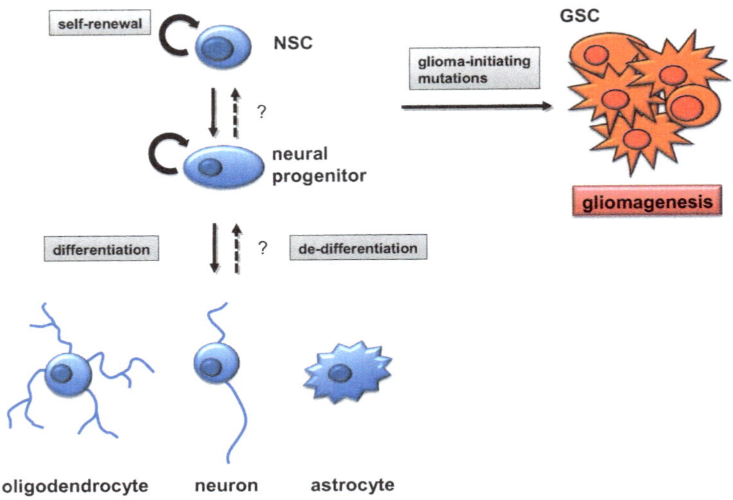

Fig. 3.2 Glioma stem cells (GSCs) may have been postulated to originate from normal neural stem cell (NSCs) and/or their progenitors. On the other hand, differentiated cells (neuron, astrocyte, and oligodendrocyte) have potential as origin of GSCs

dedifferentiation of astrocytes or oligodendrocytes. In addition, the dedifferentiated cells could be sensitized to transform GSCs in response to the appropriate oncogenic stimuli (Uhrbom et al. 2002).

GSCs and their progeny are thought to regulate different intrinsic and extrinsic factors (Fig. 3.3). Growth factor-stimulated RTK signaling, particularly EGFR and PDGFR, is involved in GSC regulation (Clark et al. 2012; Doetsch et al. 2002; Jackson et al. 2006). Overexpression or aberrant mutation of these RTKs, such as EGFRvIII, promotes self-renewal and proliferation of GSCs and therefore glioma formation (Clark et al. 2012; Dietrich et al. 2010; Doetsch et al. 2002; Jackson et al. 2006; Vescovi et al. 2006). Stimulation of these growth factor receptors activates downstream effectors, such as RAS and PI3K pathways, which are negatively regulated by the tumor suppressors NF1 and phosphatase and the tensin homolog deleted from chromosome 10 (PTEN), respectively. Inactivation of these tumor suppressors in NSCs/progenitor cells is both necessary and sufficient to induce astrocytoma formation (Alcantara Llaguno et al. 2009).

Wnt and sonic hedgehog (Shh) signaling pathways play critical roles in patterning, morphogenesis, and proliferation in GSCs (Dietrich et al. 2010). β-catenin and gliotactin (Gli) are two molecules activated by Wnt and Shh pathways, respectively, and are likely to be prognostic factors in GBM patients (Rossi et al. 2011). Wnt signaling is divided into two different pathways: the canonical, or Wnt/β-catenin pathway involved in determination of cell fate, and the noncanonical pathway, which consists of small G proteins, involved in the control of cell movement and tissue polarity (Katoh 2007; Kwiatkowska and Symons 2013). Wnt binds to its frizzled receptor (Frz) inducing a complex number of events

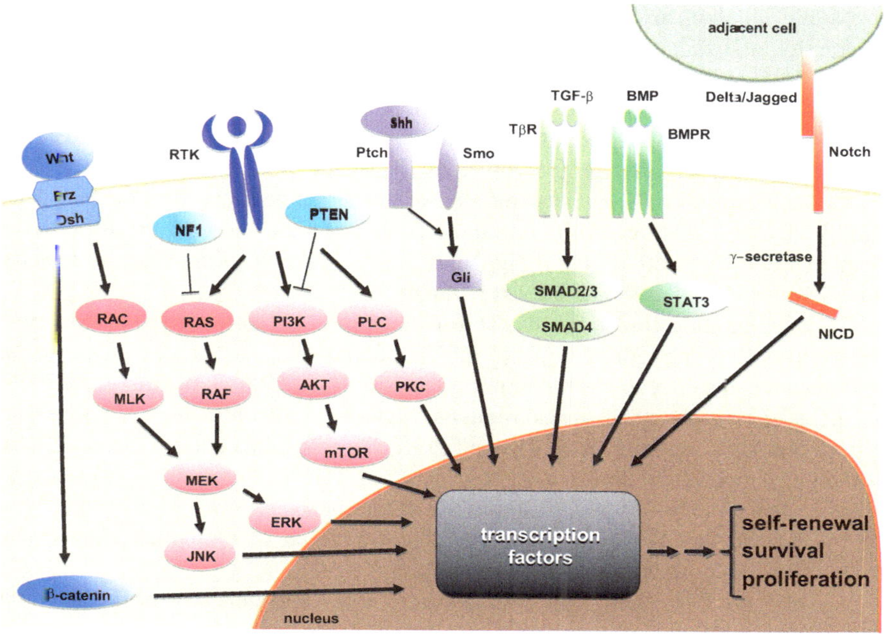

Fig. 3.3 Simplified scheme of signaling pathways in glioma stem cell. RTKs, especially EGFR and PDGFR, are frequently overactivated in malignant gliomas, stimulating its downstream effectors, such as RAS and PI3K. These signaling are negatively regulated by tumor suppressor NF1 and PTEN, respectively. TGF, Shh, Wnt, and Notch are developmentally regulated signaling pathways that are also active in normal NSCs. If dysregulation or mutation of these signaling molecules occurs, NSCs might transform into GSCs followed by gliomagenesis. *Abbreviations*: *BMP* bone morphogenic protein, *BMPR* BMP receptor, *ERK* extracellular signal-regulated kinase, *Dsh* dishevelled, *Frz* frizzled, *Gli* gliotactin, *JNK* c-Jun N-terminal kinase, *MEK* mitogen-activated protein kinase-ERK kinase, *MLK* mixed lineage kinase, *mTOR* mammalian target of rapamycin, *NCID* Notch intracellular domain, *PI3K* phosphatidylinositol 3-kinase, *PKC* protein kinase C, *PLC* protein lipase C, *PTEN* phosphatase and tensin homolog, *Ptch* patched homolog, *RTK* receptor tyrosine kinase, *Shh* sonic hedgehog, *Smo* smoothened homolog, *STAT3* signal transducer and activator of transcription 3, *TGF-β* transforming growth factor-β, *TβR* TGF-β receptor

leading to the stabilization of unbound β-catenin in the cytoplasm and translocation of this free β-catenin into the nucleus where it functions as a transcriptional activator of target genes (Peifer and Polakis 2000). These factors are involved in migration, invasion, and proliferation of malignant gliomas as well as the self-renewal of GSCs (Zhang et al. 2011; Zheng et al. 2010). SHH ligands bind to the receptor patched homolog (Ptch) to repress smoothened homolog (Smo) resulting in the activation of Gli transcription factors that are translocated into the nucleus to regulate cellular activities of GSC, such as cell stemness, survival, and proliferation (Bar et al. 2007; Onishi and Katano 2011).

Notch signaling, an evolutionarily conserved pathway mediating direct cell–cell interaction and signaling, plays a pivotal role in the stemness of NSC (Louvi and Artavanis-Tsakonas 2006). Notch cooperates with Delta/Jagged localized on the membrane of adjacent cells. When the Notch receptor is triggered by ligands on the neighboring cells, the intracellular domain of the Notch receptor (NICD) is released from the membrane after proteolytic cleavage by the γ-secretase complex. NICD then translocates into the nucleus and transcribes target genes such as the Hes gene family by associating with its transcription factor. Recent studies have demonstrated that active Notch signaling plays a role in glioma tumorigenicity and GSC self-renewal (Hu et al. 2011; Zhang et al. 2008).

TGF-β signaling, mediated by TGF-β heterodimer ligands, type 1 and type 2 receptors forming TβR complex, and Smad proteins, has crucial roles in the maintenance of self-renewal and tumorigenic activity of GSC (Ikushima et al. 2009; Penuelas et al. 2009). TGF-β also regulates signaling molecules such as p38, c-Jun N-terminal kinase (JNK), Ras/ERK, and PI3K/Akt (Ikushima and Miyazono 2010). Bone morphogenic protein (BMP), a member of TGF-β superfamily, is known to regulate the stemness of GSC differentiation mediated by STAT3 and might be a promising therapeutic agent to prevent the recurrence of malignant gliomas by inducing GSC terminal differentiation (Lee et al. 2008).

As shown in Fig. 3.3, multiple signaling pathways participate in GSC regulation and these coordinate with each other to maintain signaling. A more detailed analysis of the GSC population may lead to the identification of novel pathways important for their proliferation, self-renewal, and differentiation, thus revealing new therapeutic targets.

3.4 Major Signaling and Related Molecules

3.4.1 RTK/PI3K/Akt Pathway

RTK/PI3K/Akt pathway regulates various cellular processes, such as proliferation, growth, apoptosis, and cytoskeletal rearrangement. The pathway contains RTKs (EGFR, PDGFR, and vascular endothelial growth factor receptor [VEGFR], etc.) as well as the tumor suppressor protein phosphatase of PTEN and protein kinases PI3K, Akt, and mTOR (Fig. 3.4). Aberrant activation of this pathway is frequently observed in malignant gliomas mainly due to the alterations of these proteins (Kita et al. 2007; Ohgaki and Kleihues 2009).

3.4.1.1 RTKs

RTK activation in gliomas can be achieved not only through protein overexpression or genetic amplification of several RTKs, but also through gene mutations that lead to a constitutively active form. In GBM, *EGFR* gene amplification is the most frequent alteration among all the RTK alterations (approximately 40 %) (Ekstrand et al. 1992; Libermann et al. 1985). EGFR overexpression and/or gene alteration is frequently observed in primary (de novo) GBMs, which develop rapidly, with a

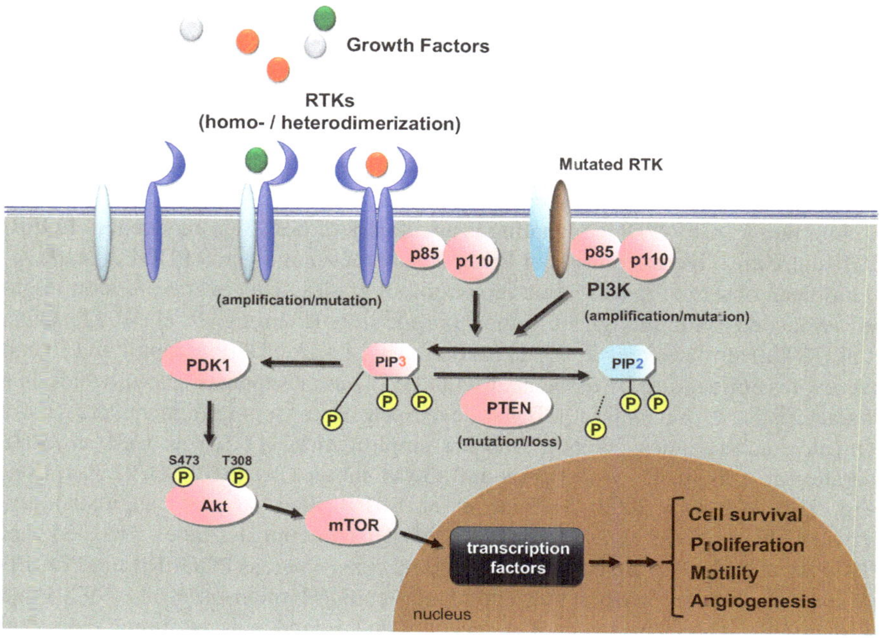

Fig. 3.4 RTK/PI3K/Akt signaling. Homo-/heterodimerization of RTKs are caused by their ligands, while certain mutants of RTKs (e.g., EGFRvIII) can be activated by ligand independent fashion. The binding of RTKs, including EGFR, PDGFR, and VEGFR, by the p85 subunit of PI3K results in activation of the catalytic subunit (p110), which then catalyzes phosphorylation of PIP2 into PIP3. PIP3 in turn activates phosphoinositide-dependent kinase-1 (PDK1), which phosphorylates Ser473 and Thr308 of Akt. Akt phosphorylates mTOR, which affects multiple cellular processes including cell survival, proliferation, and motility. *Abbreviations*: *mTOR* mammalian target of rapamycin, *PDK1* phosphoinositide-dependent kinase-1, *PI3K* phosphatidylinositide-3-kinase, *PIP2* phosphatidylinositide 3,4-bisphosphate, *PIP3* phosphatidylinositide 3,4,5-triphosphate, *RTK* receptor tyrosine kinase

short clinical history and with no evidence of a less-malignant precursor lesion (Ekstrand et al. 1992; Ohgaki et al. 2004; Wong et al. 1987). Genomic analysis by TCGA network revealed that *EGFR* aberration is related to the classical GBM subtype (Fig. 3.1) (Network 2008). Amplification of the *EGFR* gene is also associated with the expression of a constitutively activated mutation, EGFR variant III (EGFRvIII), or other members of EGFR family such as ErbB2 (Wong et al. 1987; Yamazaki et al. 1988).

The fact that EGFRvIII expression in glioma cells stimulated expression of TGF-α and heparin-binding EGF-like growth factor (HB-EGF) that act as ligands for EGFR suggests that EGFRvIII plays a role in the generation of an autocrine loop with wild-type EGFR (Ramnarain et al. 2006). Some researchers have shown that GBM patients with EGFR overexpression or EGFR mutations exhibit shorter survival, suggesting that EGFR alterations are associated with the aggressiveness of GBMs (Barker et al. 2001; Feldkamp et al. 1999; Shinojima et al. 2003). Despite the high frequency of EGFR overexpression in GBM, EGFR inhibitors and specific antibodies (e.g., gefitinib, erlotinib, and nimotuzumab) have not elicited a positive response in clinical trials (Rich et al. 2004; Strumberg et al. 2012; Van Den Bent et al. 2009). A peptide vaccine therapy against EGFRvIII (rindopepimut) instead of EGFR is now being utilized for newly diagnosed GBM patients because EGFRvIII is not expressed in normal tissues (Del Vecchio et al. 2012).

Aberrant PDGFR/PDGF signaling is a hallmark of GBM biology next to EGFR/EGF signaling. Overexpression of PDGFR subtypes α and β and PDGF ligands A–D has been observed in astrocytic tumors of all grades. This overexpression might be associated with malignant tumor progression (Fleming et al. 1992; Guha et al. 1995; Hermanson et al. 1992; Lokker et al. 2002). PDGF receptor and ligand overexpression tend to be associated with TP53 tumor suppressor loss, which is a characteristic of secondary GBMs developed from less-malignant precursors (Ohgaki and Kleihues 2009). *PDGFRA* amplification (14 %) as well as *IDH1* mutation is a hallmark of the proneural GBM subtype by the TCGA consortium (Fig. 3.1) (Network 2008; Verhaak et al. 2010). This suggests an association between this subtype and secondary GBM. Histochemical studies revealed that PDGFRα and PDGFA are expressed in glioma cells, whereas PDGFRβ and PDGFB are expressed in the surrounding endothelial cells (Hermanson et al. 1992; Plate et al. 1992). The expression of a receptor and ligand pair in tumor cells and associated blood vessels indicates that autocrine/paracrine activation also plays a role in tumor cell migration and colony formation (Hoelzinger et al. 2007; Shih and Holland 2006). Despite a deep association of this molecule with GBM, anti-PDGFR therapy (e.g., imatinib, dasatinib, and sunitinib) led to only a limited clinical response (Kreisl et al. 2013; Lu-Emerson et al. 2011; Reardon et al. 2005, 2009).

Increased VEGF and VEGFR levels in GBM have been reported and are associated with sustained angiogenesis (Lamszus et al. 2003; Salmaggi et al. 2003). As compared with external historical control, bevacizumab, a humanized monoclonal antibody against VEGF, prolonged the progression-free survival and controlled peritumoral edema (up to 50 %) in Phase II trials (Friedman et al. 2009). Recent two large randomized Phase III trials using either bevacizumab or placebo in combination with TMZ and radiation showed a statistically significant improvement in progression-free survival (3 to 4 months), however failed to show difference in overall survival (Chinot et al. 2014; Gilbert et al. 2014). One must consider that anti-VEGF/VEGFR treatment is a double-edged sword for GBM, because there is evidence suggesting that it increases tumor cell invasion in GBM by inducing the hypoxic tumor microenvironment (Ebos et al. 2009; Furuta et al. 2014; Keunen et al. 2011; Paez-Ribes et al. 2009).

3.4.1.2 PI3K

PI3Ks are widely expressed lipid kinases that promote diverse biological functions. The binding of PI3Ks and RTKs results in the activation of Akt through phosphatidylinositol 3,4,5-trisphosphate (PIP3) and phosphoinositide-dependent kinase-1 (PDK1), and thereby affects multiple cellular processes, including cell survival, proliferation, and motility (Fig. 3.4) (Vivanco and Sawyers 2002). The eight mammalian PI3Ks are classified into three classes according to their structure, regulation, and substrate specificity. Most PI3K enzymes consist of a p110 catalytic subunit that heterodimerizes with a p85 regulatory subunit. Oncogenic mutations or amplification of the *p110α* gene (*PIK3CA*) has been reported in various human neoplasms, including GBM (Samuels et al. 2004). *PIK3CA* mutations and amplification were observed in 5 % and 13 % of GBM, respectively (Kita et al. 2007). TCGA analysis revealed p85α gene (*PIK3R1*) mutations in 10 % (9/91) of GBM, although, unlike *PIK3CA*, *PIK3R1* has rarely been reported as mutated in human cancers. According to the integrated genomic classification of GBM, *PI3K* mutations (15 %) are associated with the proneural subtype (Fig. 3.1) (Network 2008; Verhaak et al. 2010). Multiple PI3K inhibitors (many of them are PI3K/mTOR dual inhibitors) such as XL147, XL765 (Exelixis and Sanofi-Aventis), BKM120, BEZ235, BGT226 (Novartis), GDC0980 (Genentech), PKI587, PF04691502 (Pfizer), and GSK2126458 (Glaxo-Smith-Kline) are currently undergoing Phase I/II trials for GBM patients (see http://clinicaltrials.gov/) (Workman et al. 2010).

3.4.1.3 PTEN

Decreased PTEN activity can activate the RTKs/PI3K/Akt pathway, because PTEN can negatively regulate the pathway by antagonizing the function of PI3Ks (Furnari et al. 1998). Homozygous deletion or mutation of *PTEN* is a common genetic feature in GBM (15–40 %) (Li et al. 1997; Network 2008; Tohma et al. 1998), resulting in constitutive activation of the RTKs/PI3K/Akt pathway. Loss of PTEN is also associated with ABC transporter activation in GBM stem-like cells, which leads to drug resistance (Bleau et al. 2009). Some reports implied that GBMs, with EGFRvIII and intact PTEN, are more likely to respond to EGFR inhibitors (Haas-Kogan et al. 1998; Mellinghoff et al. 2005). PTEN loss is also associated with both the classical and mesenchymal GBM subtypes according to the TCGA study (Fig. 3.1) (Network 2008).

3.4.1.4 Akt/mTOR

The Akt family of serine/threonine kinases consists of three members, Akt1–3, all of which have been implicated in cancer through regulation of cell growth, proliferation, and apoptosis. Simultaneous dysfunction of PTEN and activation of Akt have been found in a variety of cancers, including GBM (Choe et al. 2003). Akt activation has been reported in approximately 80 % of human GBMs (Haas-Kogan et al. 1998; Holland et al. 2000), well correlated with the fact that RTKs/PI3K/Akt signaling is altered in 88 % of GBMs (Network 2008). To date, oncogenic mutation or amplification of Akt has not been detected in GBM (Bleeker et al. 2009). Akt is well known for its anti-apoptotic activity when overexpressed; however, inhibition of Akt phosphorylation by PI3K inhibitors failed to induce apoptosis in GBM cells (Cheng et al. 2009). Phospholipid perifosine, which is a specific Akt inhibitor, is being investigated as a GBM therapeutic (see http://clinicaltrials.gov/ct2/home) (Becher et al. 2010; Gills and Dennis 2009).

mTOR is a serine/threonine kinase and can be activated by not only Akt but also the PKC and RAS pathway (Fan et al. 2010). Blockade of Akt signaling in glioma cells by Akt inhibitor or knockdown did not impact proliferation or mTOR phosphorylation levels, suggesting that Akt may not be a central regulator of mTOR for inducing proliferation (Fan et al. 2009). Rapamycin (sirolimus) and its synthesized analogs (temsirolimus and everolimus) have been intensively evaluated in clinical trials of recurrent/progressive malignant gliomas, and they demonstrate modest efficacy when combined with other anticancer agents (De Witt Hamer 2010; Hainsworth et al. 2012).

3.4.2 p14^{ARF}/MDM2/TP53 Pathway

The *p53* gene, at chromosome 17q13.1, encodes TP53 that responds to diverse cellular stresses to regulate target genes that induce cell cycle arrest, cell death, cell differentiation, senescence, DNA repair, and neovascularization (Bogler et al. 1995). Following DNA damage, TP53 is activated and induces transcription of genes such as *p21Waf1/Cip1* that function as regulators of cell cycle progression in the G1 phase (Harper et al. 1993; Sherr and Roberts 1999; Stott et al. 1998). The *MDM2* gene, at chromosome 12q14.3-q15, encodes a putative transcription factor and enhances the tumorigenic potential of cells when it is overexpressed. Forming a tight complex with the *p53* gene, the *MDM2* oncogene inhibits TP53 transcriptional activity by binding to the N-terminal transactivation domain and participates in the ubiquitination and proteasomal degradation of TP53 (Kubbutat et al. 1997; Oliner et al. 1992). Conversely, the transcription of the *MDM2* gene is induced by wild-type TP53 (Barak et al. 1994; Zauberman et al. 1995). This autoregulatory feedback loop regulates the expression of MDM2 and the activity of TP53. The *p14^{ARF}* gene (a part of the complex CDKN2A locus on chromosome 9p21) encodes a protein that directly binds to MDM2 and inhibits MDM2-mediated TP53 degradation and transactivational silencing (Kamijo et al. 1998; Mao et al. 1995; Stott et al. 1998; Zhang et al. 1998). In turn, p14^{ARF} expression is negatively regulated by TP53 (Stott et al. 1998). Therefore, inactivation of p14^{ARF}/MDM2/TP53 is caused by altered expression of any of the *TP53*, *MDM2*, or *p14^{ARF}* genes (Fig. 3.5).

The TP53 pathway plays a critical role in the development of GBMs since *p53* mutations are common in two-thirds of precursor low-grade diffuse astrocytomas. This frequency is similar to that in anaplastic astrocytomas and secondary GBMs (Ohgaki et al. 2004; Watanabe et al. 1996, 1997). *p53* mutations also occur in primary GBMs, but at a significantly lower frequency (approximately 25 %) (Ohgaki et al. 2004).

IDH1 gene mutations at chromosome 2q33, which were identified in an analysis of 20,661 protein-coding genes in GBMs (Network 2008), are early events in the development of astrocytomas and constitute a remarkably reliable molecular signature of secondary GBMs as well as their precursor lesions (Nobusawa et al. 2009; Watanabe et al. 2009; Yan et al. 2009). The additional acquisition of a *p53* mutation, which is a genetic event subsequent to the *IDH1/2* mutations, except in

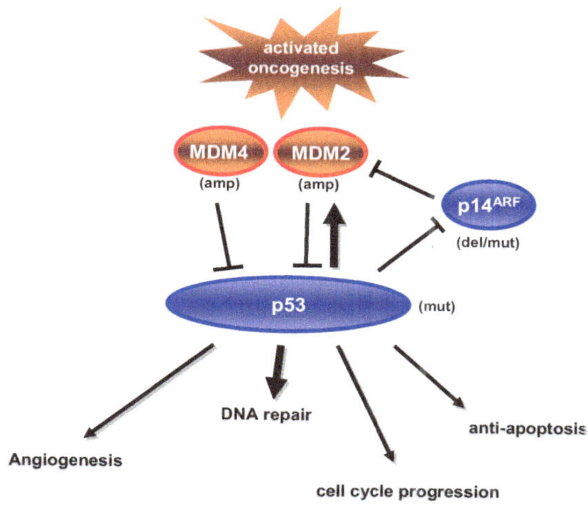

Fig. 3.5 p14^{ARF}/MDM2/MDM4/TP53 pathway. Inactivation of the pathway caused by altered expression of any of the TP53, MDM2, MDM4, or p14^{ARF} genes plays a crucial role in the progression of glioma. MDM2 and MDM4 are important negative regulators of the p53. p14^{ARF} inhibits MDM2, thus promoting TP53. DNA damage activates ATM, which leads to activation of checkpoint kinases (CHK2) and TP53. Activating genetic alterations are shown in *red circle*. Genetic alterations that lead to a loss of function are indicated in *blue circle*. Figure is modified from Ohgaki and Kleihues (2009). The p53 gene is the most commonly mutated TP53 pathway gene in glioma; however, molecular abnormalities involving other genes in the pathway—such as p14^{ARF}, MDM2, or MDM4—have also been described. MDM2 amplification is observed in about 10 % of GBMs, exclusively in primary GBMs that lack a p53 mutation (Biernat et al. 1997; Reifenberger et al. 1993). Loss of p14^{ARF} expression is frequently present in GBMs. p14^{ARF} appears to be associated mostly with aberrant promoter methylation or hemizygous deletion, whereas mutational inactivation is rare (Ichimura et al. 2000; Nakamura et al. 2001). Promoter methylation of p14^{ARF} is more frequent in secondary than primary GBMs, but there is no significant difference in the overall frequency of p14^{ARF} alterations between GBM subtypes (Nakamura et al. 2001). *Abbreviations*: *amp* amplification, *del* deletion, *mut* mutation

the case of Li-Fraumeni syndrome (Watanabe et al. 2009), may lead to astrocytic differentiation, whereas subsequent loss of 1p/19q favors the acquisition of an oligodendroglial phenotype (Fig. 3.6) (Kloosterhof et al. 2011; Watanabe et al. 2009).

GBMs with *IDH1/2* mutations frequently have *p53* mutations (Nobusawa et al. 2009). Astrocytomas typically develop from cells with IDH1/2 mutations that subsequently acquire *p53* mutations. According to the TCGA study, mutations in the *p53* gene (54 %) and *IDH1/2* (30 %) are characteristic of the proneural GBM subtype, as well as *PDGFRA* mutation (11 %) and amplification (14 %) (Network 2008) (Fig. 3.1). This finding suggests that secondary GBM belongs to the proneural subtype. Recent studies have also described mutations in the ATRX (a-thalassemia/mental-retardation-syndrome-X-linked) gene that are often concurrent with IDH1/2 mutations and TP53 mutations in diffuse astrocytomas WHO grades II/III and secondary GBMs (Jiao et al. 2012; Liu et al. 2012). On the other hand, loss of

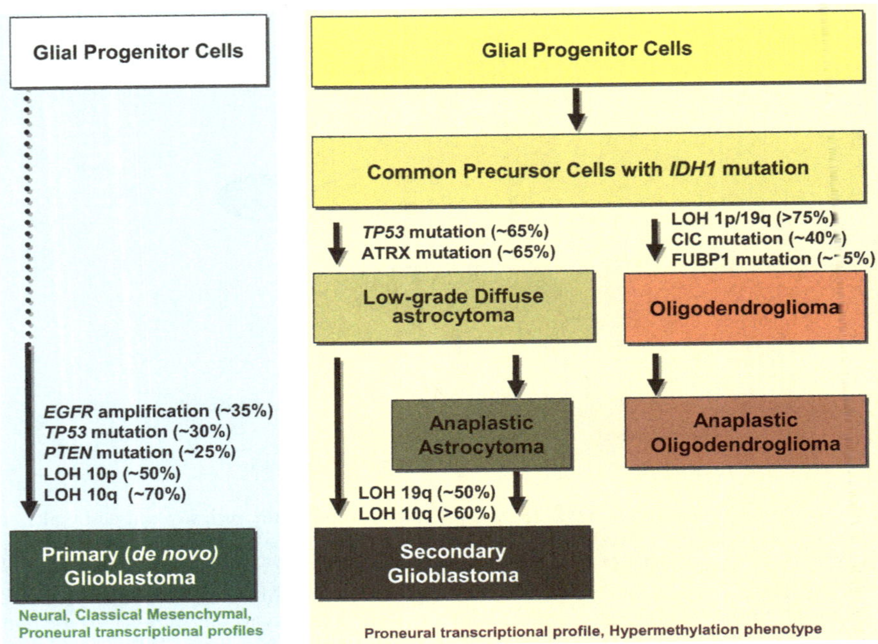

Fig. 3.6 Genetic pathways to primary and secondary GBMs. Note that only secondary GBMs share common origin of cells with oligodendrogliomas. This figure was modified from the previous paper (Ohgaki and Kleihues 2013). *Abbreviation*: *LOH* loss of heterozygosity

1p/19q in cells with IDH1/2 mutations might be the driving force toward oligodendroglial differentiation in low-grade diffuse gliomas. Recent studies indicate that the *CIC* gene (homolog of the *Drosophila* gene capicua) at 19q13.2 and the *FUBP1* gene at 1p are expressed in many oligodendrogliomas but are rare or absent in diffuse astrocytomas (Fig. 3.6) (Jiao et al. 2012; Liu et al. 2012; Yip et al. 2012).

The approaches activating wild-type TP53 have been intensively researched for years, including the use of chemo-radiation to enhance endogenous TP53, of gene therapy to introduce wild-type TP53 or modified adenovirus to kill cancer cells with mutant TP53, and of synthetic peptides or nongenotoxic small molecules to inhibit MDM2 or to activate endogenous TP53; however none of these approaches have shown efficacy in clinical trials yet (Shangary and Wang 2009; Wang and Sun 2010).

3.4.3 RB Pathway

The RB pathway suppresses cell cycle entry and progression, as well as TP53 pathway. The 107-kDa RB1 protein encoded by the *RB1* gene controls the progression through G1 into the S-phase of the cell cycle (Serrano et al. 1993). On the other hand, the CDKN2A (p16^{INK4a}) or CDKN2B (p15^{INK4b}) protein binds to CDK4 and prevents CDK4/cyclin D1 complex formation, consequently inhibiting cell cycle transition from the G1 to the S phase (Fig. 3.7). Therefore, any genetic alteration in *RB1*, *CDK4*, or *CDKN2A/CDKN2B* can cause dysregulation of the G1-S phase transition. The TCGA pilot project revealed that the frequency of genetic alterations in the RB signaling pathway was 77 %, including *CDKN2A* homozygous deletion or mutations (52 %), *CDKN2B* homozygous deletion (47 %), *CDKN2C* (p18^{INK4c}) homozygous deletion (2 %), *CDK4* amplification (18 %), *CCND2* (cyclin D2) amplification (2 %), *CDK6* amplification (1 %), and *RB1* mutation or homozygous deletion (11 %) (Network 2008). Alteration of *CDKN2A* and *CDKN2B* was observed almost concurrently, while alterations among *RB1*, *CDK4*, and *CDKN2A/CDK2NB* appeared mutually exclusive (Ciriello et al. 2012). Sole alteration in the RB pathway is not sufficient to induce tumor formation; however, combined loss of *RB* and other tumor suppressor genes (e.g., *TP53* and/or *PTEN)* leads to high incidences of malignant gliomas in an adult mouse model (Chow et al. 2011). In addition to this, EGFR amplification is associated with *CDKN2A* deletions (Hayashi et al. 1997; Hegi et al. 1997). *CDKN2A* loss is associated with the classical GBM subtype according to TCGA research (Fig. 3.1) (Network 2008).

3.4.4 RAS/ERK Pathway

3.4.4.1 NF1

The *NF1* tumor suppressor gene encodes neurofibromin, a protein that primarily functions as a negative regulator of RAS and also plays a role in adenylate cyclase- and Akt-mTOR-mediated pathways (Gottfried et al. 2010; Lee and Stephenson 2007). There is increasing evidence that the *NF1* gene is involved in the tumorigenesis of not only NF1-related but also sporadically occurring gliomas. According to the TCGA study, *NF1* mutation/homozygous deletions were identified in 18 % of GBM (Network 2008). Mesenchymal GBM showing frequent inactivation of the *NF1* (37 %), *TP53* (32 %), and *PTEN* genes responds to aggressive chemoradiation adjuvant therapies (Fig. 3.1) (Verhaak et al. 2010).

3.4.4.2 RAS

RAS proteins are small GTPases that cycle between inactive GDP-bound and active GTP-bound conformation by coupling cell membrane growth factor receptors, such as EGFR (Fig. 3.8). RAS can activate the serine/threonine kinases RAF, mitogen-activated protein kinase (MAPK: ERK1 and ERK2), PI3K, and a number of proteins that translocate to the nucleus to promote cell proliferation, differentiation, and survival. The *RAS* gene family comprises three different genes that encode HRAS, NRAS, KRAS4A, and KRAS4B (Malumbres and Barbacid 2003). Combined activation of RAS and Akt in neural progenitors induces GBM formation in mouse models (Holland et al. 2000). Although activation of *RAS* mutations is widely observed in approximately 30 % of human cancers (Bos 1989), the *RAS*

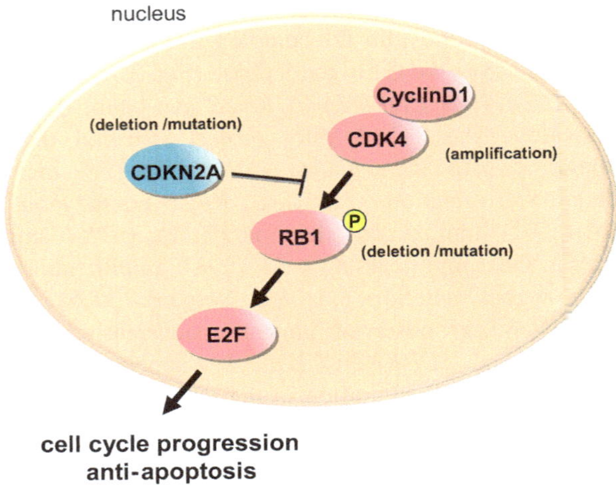

Fig. 3.7 RB tumor suppressor protein signaling. The CDK4–cyclin D1 complex phosphorylates RB1, which in turn induces release of the E2F transcription factor that can accelerate G1 to S-phase transition. CDKN2A ($p16^{INK4a}$) inhibits formation of the CDK4–cyclin D1 complex by binding to CDK4

mutations seem rare in human GBM, with a frequency of only 2 % according to the TCGA study (Network 2008). Therefore, the observed dysregulation of the RAS-RAF-ERK signaling pathway in malignant gliomas is attributed to its upstream positive regulators, including EGFR and PDGFR, known to be highly active in the majority of malignant gliomas (Feldkamp et al. 1999; Guha et al. 1997).

3.4.4.3 RAF

Activated RAS recruits RAF kinase to the membrane, where it is activated by multiple phosphorylation events. The RAF family of serine/threonine kinases, which consist of ARAF, BRAF, and CRAF1, play an important role in proliferative signaling. In pediatric low-grade glioma, a majority of which are pilocytic astrocytoma and fibrillary astrocytoma, chromosome 7q34 rearrangements result in an in-frame *KIAA1549:BRAF* fusion gene that possesses constitutive BRAF kinase activity resembling oncogenic BRAF (V600E) (Lin et al. 2012; Sievert et al. 2009). In contrast, RAF mutations or rearrangements are rare in common malignant gliomas in adults. A Phase I/II clinical trial is currently ongoing for recurring GBM by using a combination of RAF and mTOR inhibitors, and sorafenib and temsirolimus, respectively (NCT00335764).

3.4.5 MEK/ERK Pathway

Activated RAF phosphorylates and activates MAPK kinase (MAPKK), also called MEK, which in turn phosphorylates and activates MAPK or ERK (Fig. 3.8) (Adjei and Hidalgo 2005). Activated ERK can translocate into the nucleus to activate several transcription factors such as Elk1, c-Myc, Ets, STAT1/3, and PPARγ, which induce cell cycle progression and the transcription of anti-apoptotic genes (Kapoor and O'rourke 2003a, b). Aberrant signaling of this pathway consequently leads to

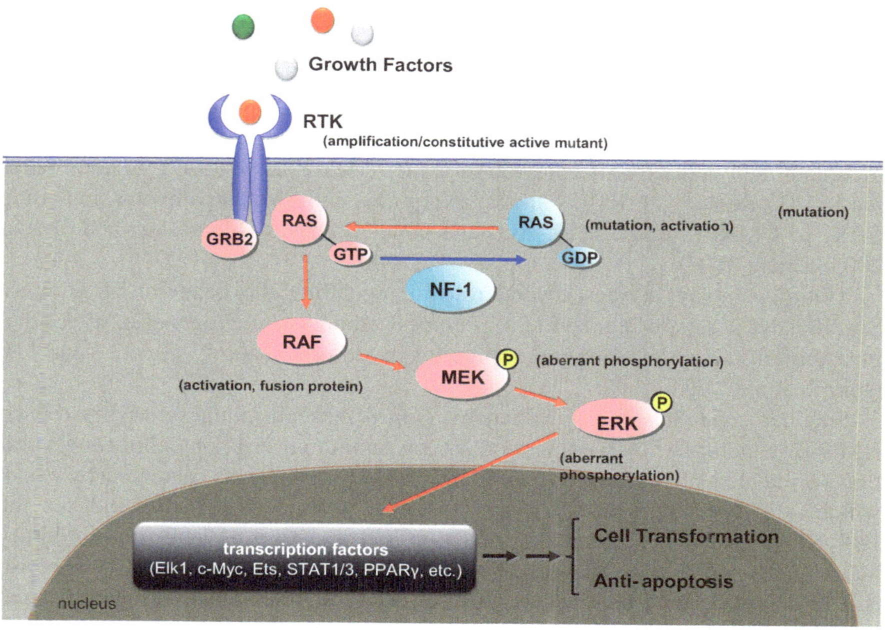

Fig 3.8 RAS signaling. RAS proteins act as on/off (RAS-GDP/RAS-GTP) switches controlled by RTKs and NF-1 protein. Activated RAS (RAS-GTP) then activates serine/threonine kinase RAF. RAF activates mitogen-activated protein kinase kinase (MAPKK), also called MEK, which in turn activates MAPK or ERK. ERK activation results in activation of various transcription factors, such as Elk1, c-Myc, Ets, STAT1/3, PPARγ, which induce cell transformation and anti-apoptosis. *Abbreviations*: *ERK* extracellular regulated kinase, *MAPKK* mitogen-activated protein kinase kinase, *MAPK* mitogen-activated protein kinase, *MEK* MAP kinase-ERK kinase, *PPARγ* peroxisome proliferator-activated receptor gamma, *RTK* receptor tyrosine kinase, *STAT* signal transducer and activator of transcription

cell transformation and resistance to apoptosis; therefore, this pathway is an attractive target for malignant gliomas (Krakstad and Chekenya 2010; Lo 2010). MEK inhibitors PD0325901 and AZD6244 are currently in clinical trials for leukemia and other cancers and have inhibited cell growth in NF1-deficient human glioma cell lines (See et al. 2012).

3.4.6 TGF Signaling

TGF-β is a family of polypeptides that regulates a wide variety of biologic functions, including cell proliferation, migration, survival, angiogenesis, immunosurveillance, and embryonic stem cell maintenance and differentiation. There are three isoforms of TGF-β in mammals, namely, TGF-β1, -β2, and -β3, all of which bind their receptors as dimers.

TGF-β activates a membrane receptor serine/threonine kinase complex (TGF-β-RI/II or TβRI/II), which phosphorylates SMAD2 and SMAD3, known as receptor-regulated (R)-SMADs (Heldin et al. 1997). The phosphorylated SMADs, together with SMAD4 or common-mediator (co)-SMAD, are translocated into the nucleus, resulting in the formation of transcriptional regulatory complexes that modulate the expression of many target genes (Heldin et al. 1997). In addition to the SMAD pathway, TGF-β is known to regulate non-SMAD pathways, including ERK, p38 MAPK, JNK, PI3K-Akt, and small GTPases (Moustakas and Heldin 2005; Zhang 2009).

TGF-β pathway plays a crucial role in regulating the behavior of gliomas. Specifically, it acts to regulate GSC stemness, abnormal angiogenesis, tumor cell invasion/migration, and immunosuppression, as described above (Figs. 3.3 and 3.9) (Joseph et al. 2013).

Stemness of GSCs is maintained by TGF-β through induction of Sry-related HMG-Box factor 2 (Sox2) expression, an important transcription factor for generation of induced pluripotent (iPS) cells. Sox2 expression was mediated by Sox4, which is a direct target of TGF-β (Ikushima et al. 2009). TGF-β induces the JAK-STAT pathway through the induction and secretion of leukemia inhibitory factor (LIF). LIF binds the heterodimeric glycoprotein 130 (gp130)/LIF receptor (LIFR) complex, and LIF-LIFR/gp130-JAK-STAT signaling pathway is implicated as a promoter of NSC self-renewal in the adult brain (Bauer and Patterson 2006; Penuelas et al. 2009). GSCs express higher levels of TGF-β2 than differentiated glioma cells do, and TGF-β2 secretion also appears to correlate with higher glioma grades (Qiu et al. 2011). Therefore, targeting TGF-β might help reduce the GSC fraction of gliomas, and is a promising therapeutic methodology to abrogate malignant gliomas (Bogdahn et al. 2011).

TGF-β signaling regulates the expression of a wide variety of genes encoding extracellular matrix (ECM) proteins (collagens, fibronectin, and laminins) and ECM-degrading enzymes, matrixmetalloproteinases (MMPs) (Bogdahn et al. 2011; Dieterich et al. 2012; Wick et al. 2001). These molecules are necessary for tumor cell migration/invasion and angiogenesis. TGF-β signaling contributes to aberrant vascular gene expression pattern in GBM through TβRI, known as activin receptor-like kinase (ALK) 5. The TGF-β1/ALK5/SMAD2 pathway is closely related with the regulation of glioma invasion and vascular proliferation, which are features of GBMs (Massague 2008; Pen et al. 2008).

Immunosuppression, which has been observed as malignant phenotype of gliomas, can also be attributed to TGF-β, a known potent immunosuppressant. TGF-β exerts its immunosuppressive effects by inhibiting proliferation and cytotoxic activity of brain tumor-infiltrating lymphocytes, natural killer cells, and lymphokine-activated killer cells (Friese et al. 2004; Uhl et al. 2004). Multiple molecular mechanisms are involved in immunosuppression: for example, inhibition of T-cell activation and effector function through downregulation of interleukin-2 (IL-2) and suppression of tumor necrosis factor-α (TNF-α), INF-γ, and IL-6, which are pro-inflammatory cytokines (Chen et al. 2005; Ksendzovsky et al. 2009). The anti-tumor immune response might be compromised during various stages of

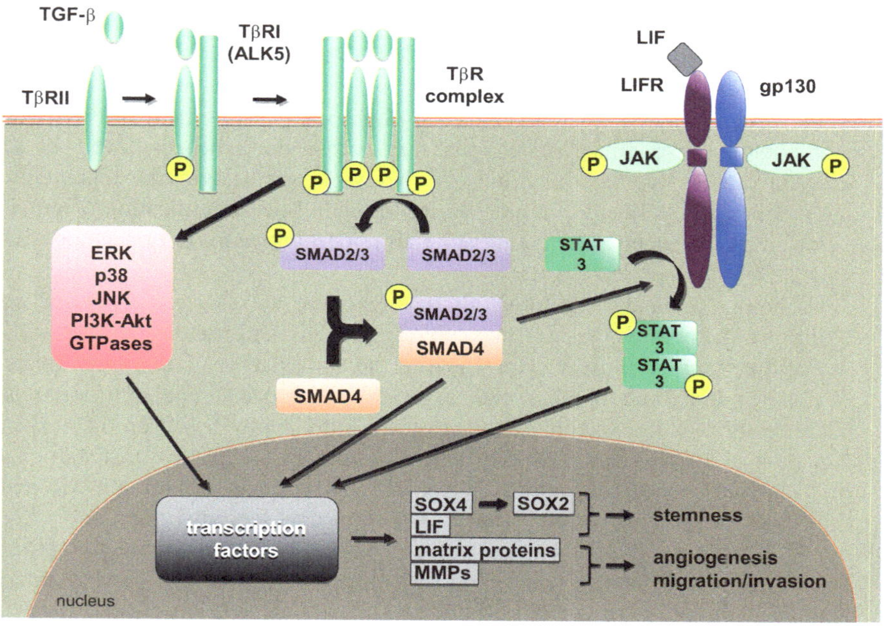

Fig. 3.9 TGF-β signaling regulates GSC self-renewal and glioma cell activities transduced through SMAD and non-SMAD pathways. TGF-β homodimer ligand binds to TβRII and then activates it. TβRII phosphorylates TβRI, or ALK5, which subsequently phosphorylates and activates R-SMAD (SMAD2 and SMAD3). Activated SMAD2 and SMAD3 form a Smad complex with co-SMAD (SMAD4) and translocate into the nucleus. In the nucleus, Smad complex drives transcription of genes, including stemness-associated transcription factors SOX and matrix-associated proteins. TGF-β also activates the JAK-STAT pathway via induction of LIF secretion acting through autocrine/paracrine loop. After binding of LIF to its cell surface receptor, LIFR, heterodimerization with another transmembrane protein gp130 occurs, followed by recruitment of JAK and STAT3. STAT3 can induce the expression of SOX2 retaining self-renewal capacity of GSCs. *Abbreviations*: *ALK5* activin receptor-like kinase 5, *gp130* glycoprotein 130, *JAK* Janus kinase, *LIF* leukemia inhibitory factor, *LIFR* LIF receptor, *MMP* matrix metalloproteinase, *SOX* Sry-related HMG-Box factor, *STAT* signal transducer and activator of transcription, *TGF-β* transforming growth factor-β, *TβR* TGF-β receptor

malignant glioma progression (Platten et al. 2001). The GBM subtypes by the TCGA examined the degree of immune infiltration and local immune deficiency via a different degree of TGF-β-mediated suppression of infiltrating immune cells (Fig. 3.1). As compared to the proneural subtype, the mesenchymal subtype showed enhanced TGF-β-induced SMAD2 phosphorylation and downstream gene expression, despite the fact that TGF-β expression did not vary between the subtypes (Beier et al. 2012).

Two compounds AP 12009 (TGF-β2 inhibitor) and LY2157299 (TβRI kinase inhibitor) are currently in Phase IIb/III and Phase IIa trials for malignant glioma patients, respectively (see http://clinicaltrials.gov/ct2/home).

Conclusions and Perspectives

The discovery of the major genetic variations in gliomas is of great significance for understanding the molecular pathways involved in gliomagenesis. Continued studies of the molecular pathogenesis of gliomas have led to new classification methodology for gliomas. Since signals are carried along these pathways by an array of protein kinases, this allows for many potential targets for therapeutic intervention using specific inhibitors. Additionally, the identification of previously unknown genetic alterations provides opportunities for drug development against new therapeutic targets.

Although decisions regarding glioma treatment are still mainly based on traditional pathology relying on histology, genetic information is becoming more important in clinical diagnosis and the formulation of treatment plans. Successful treatment requires the targeting reasonable signaling molecules. The inhibitors utilized should be determined individually based on the critical signaling molecules in GBM. To date, monotherapy with molecular-targeted drugs has not been significantly effective for GBM patients. Combining different molecular targeted drugs may lead to new concepts for targeting GBM. This would lead to successful clinical responses via control of tumor cell proliferation, survival, and invasion. However, our data suggest that combinatorial therapy with specific molecular inhibitors for two different targets has limited effects on proliferation of glioma cells compared to cells treated with their respective agents individually (Jin et al. 2013). Despite these studies, further studies regarding the molecular pathogenic mechanisms of GBMs following therapeutic intervention are still required.

The existence of GSC is a novel concept in cancer biology that must be explored further. Continued investigation of GSC development will enhance our comprehension of tumor initiation and progression, likely giving rise to innovative, preventive, and therapeutic approaches for GBM. As methods to identify CSCs improve and GSC studies advance, GSC-directed therapies might become a useful therapeutic strategy for GBM.

The continued incorporation and validation of new data by using increasingly sophisticated in vitro and in vivo models will further advance our knowledge of disease origin, progression, and treatment. There is real hope that, in the near future, survival of GBM patients will improve.

References

Adjei AA, Hidalgo M (2005) Intracellular signal transduction pathway proteins as targets for cancer therapy. J Clin Oncol 23:5386–5403

Alcantara Llaguno S, Chen J, Kwon CH, Jackson EL, Li Y, Burns DK, Alvarez-Buylla A, Parada LF (2009) Malignant astrocytomas originate from neural stem/progenitor cells in a somatic tumor suppressor mouse model. Cancer Cell 15:45–56

Bachoo RM, Maher EA, Ligon KL, Sharpless NE, Chan SS, You MJ, Tang Y, Defrances J, Stover E, Weissleder R, Rowitch DH, Louis DN, Depinho RA (2002) Epidermal growth factor receptor and Ink4a/Arf: convergent mechanisms governing terminal differentiation and transformation along the neural stem cell to astrocyte axis. Cancer Cell 1:269–277

Bar EE, Chaudhry A, Lin A, Fan X, Schreck K, Matsui W, Piccirillo S, Vescovi AL, Dimeco F, Olivi A, Eberhart CG (2007) Cyclopamine-mediated hedgehog pathway inhibition depletes stem-like cancer cells in glioblastoma. Stem Cells 25:2524–2533

Barak Y, Gottlieb E, Juven-Gershon T, Oren M (1994) Regulation of mdm2 expression by p53: alternative promoters produce transcripts with nonidentical translation potential. Genes Dev 8: 1739–1749

Barker FG II, Simmons ML, Chang SM, Prados MD, Larson DA, Sneed PK, Wara WM, Berger MS, Chen P, Israel MA, Aldape KD (2001) EGFR overexpression and radiation response in glioblastoma multiforme. Int J Radiat Oncol Biol Phys 51:410–418

Bauer S, Patterson PH (2006) Leukemia inhibitory factor promotes neural stem cell self-renewal in the adult brain. J Neurosci 26:12089–12099

Becher OJ, Hambardzumyan D, Walker TR, Helmy K, Nazarian J, Albrecht S, Hiner RL, Gall S, Huse JT, Jabado N, Macdonald TJ, Holland EC (2010) Preclinical evaluation of radiation and perifosine in a genetically and histologically accurate model of brainstem glioma. Cancer Res 70:2548–2557

Beier CP, Kumar P, Meyer K, Leukel P, Bruttel V, Aschenbrenner I, Riemenschneider MJ, Fragoulis A, Rummele P, Lamszus K, Schulz JB, Weis J, Bogdahn U, Wischhusen J, Hau P, Spang R, Beier D (2012) The cancer stem cell subtype determines immune infiltration of glioblastoma. Stem Cells Dev 21:2753–2761

Biernat W, Tohma Y, Yonekawa Y, Kleihues P, Ohgaki H (1997) Alterations of cell cycle regulatory genes in primary (de novo) and secondary glioblastomas. Acta Neuropathol 94: 303–309

Bleau AM, Hambardzumyan D, Ozawa T, Fomchenko EI, Huse JT, Brennan CW, Holland EC (2009) PTEN/PI3K/Akt pathway regulates the side population phenotype and ABCG2 activity in glioma tumor stem-like cells. Cell Stem Cell 4:226–235

Bleeker FE, Lamba S, Zanon C, Van Tilborg AA, Leenstra S, Troost D, Hulsebos T, Vandertop WP, Bardelli A (2009) Absence of AKT1 mutations in glioblastoma. PLoS One 4:e5638

Bogdahn U, Hau P, Stockhammer G, Venkataramana NK, Mahapatra AK, Suri A, Balasubramaniam A, Nair S, Oliushine V, Parfenov V, Poverennova I, Zaaroor M, Jachimczak P, Ludwig S, Schmaus S, Heinrichs H, Schlingensiepen KH (2011) Targeted therapy for high-grade glioma with the TGF-beta2 inhibitor trabedersen: results of a randomized and controlled phase IIb study. Neuro Oncol 13:132–142

Bogler O, Huang HJ, Kleihues P, Cavenee WK (1995) The p53 gene and its role in human brain tumors. Glia 15:308–327

Bos JL (1989) ras oncogenes in human cancer: a review. Cancer Res 49:4682–4689

Chen ML, Pittet MJ, Gorelik L, Flavell RA, Weissleder R, Von Boehmer H, Khazaie K (2005) Regulatory T cells suppress tumor-specific CD8 T cell cytotoxicity through TGF-beta signals in vivo. Proc Natl Acad Sci U S A 102:419–424

Cheng CK, Fan QW, Weiss WA (2009) PI3K signaling in glioma–animal models and therapeutic challenges. Brain Pathol 19:112–120

Chinot OL et al (2014) Bevacizumab plus radiotherapy–temozolomide for newly diagnosed glioblastoma. New Engl J Med 370:709–722

Choe G, Horvath S, Cloughesy TF, Crosby K, Seligson D, Palotie A, Inge L, Smith BL, Sawyers CL, Mischel PS (2003) Analysis of the phosphatidylinositol 3′-kinase signaling pathway in glioblastoma patients in vivo. Cancer Res 63:2742–2746

Chow LM, Endersby R, Zhu X, Rankin S, Qu C, Zhang J, Broniscer A, Ellison DW, Baker SJ (2011) Cooperativity within and among Pten, p53, and Rb pathways induces high-grade astrocytoma in adult brain. Cancer Cell 19:305–316

Ciriello G, Cerami E, Sander C, Schultz N (2012) Mutual exclusivity analysis identifies oncogenic network modules. Genome Res 22:398–406

Clark PA, Iida M, Treisman DM, Kalluri H, Ezhilan S, Zorniak M, Wheeler DL, Kuo JS (2012) Activation of multiple ERBB family receptors mediates glioblastoma cancer stem-like cell resistance to EGFR-targeted inhibition. Neoplasia 14:420–428

Dai C, Celestino JC, Okada Y, Louis DN, Fuller GN, Holland EC (2001) PDGF autocrine stimulation dedifferentiates cultured astrocytes and induces oligodendrogliomas and oligoastrocytomas from neural progenitors and astrocytes in vivo. Genes Dev 15:1913–1925

De Witt Hamer PC (2010) Small molecule kinase inhibitors in glioblastoma: a systematic review of clinical studies. Neuro Oncol 12:304–316

Del Vecchio CA, Li G, Wong AJ (2012) Targeting EGF receptor variant III: tumor-specific peptide vaccination for malignant gliomas. Expert Rev Vaccines 11:133–144

Dieterich LC, Mellberg S, Langenkamp E, Zhang L, Zieba A, Salomaki H, Teichert M, Huang H, Edqvist PH, Kraus T, Augustin HG, Olofsson T, Larsson E, Soderberg O, Molema G, Ponten F, Georgii-Hemming P, Alafuzoff I, Dimberg A (2012) Transcriptional profiling of human glioblastoma vessels indicates a key role of VEGF-A and TGFbeta2 in vascular abnormalization. J Pathol 228:378–390

Dietrich J, Diamond EL, Kesari S (2010) Glioma stem cell signaling: therapeutic opportunities and challenges. Expert Rev Anticancer Ther 10:709–722

Doetsch F, Petreanu L, Caille I, Garcia-Verdugo JM, Alvarez-Buylla A (2002) EGF converts transit-amplifying neurogenic precursors in the adult brain into multipotent stem cells. Neuron 36:1021–1034

Ebben JD, Treisman DM, Zorniak M, Kutty RG, Clark PA, Kuo JS (2010) The cancer stem cell paradigm: a new understanding of tumor development and treatment. Expert Opin Ther Targets 14:621–632

Ebos JM, Lee CR, Cruz-Munoz W, Bjarnason GA, Christensen JG, Kerbel RS (2009) Accelerated metastasis after short-term treatment with a potent inhibitor of tumor angiogenesis. Cancer Cell 15:232–239

Ekstrand AJ, Sugawa N, James CD, Collins VP (1992) Amplified and rearranged epidermal growth factor receptor genes in human glioblastomas reveal deletions of sequences encoding portions of the N- and/or C-terminal tails. Proc Natl Acad Sci U S A 89:4309–4313

Fan QW, Cheng C, Hackett C, Feldman M, Houseman BT, Nicolaides T, Haas-Kogan D, James CD, Oakes SA, Debnath J, Shokat KM, Weiss WA (2010) Akt and autophagy cooperate to promote survival of drug-resistant glioma. Sci Signal 3:ra81

Fan QW, Cheng C, Knight ZA, Haas-Kogan D, Stokoe D, James CD, Mccormick F, Shokat KM, Weiss WA (2009) EGFR signals to mTOR through PKC and independently of Akt in glioma. Sci Signal 2:ra4

Feldkamp MM, Lala P, Lau N, Roncari L, Guha A (1999) Expression of activated epidermal growth factor receptors, Ras-guanosine triphosphate, and mitogen-activated protein kinase in human glioblastoma multiforme specimens. Neurosurgery 45:1442–1453

Fleming TP, Saxena A, Clark WC, Robertson JT, Oldfield EH, Aaronson SA, Ali IU (1992) Amplification and/or overexpression of platelet-derived growth factor receptors and epidermal growth factor receptor in human glial tumors. Cancer Res 52:4550–4553

Friedman HS, Prados MD, Wen PY, Mikkelsen T, Schiff D, Abrey LE, Yung WK, Paleologos N, Nicholas MK, Jensen R, Vredenburgh J, Huang J, Zheng M, Cloughesy T (2009) Bevacizumab alone and in combination with irinotecan in recurrent glioblastoma. J Clin Oncol 27: 4733–4740

Friese MA, Wischhusen J, Wick W, Weiler M, Eisele G, Steinle A, Weller M (2004) RNA interference targeting transforming growth factor-beta enhances NKG2D-mediated antiglioma immune response, inhibits glioma cell migration and invasiveness, and abrogates tumorigenicity in vivo. Cancer Res 64:7596–7603

Furnari FB, Huang HJ, Cavenee WK (1998) The phosphoinositol phosphatase activity of PTEN mediates a serum-sensitive G1 growth arrest in glioma cells. Cancer Res 58:5002–5008

Furuta T, Nakada M, Misaki K, Sato Y, Hayashi Y, Nakanuma Y, Hamada JI (2014) Molecular analysis of a recurrent glioblastoma treated with bevacizumab. Brain Tumor Pathol 31(1): 32–39

Galli R, Binda E, Orfanelli U, Cipelletti B, Gritti A, De Vitis S, Fiocco R, Foroni C, Dimeco F, Vescovi A (2004) Isolation and characterization of tumorigenic, stem-like neural precursors from human glioblastoma. Cancer Res 64:7011–7021

Gilbert MR et al (2014) A randomized trial of bevacizumab for newly diagnosed glioblastoma. New Engl J Med 370(8):699–708

Gills JJ, Dennis PA (2009) Perifosine: update on a novel Akt inhibitor. Curr Oncol Rep 11: 102–110

Goodenberger ML, Jenkins RB (2012) Genetics of adult glioma. Cancer Genet 205:613–621

Gottfried ON, Viskochil DH, Couldwell WT (2010) Neurofibromatosis Type 1 and tumorigenesis: molecular mechanisms and therapeutic implications. Neurosurg Focus 28:E8

Gravendeel LA, Kouwenhoven MC, Gevaert O, De Rooi JJ, Stubbs AP, Duijm JE, Daemen A, Bleeker FE, Bralten LB, Kloosterhof NK, De Moor B, Eilers PH, Van Der Spek PJ, Kros JM, Sillevis Smitt PA, Van Den Bent MJ, French PJ (2009) Intrinsic gene expression profiles of gliomas are a better predictor of survival than histology. Cancer Res 69:9065–9072

Guha A, Dashner K, Black PM, Wagner JA, Stiles CD (1995) Expression of PDGF and PDGF receptors in human astrocytoma operation specimens supports the existence of an autocrine loop. Int J Cancer 60:168–173

Guha A, Feldkamp MM, Lau N, Boss G, Pawson A (1997) Proliferation of human malignant astrocytomas is dependent on Ras activation. Oncogene 15:2755–2765

Gupta K, Salunke P (2012) Molecular markers of glioma: an update on recent progress and perspectives. J Cancer Res Clin Oncol 138:1971–1981

Haas-Kogan D, Shalev N, Wong M, Mills G, Yount G, Stokoe D (1998) Protein kinase B (PKB/Akt) activity is elevated in glioblastoma cells due to mutation of the tumor suppressor PTEN/MMAC. Curr Biol 8:1195–1198

Hainsworth JD, Shih KC, Shepard GC, Tillinghast GW, Brinker BT, Spigel DR (2012) Phase II study of concurrent radiation therapy, temozolomide, and bevacizumab followed by bevacizumab/everolimus as first-line treatment for patients with glioblastoma. Clin Adv Hematol Oncol 10:240–246

Harper JW, Adami GR, Wei N, Keyomarsi K, Elledge SJ (1993) The p21 Cdk-interacting protein Cip1 is a potent inhibitor of G1 cyclin-dependent kinases. Cell 75:805–816

Hayashi Y, Ueki K, Waha A, Wiestler OD, Louis DN, Von Deimling A (1997) Association of EGFR gene amplification and CDKN2 (p16/MTS1) gene deletion in glioblastoma multiforme. Brain Pathol 7:871–875

Hegi ME, Zur Hausen A, Ruedi D, Malin G, Kleihues P (1997) Hemizygous or homozygous deletion of the chromosomal region containing the p16INK4a gene is associated with amplification of the EGF receptor gene in glioblastomas. Int J Cancer 73:57–63

Heldin CH, Miyazono K, Ten Dijke P (1997) TGF-beta signalling from cell membrane to nucleus through SMAD proteins. Nature 390:465–471

Hemmati HD, Nakano I, Lazareff JA, Masterman-Smith M, Geschwind DH, Bronner-Fraser M, Kornblum HI (2003) Cancerous stem cells can arise from pediatric brain tumors. Proc Natl Acad Sci U S A 100:15178–15183

Hermanson M, Funa K, Hartman M, Claesson-Welsh L, Heldin CH, Westermark B, Nister M (1992) Platelet-derived growth factor and its receptors in human glioma tissue: expression of messenger RNA and protein suggests the presence of autocrine and paracrine loops. Cancer Res 52:3213–3219

Hoelzinger DB, Demuth T, Berens ME (2007) Autocrine factors that sustain glioma invasion and paracrine biology in the brain microenvironment. J Natl Cancer Inst 99:1583–1593

Holland EC, Celestino J, Dai C, Schaefer L, Sawaya RE, Fuller GN (2000) Combined activation of Ras and Akt in neural progenitors induces glioblastoma formation in mice. Nat Genet 25:55–57

Hu YY, Zheng MH, Cheng G, Li L, Liang L, Gao F, Wei YN, Fu LA, Han H (2011) Notch signaling contributes to the maintenance of both normal neural stem cells and patient-derived glioma stem cells. BMC Cancer 11:82

Ichimura K, Bolin MB, Goike HM, Schmidt EE, Moshref A, Collins VP (2000) Deregulation of the p14ARF/MDM2/p53 pathway is a prerequisite for human astrocytic gliomas with G1-S transition control gene abnormalities. Cancer Res 60:417–424

Ignatova TN, Kukekov VG, Laywell ED, Suslov ON, Vrionis FD, Steindler DA (2002) Human cortical glial tumors contain neural stem-like cells expressing astroglial and neuronal markers in vitro. Glia 39:193–206

Ikushima H, Miyazono K (2010) TGFbeta signalling: a complex web in cancer progression. Nat Rev Cancer 10:415–424

Ikushima H, Todo T, Ino Y, Takahashi M, Miyazawa K, Miyazono K (2009) Autocrine TGF-beta signaling maintains tumorigenicity of glioma-initiating cells through Sry-related HMG-box factors. Cell Stem Cell 5:504–514

Jackson EL, Garcia-Verdugo JM, Gil-Perotin S, Roy M, Quinones-Hinojosa A, Vandenberg S, Alvarez-Buylla A (2006) PDGFR alpha-positive B cells are neural stem cells in the adult SVZ that form glioma-like growths in response to increased PDGF signaling. Neuron 51:187–199

Jiao Y, Killela PJ, Reitman ZJ, Rasheed AB, Heaphy CM, De Wilde RF, Rodriguez FJ, Rosemberg S, Oba-Shinjo SM, Nagahashi Marie SK, Bettegowda C, Agrawal N, Lipp E, Pirozzi C, Lopez G, He Y, Friedman H, Friedman AH, Riggins GJ, Holdhoff M, Burger P, Mclendon R, Bigner DD, Vogelstein B, Meeker AK, Kinzler KW, Papadopoulos N, Diaz LA, Yan H (2012) Frequent ATRX, CIC, FUBP1 and IDH1 mutations refine the classification of malignant gliomas. Oncotarget 3:709–722

Jin R, Nakada M, Teng L, Furuta T, Sabit H, Hayashi Y, Demuth T, Hirao A, Sato H, Zhao G, Hamada J (2013) Combination therapy using Notch and Akt inhibitors is effective for suppressing invasion but not proliferation in glioma cells. Neurosci Lett 534:316–321

Joseph JV, Balasubramaniyan V, Walenkamp A, Kruyt FA (2013) TGF-beta as a therapeutic target in high grade gliomas – promises and challenges. Biochem Pharmacol 85:478–485

Kamijo T, Weber JD, Zambetti G, Zindy F, Roussel MF, Sherr CJ (1998) Functional and physical interactions of the ARF tumor suppressor with p53 and Mdm2. Proc Natl Acad Sci U S A 95:8292–8297

Kapoor GS, O'rourke DM (2003a) Mitogenic signaling cascades in glial tumors. Neurosurgery 52:1425–1434, discussion 1434–1425

Kapoor GS, O'rourke DM (2003b) Receptor tyrosine kinase signaling in gliomagenesis: pathobiology and therapeutic approaches. Cancer Biol Ther 2:330–342

Katoh M (2007) WNT signaling pathway and stem cell signaling network. Clin Cancer Res 13: 4042–4045

Keunen O, Johansson M, Oudin A, Sanzey M, Rahim SA, Fack F, Thorsen F, Taxt T, Bartos M, Jirik R, Miletic H, Wang J, Stieber D, Stuhr L, Moen I, Rygh CB, Bjerkvig R, Niclou SP (2011) Anti-VEGF treatment reduces blood supply and increases tumor cell invasion in glioblastoma. Proc Natl Acad Sci U S A 108:3749–3754

Kita D, Yonekawa Y, Weller M, Ohgaki H (2007) PIK3CA alterations in primary (de novo) and secondary glioblastomas. Acta Neuropathol 113:295–302

Kloosterhof NK, Bralten LB, Dubbink HJ, French PJ, Van Den Bent MJ (2011) Isocitrate dehydrogenase-1 mutations: a fundamentally new understanding of diffuse glioma? Lancet Oncol 12:83–91

Krakstad C, Chekenya M (2010) Survival signalling and apoptosis resistance in glioblastomas: opportunities for targeted therapeutics. Mol Cancer 9:135

Kreisl TN, Smith P, Sul J, Salgado C, Iwamoto FM, Shih JH, Fine HA (2013) Continuous daily sunitinib for recurrent glioblastoma. J Neurooncol 111(1):41–48

Ksendzovsky A, Feinstein D, Zengou R, Sharp A, Polak P, Lichtor T, Glick RP (2009) Investigation of immunosuppressive mechanisms in a mouse glioma model. J Neurooncol 93:107–114

Kubbutat MH, Jones SN, Vousden KH (1997) Regulation of p53 stability by Mdm2. Nature 387: 299–303
Kwiatkowska A, Symons M (2013) Signaling determinants of glioma cell invasion. Adv Exp Med Biol 986:121–141
Lamszus K, Ulbricht U, Matschke J, Brockmann MA, Fillbrandt R, Westphal M (2003) Levels of soluble vascular endothelial growth factor (VEGF) receptor 1 in astrocytic tumors and its relation to malignancy, vascularity, and VEGF-A. Clin Cancer Res 9:1399–1405
Lee J, Son MJ, Woolard K, Donin NM, Li A, Cheng CH, Kotliarova S, Kotliarov Y, Walling J, Ahn S, Kim M, Totonchy M, Cusack T, Ene C, Ma H, Su Q, Zenklusen JC, Zhang W, Maric D, Fine HA (2008) Epigenetic-mediated dysfunction of the bone morphogenetic protein pathway inhibits differentiation of glioblastoma-initiating cells. Cancer Cell 13:69–80
Lee MJ, Stephenson DA (2007) Recent developments in neurofibromatosis type 1. Curr Opin Neurol 20:135–141
Li J, Yen C, Liaw D, Podsypanina K, Bose S, Wang SI, Puc J, Miliaresis C, Rodgers L, Mccombie R, Bigner SH, Giovanella BC, Ittmann M, Tycko B, Hibshoosh H, Wigler MH, Parsons R (1997) PTEN, a putative protein tyrosine phosphatase gene mutated in human brain, breast, and prostate cancer. Science 275:1943–1947
Libermann TA, Nusbaum HR, Razon N, Kris R, Lax I, Soreq H, Whittle N, Waterfield MD, Ullrich A, Schlessinger J (1985) Amplification and overexpression of the EGF receptor gene in primary human glioblastomas. J Cell Sci Suppl 3:161–172
Lin A, Rodriguez FJ, Karajannis MA, Williams SC, Legault G, Zagzag D, Burger PC, Allen JC, Eberhart CG, Bar EE (2012) BRAF alterations in primary glial and glioneuronal neoplasms of the central nervous system with identification of 2 novel KIAA1549:BRAF fusion variants. J Neuropathol Exp Neurol 71:66–72
Liu XY, Gerges N, Korshunov A, Sabha N, Khuong-Quang DA, Fontebasso AM, Fleming A, Hadjadj D, Schwartzentruber J, Majewski J, Dong Z, Siegel P, Albrecht S, Croul S, Jones DT, Kool M, Tonjes M, Reifenberger G, Faury D, Zadeh G, Pfister S, Jabado N (2012) Frequent ATRX mutations and loss of expression in adult diffuse astrocytic tumors carrying IDH1/IDH2 and TP53 mutations. Acta Neuropathol 124:615–625
Lo HW (2010) Targeting Ras-RAF-ERK and its interactive pathways as a novel therapy for malignant gliomas. Curr Cancer Drug Targets 10(8):840–848
Lokker NA, Sullivan CM, Hollenbach SJ, Israel MA, Giese NA (2002) Platelet-derived growth factor (PDGF) autocrine signaling regulates survival and mitogenic pathways in glioblastoma cells: evidence that the novel PDGF-C and PDGF-D ligands may play a role in the development of brain tumors. Cancer Res 62:3729–3735
Louvi A, Artavanis-Tsakonas S (2006) Notch signalling in vertebrate neural development. Nat Rev Neurosci 7:93–102
Lu-Emerson C, Norden AD, Drappatz J, Quant EC, Beroukhim R, Ciampa AS, Doherty LM, Lafrankie DC, Ruland S, Wen PY (2011) Retrospective study of dasatinib for recurrent glioblastoma after bevacizumab failure. J Neurooncol 104:287–291
Malumbres M, Barbacid M (2003) RAS oncogenes: the first 30 years. Nat Rev Cancer 3:459–465
Mao L, Merlo A, Bedi G, Shapiro GI, Edwards CD, Rollins BJ, Sidransky D (1995) A novel p16INK4A transcript. Cancer Res 55:2995–2997
Marumoto T, Saya H (2012) Molecular biology of glioma. Adv Exp Med Biol 746:2–11
Massague J (2008) TGFbeta in cancer. Cell 134:215–230
Mellinghoff IK, Wang MY, Vivanco I, Haas-Kogan DA, Zhu S, Dia EQ, Lu KV, Yoshimoto K, Huang JH, Chute DJ, Riggs BL, Horvath S, Liau LM, Cavenee WK, Rao PN, Beroukhim R, Peck TC, Lee JC, Sellers WR, Stokoe D, Prados M, Cloughesy TF, Sawyers CL, Mischel PS (2005) Molecular determinants of the response of glioblastomas to EGFR kinase inhibitors. N Engl J Med 353:2012–2024
Moustakas A, Heldin CH (2005) Non-Smad TGF-beta signals. J Cell Sci 118:3573–3584
Nakada M, Kita D, Teng L, Pyko IV, Watanabe T, Hayashi Y, Hamada J (2013) Receptor tyrosine kinases: principles and functions in glioma invasion. Adv Exp Med Biol 986:143–170

Nakada M, Nakada S, Demuth T, Tran NL, Hoelzinger DB, Berens ME (2007) Molecular targets of glioma invasion. Cell Mol Life Sci 64:458–478

Nakada M, Kita D, Watanabe T, Hayashi Y, Teng L, Pyko IV, Hamada J-I (2011) Aberrant signaling pathways in glioma. Cancers 3:3242–3278

Nakamura M, Watanabe T, Klangby U, Asker C, Wiman K, Yonekawa Y, Kleihues P, Ohgaki H (2001) p14ARF deletion and methylation in genetic pathways to glioblastomas. Brain Pathol 11:159–168

Natsume A, Kinjo S, Yuki K, Kato T, Ohno M, Motomura K, Iwami K, Wakabayashi T (2011) Glioma-initiating cells and molecular pathology: implications for therapy. Brain Tumor Pathol 28:1–12

Network CGaR (2008) Comprehensive genomic characterization defines human glioblastoma genes and core pathways. Nature 455:1061–1068

Nobusawa S, Watanabe T, Kleihues P, Ohgaki H (2009) IDH1 mutations as molecular signature and predictive factor of secondary glioblastomas. Clin Cancer Res 15:6002–6007

Ohgaki H, Dessen P, Jourde B, Horstmann S, Nishikawa T, Di Patre PL, Burkhard C, Schuler D, Probst-Hensch NM, Maiorka PC, Baeza N, Pisani P, Yonekawa Y, Yasargil MG, Lutolf UM, Kleihues P (2004) Genetic pathways to glioblastoma: a population-based study. Cancer Res 64: 6892–6899

Ohgaki H, Kleihues P (2013) The definition of primary and secondary glioblastoma. Clin Cancer Res 19:764–772

Ohgaki H, Kleihues P (2009) Genetic alterations and signaling pathways in the evolution of gliomas. Cancer Sci 100:2235–2241

Oliner JD, Kinzler KW, Meltzer PS, George DL, Vogelstein B (1992) Amplification of a gene encoding a p53-associated protein in human sarcomas. Nature 358:80–83

Onishi H, Katano M (2011) Hedgehog signaling pathway as a therapeutic target in various types of cancer. Cancer Sci 102:1756–1760

Paez-Ribes M, Allen E, Hudock J, Takeda T, Okuyama H, Vinals F, Inoue M, Bergers G, Hanahan D, Casanovas O (2009) Antiangiogenic therapy elicits malignant progression of tumors to increased local invasion and distant metastasis. Cancer Cell 15:220–231

Parsons DW, Jones S, Zhang X, Lin JC, Leary RJ, Angenendt P, Mankoo P, Carter H, Siu IM, Gallia GL, Olivi A, Mclendon R, Rasheed BA, Keir S, Nikolskaya T, Nikolsky Y, Busam DA, Tekleab H, Diaz LA Jr, Hartigan J, Smith DR, Strausberg RL, Marie SK, Shinjo SM, Yan H, Riggins GJ, Bigner DD, Karchin R, Papadopoulos N, Parmigiani G, Vogelstein B, Velculescu VE, Kinzler KW (2008) An integrated genomic analysis of human glioblastoma multiforme. Science 321:1807–1812

Peifer M, Polakis P (2000) Wnt signaling in oncogenesis and embryogenesis–a look outside the nucleus. Science 287:1606–1609

Pen A, Moreno MJ, Durocher Y, Deb-Rinker P, Stanimirovic DB (2008) Glioblastoma-secreted factors induce IGFBP7 and angiogenesis by modulating Smad-2-dependent TGF-beta signaling. Oncogene 27:6834–6844

Penuelas S, Anido J, Prieto-Sanchez RM, Folch G, Barba I, Cuartas I, Garcia-Dorado D, Poca MA, Sahuquillo J, Baselga J, Seoane J (2009) TGF-beta increases glioma-initiating cell self-renewal through the induction of LIF in human glioblastoma. Cancer Cell 15:315–327

Plate KH, Breier G, Farrell CL, Risau W (1992) Platelet-derived growth factor receptor-beta is induced during tumor development and upregulated during tumor progression in endothelial cells in human gliomas. Lab Invest 67:529–534

Platten M, Wick W, Weller M (2001) Malignant glioma biology: role for TGF-beta in growth, motility, angiogenesis, and immune escape. Microsc Res Tech 52:401–410

Qiu B, Zhang D, Wang C, Tao J, Tie X, Qiao Y, Xu K, Wang Y, Wu A (2011) IL-10 and TGF-beta2 are overexpressed in tumor spheres cultured from human gliomas. Mol Biol Rep 38:3585–3591

Ramnarain DB, Park S, Lee DY, Hatanpaa KJ, Scoggin SO, Otu H, Libermann TA, Raisanen JM, Ashfaq R, Wong ET, Wu J, Elliott R, Habib AA (2006) Differential gene expression analysis reveals generation of an autocrine loop by a mutant epidermal growth factor receptor in glioma cells. Cancer Res 66:867–874

Reardon DA, Dresemann G, Taillibert S, Campone M, Van Den Bent M, Clement P, Blomquist E, Gordower L, Schultz H, Raizer J, Hau P, Easaw J, Gil M, Tonn J, Gijtenbeek A, Schlegel U, Bergstrom P, Green S, Weir A, Nikolova Z (2009) Multicentre phase II studies evaluating imatinib plus hydroxyurea in patients with progressive glioblastoma. Br J Cancer 101:1995–2004

Reardon DA, Egorin MJ, Quinn JA, Rich JN, Gururangan S, Vredenburgh JJ, Desjardins A, Sathornsumetee S, Provenzale JM, Herndon JE II, Dowell JM, Badruddoja MA, Mclendon RE, Lagattuta TF, Kicielinski KP, Dresemann G, Sampson JH, Friedman AH, Salvado AJ, Friedman HS (2005) Phase II study of imatinib mesylate plus hydroxyurea in adults with recurrent glioblastoma multiforme. J Clin Oncol 23:9359–9368

Reifenberger G, Liu L, Ichimura K, Schmidt EE, Collins VP (1993) Amplification and overexpression of the MDM2 gene in a subset of human malignant gliomas without p53 mutations. Cancer Res 53:2736–2739

Rich JN, Rasheed BK, Yan H (2004) EGFR mutations and sensitivity to gefitinib. N Engl J Med 351:1260–1261, author reply 1260–1261

Rossi M, Magnoni L, Miracco C, Mori E, Tosi P, Pirtoli L, Tini P, Oliveri G, Cosci E, Bakker A (2011) beta-catenin and Gli1 are prognostic markers in glioblastoma. Cancer Biol Ther 11:753–761

Salmaggi A, Eoli M, Frigerio S, Silvani A, Gelati M, Corsini E, Broggi G, Boiardi A (2003) Intracavitary VEGF, bFGF, IL-8, IL-12 levels in primary and recurrent malignant glioma. J Neurooncol 62:297–303

Sampetrean O, Saya H (2013) Characteristics of glioma stem cells. Brain Tumor Pathol 30(4):209–214

Samuels Y, Wang Z, Bardelli A, Silliman N, Ptak J, Szabo S, Yan H, Gazdar A, Powell SM, Riggins GJ, Willson JK, Markowitz S, Kinzler KW, Vogelstein B, Velculescu VE (2004) High frequency of mutations of the PIK3CA gene in human cancers. Science 304:554

See WL, Tan IL, Mukherjee J, Nicolaides T, Pieper RO (2012) Sensitivity of glioblastomas to clinically available MEK inhibitors is defined by neurofibromin 1 deficiency. Cancer Res 72:3350–3359

Serrano M, Hannon GJ, Beach D (1993) A new regulatory motif in cell-cycle control causing specific inhibition of cyclin D/CDK4. Nature 366:704–707

Shangary S, Wang S (2009) Small-molecule inhibitors of the MDM2-p53 protein-protein interaction to reactivate p53 function: a novel approach for cancer therapy. Annu Rev Pharmacol Toxicol 49:223–241

Sherr CJ, Roberts JM (1999) CDK inhibitors: positive and negative regulators of G1-phase progression. Genes Dev 13:1501–1512

Shih AH, Holland EC (2006) Platelet-derived growth factor (PDGF) and glial tumorigenesis. Cancer Lett 232:139–147

Shinojima N, Tada K, Shiraishi S, Kamiryo T, Kochi M, Nakamura H, Makino K, Saya H, Hirano H, Kuratsu J, Oka K, Ishimaru Y, Ushio Y (2003) Prognostic value of epidermal growth factor receptor in patients with glioblastoma multiforme. Cancer Res 63:6962–6970

Sievert AJ, Jackson EM, Gai X, Hakonarson H, Judkins AR, Resnick AC, Sutton LN, Storm PB, Shaikh TH, Biegel JA (2009) Duplication of 7q34 in pediatric low-grade astrocytomas detected by high-density single-nucleotide polymorphism-based genotype arrays results in a novel BRAF fusion gene. Brain Pathol 19(3):449–458

Singh SK, Hawkins C, Clarke ID, Squire JA, Bayani J, Hide T, Henkelman RM, Cusimano MD, Dirks PB (2004) Identification of human brain tumour initiating cells. Nature 432:396–401

Specenier P (2012) Bevacizumab in glioblastoma multiforme. Expert Rev Anticancer Ther 12:9–18

Stott FJ, Bates S, James MC, Mcconnell BB, Starborg M, Brookes S, Palmero I, Ryan K, Hara E, Vousden KH, Peters G (1998) The alternative product from the human CDKN2A locus, p14

(ARF), participates in a regulatory feedback loop with p53 and MDM2. EMBO J 17: 5001–5014

Strumberg D, Schultheis B, Scheulen ME, Hilger RA, Krauss J, Marschner N, Lordick F, Bach F, Reuter D, Edler L, Mross K (2012) Phase II study of nimotuzumab, a humanized monoclonal anti-epidermal growth factor receptor (EGFR) antibody, in patients with locally advanced or metastatic pancreatic cancer. Invest New Drugs 30:1138–1143

Tamase A, Muraguchi T, Naka K, Tanaka S, Kinoshita M, Hoshii T, Ohmura M, Shugo H, Ooshio T, Nakada M, Sawamoto K, Onodera M, Matsumoto K, Oshima M, Asano M, Saya H, Okano H, Suda T, Hamada J, Hirao A (2009) Identification of tumor-initiating cells in a highly aggressive brain tumor using promoter activity of nucleostemin. Proc Natl Acad Sci U S A 106:17163–17168

Tanaka S, Louis DN, Curry WT, Batchelor TT, Dietrich J (2013) Diagnostic and therapeutic avenues for glioblastoma: no longer a dead end? Nat Rev Clin Oncol 10:14–26

Tohma Y, Gratas C, Biernat W, Peraud A, Fukuda M, Yonekawa Y, Kleihues P, Ohgaki H (1998) PTEN (MMAC1) mutations are frequent in primary glioblastomas (de novo) but not in secondary glioblastomas. J Neuropathol Exp Neurol 57:684–689

Uhl M, Aulwurm S, Wischhusen J, Weiler M, Ma JY, Almirez R, Mangadu R, Liu YW, Platten M, Herrlinger U, Murphy A, Wong DH, Wick W, Higgins LS, Weller M (2004) SD-208, a novel transforming growth factor beta receptor I kinase inhibitor, inhibits growth and invasiveness and enhances immunogenicity of murine and human glioma cells in vitro and in vivo. Cancer Res 64:7954–7961

Uhrbom L, Dai C, Celestino JC, Rosenblum MK, Fuller GN, Holland EC (2002) Ink4a-Arf loss cooperates with KRas activation in astrocytes and neural progenitors to generate glioblastomas of various morphologies depending on activated Akt. Cancer Res 62:5551–5558

Van Den Bent MJ, Brandes AA, Rampling R, Kouwenhoven MC, Kros JM, Carpentier AF, Clement PM, Frenay M, Campone M, Baurain JF, Armand JP, Taphoorn MJ, Tosoni A, Kletzl H, Klughammer B, Lacombe D, Gorlia T (2009) Randomized phase II trial of erlotinib versus temozolomide or carmustine in recurrent glioblastoma: EORTC brain tumor group study 26034. J Clin Oncol 27:1268–1274

Van Meir EG, Hadjipanayis CG, Norden AD, Shu HK, Wen PY, Olson JJ (2010) Exciting new advances in neuro-oncology: the avenue to a cure for malignant glioma. CA Cancer J Clin 60: 166–193

Verhaak RG, Hoadley KA, Purdom E, Wang V, Qi Y, Wilkerson MD, Miller CR, Ding L, Golub T, Mesirov JP, Alexe G, Lawrence M, O'kelly M, Tamayo P, Weir BA, Gabriel S, Winckler W, Gupta S, Jakkula L, Feiler HS, Hodgson JG, James CD, Sarkaria JN, Brennan C, Kahn A, Spellman PT, Wilson RK, Speed TP, Gray JW, Meyerson M, Getz G, Perou CM, Hayes DN (2010) Integrated genomic analysis identifies clinically relevant subtypes of glioblastoma characterized by abnormalities in PDGFRA, IDH1, EGFR, and NF1. Cancer Cell 17: 98–110

Vescovi AL, Galli R, Reynolds BA (2006) Brain tumour stem cells. Nat Rev Cancer 6:425–436

Vitucci M, Hayes DN, Miller CR (2011) Gene expression profiling of gliomas: merging genomic and histopathological classification for personalised therapy. Br J Cancer 104:545–553

Vivanco I, Sawyers CL (2002) The phosphatidylinositol 3-Kinase AKT pathway in human cancer. Nat Rev Cancer 2:489–501

Wang Z, Sun Y (2010) Targeting p53 for novel anticancer therapy. Transl Oncol 3:1–12

Watanabe K, Sato K, Biernat W, Tachibana O, Von Ammon K, Ogata N, Yonekawa Y, Kleihues P, Ohgaki H (1997) Incidence and timing of p53 mutations during astrocytoma progression in patients with multiple biopsies. Clin Cancer Res 3:523–530

Watanabe K, Tachibana O, Sata K, Yonekawa Y, Kleihues P, Ohgaki H (1996) Overexpression of the EGF receptor and p53 mutations are mutually exclusive in the evolution of primary and secondary glioblastomas. Brain Pathol 6:217–223, discussion 223–214

Watanabe T, Nobusawa S, Kleihues P, Ohgaki H (2009) IDH1 mutations are early events in the development of astrocytomas and oligodendrogliomas. Am J Pathol 174:1149–1153

Wick W, Platten M, Weller M (2001) Glioma cell invasion: regulation of metalloproteinase activity by TGF-beta. J Neurooncol 53:177–185

Wong AJ, Bigner SH, Bigner DD, Kinzler KW, Hamilton SR, Vogelstein B (1987) Increased expression of the epidermal growth factor receptor gene in malignant gliomas is invariably associated with gene amplification. Proc Natl Acad Sci U S A 84:6899–6903

Workman P, Clarke PA, Raynaud FI, Van Montfort RL (2010) Drugging the PI3 kinome: from chemical tools to drugs in the clinic. Cancer Res 70:2146–2157

Yamazaki H, Fukui Y, Ueyama Y, Tamaoki N, Kawamoto T, Taniguchi S, Shibuya M (1988) Amplification of the structurally and functionally altered epidermal growth factor receptor gene (c-erbB) in human brain tumors. Mol Cell Biol 8:1816–1820

Yan H, Parsons DW, Jin G, Mclendon R, Rasheed BA, Yuan W, Kos I, Batinic-Haberle I, Jones S, Riggins GJ, Friedman H, Friedman A, Reardon D, Herndon J, Kinzler KW, Velculescu VE, Vogelstein B, Bigner DD (2009) IDH1 and IDH2 mutations in gliomas. N Engl J Med 360: 765–773

Yip S, Butterfield YS, Morozova O, Chittaranjan S, Blough MD, An J, Birol I, Chesnelong C, Chiu R, Chuah E, Corbett R, Docking R, Firme M, Hirst M, Jackman S Karsan A, Li H, Louis DN, Maslova A, Moore R, Moradian A, Mungall KL, Perizzolo M, Qian J, Roldan G, Smith EE, Tamura-Wells J, Thiessen N, Varhol R, Weiss S, Wu W, Young S, Zhao Y, Mungall AJ, Jones SJ, Morin GB, Chan JA, Cairncross JG, Marra MA (2012) Concurrent CIC mutations, IDH mutations, and 1p/19q loss distinguish oligodendrogliomas from other cancers. J Pathol 226:7–16

Zauberman A, Flusberg D, Haupt Y, Barak Y, Oren M (1995) A functional p53-responsive intronic promoter is contained within the human mdm2 gene. Nucleic Acids Res 23:2584–2592

Zhang N, Wei P, Gong A, Chiu WT, Lee HT, Colman H, Huang H, Xue J, Liu M, Wang Y, Sawaya R, Xie K, Yung WK, Medema RH, He X, Huang S (2011) FoxM1 promotes beta-catenin nuclear localization and controls Wnt target-gene expression and glioma tumorigenesis. Cancer Cell 20:427–442

Zhang XP, Zheng G, Zou L, Liu HL, Hou LH, Zhou P, Yin DD, Zheng QJ, Liang L, Zhang SZ, Feng L, Yao LB, Yang AG, Han H, Chen JY (2008) Notch activation promotes cell proliferation and the formation of neural stem cell-like colonies in human glioma cells. Mol Cell Biochem 307:101–108

Zhang Y, Xiong Y, Yarbrough WG (1998) ARF promotes MDM2 degradation and stabilizes p53: ARF-INK4a locus deletion impairs both the Rb and p53 tumor suppression pathways. Cell 92:725–734

Zhang YE (2009) Non-Smad pathways in TGF-beta signaling. Cell Res 19:128–139

Zheng H, Ying H, Wiedemeyer R, Yan H, Quayle SN, Ivanova EV, Paik JH, Zhang H, Xiao Y, Perry SR, Hu J, Vinjamoori A, Gan B, Sahin E, Chheda MG, Brennan C, Wang YA, Hahn WC, Chin L, Depinho RA (2010) PLAGL2 regulates Wnt signaling to impede differentiation in neural stem cells and gliomas. Cancer Cell 17:497–509

MicroRNAs in the Molecular Pathology of Gliomas

4

Jiri Sana, Andrej Besse, and Ondrej Slaby

Abstract

MicroRNAs (miRNAs) are endogenously expressed small noncoding RNAs that act as posttranscriptional regulators of gene expression. Dysregulation of these molecules has been observed in many types of cancers. Altered expression levels of several miRNAs were identified also in gliomas. It was been frequently shown that miRNAs are involved in core signaling pathways, which play key roles in cellular processes, such as proliferation, apoptosis, cell cycle regulation, invasion, angiogenesis, and stem cell behavior. Therefore, miRNAs have a great potential to act as new class of diagnostic, prognostic, and predictive biomarkers as well as promising therapeutic targets in gliomas. Here, we summarize the current knowledge about miRNAs significance in glioma molecular pathology, with special focus on their involvement in core signaling pathways, their roles in drug resistance, and their potential clinical implications.

Keywords

MicroRNA • Transcription • Signal transduction

Abbreviations

BMI1	BMI1 polycomb ring finger oncogene
Cdk	Cyclin-dependent kinase
c-MYC	v-myc avian myelocytomatosis viral oncogene homolog
EGFR	Epidermal growth factor receptor
GBM	*Glioblastoma multiforme*

J. Sana • A. Besse • O. Slaby (✉)
Masaryk Memorial Cancer Institute, Zluty kopec 7, 656 53 Brno, Czech Republic

Central European Institute of Technology, Kamenice 5, 625 00, Brno, Czech Republic
e-mail: on.slaby@gmail.com

A. Sedo and R. Mentlein (eds.), *Glioma Cell Biology*,
DOI 10.1007/978-3-7091-1431-5_4, © Springer-Verlag Wien 2014

GSC	Glioma stem cell
IFN	Interferon
IKK$\alpha/\beta/\gamma$	Conserved helix-loop-helix ubiquitous kinase $\alpha/\beta/\gamma$
IRS	Insulin receptor substrate
JAK-STAT	Janus kinase-signal transducer and activator of transcription
miRNA	MicroRNA
MMP	Matrix metalloproteinase
NF-κB	Nuclear factor of kappa light polypeptide gene enhancer in B-cells
p21CIP1	Cyclin-dependent kinase inhibitor 1A
p27	p27 protein
PDGFR	Platelet-derived growth factor receptor
PI3K/AKT	Phosphatidylinositol-4,5-bisphosphate 3-kinase/v-akt murine thymoma viral oncogene homolog)
PLD2	Phospholipase D2
PTEN	Phosphatase and tensin homolog
RAS	Rat sarcoma viral oncogene homolog
TMZ	Temozolomide
VEGF	Vascular endothelial cell growth factor
VEGFR	VEGF receptor
WHO	World Health Organization

4.1 Introduction

Glial tumors of adults are the most frequently occurring primary tumors of central nervous system with a tendency to invade the surrounding brain tissue, originating in astrocytic glial cells. They are traditionally divided into histopathological subtypes defined by World Health Organization (WHO) classification. Numerous investigations have indicated that the improved understanding of the biology and molecular factors involved in the development, progression, and drug resistance of gliomas will be necessary for prediction of clinical outcome and therapy response, and mainly for development of new therapeutic strategies. Only a few molecular biomarkers have been validated and widely accepted in clinical practice till now and despite introduction of novel therapeutic approaches (bevacizumab, cilengitide) this cancer remains associated with very poor prognosis.

MicroRNAs (miRNAs) are endogenously expressed small noncoding RNAs that act as posttranscriptional regulators of gene expression. Deregulation of miRNAs has been observed in various types of cancers, including gliomas. There is continuously increasing evidence showing miRNAs involvement in core signaling pathways of glioma pathogenesis and their great potential as biomarkers and therapeutic targets.

4.2 MicroRNAs Function and Biogenesis

MiRNAs are a class of small 18–25 nucleotides long noncoding RNA molecules, whose function is to regulate gene expression on the posttranscriptional level. This regulation is mediated by the binding of miRNA to the 3′-untranslated region (3′UTR) of its target mRNA that induces either direct mRNA chain degradation or translational inhibition followed by deadenylation and mRNA decay. The decision on the fate of the target molecule is dependent on the degree of complementarity of both molecules. Perfect base pairing leads to the degradation in RISC, whereas imperfect bond leads first to the inhibition of mRNA translation subsequently followed by the degradation of target molecule due to its instability. Being able to regulate the translation even in case of imperfect complementarity causes that a single miRNA can regulate multiple mRNAs while a single mRNA can be targeted by different miRNAs. To date, over 1,400 human miRNAs have been identified (Griffiths-Jones et al. 2008). MiRNA expression is largely tissue and cell type specific and, moreover, these molecules play important role in a wide range of biological processes, such as development, differentiation, proliferation, and apoptosis (Niyazi et al. 2011). It is not surprising that deregulation of miRNAs was described as an important step in initiation as well as in progression and metastasis of many cancers including gliomas. On the basis of the target molecules, miRNAs can be classified as onco-miRs or tumor-suppressive miRNAs; some miRNAs may exhibit both features dependently on cellular context in various cancers (Chen 2005).

Most miRNA genes have their own promoters and are transcribed as autonomous transcription units (Carthew and Sontheimer 2009). Primary transcripts (pri-miRNAs) are generally transcribed by RNA polymerase II. Pri-miRNA can be up to 10 kb long and is characterized by the formation of stem loop structures. Single pri-miRNA frequently contains polycistronic cluster of miRNAs that are co-transcribed (Lawler and Chiocca 2009). In the nucleus, the pri-miRNA precursor is further cleaved through the activity of an enzyme called Drosha and cofactor DGCR8/Pasha. This processing results in a precursor miRNA (pre-miRNA) that is usually between 60 and 100 nucleotides in length. The pre-miRNA is exported from the nucleus in a RanGTP-dependent process after binding to transporter Exportin-5. In the cytoplasm, pre-miRNA is processed by the enzyme Dicer and, subsequently, one (passenger) of two chains of mature duplex miRNA is destroyed while guide chain is stabilized by the Argonaute proteins and incorporated into RISC (RNA-induced silencing complex). Recent findings show that guide chain can be derived from both the 5′ and 3′ ends of pre-miRNA hairpin depending on the thermodynamic stability on the 5′ end of the mature miRNA (Schwarz et al. 2003; Ladomery et al. 2011). The RISC complex with a miRNA guide strand is known as a miRISC complex. MiRNA involved in this macromolecular complex is prepared to perform its regulatory effects.

4.3 MicroRNAs in Glioma Pathogenesis

Tumor behavior is largely dependent on the abilities of the individual tumor cells to have the capacity of uncontrolled proliferation, to regulate own cell cycle, to escape an apoptosis, to regulate angiogenesis, and to migrate and invade in the surrounding tissues. The other specific properties such as the capability to self-renewal, unlimited proliferation potential, and differentiation have been observed in glioma stem cells. It has been described many times that miRNAs are involved in the regulation of these processes in many cancers including gliomas and, thus, these molecules can significantly influence glioma pathogenesis.

4.3.1 MicroRNAs in Proliferation, Migration, and Invasiveness

Among the key signaling pathways regulating cell proliferation, migration, and invasiveness belong EGFR (epidermal growth factor receptor), PI3K/AKT (phosphatidylinositol-4,5-bisphosphate 3-kinase/v-akt murine thymoma viral oncogene homolog), and NF-κB (nuclear factor of kappa light polypeptide gene enhancer in B-cells) (Nikaki et al. 2012). These signalizations have been many times described as deregulated in gliomas and, thus, deeper understanding of them could help discover new therapeutic approaches in initial as well as recurrent diseases of glial origin. The EGFR signaling network contributes to promotion, progression, and metastasis of the wide range of solid cancers, including gliomas. It is not surprising that this pathway is regulated by many miRNAs, which are therefore promising therapeutic targets (Fig. 4.1).

Mechanistic studies identified gene targets of miR-21 among important components of the EGFR signaling pathway. GBM (*glioblastoma multiforme*) cell lines U251 and LN229 characterized by mutated and wild-type PTEN (phosphatase and tensin homolog), respectively, showed a decreased expression of EGFR, activated AKT, Cyclin D, and Bcl-2 (B-cell CLL/lymphoma 2) after silencing of miR-21 expression (Zhou et al. 2010a; Sana et al. 2011). Although miR-21 is known to regulate PTEN and downregulation of miR-21 led to increased PTEN expression, the GBM suppressor effect of antisense miR-21 is most likely independent of the PTEN status because U251 has mutated PTEN (Ren et al. 2010b; Zhou et al. 2010a; Sana et al. 2011). In addition to the PTEN, miR-21 regulates proliferation, migration, invasiveness, and other cell processes through direct or indirect repression of many other cancer genes (Gabriely et al. 2008; Gaur et al. 2011a). Interestingly, significant synergistic inhibition of proliferation and invasiveness was observed after co-inhibition of miR-21 and miR-10b. It was hypothesized that miR-21 inhibitor may interrupt the activity of EGFR pathways, increasing PDCD4 [programmed cell death 4 (neoplastic transformation inhibitor)] and TPM1 [tropomyosin 1 (alpha)] expression and reducing MMP (matrix metalloproteinase) activities, independently of PTEN status. Meanwhile, miR-10b inhibitor induces translational suppression of the mRNA encoding HOXD10 (homeobox D10) leading to the increase of the expression of the well-

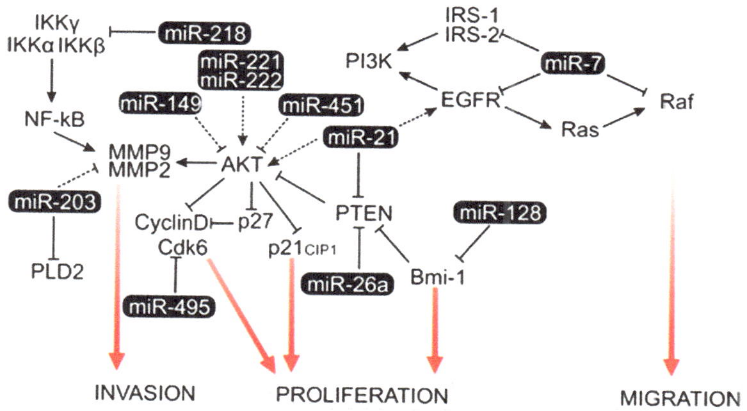

Fig. 4.1 MicroRNAs involved in regulation of glioma proliferation, migration, and invasion. *Bmi-1* BMI1 polycomb ring finger oncogene; *Cdk6* cyclin-dependent kinase 6; *EGFR* epidermal growth factor receptor; *IKKα/β/γ* conserved helix-loop-helix ubiquitous kinase α/β/γ; *IRS-1/2* insulin receptor substrate 1/2; *MMP2/9* matrix metalloproteinase 2/9; *NF-κB* nuclear factor of kappa light polypeptide gene enhancer in B-cells; *p21CIP1* cyclin-dependent kinase inhibitor 1A; *p27* p27 protein; *PI3K* phosphatidylinositol-4,5-bisphosphate 3-kinase; *PLD2* phospholipase D2; *PTEN* phosphatase and tensin homolog; *Raf* zinc fingers and homeoboxes 2; *Ras* rat sarcoma viral oncogene homolog. —| direct suppression; —> direct activation; ---| indirect suppression; --> indirect activation

characterized pro-metastatic gene RHOC (ras homolog family member C). Taken together, combination of miR-21 inhibitor and miR-10b inhibitor could be an effective GBM therapeutic strategy (Dong et al. 2012).

PTEN downregulation followed by AKT activation was described after transfection of GBM cells with the miR-26a-2 primary transcript. Similarly, the miR-26a mimics decreased PTEN protein levels and increased AKT phosphorylation (Huse et al. 2009; Kim et al. 2010; Sana et al. 2011). Another research group also showed that ectopic expression of miR-26a influenced cell proliferation by targeting PTEN and suggested that miR-26a is regulated by transcription factor c-MYC (v-myc avian myelocytomatosis viral oncogene homolog) (Guo et al. 2013).

Modulation of AKT signaling cascade using miRNAs in GBM cell lines was described also by Nan et al. In this study, transfection of miR-451 mimic reduced expression levels of many molecules including AKT1, MMP-2, and MMP-9. According to phenotypic experiments, miR-451 inhibited invasive ability and cell proliferation in GBM cells in vitro (Nan et al. 2010; Sana et al. 2011). Another miRNA involved in the EGFR signaling pathway is miR-7. Kefas et al. showed that miR-7 directly inhibits EGFR expression and independently suppressed the AKT pathway via targeting upstream regulators IRS-1 (insulin receptor substrate) and IRS-2. Moreover, transfection with miR-7 oligonucleotides decreased viability and invasiveness of primary GBM cell lines (Kefas et al. 2008; Webster et al. 2009). Webster et al. also described Raf1, another member of the EGFR signaling pathway, as a direct target of miR-7 in cancer cells (Webster et al. 2009).

Godlewski et al. published that miR-128 expression significantly reduced glioma cell proliferation in vitro and glioma growth in vivo. This effect was explained by direct regulation of the BMI1 (BMI1 polycomb ring finger oncogene), which is significantly upregulated in GBM compared to normal brain while miR-128 showed an opposite regulation. Moreover, miR-128 expression caused a decrease in histone methylation and AKT phosphorylation, and upregulation of p21(CIP1) (cyclin-dependent kinase inhibitor 1A) levels, consistent with BMI1 downregulation (Godlewski et al. 2008).

Finally, miR-221 and miR-222 were revealed as potential regulators of many target genes involved in AKT signaling pathway. Upregulation of miR-221/222 resulted in remarkable increase of p-AKT (phosphorylated AKT) and significant changes in the expression of *AKT*-related genes in glioma cells. Consequently, miR-221/222 overexpression increased glioma cell proliferation and invasion in vitro and induced glioma growth in mouse model. These results suggest that miR-221/222 enhance glioma malignant phenotype via activation of the AKT signaling pathway (Zhang et al. 2010b; Sana et al. 2011). Another study revealed that phenotypic effect of miR-221/222 is caused at least in part by targeting Cx43 (gap junction protein, alpha 1, 43 kDa), which has been identified as a tumor suppressor and major component for the establishment of GJIC (gap junction intercellular communication) in glial cells (Hao et al. 2012).

Another signaling pathway mentioned at the beginning of this chapter is the NF-κB pathway. NF-κB is the transcription factor with pleiotropic activity owing to its central roles in various biological processes. Aberrant activation of NF-κB signaling pathway has been proved to be important for invasiveness and metastatic capacity of tumors through upregulation of MMPs and transcription factors regulating E-cadherin, such as Snail (snail family zinc finger 1), Twist (twist family bHLH transcription factor), or Slug (snail family zinc finger 2). A critical component in NF-κB regulation is the IKK-β (conserved helix-loop-helix ubiquitous kinase) complex (Ghosh and Karin 2002; Song et al. 2010). Song et al. identified miR-218 expression in glioma cell lines and in human primary glioma tissues was substantially downregulated, when compared to miR-218 expression in normal human astrocytes and normal brain tissues. Forced upregulation of miR-218 dramatically reduced the migratory speed and invasive ability of analyzed cells. Ectopic expression of miR-218 downregulated MMP-9 and reduced NF-κB transactivity at transcriptional level, whereas inhibition of miR-218 has the opposite effect. Authors also demonstrated that miR-218 could inactivate NF-κB/MMP-9 signaling by directly targeting IKK-β (Song et al. 2010; Sana et al. 2011).

High-throughput analysis of the effects of 319 miRNA precursor molecules on cell proliferation in three (A172, LN405, and U87MG) GBM cell lines revealed nine miRNAs (miR-129, miR-136, miR-145, miR-155, miR-181b, miR-342-5p, miR-342-3p, miR-376a, and miR-376b), which had an antiproliferative effect in glioma cells. Moreover, six miRNA target genes [*ROCK1* (Rho-associated, coiled-coil containing protein kinase 1), *RHOA* (ras homolog family member A), *MET* (met proto-oncogene), *CSF1R* (colony-stimulating factor 1 receptor), *EIF2AK1* (eukaryotic translation initiation factor 2-alpha kinase 1), and *FGF7* (fibroblast growth factor 7)] were subsequently validated for the similar effects (Haapa-Paananen et al. 2013).

MiR-124 is another miRNA that plays an important role in regulating proliferation and invasiveness of GBM cells. As one of the possible mechanisms of these effects was revealed signalization through PPP1R3L, an inhibitory member of the apoptosis-stimulating protein of p53 family (IASPP), which is also able to affect growth, cell cycle progression, apoptosis, and metastasis of various types of cancer (Zhao et al. 2013b). Similarly, miR-203 inhibits the proliferation and invasion of U251 GBM cells partially via direct targeting of PLD2 (phospholipase D2) and/or ROBO1 [roundabout, axon guidance receptor, homolog 1 (Drosophila)]/ERK (mitogen-activated protein kinase 1)/MMP-9 signaling pathway (Dontula et al. 2013; Chen et al. 2013d). The ability of miR-100 to reduce proliferation in vitro was subsequently demonstrated in vivo where this miRNA treatment improved overall survival of treated animals. As a direct target of miR-100 was later confirmed SMRT/NCOR2 (silencing mediator of retinoid or thyroid hormone receptor-2/nuclear receptor corepressor 2) (Alrfaei et al. 2013).

Recently, many other miRNAs and their targets were described as involved in the regulation of proliferation, migration, and invasiveness in gliomas. Among them belong miR-326/NOB1 (Nin one binding protein) (Zhou et al. 2013); miR-125/PIAS3 (protein inhibitor of activated STAT, 3), which contributed to reduced STAT3 [signal transducer and activator of transcription 3 (acute-phase response factor)] transcriptional activity and subsequent decreased expression of MMP-2/9 (Shi et al. 2013); miR-138/ETH2 (methionine adenosyltransferase SAM2)–CDK4/6 (Cyclin-dependent kinase 4/6)–pRb (protein retinoblastoma 1)–E2F1 (E2F transcription factor 1) signal loop (Qiu et al. 2013); miR-495/CDK6 (Chen et al. 2013c); miR-149/AKT1 signaling (Pan et al. 2012); miR-330/SH3GL2 (SH3-domain GRB2-like 2) (Qu et al. 2012); miR-195/CCND3 (cyclin D3) (Zhang et al. 2012b); miR-34a/NOTCH1 (Li et al. 2011); miR-128/BMI1 (Godlewski et al. 2008); and miR-124, miR-137, miR-143, miR-145, and let-7 (Silber et al. 2008; Lee et al. 2011b; Koo et al. 2012).

4.3.2 MicroRNAs in Cell Cycle

Many miRNAs directly or indirectly target molecules such as cyclins, cyclin-dependent kinases, CDK regulators, and others, which are crucial for a progression of the cell through individual cell cycle checkpoints (Fig. 4.2).

The best-described miRNAs in gliomas with regard to the cell cycle are undoubtedly members of miR-34 family. Initially, overexpression of miR-34a in U251 GBM cells was described resulting in inhibition of cell growth and arrest in G0/G1 phase, which is inter alia accompanied by increased apoptosis. Moreover, authors supposed that this effect is due to the regulation of SIRT1 (sirtuin 1), a direct target of miR-34a (Luan et al. 2010). Two other studies confirmed c-Met (met proto-oncogene), NOTCH1, NOTCH2, and CDK6 also as direct targets of miR-34a and observed inhibition of G1/S progression after miR-34a artificial upregulation.

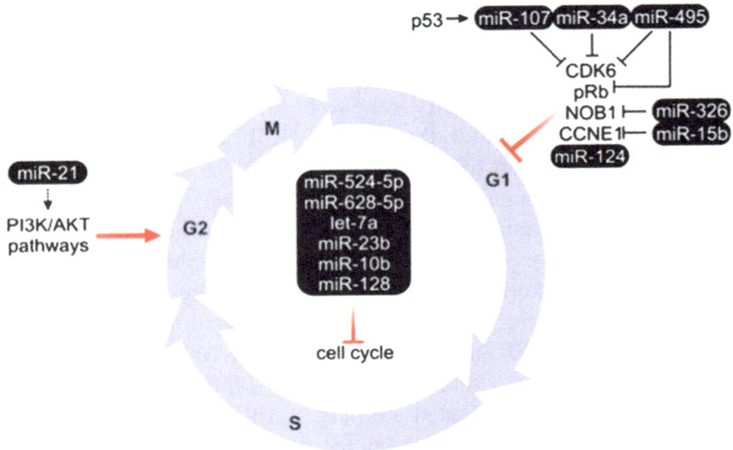

Fig. 4.2 MicroRNAs involved in regulation of glioma cell cycle. *CCNE1* cyclin E1; *CDK6* cyclin-dependent kinase 6; *NOB1* NIN1/RPN12 binding protein 1 homolog (*S. cerevisiae*); *p53* protein 53; *PI3K/AKT* phosphatidylinositol-4,5-bisphosphate 3-kinase/v-akt murine thymoma viral oncogene homolog; *pRB1* protein retinoblastoma 1. —| direct suppression; —> direct activation; ---> indirect activation

It seems that miR-34a regulates cell cycle via more important targets and, thus, presents a promising therapeutic target for brain tumors (Li et al. 2009a; Guessous et al. 2010b). CDK6 is confirmed as target molecule also for miR-495 and miR-107. Similar to miR-34a, cell cycle analysis revealed that upregulation of both miRNAs resulted in cell cycle arrest at the G1/S checkpoint. Interestingly, miR-107 inhibits also translation of NOTCH2 and its upstream regulator is p53 (protein 53) (Chen et al. 2013b). MiR-495 inhibits another crucial molecule of G1/S transition, tumor suppressor pRB, highlighting its potential in regulation of cell cycle (Chen et al. 2013c). Further, G1 cell cycle arrest was associated also with overexpression of miR-326, miR-15b, and miR-124. Nevertheless, the mechanisms seem to be different. While miR-326 targets NOB1, miR-15b decreases expression of downstream *CCNE1* gene encoding cyclin E1 (Xia et al. 2009; Zhao et al. 2013b; Zhou et al. 2013).

But miRNAs are not involved just in the disruption of G1/S transition; it seems that miR-21 decreases G2/M transition. Gwak et al. showed that transfection of malignant glioma cells by anti-miR-21 resulted in the increased transition at this phase. The mechanism of miR-21 modulation of G2/M transition is probably similar to its regulation of proliferation, through the activation of the PI3K/AKT pathway (Gwak et al. 2012).

Finally, the ability to arrest a cell cycle progression in gliomas is linked also with miR-524-5p, miR-628-5p, let-7a, miR-23b, miR-10b, and miR-128 (Zhang et al. 2009; Gabriely et al. 2011; Jiang et al. 2013; Li et al. 2013c; Wang et al. 2013c).

4.3.3 MicroRNAs in Apoptosis

MiRNAs are important regulators of apoptotic pathways. Alterations in functioning of these apoptotic miRNAs are frequently observed also in the glial tumors (Fig. 4.3). One of the most important regulatory mechanisms of apoptosis in glioma cells is cross talk between TGF-β (transforming growth factor, beta) and p53 signaling, which are regulated by the well-known onco-miR, miR-21. Papagiannakopoulos et al. reported that p53, TGF-β, and mitochondrial apoptotic networks are derepressed in response to miR-21 knockdown. They published a panel of genes, which are involved in particular pathways and are simultaneously modulated by miR-21 treatment. Some of these genes were subsequently predicted to be direct targets of miR-21 that can stabilize p53 protein levels by interfering with MDM2 (MDM2 oncogene, E3 ubiquitin protein ligase) and/or act as p53 transcriptional cofactors (Papagiannakopoulos et al. 2008; Sana et al. 2011). Inhibition of miR-21 increased also endogenous levels of PDCD4 in human glioma cell lines and activated caspases 9/3, which may be mediated by modulating multiple potential target genes, such as *TIMP3* (tissue inhibitor of matrix metalloproteinase 3) (Chen et al. 2008; Zhou et al. 2010b; Sana et al. 2011). The upregulation of PDCD4 and caspase 3 is in contrast to the low level of miR-21 and inactivation of TGF-β1/SMAD signaling, which are the critical upstream regulators of miR-21 in dose- and time-dependent manner in human GBM cell line U251 (Wang et al. 2012a).

Other direct targets of miR-21 involved in apoptosis were confirmed molecules ANP32A (acidic (leucine-rich) nuclear phosphoprotein 32 family, member A) and SMARCA4 (SWI/SNF-related, matrix-associated, actin-dependent regulator of chromatin, subfamily a, member 4). In A172 GBM cells, enhanced ANP32A expression compensated for the positive effects of anti-miR-21 treatment on cell viability and apoptosis (Schramedei et al. 2011). MiR-21 silencing also enhances apoptosis of GBM cells also after treatment by the antiangiogenic drug sunitinib and alkylation agent temozolomide (TMZ) (Zhang et al. 2012c; Costa et al. 2013). In relation to the enhancement of chemosensitivity was described also miR-155. This miRNA indicated oncogenic character and its silencing led to the elevation of apoptosis after taxol treatment via EAG1 (potassium voltage-gated channel, subfamily H (eag-related), member 1) expression (Meng et al. 2012).

Interestingly, miR-211 overexpression and MMP-9 treatments led to the activation of the intrinsic mitochondrial/caspase-9/3-mediated apoptotic pathway in both glioma cells and GSCs (Asuthkar et al. 2012). MiR-34a was identified also as regulator of TGF-β signaling in GBM (Genovese et al. 2012). This miRNA was suggested to be promoter of mitochondrial apoptosis in several cancers including gliomas (Li et al. 2011; Sasaki et al. 2012; Bienertova-Vasku et al. 2013). Therefore, it is probable that miR-34a regulates apoptosis in glioma also through the TGF-β signaling pathway.

There are also other miRNAs described as regulators of apoptosis in glioma cells. Overexpression of miR-326 in human glioma cell lines A172 and U373 caused cell cycle arrest at the G1 phase, delayed cell proliferation, and mainly enhanced apoptosis (Zhou et al. 2013). Interestingly, miR-326 targets PKM2

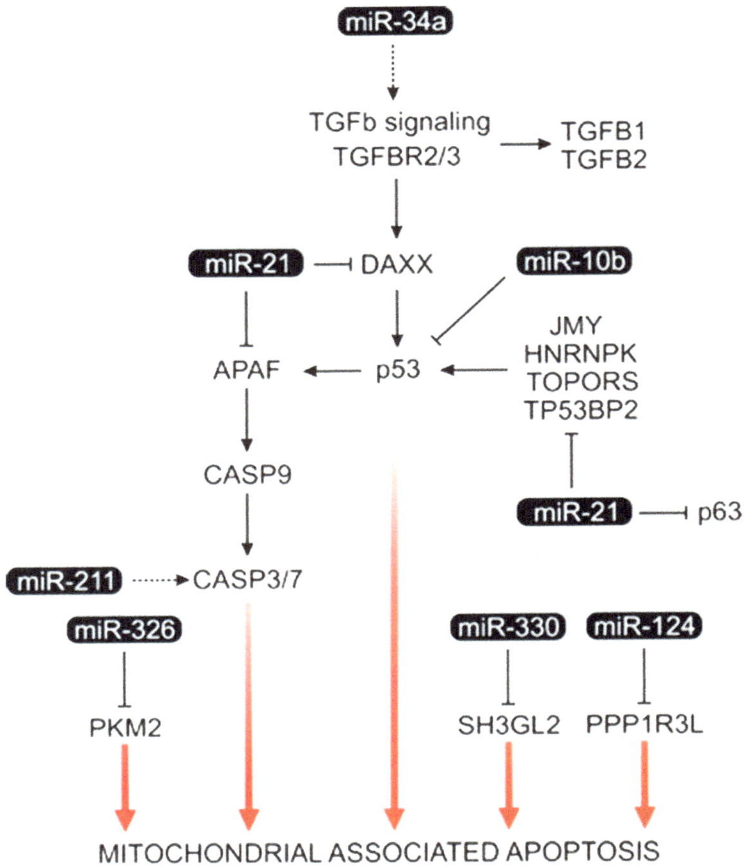

Fig. 4.3 MicroRNAs involved in regulation of glioma mitochondrial associated apoptosis. *APAF* apoptotic peptidase activating factor; *CASP3/7/9* caspase 3/7/9, apoptosis-related cysteine peptidase; *DAXX* death-domain associated protein; *HNRNPK* heterogeneous nuclear ribonucleoprotein K; *JMY* unction mediating and regulatory protein, p53 cofactor; *p53* protein 53; *p63* protein 63; *PKM2* pyruvate kinase 2; *PPP1R3L* an inhibitory member of the apoptosis-stimulating protein of p53 family (IASPP); *SH3GL2* SH3-domain GRB2-like 2; *TGFB1/2* transforming growth factor, beta 1; *TGFBR2/3* transforming growth factor, beta receptor 2/3; *TOPORS* topoisomerase I binding, arginine/serine-rich, E3 ubiquitin protein ligase; *TP53BP2* tumor protein p53 binding protein, 2. —| direct suppression; —> direct activation; ---> indirect activation

(pyruvate kinase M2), which plays a key role in cancer cell metabolism. This could be a mechanism how miR-326 regulates glioma cell survival (Kefas et al. 2010). MiR-10b suppresses many tumor suppressors including p53, FOXO3 (forkhead box O3), CYLD [cylindromatosis (turban tumor syndrome)], PAX6 (paired box 6), PTCH1 (patched 1), HOXD10, and NOTCH1. Hence, this miRNA shows pleiotropic feature and regulates many key cellular processes (Lin et al. 2012) and its downregulation; it is not expressed in human brain and strongly upregulated in both low-grade and high-grade gliomas and reduces glioma cell growth through cell cycle arrest and activation of apoptosis (Gabriely et al. 2011). MiR-124 regulates apoptosis and other cell processes via PPP1R3L (Zhao et al. 2013b). Similar pleotropic nature has also onco-miR miR-330 targeting SH3GL2 (Qu et al. 2012).

4.3.4 MicroRNAs in Angiogenesis

Angiogenesis is essential for many physiological processes as well as for tumor development. It can be triggered by extracellular signals such as vascular endothelial growth factors and by genetic alterations such as activation of oncogenes (Sternlicht et al. 1999; Folkman 2006; Zheng et al. 2013). In gliomas, aberrant activation of angiogenic signaling pathways EGFR/PI3K/AKT, NF-κB, VEGF (vascular endothelial growth factor), and PDGF (platelet-derived growth factor) was repeatedly described (Kozomara and Griffiths-Jones 2011; Mizoguchi et al. 2012).

Glioma is one of the first cancers in which angiogenesis was found to be a key phenotypic feature to disease progression. Indeed, GBM is one of the most highly vascularized human cancers. It is well recognized that the progression of human glioma is angiogenesis dependent. Accumulated evidence has shown that high vascular density and overexpression of angiogenic factors correlate with poor prognosis of glioma patients (Mizoguchi et al. 2012).

Recently, several miRNAs were described to be involved in glioma angiogenesis regulation (Fig. 4.4). In addition to miR-30e* and miR-182 activating NF-κB signaling and subsequent expression of downstream angiogenic factors such as VEGF-C and MMPs (Mizoguchi et al. 2012), tumor-suppressive miR-205 can specifically suppress expression of the key angiogenic factor, VEGF-A. The effect of the miR-205 ectopic expression has not only anti-angiogenic impact, but it is accompanied by the induction of apoptosis, cell cycle arrest, impairing of cell viability, clonability, and invasiveness (Yue et al. 2012). Angiogenic function was supposed also for miR-15b and miR-152, but subsequent evaluation of tube formation in cultured endothelial cells with culture supernatant from 9L cells, rat glial cell line derived from gliosarcoma, treated with these miRNAs revealed that only miR-15b significantly reduced capillary-like tube formation. Authors have suggested that this effect is caused by NRP2 (neuropilin 2), which was confirmed to be a direct target of miR-15b. NRPs are receptors for the SEMA (class-3 semaphorin) family of axon guidance molecules and also for VEGF family of angiogenic factors. VEGF–NRP interactions promote developmental angiogenesis and also metastases (Geretti and Klagsbrun 2007; Zheng et al. 2013).

Glioma cells and angiogenic growth factors elevate the level of miR-296 in primary human brain microvascular endothelial cells in culture. Subsequently, growth factor-induced miR-296 significantly contributes to angiogenesis by directly targeting the HGS (hepatocyte growth factor-regulated tyrosine kinase substrate) leading to reduced HGS-mediated degradation of the VEGFR2 (vascular endothelial growth factor receptor 2) and PDGFRβ (platelet-derived growth factor

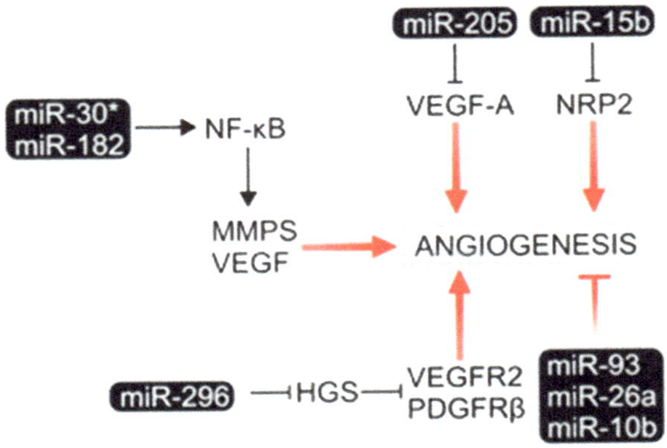

Fig. 4.4 MicroRNAs involved in regulation of glioma angiogenesis. *HGS* hepatocyte growth factor-regulated tyrosine kinase substrate; *MMPs* matrix metalloproteinases; *NF-κB* nuclear factor of kappa light polypeptide gene enhancer in B-cells; *NRP2* neuropilin 2; *PDGFRB* platelet-derived growth factor receptor, beta polypeptide; *VEGF-A* vascular endothelial cell growth factor A; *VEGFR2* VEGF kinase insert domain receptor (a type III receptor tyrosine kinase). —| direct suppression; —> direct activation; ---> indirect activation

receptor, beta). Furthermore, inhibition of miR-296 reduces angiogenesis in tumor xenografts in vivo (Wurdinger et al. 2008). Similarly, when miR-93-overexpressing U87 glioma cells were cocultured with endothelial cells, they supported endothelial cell spreading, growth, migration, and tube formation. In vivo studies revealed that miR-93-expressing cells induced blood vessel formation allowing blood vessels to extend to tumor tissues in high densities. This effect is explained by suppressing, at least in part, integrin-β8 expression (Fang et al. 2011).

MiR-26a has been described above as a regulator of proliferation. In addition, forced expression of miR-26a in glioma significantly increased tumor growth and angiogenesis in vivo, while reduced expression of this miRNA played opposite role (Qian et al. 2013). Functionally pleiotropic miR-10b is able to regulate angiogenicity in tumor cells resembling mesenchymal subtype of GBM (Lin et al. 2012).

4.3.5 MicroRNAs in Immune Response

Interferons (IFNs) are cytokines released by lymphocytes that have antiviral, antiproliferative, and immunomodulatory effects. They are connected with the JAK-STAT (Janus kinase-Signal Transducer and Activator of Transcription) signaling pathway and allow communication between cells to trigger protective defenses of the immune system leading to eradication of affected cells. Therefore, deregulation of JAK-STAT cascade is the key step to mediating immunosuppression in the tumor microenvironment (Platanias 2005; Sana et al. 2011). Recently, the possibility was investigated if IFN-β may induce or reduce cellular miRNAs in human gliomas. Analysis of IFN-β treatment on miR-21 expression in glioma cells and intracranial glioma xenografts revealed that systemic delivery of this cytokine markedly reduced the level of miR-21 in glioma cells. In contrast, the addition of the STAT3-specific inhibitor increased the level of miR-21 (Ohno et al. 2009; Sana et al. 2011). Another study revealed miR-221 and miR-222 as possible regulators of IFN pathways. Zhang et al. found that the IFN-α signaling pathway is the most significant pathway modulated by genes with the different expression after knockdown of miR-221 and miR-222. The authors showed that STAT1 and STAT2 expression and phosphorylation were upregulated in U251 cells with silenced miR-221/222. Tyrosine phosphorylation of STAT1 and STAT2 was present in the nucleus after repression of the same miRNAs. These data illustrate a mechanism of STAT1/2 upregulation under the transcriptional control of IFN-α signaling after knockdown of miR-221/222 cluster in U251 glioma cells (Zhang et al. 2010a; Sana et al. 2011).

Interestingly, on the basis of miRNA expression in gliomas using tissue microarrays, in situ hybridization, and molecular modeling, miR-124 was identified as another candidate for modulating STAT3 signaling pathway. MiR-124 is absent in all grades and pathologic types of gliomas. Upon replacement of miR-124 in GSCs (glioma stem cells), the STAT3 pathway was inhibited, and miR-124 reversed GSC-mediated immunosuppression of T-cell proliferation and induction of Foxp3 (forkhead box P3) regulatory T cells (Treg). Treatment of T cells from immunosuppressed GBM patients with miR-124 induced remarkable effector response including upregulation of IL-2 (interleukin 2), IFN-γ, and TNF-α (tumor necrosis factor, alpha). Both systemic administration of miR-124 and adoptive miR-124-transfected T-cell transfers caused strong anti-glioma therapeutic effects and enhanced effector responses in the local tumor microenvironment in vivo. These findings highlight the potential application of miR-124 as a novel immunotherapeutic agent for gliomas (Shi et al. 2013).

4.3.6 MicroRNAs in Glioma Stem Cells

During the last years there is an increasing interest in the small population of glioma cells, later termed as glioma stem cells (GSCs) or glioma initiating cells (GICs), which have the ability to self-renew (Koshkin et al. 2013). Conventional therapies target the cells that rapidly divide, while the benefit of GICs is to divide slowly but infinitely. These properties enable them to initiate new tumors and also to serve as a source of recurrence (Martin et al. 2013). These cells, which maintain stem-like phenotype, show generally much more resistance to radiation or chemotherapy in comparison to "normal" glioma cells (Wang et al. 2010; Yamada and Nakano 2012). Similar to normal stem cells, GSCs express CD133 (prominin 1), NESTIN, OCT4 (POU class 5 homeobox 1), SOX2 (SRY (sex determining region Y)-box 2), and NANOG (Nanog homeobox) markers and are capable of self-renewal as well as they have the ability to initiate neurosphere growth in vitro and generation of highly

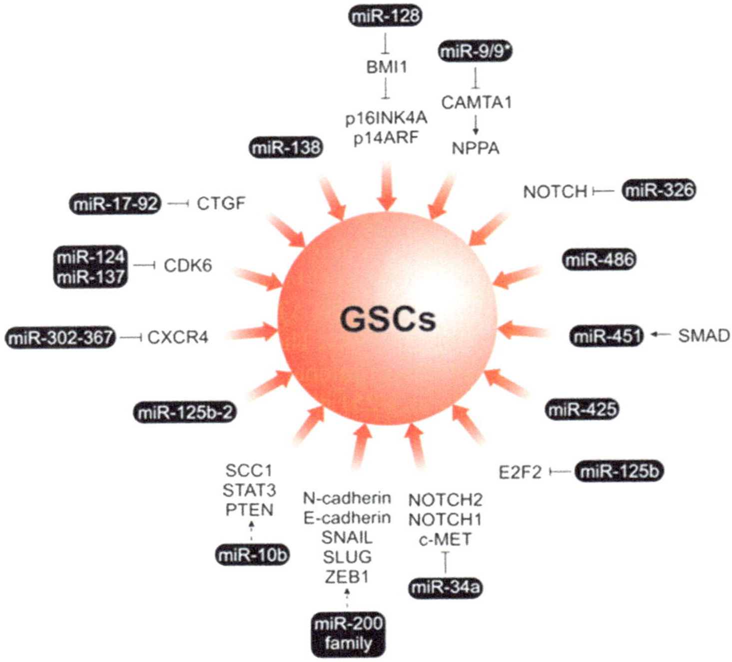

Fig. 4.5 MicroRNAs involved in regulation of glioma stem cells. *BMI1* BMI1 polycomb ring finger oncogene; *c-MET* met proto-oncogene; *CAMTA1* calmodulin binding transcription activator 1; *CDK6* cyclin-dependent kinase 6; *CTGF* connective tissue growth factor; *CXCR4* chemokine (C-X-C motif) receptor 4; *E2F2* E2F transcription factor 2; *NPPA* natriuretic peptide A; *p14ARF* cyclin-dependent kinase inhibitor 2A; *p16INK4A* cyclin-dependent kinase inhibitor 2A; *PTEN* phosphatase and tensin homolog; *SCC1* protein tyrosine phosphatase; *SLUG* snail family zinc finger 2; *SMAD* SMAD family member; *SNAIL* snail family zinc finger 1; *STAT3* signal transducer and activator of transcription 3 (acute-phase response factor); *ZEB1* zinc finger E-box binding homeobox 1. —| direct suppression; —> direct activation; ---> indirect activation

malignant tumor in NOD/SCID (nonobese diabetic/severe combined immunodeficient) mice (Altaner 2008; Chen et al. 2010b). There is increasing amount of evidence suggesting involvement of miRNAs in the regulation of stem-like phenotype of GSCs (Fig. 4.5).

Two independent research groups found out that miR-128 inhibits self-renewal and proliferation capacities of GSCs through the direct targeting of BMI1 (Godlewski et al. 2008; Fu et al. 2013), which negatively regulates expression of tumor-suppressor proteins p16INK4a (cyclin-dependent kinase inhibitor 2A) and p14ARF (CDKN2A; cyclin-dependent kinase inhibitor 2A). These two molecules subsequently regulate cyclin D and p53 leading to the cell cycle arrest (Siddique and Saleem 2012). Fu et al. described capability of miR-200 family to suppress an epithelial–mesenchymal transition (EMT) in GSCs via E-cadherin, N-cadherin, SNAIL, SLUG, and ZEB1 (zinc finger E-box binding homeobox 1)

(Fu et al. 2013). It was also discovered that miR-200c is involved in the regulation of BMI1 in breast cancer and, thus, this miRNA could function analogically to miR-128 in GSCs (Siddique and Saleem 2012).

Chan et al. published data showing miR-138 being an important regulator of self-renewal, proliferation, and apoptosis in GSCs. Interestingly, this miRNA did not affect healthy neural stem cells. This fact is remarkable accordingly to potential use of miR-138 in therapy (Chan et al. 2012). The similar effects indicated also miR-10b, the inhibition of which decreased proliferation, migration, and invasiveness in vitro and, moreover, the ability of GSCs to initiate tumorigenesis in vivo (Guessous et al. 2013). Oncogenic properties have also miR-9/9* that inhibit expression tumor suppressor CAMTA1 (calmodulin-binding transcription activator 1) and thus support tumorigenic potential of CD133 positive cells. CAMTA1 activates transcription of NPPA (antiproliferative peptide), protein with proven antiproliferative effects in GBM. Furthermore, expression levels of both CAMTA1 and NPPA correlate with prognosis and overall survival of GBM patients (Schraivogel et al. 2011).

On the contrary, miR-34a was observed as tumor-suppressive miRNA in GBM. This molecule targets genes involved in tumorigenesis as well as in regulation of GSCs (c-MET, NOTCH1, NOTCH2). It was demonstrated that miR-34a has significantly lower expression in glioma cells and, further, decreases cell proliferation, migration, invasiveness, and viability when replaced in vitro (Li et al. 2009a). This replacement approach led in GSCs to the induction of apoptosis and differentiation, which was connected with significant reduction of stem cell markers CD133 and nestin (Guessous et al. 2010a). Another regulator of NOTCH signaling is miR-326 that also decreases viability, invasiveness, and tumorigenic potential in GBM when its levels are upregulated (Kefas et al. 2009). MiR-125 is another molecule involved in the regulation of neural differentiation (Le et al. 2009), which is downregulated in GBM tissue and CD133 positive GBM cells. Ectopic expression of miR-125 led to the inhibition of proliferation and sphere formation, to cell cycle arrest in G0–G1 phase, and to the decreased expression of stemness markers indicating cell differentiation. Further, E2F2 (E2F transcription factor 2), critical regulator of cell cycle, is the direct target of miR-125b (Wu et al. 2012). Moreover, miR-124 and miR-137 are involved in a regulation of the cell cycle progression through G1 phase into early S phase in GSCs via inhibition of CDK6. This results in decreased proliferation and enhanced differentiation (Silber et al. 2008). In another study was observed that miR-125b-2 is overexpressed in initial GBMs as well in relapsed tumors after concomitant RT/TMZ (radiochemotherapy with temozolomide) therapy and, thus, this miRNA indicates oncogenic character. Increased levels of miR-125b-2 were also detected in CD133 positive GBM cells. Inhibition of this miRNA leads to the induction of mitochondrial apoptosis after TMZ treatment in CD133 positive cells Shi et al. 2012.

As an important regulators of glioblastoma tumor initiating cells (GICs; this term is sometimes used by authors, who decline to use the term glioma stem cell) have been described also members of cluster miR-302-367. This cluster is able to suppress the self-renewal, infiltration, and proliferation potential of GICs through a repression of CXCR4 (chemokine C-X-C motif, receptor 4) leading to the suppression of SHH-GLI-NANOG signaling pathway (Card et al. 2008; Zbinden et al. 2010; Fareh et al. 2012). Similarly, another cluster miR-17-92 regulates neurosphere formation, differentiation, proliferation, and apoptosis in vitro in GICs. Supposed mediator of these effects is the gene CTFG (connective tissue growth factor), a direct target of this cluster (Ernst et al. 2010). Global expression analysis of miRNAs revealed changes during differentiation of GICs, including overexpression of miR-21, miR-29a, miR-29b, miR-221, and miR-222 and downregulation of miR-93 and miR-106a. Functional studies showed that miR-21 overexpression in GICs induced comparable cell differentiation features and targeted SPRY1 [sprouty homolog 1, antagonist of FGF signaling (Drosophila)] mRNA, which encodes for a negative regulator of neural stem cell differentiation. In addition, miR-221 and miR-222 inhibition in differentiated cells restored the expression of stem cell markers while reducing differentiation markers. Finally, miR-29a and miR-29b targeted MCL1 [myeloid cell leukemia sequence 1 (BCL2-related)] mRNA in GICs and increased apoptosis (Aldaz et al. 2013).

Analysis of the expression profiles of both CD133 positive and negative GBM cells revealed several differentially expressed miRNAs including miR-451, miR-486, and miR-425. Ectopic expression of these miRNAs led to the inhibition of cell growth and neurosphere formation in GBM cells. As positive regulator of miR-451 was subsequently revealed SMAD (Gal et al. 2008). Another analysis showed that miR-153 expression was downregulated in GBM tissues relative to normal brain tissues and in CD133 positive cells relative to CD133 negative cells (Zhao et al. 2013a).

4.4 MicroRNAs as Biomarkers in Gliomas

Recent research suggests that deregulation of miRNAs is involved in initiation and progression of many cancers, including gliomas, and that miRNAs hold great potential as future diagnostic, prognostic, and predictive biomarkers in cancer. The rationale of this potential is mainly based on the miRNAs' increased stability when compared to mRNAs and the ability of a single miRNA molecule to affect numerous targets, oncogenes and tumor-suppressor genes including (Hermansen and Kristensen 2013).

4.4.1 Diagnostic Biomarkers

Study of miRNA expression profiles in tumor tissue of gliomas of different grades and their comparison to miRNA profiles in non-tumoral brain tissue is not only the first step in the determination of miRNAs involved in glioma pathogenesis, but also important approach enabling new level of glioma molecular taxonomy, which is not based on morphology but on concrete molecular alterations which are present in individual tumors. This molecular classification has potential to be more accurate in

Table 4.1 MicroRNAs significantly deregulated in glioma tissue, which were observed in at least two independent studies

miRNA | study	Chan	Sasayama	Conti	Wang	Ciafrè	Slaby	Silber	Rao	Others
miR-17								↑	↑
miR-19a								↑	↑
miR-19b								↑	↑
miR-21	↑		↑	↑	↑	↑	↑	↑	
miR-25					↑			↑	
miR-92b		↑						↑	
miR-106b		↑						↑	
miR-125b					↑	↑			
miR-130a					↑			↑	
miR-155							↑	↑	
miR-182				↑				↑	
miR-335				↑					↑
miR-10b					↑			↓	
miR-138	↑							↓	
miR-221			↑		↑	↓			
miR-34a									2x↓
miR-128a					↓	↓	↓		
miR-132							↓	↓	
miR-181a					↓	↓			
miR-181b			↓		↓	↓			
miR-181c					↓	↓			
miR-198	↓							↓	
miR-219-5p				↓				↓	
miR-329		↓						↓	
miR-338-3p				↓				↓	
miR-483-5p				↓				↓	

↑ upregulated in gliomas, ↓ downregulated in gliomas

stratification of the patients accordingly to their prognosis and therapeutic response in comparison to conventional histopathological classification and open new diagnostic possibilities for glioma patients. MiRNAs which were repeatedly found to be differentially expressed between glioma tumor tissue and non-tumoral brain tissue are summarized in Table 4.1.

The first work aiming miRNA profiling in glioblastoma tissue was published by Ciafre et al. (2005). These authors analyzed 254 miRNAs in nine paired GBM and non-tumor brain tissue samples. The findings showed that nine miRNAs (miR-10b, miR-130a, miR-221, miR-125b-1, miR-125b-2, miR-9-2, miR-21, miR-25, and miR-123) were upregulated and four miRNAs (miR-128a and miR-181a/b/c) downregulated in GBM. The greatest expression changes were observed in miR-221, miR-128a, and miR-181a/b/c (Ciafre et al. 2005). A similar analysis was performed by Chan et al., which compared three primary high-grade gliomas with eight samples of fetal and adult brain tissues. In this study, five miRNAs (miR-21, miR-138, miR-347, miR-135, and miR-291-5) were significantly overexpressed and three miRNAs (miR-198, miR-188, and miR-202) showed lower expression levels in tumor tissue. MiR-21 was almost nine times more expressed in GBM compared to control samples (Chan et al. 2005). Silber et al. analyzed 192 matured miRNAs in four anaplastic astrocytomas (grade III), four GBMs (grade IV), and four non-tumor tissues obtained from surgically resected tissue of temporal lobes of epileptic patients. Statistical analysis revealed 13 downregulated and three upregulated miRNAs in GBM compared with the non-tumor tissue. Furthermore, only six deregulated miRNAs were observed between grade III and IV gliomas (Silber et al. 2008).

Similar study was published comparing global expression profiles of miRNAs in 39 gliomas (13 primary and 13 secondary GBMs, 13 anaplastic astrocytomas) and seven normal brain tissues and revealed 55 upregulated and 29 downregulated miRNAs in glioma tissue. Moreover, 67 differentially expressed miRNAs were identified by a comparison of anaplastic astrocytomas and GBMs. The comparison of primary and secondary GBMs revealed seven differentially expressed miRNAs. Several miRNAs significantly differed also between sequentially progressive astrocytomas (anaplastic astrocytomas and secondary GBM) and primary GBMs (21 miRNAs), anaplastic astrocytomas and primary GBMs (76 miRNAs), as well as anaplastic astrocytomas and secondary GBMs (68 miRNAs) (Rao et al. 2010). Indeed, D'Urso et al. found that a set of 10 miRNAs is able to classify primary and secondary GBMs (D'Urso et al. 2012).

Skalsky and Cullen investigated miRNA expression profiles in six GBMs and three samples of non-tumor brain tissues using high-throughput sequencing technology. They published 68 differentially expressed miRNAs between both examined sets ($p < 0.05$, Student's t-test). Nevertheless, when they applied the standard FDR (false discovery rate)-adjusted p-value < 0.05 cutoff, only nine miRNAs (miR-124, miR-10b*, miR-139-5p, miR-7, miR-10b, miR-132, miR-95, miR-543, miR-7d) were identified that met the criteria. Given the much broader dynamic range of deep sequencing compared to finite platforms such as miRNA microarrays or real-time PCR (Polymerase Chain Reaction) arrays, they therefore used an FDR cutoff of $p < 0.1$. Under these conditions 12 more miRNAs (miR-323-3p, miR-128, miR-139-3p, miR-598, miR-103a, miR-103b, miR-873, miR-891a, miR-487b, miR-323b-3p, miR-138-1*, miR-93) met the criteria (Skalsky and Cullen 2011).

Sasayama et al. analyzed three paired samples of GBM and non-tumor brain tissues. They found that five miRNAs (miR-10b, miR-21, miR-183, miR-92b, and miR-106b) were upregulated and five miRNAs (miR-302c*, miR-379, miR-329, miR-134, and miR-369-3p) were downregulated in GBMs (Sasayama et al. 2009). Wang et al. also analyzed miRNA expression profiles in three GBMs and three paired non-tumor adjacent tissues. They identified 91 miRNAs, whose expressions were at least twofold changed between both sets of samples. Interestingly, an expression of miR-483-5p was almost 100-fold decreased in tumors (Wang et al. 2012b). Another study validated expression of eight selected miRNAs in 10 GBM tissues in comparison to the four non-tumor samples of brain tissues obtained from epileptic patients. MiR-21 and miR-221 showed higher and miR-181b lower expression levels in tumor samples (Conti et al. 2009). Slaby et al. compared expressions of eight selected miRNAs in GBMs and non-tumor brain tissues obtained from surgeries of arteriovenous malformations (AVM). Results showed that while miR-21 and miR-125b were overexpressed in GBMs, miR-128a, miR-181a/b/c, and miR-221/222 were downregulated in tumors (Slaby et al. 2010). Other studies show that miR-31 (Hua et al. 2012), miR-205 (Yue et al. 2012), miR-124a (Fowler et al. 2011), and miR-34a (Li et al. 2011) have lower expression levels in GBM in comparison with normal brain tissue.

Malzkorn et al. analyzed miRNA expression profiles among different grades of gliomas (low-grade astrocytomas vs. anaplastic astrocytomas vs. secondary GBMs). Authors studied expression profiles of 157 miRNAs in patients affected by low-grade astrocytomas (grade II) that gradually progressed to secondary GBMs (grade IV). It was found that 12 miRNAs (miR-9, miR-15a, miR-16, miR-17, miR-19a, miR-20a, miR-21, miR-25, miR-28, miR-130b, miR-140, and miR-210) are upregulated and two miRNAs (miR-184 and miR-328) are downregulated dependently on the grade of glioma (Malzkorn et al. 2010).

There are also many studies characterizing expression pattern of only one miRNA in gliomas and their different clinicopathological features. From these studies it implies that miR-34a, miR-203, miR-326, and miR-375 are reduced in tumors (Chang et al. 2012; Gao et al. 2013; He et al. 2013; Wang et al. 2013b) and miR-17, miR-19a/b, miR-224, miR-335, and miR-372 are overexpressed in gliomas (Jiang et al. 2012; Lu et al. 2012, 2013; Jia et al. 2013; Li et al. 2013a) compared to non-tumor brain tissues. Moreover, most of these miRNAs were significantly correlated with grading of gliomas.

4.4.2 Prognostic and Predictive Biomarkers

One of the main aims in glioma research is the discovery of highly sensitive prognostic and predictive biomarkers enabling stratification of the patients accordingly to their risk of progression and predicted therapeutic response. The importance of this challenge increases with the advent of new therapeutic possibilities in the treatment of gliomas, e.g., TMZ, bevacizumab, and cilengitide (Chamberlain 2011). MiRNAs with potential as prognostic and/or predictive biomarkers are summarized in Table 4.2.

Srinivasan et al. analyzed 10 selected miRNAs in 222 GBM patients. The aim of this study was to find differently expressed miRNAs between short-time and long-time surviving GBM patients. Tumor samples of the patients with long-time OS (overall survival) showed higher expression levels of protective miRNAs (miR-20a, miR-106a, miR-17-5p). Conversely, overexpression of oncogenic miRNAs (miR-31, miR-221, miR-222, miR-148a, miR-146b, miR-200b, miR-193a) was observed in tumors of patients with short-time OS (Srinivasan et al. 2011). Further, Niyazi et al. performed retrospective study analyzing 1,100 miRNAs in 35 GBM patients, who underwent adjuvant concomitant chemoradiotherapy. Subsequently, 30 miRNAs, which mostly differ between short- and long-time surviving patients,

Table 4.2 MicroRNAs with prognostic and/or predictive significance in gliomas

MiRNA	Glioma (WHO)	Prognostic and/or predictive significance	References
let-7a	IV	Correlation with survival (+)	Niyazi et al. (2011)
miR-106a	IV	Correlation with survival (+)	Srinivasan et al. (2011)
miR-1260	IV	Correlation with survival (+)	Niyazi et al. (2011)
miR-1305	IV	Correlation with survival (+)	Niyazi et al. (2011)
miR-146b	IV	Correlation with survival (−)	Srinivasan et al. (2011)
miR-148	IV	Correlation with survival (−)	Srinivasan et al. (2011)
miR-17	I–IV	Correlation with survival (−)	Srinivasan et al. (2011)
miR-17-5p	IV	Correlation with survival (+)	Srinivasan et al. (2011)
miR-181b	IV	Association with poor response to RT/TMZ	Slaby et al. (2010)
miR-181c	IV	Association with poor response to RT/TMZ	Slaby et al. (2010)
miR-193a	IV	Correlation with survival (−)	Srinivasan et al. (2011)
miR-195	IV	Correlation with survival (+)	Lakomy et al. (2011)
miR-196a	III–IV	Correlation with survival (−)	Ujifuku et al. (2010)
miR-196b	IV	Correlation with survival (+)	Lakomy et al. (2011)
miR-196b	III–IV	Correlation with survival (−)	Ujifuku et al. (2010)
miR-200b	IV	Correlation with survival (−)	Srinivasan et al. (2011)
miR-203	I–IV	Correlation with survival (+)	He et al. (2013)
miR-20a	IV	Correlation with survival (+)	Srinivasan et al. (2011)
miR-21	I–IV	Correlation with survival (−)	Hermansen et al. (2013)
miR-214	I–IV	Correlation with survival (+)	Wang et al. (2014)
miR-221	IV	Correlation with survival (−)	Srinivasan et al. (2011)
miR-222	IV	Correlation with survival (−)	Srinivasan et al. (2011)
miR-224	I–IV	Correlation with survival (−)	Lu et al. (2013)
miR-31	IV	Correlation with survival (−)	Srinivasan et al. (2011)
miR-3163	IV	Correlation with survival (−)	Niyazi et al. (2011)
miR-326	I–IV	Correlation with survival (+)	Wang et al. (2013b)
miR-335	I–IV	Correlation with survival (−)	Jiang et al. (2012)
miR-34a	I–IV	Correlation with survival (+)	Gao et al. (2013)
miR-372	I–IV	Correlation with survival (−)	Li et al. (2013a)
miR-375	I–IV	Correlation with survival (+)	Chang et al. (2012)
miR-539	IV	Correlation with survival (+)	Niyazi et al. (2011)
miR-650	I–IV	Correlation with survival (−)	Sun et al. (2013)
miR-9	I–IV	Correlation with survival (−)	Wu et al. (2013)

were selected for the following analysis. However, only five miRNAs were significantly deregulated: miR-3163 (fold change (FC) = 2.0, $p = 0.05$), miR-539 (FC = 0.5, $p = 0.001$), miR-1305 (FC = 0.5, $p = 0.05$), miR-1260 (FC = 0.5, $p = 0.03$), and let-7a (FC = 0.3, $p = 0.02$). Nevertheless, an application of all

30 miRNAs classified tested patients into two groups according to overall survival. The miRNA pattern and the prognostic power were both independent of the MGMT (O-6-methylguanine-DNA methyltransferase) methylation status. When performing a multivariate analysis for all patients including the factors MGMT status, miRNA pattern, adjuvant TMZ, age category, and RPA (recursive partitioning analysis) class, it turns out that only adjuvant TMZ remains a prognostic factor ($p = 0.01$) and MGMT status and miRNA pattern lose their prognostic significance ($p = 0.17$ and $p = 0.22$) (Niyazi et al. 2011).

Recent study published by Slaby et al. described significant negative impact of miR-181c and miR-181b on the response to the concomitant chemoradiotherapy with TMZ (Slaby et al. 2010). Lakomy et al. analyzed expression of eight miRNAs (miR-21, miR-128a, miR-181c, miR-195, miR-196a, miR-196b, miR-221, and miR-222) in GBM patients treated with concomitant chemoradiotherapy with TMZ and results subsequently correlated with MGMT status and clinical data. It was found that miR-195 ($p = 0.0124$) and miR-196b ($p = 0.0492$) had a positive impact on overall survival of the patients. Moreover, combination of miR-181c and miR-21 enabled to identify patients, which progressed in the 6 months after diagnosis (sensitivity 92 %, specificity 81 %, $p < 0.0001$) (Lakomy et al. 2011). These results are partially conflicting with conclusions of Ujifuku at al., who identified miR-195, miR-455-3p, and miR-10a* to be three most increased miRNAs in the TMZ-resistant cell lines. Moreover, miR-195 silencing led to the increased sensitivity to TMZ in these cells (Ujifuku et al. 2010). Similarly, Guan et al. described significant relationship between high expression of miR-196a/b and a worse prognosis of GBM and anaplastic astrocytoma patients (Guan et al. 2010). However, it is interesting that miR-195 and miR-196b have the same impact on prognosis in colorectal, hepatocellular, and adrenocortical carcinomas as described by Lakomy et al. (Soon et al. 2009; Xu et al. 2009; Liu et al. 2010; Wang et al. 2012c).

Wang et al. evaluated the prognostic value of miR-214 in overall survival of 108 glioma patients (WHO I—18, WHO II—12, WHO III—32, WHO IV—46). The overall survivals of patients, whose tumors expressed low level of miR-214, were significantly shorter than those with high level of miR-214 ($p < 0.001$). Moreover miR-214 was significantly downregulated compared to 20 normal brain tissues ($p = 0.001$) (Wang et al. 2014). Similarly, miR-34a, miR-203, miR-326, and miR-375 were reduced in tumors and their lower expression levels were significantly associated with worse progression-free survival and overall survival of glioma patients. Furthermore, multivariate Cox regression analyses indicated that all these miRNAs have been independent prognostic factors (Chang et al. 2012; Gao et al. 2013; He et al. 2013; Wang et al. 2013b).

On the contrary, miR-9 expression in 128 glioma tissues (WHO I—18, WHO II—14, WHO III—38, WHO IV—58) was significantly higher in comparison to 10 corresponding non-neoplastic brain tissues ($p < 0.001$). The increased expression of miR-9 was more frequently observed in the tissue of high-grade gliomas ($p = 0.001$). The expression levels of miR-9 in glioma tissues with low Karnofsky performance score (KPS) were also significantly higher than those with high KPS

($p = 0.008$). Moreover, the overall survival of glioma patients with high miR-9 expression was obviously lower than that with low miR-9 expression ($p < 0.001$). Multivariate analysis further showed that high miR-9 expression was an independent prognostic factor for overall survival in glioma patients ($p = 0.01$). More importantly, the subgroup analyses indicated that the overall survival of patients with high-grade gliomas was significantly worse for high miR-9 expression group than for low miR-9 expression group ($p < 0.001$), but no significant difference was found for patients with low-grade (I–II) gliomas (Wu et al. 2013). Furthermore, miR-650 expression was increased in 168 gliomas (WHO I—35, WHO II—40, WHO III—41, WHO IV—52) compared with 21 normal control specimens ($p < 0.001$). It was also found that miR-650 expression was related to grade and KPS, whereas high expression was more frequently detected in gliomas of high grade or low KPS score ($p < 0.001$). The prognosis of glioma with high miR-650 expression was significantly worse compared with gliomas with low miR-650 expression (Sun et al. 2013).

MiR-21 is the most consistently overexpressed miRNA in many cancers including gliomas. To better understand the role of miR-21 in gliomas, paraffin-embedded glioma tissue samples from 193 patients with WHO I–IV tumors were analyzed by in situ hybridization. It was found that miR-21 expression is localized in tumor cells and tumor-associated blood vessels, whereas no expression was seen in adjacent normal brain parenchyma. Only tumor cell miR-21 was associated with poor prognosis when adjusting for known clinical parameters in a multivariate analysis (Hermansen et al. 2013). Similarly, miR-335 expression level was also described to be significantly higher than that in corresponding non-neoplastic brain tissues ($p < 0.001$) and, in addition, survival analysis demonstrated that patients with high miR-335 expression tumors had significantly shorter survival times ($p = 0.01$) and that miR-335 is an independent prognostic factor ($p = 0.02$) (Jiang et al. 2012). Similarly, miR-17, miR-224, and miR-372 are overexpressed in gliomas and their higher expression levels are significantly associated with worse progression-free survival and overall survival of glioma patients. Multivariate Cox regression analyses indicated that all these miRNAs are independent prognostic factors (Lu et al. 2012, 2013; Li et al. 2013a).

4.5 MicroRNAs Involved in Chemoradioresistance of Gliomas

DNA damage, caused by radiation, activates pathways that protect cells from apoptosis and give them the ability to grow and proliferate. There are two well-known pathways: PI3K/AKT and ATM (ataxia telangiectasia mutated)/Chk2 (checkpoint kinase 2)/p53 that are activated as a response to the treatment. These pathways are regulated by several miRNAs. As for radiation, DNA damage is the main reason for cell death in normal cells, where AKT is important activator of multiple proteins, such as DNA-PK (protein kinase, DNA-activated, catalytic polypeptide), which subsequently leads to DNA repair, BCL-2, or mTOR [mechanistic target of rapamycin (serine/threonine kinase)], which ensures cell growth and proliferation. Next possibility of cells to recognize DSB (double strand breaks) or DNA damage is via BRCA1 (breast cancer 1, early onset) and activation of ATM/CHK2/p53 signaling pathway. The survival and cell apoptosis is dependent on activation of downstream proteins of this pathway.

Among treatment approaches that modulate and target DNA, TMZ is one of the currently frequently used. TMZ is an alkylating agent, which methylates nucleotides and so causes inhibition of DNA replication. Resistant cancer cells have evolved ways how to overcome this process. First mechanism is based on the MGMT enzyme that is able to excise nucleotides in DNA base-pair mismatch, and so neutralize the effect of TMZ. Other mechanism is efflux of TMZ by ATP-binding cassette transporters, which decrease the concentration of the drug within the cell. ATP-binding cassettes are probably the main reason for multidrug resistance. These processes were experimentally proved to be re-regulated by miRNAs (Fig. 4.6).

4.5.1 MicroRNAs Involved in Radioresistance

PI3K/AKT (Lee et al. 2011a) and ATM/CHk2/p53 (Squatrito et al. 2010) are two main pathways involved in radioresistance. These pathways are responsible for either activation of downstream proteins involved in repair of damaged DNA caused by radiation or regulation of apoptosis and survival rate (Li et al. 2009a). EGFR which triggers PI3K/AKT signaling pathways is important for proliferation activity of cells and is highly expressed in various types of cancer including gliomas. Furthermore, there is substantial experimental evidence supporting a causal role for aberrant EGFR signaling in cancer pathogenesis and resistance to treatment (Huang et al. 2009; Hatanpaa et al. 2010). PI3K is crucial for normal brain development and function. It was found to be hyperactivated in brain tumors (Guillamo et al. 2009; Kwiatkowska and Symons 2013). Studies concerning miRNAs and resistance revealed a connection between miRNAs and downstream proteins of EGFR/PI3K/AKT signaling.

PI3K/AKT signaling pathway is usually activated in GBM. Inactivation of this pathway impairs mechanisms of DNA repair, which can consequently enhance radiosensitivity. MiR-21 was shown to activate this pathway through suppression of PI3K/AKT inhibitor PTEN (Chakravarti et al. 2004; Kao et al. 2007; Gwak et al. 2012). MiR-21 belongs to one of the most studied miRNAs across many tumors including gliomas and is classified as an onco-miR. (Hong et al. 2013). It seems than miR-21 act as a major player in radioresistance, as decrease of this miRNA sensitizes GBM cell lines to irradiation. Sensitization of cells is mediated via γ-H2AX (H2A histone family, member X, gamma) blocking, which serves as an indicator of DSB and subsequently leads to prolongation of DNA repair (Gwak et al. 2012).

Further studies involved miR-7 in the regulation of PI3K/AKT signaling (Kefas et al. 2008). The involvement of miR-7 in this pathway was evaluated on two glioma cell lines U251 and U87, where increased level or miR-7 induced attenuation of EGFR and AKT, and caused a radio-sensitization (Lee et al. 2011a). AKT is

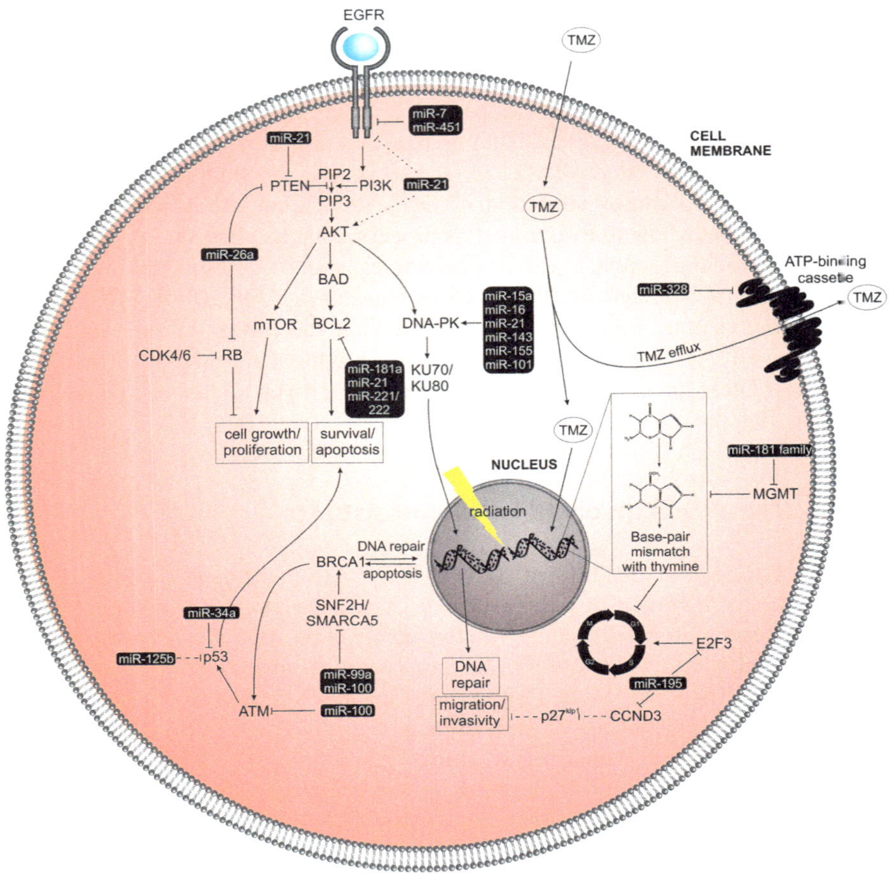

Fig. 4.6 MicroRNAs involved in radiotherapy and chemotherapy resistance. *AKT* v-akt murine thymoma viral oncogene homolog; *ATM* ataxia telangiectasia mutated; *BAD* BCL2-associated agonist of cell death; *BCL2* B-cell CLL/lymphoma 2; *BRCA1* breast cancer 1; *CCND3* cyclin D3; *CDK4/6* cyclin-dependent kinase 4/6; *DNA-PK* protein kinase, DNA-activated, catalytic polypeptide; *E2F3* E2F transcription factor 3; *EGFR* epidermal growth factor receptor; *KU70/80* X-ray repair complementing defective repair in Chinese hamster cells 5/6; *MGMT* O-6-methylguanine-DNA methyltransferase; *mTOR* mechanistic target of rapamycin (serine/threonine kinase); *p27* protein 27; *p53* protein 53; *PI3K* phosphatidylinositol-4,5-bisphosphate 3-kinase; *PIP2* phosphatidylinositol 4,5-bisphosphate; *PIP3* phosphatidylinositol (3,4,5)-triphosphate; *PTEN* phosphatase and tensin homolog; *RB* retinoblastoma; *SMARCA5 (SNF2H)* SWI/SNF-related, matrix-associated, actin-dependent regulator of chromatin, subfamily a, member 5; *TMZ* temozolomide. —| direct suppression; —> direct activation; ---| indirect suppression; ---> indirect activation

also important player in DNA damage repair as it controls expression of DNA-PK. DNA-PK serves as an activator of heterodimers Ku70/Ku80 (X-ray repair complementing defective repair in Chinese hamster cells 5/6), which are able to recognize DNA DSB during NHEJ (non-homologous end joining), and so it enables to repair DNA after DNA damage (Fell and Schild-Poulter 2012). As by decrease of miR-21, also miR-7 was able to regulate response of cell to DNA damage, through regulation of DNA-PK. This indicates the relationship between delayed DNA damage repair and increased radiation-induced cell killing by miR-7 overexpression. It seems that miR-7 could be a useful therapeutic target for overcoming the radioresistance of GBM and other tumors with activated EGFR-PI3K-AKT signaling, as it had an effect in different cancer types treated with radiation (Lee et al. 2011a).

MiR-181a is next miRNA which has been described in connection with GBM and radiation treatment. It affects anti-apoptotic gene *BCL2* and it was shown that irradiation exhibited and induced expression of miR-181a, while overexpression of miR-181a caused sensitization to radiation (Chen et al. 2010a).

ATM/Chk2/p53 pathway is closely connected to cellular processes such as regulation of apoptosis, cell cycle, or senescence (Cao et al. 2006). The importance of ATM level in radioresistance was shown on two GBM cell lines, which differ in ATM expression. An M059J radiosensitive cell line is characterized by lower level of ATM and M059K radioresistant cell line, which show higher expression of ATM due to deficiency in DNA-PK. There is an evidence that miR-100 could be involved in this process, as it is differently expressed in these cell lines and accordingly to the computational analyses is a direct regulator of ATM (Ng et al. 2010). Irradiation of M059K and M059J cell lines caused upregulation of several miRNAs: miR-17-3p, miR-17-5p, miR-19a, miR-19b miR-142-3p, and miR142-5p. However, cell line with lower level of ATM and normal DNA-PK activity exhibited induction of miR-15a, miR-16, miR-21, miR-143, and miR-155 level (Chaudhry et al. 2010b). Establishment of these miRNAs in radiation response was confirmed by other studies. Study focused on lymphoblastoid cell revealed that low dosage of IR elevates expression level of miR-15a, miR-16, and miR-21 (Chaudhry et al. 2010a). Mir-143 was found to directly target gene *FHIT* (fragile histidine triad) the deregulation of which is often seen in epithelial tumors and homozygous deletion contributes to higher radioresistance. Regulation of FHIT was shown via miR-143 overexpression which consequently leads to cell cycle arrest in G2 phase (Lin et al. 2011). It was also shown that elevated level of miR-155 serves as a radioprotectant in lung cancer cells. Inhibition of miR-155 led to radio-sensitization (Babar et al. 2011). ATM and DNA-PK were confirmed as targets of miR-101. These miRNAs sensitized U87MG glioma cells to radiation after lentiviral transduction in vitro and in vivo (Yan et al. 2010).

For radioresistance of cells it is very important to recognize and repair DSB or DNA damage, chromatin remodeling protein complex. A positive correlation between miR-99 and SWI/SNF (SWItch/Sucrose NonFermentable) chromatin remodeling factor SNF2H/SMARCA5 (SWI/SNF-related, matrix-associated, actin-dependent regulator of chromatin, subfamily a, member 5), a component of the ACF1 (bromodomain adjacent to zinc finger domain, 1A) complex, was found. Moreover, it has been elucidated that reduction of BRCA1 level on DNA damage site was arranged due to the downregulation of SNF2H, miR-99a, and miR-100. Ectopic expression of the miR-99 family in cells reduced the rate and overall

Table 4.3 Effects of microRNAs involved in regulation of radioresistance in gliomas

miRNA	Up/ downregulation	Effect on radioresistance	Targets	References
miR-181a	↓	+	BCL-2	Chen et al. (2010a)
miR-7	↓	+	EGFR, AKT, DNA-PKs	Lee et al. (2011b)
miR-101	↓	+	ATM, DNA-PK	Yan et al. (2010)
miR-99	↑	–	SNF2H, SMARCA5	Mueller et al. (2013)
miR-100	↑	–	SNF2H, SMARCA5	Ng et al. (2010)
miR-21	↑	–	γ-H2AX, AKT	Gwak et al. (2012), Hong et al. (2013)
miR-143	↑	+	FHIT	Lin et al. (2011)
miR-155	↑	+	FOXO3a	Ling et al. (2013)
miR17-3p	Induction after IR		MDM2	Chaudhry et al. (2010b), Wang et al. (2012a)
miR-17-5p	Induction after IR		PTEN	Chaudhry et al. (2010b), Wang et al. (2012a)
miR-19a	Induction after IR			Chaudhry et al. (2010b), Jia et al. (2013)
miR-19b	Induction after IR			Jia et al. (2013)
miR142-3p	Induction after IR			Chaudhry et al. (2010b)
miR142-5p	Induction after IR			Chaudhry et al. (2010b)

Adapted from Besse et al. (2013)

efficiency of repair by both homologous recombination and non-homologous end joining, which led to sensitization of cells to radiation (Nakano and Kornblum 2006; Mueller et al. 2013; Besse et al. 2013). Effects of miRNAs and their targets are summarized in Table 4.3 and Fig. 4.6.

4.5.2 MiRNAs Involved in Chemoresistance

Temozolomide is a cytotoxic prodrug agent, which after hydrolyzation methylates nucleotides and causes inhibition of DNA replication. The resistance after long-time exposure is nearly universal. This is caused by three main mechanisms involved in DNA repair: increased levels of MGMT; a deficient mismatch repair process (MMR); and activation of the PARP (poly(ADP)-ribose polymerase) pathway (Darkes et al. 2002; Carmo et al. 2011).

MGMT repairs mutagenic DNA lesions, caused by TMZ guanines methylation on O^6 position that subsequently leads to base-pair mismatch with thymine. Further,

MGMT prevents mismatches and errors during DNA replication and transcription. Cells with increased level of MGMT become resistant to the exposure of TMZ, while in nonresistant glioma cells the pathway induced by TMZ is active and is able to induce cell death (Darkes et al. 2002; Sharma et al. 2009).

As described previously, miRNAs have the possibility to regulate multiple genes, as was observed in miR-181, which is involved in the regulation of BCL2 and MGMT. Silencing of miR-181 family led to increased level of MGMT. Therefore, miR-181 could be used as a predictive factor to TMZ response (Fig. 4.6) (Chen et al. 2010a; Zhang et al. 2012c).

Correlation of TMZ chemoresistance and survival pathways PI3K/AKT and ERK1/2/MAPK (mitogen-activated protein kinase) was confirmed on U118 cell line, which evinces chemoresistance, while after inhibition of these pathways chemoresistance was partially eradicated (Carmo et al. 2011).

Regulation of PI3K/AKT by miR-21 was described also by TMZ-treated glioma cell lines. Decreased level of this miRNAs caused inhibition of proliferation, increased apoptosis, and cell cycle arrest. This was particularly caused by the decrease of BAX (BCL2-associated X protein)/BCL-2 ratio and decreased caspase-3 activity (Ren et al. 2010b; Shi et al. 2010). Among miRNAs targeting apoptotic factors such as BAX, cytochrome c, and cleaved caspase-3 belongs miR-221/222 as well. Their decreased levels enhanced sensitivity to TMZ, but independently of the p53 status (Chen et al. 2012).

To evaluate differences in cell resistant to TMZ, two cell lines U251R, U251 were compared on their miRNA profile. This revealed an elevated level of miR-195, miR-455-3p, and miR-10a* in U251R (TMZ-resistant) cells, which could be important in TMZ resistance. Inhibition of miR-455-3p or miR-10a(*) showed modest cell killing effect in the presence of TMZ. However, miR-195 with TMZ in combination strongly increased cell death (Ujifuku et al. 2010). This was caused by inverse correlation of E2F3 (E2F transcription factor 3) and CCND3 (cyclin D3) and expression level of miR-195. These results indicate that E2F3 and CCND3 are targets of this miRNA. E2F3 is able to suppress transcription of genes related to cell cycle progression and so causes G1 phase arrest. On the other hand CCND3 suppression mediated by miR-195 allows an increase of p27Kip1 (cyclin-dependent kinase inhibitor 1B) expression level, which consequently influences proteins involved in cell migration, resulting in repression of GBM cell invasion (Zhang et al. 2012a).

Although TMZ is the most often used chemotherapeutical agent, there are other chemotherapeutics used in GBM treatment. VM-26 (Teniposide) was tested on U373MG cell line in combination with miR-21 inhibitor, which led to enhancement of cytotoxicity. As a direct target LRRFIP1 (leucine-rich repeat (in FLII) interacting protein 1) was validated, whose product is an inhibitor of NF-kB, a downstream effector of PI3K/AKT signaling (Li et al. 2009b). MiR-21 effect was also examined in cells treated by 5-FU (5-fluorouracil) and showed increased apoptosis and decreased ability to migrate (Ren et al. 2010a). Not only deregulation of PI3K pathway and its downstream factor is responsible for resistance, but also family of proteins called *ABCG2* (ATP-binding cassettes, a subfamily of G member

Table 4.4 Effects of miRNA involved in regulation of chemoresistance in gliomas

miRNA	Up/ downregulation	Effect on chemoresistance	Targets	References
miR-181	↓	–	MGMT	Zhang et al. (2012c)
miR-21	↑	–	BAX, BCL-2	Shi et al. (2010)
miR-221/ 222	↓	–	BAX, cyt c, Caspase3	Li et al. (2009b)
miR-195	↑	+	ENF3, CCND3	Chen et al. (2012)
miR-455-3p	↑	+		Zhang et al. (2012a)
miR-10a*	↑	+	EPHX1, BED7	Ujifuku et al. (2010)
miR-328	↑	–	ABCG2	Li et al. (2010)

Adapted from Besse et al. (2013)

2 protein family), which contribute to multidrug resistance. These transporters cause selective survival of cells through the efflux of drugs. MiR-328 contributes to the regulation of *ABCG2* expression, while increased level of this miRNA led to decrease of *ABCG2* expression (Li et al. 2010). Regulatory effects of miRNAs involved in chemoresistance of gliomas are summarized in Table 4.4 and Fig. 4.6.

4.5.3 Glioma Initiating Cells/Stem Cells and Their Role in Tumor Resistance

GSCs, which maintain stem-like phenotype, show generally much more resistance to radiation or chemotherapy in comparison to "normal" glioma cells (Wang et al. 2010; Yamada and Nakano 2012). Therefore, it seems that the best way how to decrease the number of recurrences which is connected with high treatment resistance is to target these cells and their signaling pathways. This population of CD133 positive (CD133+) cancer stem cells contributes to the radio- and chemoresistance and bears the activation of NOTCH and SHH (Sonic hedgehog) pathways. Inhibition of these pathways led to enhancement of CD133+ glioma cells to TMZ treatment (Ulasov et al. 2011). Further, SHH has been found to play a role in maintaining undifferentiated stem cell state in GSCs and Notch signaling in active GSCs as well as in GBM pathogenesis (Yamada and Nakano 2012). In fact, NOTCH and SHH were shown to be upregulated in GSCs and their inhibition led to increased susceptibility of these cells to TMZ. Involvement of miRNAs in regulation of stem-like phenotype of GSCs is described in detail in Sect. 4.3.6.

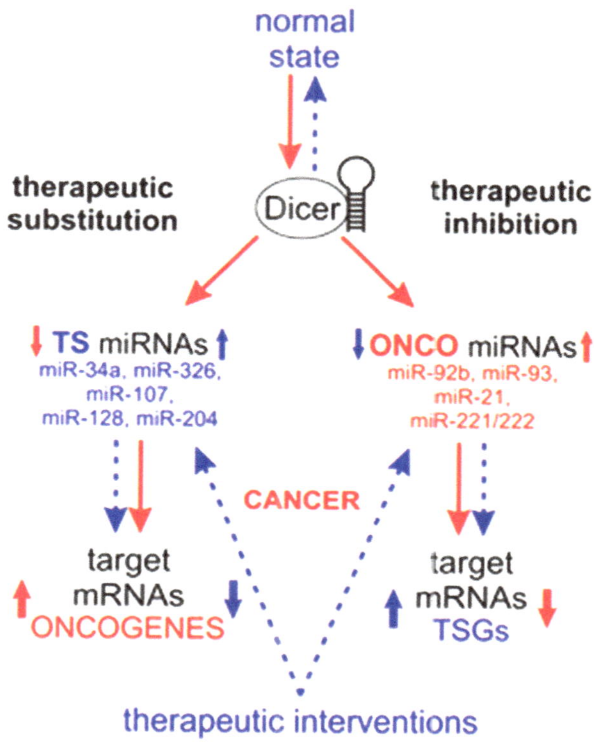

Fig. 4.7 MicroRNAs as a potential therapeutic target in gliomas; see text

4.6 MicroRNA as Therapeutic Targets

The role of miRNA in carcinogenesis depends on the functions of their targets (oncogenes or tumor suppressors). Downregulated miRNAs, which regulate oncogenes, are tumor-suppressive miRNAs (TS-miRNAs), while miRNAs overexpressed in tumors and inhibiting tumor suppressors are classified as an onco-miRs. Therefore, there are two possible miRNA-based strategies: therapeutic inhibition (in case of onco-miRs) or substitution (in case of TS-miRNAs) (Fig. 4.7).

Specific inhibitory effect of miRNAs can be achieved by use of their complementary antagonists (anti-miRs). Whereas in the most cancers tumor tissues indicate global decrease in miRNA expression levels, therapeutic substitution presents the most promising approach. So far, many studies have described oncogenic/tumor-suppressive miRNAs and their contribution to development and progression of gliomas. Most of these studies were performed under in vitro conditions (Li et al. 2013b). In vitro models are suitable for studies on specific targets and regulatory pathways; however, to evaluate the real therapeutic potential, it is necessary to study miRNAs on animal models. MiRNAs in vivo studies are mainly focused on the evaluation of particular miRNA effects on the tumor growth and confirmation of phenotypic observations from in vitro studies.

It is presumed that GSCs may stand for the cells responsible for tumor formation and treatment failure. Notch signaling is important for maintaining this stem-like phenotype and has been found to be aberrantly expressed in GSCs. Inhibition of this signaling pathway decreases proliferation, increases neuronal differentiation and reduces CD133+ cell fraction in vitro; moreover, in vivo such inhibition attenuates tumorigenicity (Stockhausen et al. 2010). MiR-34a is often described as direct regulator of Notch signaling and could serve as potential tumor suppressor in brain tumors. This was demonstrated by transfection of pre-miR-34a to cells, which were consequently implanted to immunocompromised mice. Four weeks after implantation, xenografts of pre-miR-34a-transfected cells developed tumor of significantly smaller volume compared to control cells (Li et al. 2009a, 2011). Moreover, other miRNAs were confirmed as direct regulators of Notch signaling, such as miR-326 and miR-107. Similarly to miR-34a, exogenous expression of these miRNAs in glioma cells caused decrease of the tumor bearing in mice (Kefas et al. 2009; Chen et al. 2013a). Inhibition of NOTCH can be also achieved by miR-92b. It was shown that this miRNA regulates NLK (Nemo-like kinase), which negatively regulates NOTCH–dependent transcriptional activation. Furthermore, expression of NLK was inversely correlated with miR-92b in clinical glioma samples and enables prediction of clinical outcome. An effect of miR-92b on glioma tumors was evaluated in vivo by using nude mice. Glioma cell lines were subcutaneously injected to mice and after reaching 100 mm^3 tumor volumes were animals treated with anti-miR-92b oligonucleotide. Anti-miR-92b triggered growth inhibition, induced apoptosis, and suppressed invasion of glioma (Ishitani et al. 2010; Wang et al. 2013a).

Consistently, miR-204 was found to be downregulated in glioma cells and also in neural stem cells. Inhibition of miR-204 promoted not only cellular stemness, but also the biological properties essential to lethality of the disease; however, restoration of miR-204 greatly abrogates the aggressiveness of glioma cells. This observation is supported by Kaplan–Meier survival analysis of mice transduced by miR-204 compared to control. MiR-204 prolonged the survival of mice by approx. 15 %. This is probably caused by miR-204 targeting of *SOX4* gene, which is the core regulator governing the stemness of both glioma and neural stem cells (Ying et al. 2013).

Aggressiveness of stem-like cells and ability to form new in situ tumors is closely related to proliferation and invasive ability. Receptor tyrosine kinases such as EGFR and PDGFR play a major role in glioma proliferation and invasion (Nakada et al. 2013). In gliomas, regulation of these receptors by miRNAs was observed in many in vitro studies, which were consequently confirmed in animal models. It was shown that miR-128 represses GSC growth by enhancing neuronal differentiation and mediates differentiation by targeting EGFR and PDGFRα. Authors of this study proved this by utilization and autochthonous glioma mouse model, which arises from activation of oncogenic H-RasV12 (Harvey-RasV12) and loss of p53. Developed glioma tumors in mice showed significant decrease of miR-128 and high levels of PDGFRα compared to normal brain tissue. The introduction of miR-128 led to remarkable improvement of survival in treated animals. This study validated miR-128 as effective suppressor of glioma growth (Papagiannakopoulos et al. 2012).

Another interesting example of onco-miR is miR-93. It enhances tumor cell survival, blood vessel expansion, and tumor growth, by targeting, at least in part, integrin-β8 expression. Consistent with this was the decreased survival rate of mice bearing the miR-93 positive tumors compared with the controls and Kaplan–Maier survival curves which indicated that miR-93 expression decreased rates of mouse survival. Further, higher levels of integrin-β8 are associated with cell death in tumor mass and in human glioblastoma (Fang et al. 2011). Accelerated proliferation is a characteristic for tumor cells and contributes to excessive growth and tumor progression. *PDCD4* is a well-known tumor-suppressor gene. Overexpression of PDCD4 caused by downregulation of miR-21 led to decreased proliferation and apoptotic rate and in the xenograft model decreased tumor formation and growth (Gaur et al. 2011b). Clinical relevance of miR-21 was further studied by Xuan Zhou et al. In the in vivo pharmacological study 30 mice were included and screened for tumor volume of xenografts treated with PBS, negative control, and anti-sense-miR-21 (as-miR-21). Significant decrease in tumor volume was observed only in as-miR-21-treated group. These data suggest miR-21 as a promising therapeutic target for malignant gliomas (Zhou et al. 2010a).

Even if miRNA-based therapies were not clinically implemented to the therapy, they have several important advantages to conventional drugs and targeted therapy in oncology. MiRNAs have ability to target multiple genes, which give them a priority in cancer treatment, considering cancer as a heterogeneous disease with multiple molecular alterations. In addition, miRNAs are not detected by immune system and so they could be an attractive option for clinical drug development.

Conclusions and Future Perspectives

The discovery of miRNA function has markedly spread the view on regulation of gene expression. Its remarkable ability to regulate large number of genes, including oncogenes and tumor-suppressor genes, has catapulted miRNAs into the center of cancer molecular biology over the past 5 years. It is now evident that dysregulation of miRNAs is an important step in the development of many cancers, including gliomas. Several studies based on expression profiling have proved that there are significant changes of miRNA expression levels in gliomas compared to non-tumor brain tissue. These studies also identified groups of miRNAs with potential for prognostic stratification and prediction of responses to chemoradiotherapy in glioma patients. But much more studies have been focused on the improvement of our knowledge about miRNAs involvement in glioma core signaling pathways. The results of these studies suggest a great potential and relevance of miRNAs as a novel class of therapeutic targets and possibly powerful intervention tools in glioma.

Acknowledgment This work was supported by grants of Internal Grant Agency of the Czech Ministry of Health no. NT13514-4/2012 and NT11214-4/2010; project "CEITEC—Central European Institute of Technology" (CZ.1.05/1.1.00/02.0068).

References

Aldaz B, Sagardoy A, Nogueira L, Guruceaga E, Grande L, Huse JT et al (2013) Involvement of miRNAs in the differentiation of human glioblastoma multiforme stem-like cells. PLoS One 8: e77098

Alrfaei BM, Vemuganti R, Kuo JS (2013) microRNA-100 targets SMRT/NCOR2, reduces proliferation, and improves survival in glioblastoma animal models. PLoS One 8:e80865

Altaner C (2008) Glioblastoma and stem cells. Neoplasma 55:369–374

Asuthkar S, Velpula KK, Chetty C, Gorantla B, Rao JS (2012) Epigenetic regulation of miRNA-211 by MMP-9 governs glioma cell apoptosis, chemosensitivity and radiosensitivity. Oncotarget 3:1439–1454

Babar IA, Czochor J, Steinmetz A, Weidhaas JB, Glazer PM, Slack FJ (2011) Inhibition of hypoxia-induced miR-155 radiosensitizes hypoxic lung cancer cells. Cancer Biol Ther 12:908–914

Besse A, Sana J, Fadrus P, Slaby O (2013) MicroRNAs involved in chemo- and radioresistance of high-grade gliomas. Tumour Biol 34(4):1969–1978

Bienertova-Vasku J, Sana J, Slaby O (2013) The role of microRNAs in mitochondria in cancer. Cancer Lett 336:1–7

Cao L, Kim S, Xiao C, Wang RH, Coumoul X, Wang X et al (2006) ATM-Chk2-p53 activation prevents tumorigenesis at an expense of organ homeostasis upon Brca1 deficiency. EMBO J 25:2167–2177

Card DA, Hebbar PB, Li L, Trotter KW, Komatsu Y, Mishina Y et al (2008) Oct4/Sox2-regulated miR-302 targets cyclin D1 in human embryonic stem cells. Mol Cell Biol 28:6426–6438

Carmo A, Carvalheiro H, Crespo I, Nunes I, Lopes MC (2011) Effect of temozolomide on the U-118 glioma cell line. Oncol Lett 2:1165–1170

Carthew RW, Sontheimer EJ (2009) Origins and mechanisms of miRNAs and siRNAs. Cell 136:642–655

Ciafre SA, Galardi S, Mangiola A, Ferracin M, Liu CG, Sabatino G et al (2005) Extensive modulation of a set of microRNAs in primary glioblastoma. Biochem Biophys Res Commun 334:1351–1358

Conti A, Aguennouz M, La Torre D, Tomasello C, Cardali S, Angileri FF et al (2009) miR-21 and 221 upregulation and miR-181b downregulation in human grade II-IV astrocytic tumors. J Neurooncol 93:325–332

Costa PM, Cardoso AL, Nobrega C, Pereira de Almeida LF, Bruce JN, Canoll P et al (2013) MicroRNA-21 silencing enhances the cytotoxic effect of the antiangiogenic drug sunitinib in glioblastoma. Hum Mol Genet 22:904–918

D'Urso PI, D'Urso OF, Storelli C, Mallardo M, Gianfreda CD, Montinaro A et al (2012) miR-155 is up-regulated in primary and secondary glioblastoma and promotes tumour growth by inhibiting GABA receptors. Int J Oncol 41:228–234

Darkes MJM, Plosker GL, Jarvis B (2002) Temozolomide: a review of its use in the treatment of malignant gliomas, malignant melanoma and other advanced cancers. Am J Cancer 1:55–80

Dong CG, Wu WK, Feng SY, Wang XJ, Shao JF, Qiao J (2012) Co-inhibition of microRNA-10b and microRNA-21 exerts synergistic inhibition on the proliferation and invasion of human glioma cells. Int J Oncol 41:1005–1012

Dontula R, Dinasarapu A, Chetty C, Pannuru P, Herbert E, Ozer H et al (2013) MicroRNA 203 modulates glioma cell migration via Robo1/ERK/MMP-9 signaling. Genes Cancer 4:285–296

Ernst A, Campos B, Meier J, Devens F, Liesenberg F, Wolter M et al (2010) De-repression of CTGF via the miR-17-92 cluster upon differentiation of human glioblastoma spheroid cultures. Oncogene 29:3411–3422

Fang L, Deng Z, Shatseva T, Yang J, Peng C, Du WW et al (2011) MicroRNA miR-93 promotes tumor growth and angiogenesis by targeting integrin-beta8. Oncogene 30:806–821

Fareh M, Turchi L, Virolle V, Debruyne D, Almairac F, de-la-Forest Divonne S et al (2012) The miR 302-367 cluster drastically affects self-renewal and infiltration properties of glioma-initiating cells through CXCR4 repression and consequent disruption of the SHH-GLI-NANOG network. Cell Death Differ 19:232–244

Fell VL, Schild-Poulter C (2012) Ku regulates signaling to DNA damage response pathways through the Ku70 von Willebrand A domain. Mol Cell Biol 32:76–87

Folkman J (2006) Angiogenesis. Annu Rev Med 57:1–18

Fowler A, Thomson D, Giles K, Maleki S, Mreich E, Wheeler H et al (2011) miR-124a is frequently down-regulated in glioblastoma and is involved in migration and invasion. Eur J Cancer 47:953–963

Fu J, Rodova M, Nanta R, Meeker D, Van Veldhuizen PJ, Srivastava RK et al (2013) NPV-LDE-225 (Erismodegib) inhibits epithelial mesenchymal transition and self-renewal of glioblastoma initiating cells by regulating miR-21, miR-128, and miR-200. Neuro Oncol 15:691–706

Gabriely G, Wurdinger T, Kesari S, Esau CC, Burchard J, Linsley PS et al (2008) MicroRNA 21 promotes glioma invasion by targeting matrix metalloproteinase regulators. Mol Cell Biol 28:5369–5380

Gabriely G, Yi M, Narayan RS, Niers JM, Wurdinger T, Imitola J et al (2011) Human glioma growth is controlled by microRNA-10b. Cancer Res 71:3563–3572

Gal H, Pandi G, Kanner AA, Ram Z, Lithwick-Yanai G, Amariglio N et al (2008) MIR-451 and Imatinib mesylate inhibit tumor growth of Glioblastoma stem cells. Biochem Biophys Res Commun 376:86–90

Gao H, Zhao H, Xiang W (2013) Expression level of human miR-34a correlates with glioma grade and prognosis. J Neurooncol 113:221–228

Gaur AB, Holbeck SL, Colburn NH, Israel MA (2011) Downregulation of Pdcd4 by mir-21 facilitates glioblastoma proliferation in vivo. Neuro Oncol 13:580–590

Genovese G, Ergun A, Shukla SA, Campos B, Hanna J, Ghosh P et al (2012) microRNA regulatory network inference identifies miR-34a as a novel regulator of TGF-beta signaling in glioblastoma. Cancer Discov 2:736–749

Geretti E, Klagsbrun M (2007) Neuropilins: novel targets for anti-angiogenesis therapies. Cell Adh Migr 1:56–61

Ghosh S, Karin M (2002) Missing pieces in the NF-kappaB puzzle. Cell 109(Suppl):S81–S96

Godlewski J, Nowicki MO, Bronisz A, Williams S, Otsuki A, Nuovo G et al (2008) Targeting of the Bmi-1 oncogene/stem cell renewal factor by microRNA-128 inhibits glioma proliferation and self-renewal. Cancer Res 68:9125–9130

Griffiths-Jones S, Saini HK, van Dongen S, Enright AJ (2008) miRBase: tools for microRNA genomics. Nucleic Acids Res 36:D154–D158

Guan Y, Mizoguchi M, Yoshimoto K, Hata N, Shono T, Suzuki SO et al (2010) MiRNA-196 is upregulated in glioblastoma but not in anaplastic astrocytoma and has prognostic significance. Clin Cancer Res 16:4289–4297

Guessous F, Alvarado-Velez M, Marcinkiewicz L, Zhang Y, Kim J, Heister S et al (2013) Oncogenic effects of miR-10b in glioblastoma stem cells. J Neurooncol 112:153–163

Guessous F, Zhang Y, Kofman A, Catania A, Li Y, Schiff D et al (2010) microRNA-34a is tumor suppressive in brain tumors and glioma stem cells. Cell Cycle (Georgetown, TX) 9:1031–1036

Guillamo JS, de Bouard S, Valable S, Marteau L, Leuraud P, Marie Y et al (2009) Molecular mechanisms underlying effects of epidermal growth factor receptor inhibition on invasion, proliferation, and angiogenesis in experimental glioma. Clin Cancer Res 15:3697–3704

Guo P, Nie Q, Lan J, Ge J, Qiu Y, Mao Q (2013) C-Myc negatively controls the tumor suppressor PTEN by upregulating miR-26a in glioblastoma multiforme cells. Biochem Biophys Res Commun 441:186–190

Gwak HS, Kim TH, Jo GH, Kim YJ, Kwak HJ, Kim JH et al (2012) Silencing of microRNA-21 confers radio-sensitivity through inhibition of the PI3K/AKT pathway and enhancing autophagy in malignant glioma cell lines. PLoS One 7:e47449

Haapa-Paananen S, Chen P, Hellstrom K, Kohonen P, Hautaniemi S, Kallioniemi O et al (2013) Functional profiling of precursor MicroRNAs identifies MicroRNAs essential for glioma proliferation. PLoS One 8:e60930

Hao J, Zhang C, Zhang A, Wang K, Jia Z, Wang G et al (2012) miR-221/222 is the regulator of Cx43 expression in human glioblastoma cells. Oncol Rep 27:1504–1510

Hatanpaa KJ, Burma S, Zhao D, Habib AA (2010) Epidermal growth factor receptor in glioma: signal transduction, neuropathology, imaging, and radioresistance. Neoplasia 12:675–684

He J, Deng Y, Yang G, Xie W (2013) MicroRNA-203 down-regulation is associated with unfavorable prognosis in human glioma. J Surg Oncol 108:121–125

Hermansen SK, Dahlrot RH, Nielsen BS, Hansen S, Kristensen BW (2013) MiR-21 expression in the tumor cell compartment holds unfavorable prognostic value in gliomas. J Neurooncol 111:71–81

Hermansen SK, Kristensen BW (2013) MicroRNA biomarkers in glioblastoma. J Neurooncol 114:13–23

Hong L, Han Y, Zhang Y, Zhang H, Zhao Q, Wu K et al (2013) MicroRNA-21: a therapeutic target for reversing drug resistance in cancer. Expert Opin Ther Targets 17:1073–1080

Hua D, Ding D, Han X, Zhang W, Zhao N, Foltz G et al (2012) Human miR-31 targets radixin and inhibits migration and invasion of glioma cells. Oncol Rep 27:700–706

Huang PH, Xu AM, White FM (2009) Oncogenic EGFR signaling networks in glioma. Sci Signal 2:re6

Huse JT, Brennan C, Hambardzumyan D, Wee B, Pena J, Rouhanifard SH et al (2009) The PTEN-regulating microRNA miR-26a is amplified in high-grade glioma and facilitates gliomagenesis in vivo. Genes Dev 23:1327–1337

Chakravarti A, Zhai G, Suzuki Y, Sarkesh S, Black PM, Muzikansky A et al (2004) The prognostic significance of phosphatidylinositol 3-kinase pathway activation in human gliomas. J Clin Oncol 22:1926–1933

Chamberlain MC (2011) Bevacizumab for the treatment of recurrent glioblastoma. Clin Med Insights Oncol 5:117–129

Chan JA, Krichevsky AM, Kosik KS (2005) MicroRNA-21 is an antiapoptotic factor in human glioblastoma cells. Cancer Res 65:6029–6033

Chan XH, Nama S, Gopal F, Rizk P, Ramasamy S, Sundaram G et al (2012) Targeting glioma stem cells by functional inhibition of a prosurvival oncomiR-138 in malignant gliomas. Cell Rep 2:591–602

Chang C, Shi H, Wang C, Wang J, Geng N, Jiang X et al (2012) Correlation of microRNA-375 downregulation with unfavorable clinical outcome of patients with glioma. Neurosci Lett 531:204–208

Chaudhry MA, Kreger B, Omaruddin RA (2010a) Transcriptional modulation of micro-RNA in human cells differing in radiation sensitivity. Int J Radiat Biol 86:569–583

Chaudhry MA, Sachdeva H, Omaruddin RA (2010b) Radiation-induced micro-RNA modulation in glioblastoma cells differing in DNA-repair pathways. DNA Cell Biol 29:553–561

Chen C-Z (2005) MicroRNAs as oncogenes and tumor suppressors. New Engl J Med 353:1768–1771

Chen G, Zhu W, Shi D, Lv L, Zhang C, Liu P et al (2010a) MicroRNA-181a sensitizes human malignant glioma U87MG cells to radiation by targeting Bcl-2. Oncol Rep 23:997–1003

Chen L, Chen XR, Chen FF, Liu Y, Li P, Zhang R et al (2013a) MicroRNA-107 inhibits U87 glioma stem cells growth and invasion. Cell Mol Neurobiol 33:651–657

Chen L, Zhang J, Han L, Zhang A, Zhang C, Zheng Y et al (2012) Downregulation of miR-221/222 sensitizes glioma cells to temozolomide by regulating apoptosis independently of p53 status. Oncol Rep 27:854–860

Chen L, Zhang R, Li P, Liu Y, Qin K, Fa ZQ et al (2013b) P53-induced microRNA-107 inhibits proliferation of glioma cells and down-regulates the expression of CDK6 and Notch-2. Neurosci Lett 534:327–332

Chen R, Nishimura MC, Bumbaca SM, Kharbanda S, Forrest WF, Kasman IM et al (2010b) A hierarchy of self-renewing tumor-initiating cell types in glioblastoma. Cancer Cell 17:362–375

Chen SM, Chen HC, Chen SJ, Huang CY, Chen PY, Wu TW et al (2013c) MicroRNA-495 inhibits proliferation of glioblastoma multiforme cells by downregulating cyclin-dependent kinase 6 World J Surg Oncol 11:87

Chen Y, Liu W, Chao T, Zhang Y, Yan X, Gong Y et al (2008) MicroRNA-21 down-regulates the expression of tumor suppressor PDCD4 in human glioblastoma cell T98G. Cancer Lett 272:197–205

Chen Z, Li D, Cheng Q, Ma Z, Jiang B, Peng R et al (2013d) MicroRNA-203 inhibits the proliferation and invasion of U251 glioblastoma cells by directly targeting PLD2. Mol Med Rep 9(2):503–508

Ishitani T, Hirao T, Suzuki M, Isoda M, Ishitani S, Harigaya K et al (2010) Nemo-like kinase suppresses Notch signalling by interfering with formation of the Notch active transcriptional complex. Nat Cell Biol 12:278–285

Jia Z, Wang K, Zhang A, Wang G, Kang C, Han L et al (2013) miR-19a and miR-19b overexpression in gliomas. Pathol Oncol Res 19:847–853

Jiang J, Sun X, Wang W, Jin X, Bo X, Li Z et al (2012) Tumor microRNA-335 expression is associated with poor prognosis in human glioma. Med Oncol 29:3472–3477

Jiang J, Yang J, Wang Z, Wu G, Liu F (2013) TFAM is directly regulated by miR-23b in glioma. Oncol Rep 30:2105–2110

Kao GD, Jiang Z, Fernandes AM, Gupta AK, Maity A (2007) Inhibition of phosphatidylinositol-3-OH kinase/Akt signaling impairs DNA repair in glioblastoma cells following ionizing radiation. J Biol Chem 282:21206–21212

Kefas B, Comeau L, Erdle N, Montgomery E, Amos S, Purow B (2010) Pyruvate kinase M2 is a target of the tumor-suppressive microRNA-326 and regulates the survival of glioma cells. Neuro Oncol 12:1102–1112

Kefas B, Comeau L, Floyd DH, Seleverstov O, Godlewski J, Schmittgen T et al (2009) The neuronal microRNA miR-326 acts in a feedback loop with notch and has therapeutic potential against brain tumors. J Neurosci 29:15161–15168

Kefas B, Godlewski J, Comeau L, Li Y, Abounader R, Hawkinson M et al (2008) microRNA-7 inhibits the epidermal growth factor receptor and the Akt pathway and is down-regulated in glioblastoma. Cancer Res 68:3566–3572

Kim H, Huang W, Jiang X, Pennicooke B, Park PJ, Johnson MD (2010) Integrative genome analysis reveals an oncomir/oncogene cluster regulating glioblastoma survivorship. Proc Natl Acad Sci U S A 107:2183–2188

Koo S, Martin GS, Schulz KJ, Ronck M, Toussaint LG (2012) Serial selection for invasiveness increases expression of miR-143/miR-145 in glioblastoma cell lines. BMC Cancer 12:143

Koshkin PA, Chistiakov DA, Chekhonin VP (2013) Role of microRNAs in mechanisms of glioblastoma resistance to radio- and chemotherapy. Biochemistry (Mosc) 78:325–334

Kozomara A, Griffiths-Jones S (2011) miRBase: integrating microRNA annotation and deep-sequencing data. Nucleic Acids Res 39:D152–D157

Kwiatkowska A, Symons M (2013) Signaling determinants of glioma cell invasion. Adv Exp Med Biol 986:121–141

Ladomery MR, Maddocks DG, Wilson ID (2011) MicroRNAs: their discovery, biogenesis, function and potential use as biomarkers in non-invasive prenatal diagnostics. Int J Mol Epidemiol Genet 2:253–260

Lakomy R, Sana J, Hankeova S, Fadrus P, Kren L, Lzicarova E et al (2011) MiR-195, miR-196b, miR-181c, miR-21 expression levels and O-6-methylguanine-DNA methyltransferase methylation status are associated with clinical outcome in glioblastoma patients. Cancer Sci 102:2186–2190

Lawler S, Chiocca EA (2009) Emerging functions of microRNAs in glioblastoma. J Neurooncol 92:297–306

Le MT, Xie H, Zhou B, Chia PH, Rizk P, Um M et al (2009) MicroRNA-125b promotes neuronal differentiation in human cells by repressing multiple targets. Mol Cell Biol 29:5290–5305

Lee KM, Choi EJ, Kim IA (2011a) microRNA-7 increases radiosensitivity of human cancer cells with activated EGFR-associated signaling. Radiother Oncol 101:171–176

Lee ST, Chu K, Oh HJ, Im WS, Lim JY, Kim SK et al (2011b) Let-7 microRNA inhibits the proliferation of human glioblastoma cells. J Neurooncol 102:19–24

Li G, Zhang Z, Tu Y, Jin T, Liang H, Cui G et al (2013a) Correlation of microRNA-372 upregulation with poor prognosis in human glioma. Diagn Pathol 8:1

Li M, Li J, Liu L, Li W, Yang Y, Yuan J (2013b) MicroRNA in human glioma. Cancers (Basel) 5:1306–1331

Li W-Q, Li Y-M, Tao B-B, Lu Y-C, Hu G-H, Liu H-M et al (2010) Downregulation of ABCG2 expression in glioblastoma cancer stem cells with miRNA-328 may decrease their chemoresistance. Med Sci Monit 16:HY27–HY30

Li WB, Ma MW, Dong LJ, Wang F, Chen LX, Li XR (2011) MicroRNA-34a targets notch1 and inhibits cell proliferation in glioblastoma multiforme. Cancer Biol Ther 12:477–483

Li Y, Guessous F, Zhang Y, Dipierro C, Kefas B, Johnson E et al (2009a) MicroRNA-34a inhibits glioblastoma growth by targeting multiple oncogenes. Cancer Res 69:7569–7576

Li Y, Li W, Yang Y, Lu Y, He C, Hu G et al (2009b) MicroRNA-21 targets LRRFIP1 and contributes to VM-26 resistance in glioblastoma multiforme. Brain Res 1286:13–18

Li Y, Xu J, Chen H, Bai J, Li S, Zhao Z et al (2013c) Comprehensive analysis of the functional microRNA-mRNA regulatory network identifies miRNA signatures associated with glioma malignant progression. Nucleic Acids Res 41(22):e203

Lin J, Teo S, Lam DH, Jeyaseelan K, Wang S (2012) MicroRNA-10b pleiotropically regulates invasion, angiogenicity and apoptosis of tumor cells resembling mesenchymal subtype of glioblastoma multiforme. Cell Death Dis 3:e398

Lin YX, Yu F, Gao N, Sheng JP, Qiu JZ, Hu BC (2011) microRNA-143 protects cells from DNA damage-induced killing by downregulating FHIT expression. Cancer Biother Radiopharm 26:365–372

Ling N, Gu J, Lei Z, Li M, Zhao J, Zhang HT et al (2013) microRNA-155 regulates cell proliferation and invasion by targeting FOXO3a in glioma. Oncol Rep 30:2111–2118

Liu L, Chen L, Xu Y, Li R, Du X (2010) microRNA-195 promotes apoptosis and suppresses tumorigenicity of human colorectal cancer cells. Biochem Biophys Res Commun 400:236–240

Lu S, Wang S, Geng S, Ma S, Liang Z, Jiao B (2012) Increased expression of microRNA-17 predicts poor prognosis in human glioma. J Biomed Biotechnol 2012:970761

Lu S, Wang S, Geng S, Ma S, Liang Z, Jiao B (2013) Upregulation of microRNA-224 confers a poor prognosis in glioma patients. Clin Transl Oncol 15:569–574

Luan S, Sun L, Huang F (2010) MicroRNA-34a: a novel tumor suppressor in p53-mutant glioma cell line U251. Arch Med Res 41:67–74

Malzkorn B, Wolter M, Liesenberg F, Grzendowski M, Stuhler K, Meyer HE et al (2010) Identification and functional characterization of microRNAs involved in the malignant progression of gliomas. Brain Pathol 20:539–550

Martin V, Sanchez-Sanchez AM, Herrera F, Gomez-Manzano C, Fueyo J, Alvarez-Vega MA et al (2013) Melatonin-induced methylation of the ABCG2/BCRP promoter as a novel mechanism to overcome multidrug resistance in brain tumour stem cells. Br J Cancer 108:2005–2012

Meng W, Jiang L, Lu L, Hu H, Yu H, Ding D et al (2012) Anti-miR-155 oligonucleotide enhances chemosensitivity of U251 cell to taxol by inducing apoptosis. Cell Biol Int 36:653–659

Mizoguchi M, Guan Y, Yoshimoto K, Hata N, Amano T, Nakamizo A et al (2012) MicroRNAs in human malignant gliomas. J Oncol 2012:732874

Mueller AC, Sun D, Dutta A (2013) The miR-99 family regulates the DNA damage response through its target SNF2H. Oncogene 32(9):1164–1172

Nakada M, Kita D, Teng L, Pyko IV, Watanabe T, Hayashi Y et al (2013) Receptor tyrosine kinases: principles and functions in glioma invasion. Adv Exp Med Biol 986:143–170

Nakano I, Kornblum HI (2006) Brain tumor stem cells. Pediatr Res 59:54R–58R

Nan Y, Han L, Zhang A, Wang G, Jia Z, Yang Y et al (2010) MiRNA-451 plays a role as tumor suppressor in human glioma cells. Brain Res 1359:14–21

Ng WL, Yan D, Zhang X, Mo YY, Wang Y (2010) Over-expression of miR-100 is responsible for the low-expression of ATM in the human glioma cell line: M059J. DNA Repair (Amst) 9:1170–1175

Nikaki A, Piperi C, Papavassiliou AG (2012) Role of microRNAs in gliomagenesis: targeting miRNAs in glioblastoma multiforme therapy. Expert Opin Investig Drugs 21:1475–1488

Niyazi M, Zehentmayr F, Niemoller OM, Eigenbrod S, Kretzschmar H, Schulze-Osthoff K et al (2011) MiRNA expression patterns predict survival in glioblastoma. Radiat Oncol 6:153

Ohno M, Natsume A, Kondo Y, Iwamizu H, Motomura K, Toda H et al (2009) The modulation of microRNAs by type I IFN through the activation of signal transducers and activators of transcription 3 in human glioma. Mol Cancer Res 7:2022–2030

Pan SJ, Zhan SK, Pei BG, Sun QF, Bian LG, Sun BM (2012) MicroRNA-149 inhibits proliferation and invasion of glioma cells via blockade of AKT1 signaling. Int J Immunopathol Pharmacol 25:871–881

Papagiannakopoulos T, Friedmann-Morvinski D, Neveu P, Dugas JC, Gill RM, Huillard E et al (2012) Pro-neural miR-128 is a glioma tumor suppressor that targets mitogenic kinases. Oncogene 31:1884–1895

Papagiannakopoulos T, Shapiro A, Kosik KS (2008) MicroRNA-21 targets a network of key tumor-suppressive pathways in glioblastoma cells. Cancer Res 68:8164–8172

Platanias LC (2005) Mechanisms of type-I- and type-II-interferon-mediated signalling. Nat Rev Immunol 5:375–386

Qian X, Zhao P, Li W, Shi ZM, Wang L, Xu Q et al (2013) MicroRNA-26a promotes tumor growth and angiogenesis in glioma by directly targeting prohibitin. CNS Neurosci Ther 19:804–812

Qiu S, Huang D, Yin D, Li F, Li X, Kung HF et al (2013) Suppression of tumorigenicity by microRNA-138 through inhibition of EZH2-CDK4/6-pRb-E2F1 signal loop in glioblastoma multiforme. Biochim Biophys Acta 1832:1697–1707

Qu S, Yao Y, Shang C, Xue Y, Ma J, Li Z et al (2012) MicroRNA-330 is an oncogenic factor in glioblastoma cells by regulating SH3GL2 gene. PLoS One 7:e46010

Rao SA, Santosh V, Somasundaram K (2010) Genome-wide expression profiling identifies deregulated miRNAs in malignant astrocytoma. Mod Pathol 23:1404–1417

Ren Y, Kang CS, Yuan XB, Zhou X, Xu P, Han L et al (2010a) Co-delivery of as-miR-21 and 5-FU by poly(amidoamine) dendrimer attenuates human glioma cell growth in vitro. J Biomater Sci Polym Ed 21:303–314

Ren Y, Zhou X, Mei M, Yuan XB, Han L, Wang GX et al (2010b) MicroRNA-21 inhibitor sensitizes human glioblastoma cells U251 (PTEN-mutant) and LN229 (PTEN-wild type) to taxol. BMC Cancer 10:27

Sana J, Hajduch M, Michalek J, Vyzula R, Slaby O (2011) MicroRNAs and glioblastoma: roles in core signalling pathways and potential clinical implications. J Cell Mol Med 15:1636–1644

Sasaki A, Udaka Y, Tsunoda Y, Yamamoto G, Tsuji M, Oyamada H et al (2012) Analysis of p53 and miRNA expression after irradiation of glioblastoma cell lines. Anticancer Res 32:4709–4713

Sasayama T, Nishihara M, Kondoh T, Hosoda K, Kohmura E (2009) MicroRNA-10b is overexpressed in malignant glioma and associated with tumor invasive factors, uPAR and RhoC. Int J Cancer 125:1407–1413

Sharma S, Salehi F, Scheithauer BW, Rotondo F, Syro LV, Kovacs K (2009) Role of MGMT in tumor development, progression, diagnosis, treatment and prognosis. Anticancer Res 29:3759–3768

Shi L, Chen J, Yang J, Pan T, Zhang S, Wang Z (2010) MiR-21 protected human glioblastoma U87MG cells from chemotherapeutic drug temozolomide induced apoptosis by decreasing Bax/Bcl-2 ratio and caspase-3 activity. Brain Res 1352:255–264

Shi L, Zhang S, Feng K, Wu F, Wan Y, Wang Z, Zhang J, Wang Y, Yan W, Fu Z, You Y (2012) MicroRNA-125b-2 confers human glioblastoma stem cells resistance to temozolomide through the mitochondrial pathway of apoptosis. Int J Oncol 40:119–129

Shi L, Wan Y, Sun G, Zhang S, Wang Z, Zeng Y (2013) miR-125b inhibitor may enhance the invasion-prevention activity of temozolomide in glioblastoma stem cells by targeting PIAS3. BioDrugs 28(1):41–54

Schraivogel D, Weinmann L, Beier D, Tabatabai G, Eichner A, Zhu JY et al (2011) CAMTA1 is a novel tumour suppressor regulated by miR-9/9* in glioblastoma stem cells. EMBO J 30:4309–4322

Schramedei K, Morbt N, Pfeifer G, Lauter J, Rosolowski M, Tomm JM et al (2011) MicroRNA-21 targets tumor suppressor genes ANP32A and SMARCA4. Oncogene 30:2975–2985

Schwarz DS, Hutvagner G, Du T, Xu Z, Aronin N, Zamore PD (2003) Asymmetry in the assembly of the RNAi enzyme complex. Cell 115:199–208

Siddique HR, Saleem M (2012) Role of BMI1, a stem cell factor, in cancer recurrence and chemoresistance: preclinical and clinical evidences. Stem Cells 30:372–378

Silber J, Lim DA, Petritsch C, Persson AI, Maunakea AK, Yu M et al (2008) miR-124 and miR-137 inhibit proliferation of glioblastoma multiforme cells and induce differentiation of brain tumor stem cells. BMC Med 6:14

Skalsky RL, Cullen BR (2011) Reduced expression of brain-enriched microRNAs in glioblastomas permits targeted regulation of a cell death gene. PLoS One 6:e24248

Slaby O, Lakomy R, Fadrus P, Hrstka R, Kren L, Lzicarova E et al (2010) MicroRNA-181 family predicts response to concomitant chemoradiotherapy with temozolomide in glioblastoma patients. Neoplasma 57:264–269

Song L, Huang Q, Chen K, Liu L, Lin C, Dai T et al (2010) miR-218 inhibits the invasive ability of glioma cells by direct downregulation of IKK-beta. Biochem Biophys Res Commun 402:135–140

Soon PS, Tacon LJ, Gill AJ, Bambach CP, Sywak MS, Campbell PR et al (2009) miR-195 and miR-483-5p Identified as Predictors of Poor Prognosis in Adrenocortical Cancer. Clin Cancer Res 15:7684–7692

Squatrito M, Brennan CW, Helmy K, Huse JT, Petrini JH, Holland EC (2010) Loss of ATM/Chk2/p53 pathway components accelerates tumor development and contributes to radiation resistance in gliomas. Cancer Cell 18:619–629

Srinivasan S, Patric IR, Somasundaram K (2011) A ten-microRNA expression signature predicts survival in glioblastoma. PLoS One 6:e17438

Sternlicht MD, Lochter A, Sympson CJ, Huey B, Rougier JP, Gray JW et al (1999) The stromal proteinase MMP3/stromelysin-1 promotes mammary carcinogenesis. Cell 98:137–146

Stockhausen MT, Kristoffersen K, Poulsen HS (2010) The functional role of Notch signaling in human gliomas. Neuro Oncol 12:199–211

Sun B, Pu B, Chu D, Chu X, Li W, Wei D (2013) MicroRNA-650 expression in glioma is associated with prognosis of patients. J Neurooncol 115:375–380

Ujifuku K, Mitsutake N, Takakura S, Matsuse M, Saenko V, Suzuki K et al (2010) miR-195, miR-455-3p and miR-10a(*) are implicated in acquired temozolomide resistance in glioblastoma multiforme cells. Cancer Lett 296:241–248

Ulasov IV, Nandi S, Dey M, Sonabend AM, Lesniak MS (2011) Inhibition of Sonic hedgehog and Notch pathways enhances sensitivity of CD133(+) glioma stem cells to temozolomide therapy. Mol Med 17:103–112

Wang J, Li Y, Wang X, Jiang C (2012a) Ursolic acid inhibits proliferation and induces apoptosis in human glioblastoma cell lines U251 by suppressing TGF-beta1/miR-21/PDCD4 pathway. Basic Clin Pharmacol Toxicol 111:106–112

Wang J, Wakeman TP, Lathia JD, Hjelmeland AB, Wang XF, White RR et al (2010) Notch promotes radioresistance of glioma stem cells. Stem Cells 28:17–28

Wang K, Wang X, Zou J, Zhang A, Wan Y, Pu P et al (2013a) miR-92b controls glioma proliferation and invasion through regulating Wnt/beta-catenin signaling via Nemo-like kinase. Neuro Oncol 15:578–588

Wang L, Shi M, Hou S, Ding B, Liu L, Ji X et al (2012b) MiR-483-5p suppresses the proliferation of glioma cells via directly targeting ERK1. FEBS Lett 586:1312–1317

Wang S, Jiao B, Geng S, Ma S, Liang Z, Lu S (2014) Combined aberrant expression of microRNA-214 and UBC9 is an independent unfavorable prognostic factor for patients with gliomas. Med Oncol 31:767

Wang S, Lu S, Geng S, Ma S, Liang Z, Jiao B (2013b) Expression and clinical significance of microRNA-326 in human glioma miR-326 expression in glioma. Med Oncol 30:373

Wang X, Wang J, Ma H, Zhang J, Zhou X (2012c) Downregulation of miR-195 correlates with lymph node metastasis and poor prognosis in colorectal cancer. Med Oncol 29:919–927

Wang XR, Luo H, Li HL, Cao L, Wang XF, Yan W et al (2013c) Overexpressed let-7a inhibits glioma cell malignancy by directly targeting K-ras, independently of PTEN. Neuro Oncol 15:1491–1501

Webster RJ, Giles KM, Price KJ, Zhang PM, Mattick JS, Leedman PJ (2009) Regulation of epidermal growth factor receptor signaling in human cancer cells by microRNA-7. J Biol Chem 284:5731–5741

Wu N, Xiao L, Zhao X, Zhao J, Wang J, Wang F et al (2012) miR-125b regulates the proliferation of glioblastoma stem cells by targeting E2F2. FEBS Lett 586:3831–3839

Wu Z, Wang L, Li G, Liu H, Fan F, Li Z et al (2013) Increased expression of microRNA-9 predicts an unfavorable prognosis in human glioma. Mol Cell Biochem 384(1–2):263–268

Wurdinger T, Tannous BA, Saydam O, Skog J, Grau S, Soutschek J et al (2008) miR-296 regulates growth factor receptor overexpression in angiogenic endothelial cells. Cancer Cell 14:382–393

Xia H, Qi Y, Ng SS, Chen X, Chen S, Fang M et al (2009) MicroRNA-15b regulates cell cycle progression by targeting cyclins in glioma cells. Biochem Biophys Res Commun 380:205–210

Xu T, Zhu Y, Xiong Y, Ge YY, Yun JP, Zhuang SM (2009) MicroRNA-195 suppresses tumorigenicity and regulates G1/S transition of human hepatocellular carcinoma cells. Hepatology 50:113–121

Yamada R, Nakano I (2012) Glioma stem cells: their role in chemoresistance. World Neurosurg 77:237–240

Yan D, Ng WL, Zhang X, Wang P, Zhang Z, Mo YY et al (2010) Targeting DNA-PKcs and ATM with miR-101 sensitizes tumors to radiation. PLoS One 5:e11397

Ying Z, Li Y, Wu J, Zhu X, Yang Y, Tian H et al (2013) Loss of miR-204 expression enhances glioma migration and stem cell-like phenotype. Cancer Res 73:990–999

Yue X, Wang P, Xu J, Zhu Y, Sun G, Pang Q et al (2012) MicroRNA-205 functions as a tumor suppressor in human glioblastoma cells by targeting VEGF-A. Oncol Rep 27:1200–1206

Zbinden M, Duquet A, Lorente-Trigos A, Ngwabyt SN, Borges I, Ruiz i Altaba A (2010) NANOG regulates glioma stem cells and is essential in vivo acting in a cross-functional network with GLI1 and p53. EMBO J 29:2659–2674

Zhang C, Han L, Zhang A, Yang W, Zhou X, Pu P et al (2010a) Global changes of mRNA expression reveals an increased activity of the interferon-induced signal transducer and activator of transcription (STAT) pathway by repression of miR-221/222 in glioblastoma U251 cells. Int J Oncol 36:1503–1512

Zhang J, Han L, Ge Y, Zhou X, Zhang A, Zhang C et al (2010b) miR-221/222 promote malignant progression of glioma through activation of the Akt pathway. Int J Oncol 36:913–920

Zhang QQ, Xu H, Huang MB, Ma LM, Huang QJ, Yao Q et al (2012a) MicroRNA-195 plays a tumor-suppressor role in human glioblastoma cells by targeting signaling pathways involved in cellular proliferation and invasion. Neuro Oncol 14:278–287

Zhang S, Wan Y, Pan T, Gu X, Qian C, Sun G et al (2012b) MicroRNA-21 inhibitor sensitizes human glioblastoma U251 stem cells to chemotherapeutic drug temozolomide. J Mol Neurosci 47:346–356

Zhang W, Zhang J, Hoadley K, Kushwaha D, Ramakrishnan V, Li S et al (2012c) miR-181d: a predictive glioblastoma biomarker that downregulates MGMT expression. Neuro Oncol 14:712–719

Zhang Y, Chao T, Li R, Liu W, Chen Y, Yan X et al (2009) MicroRNA-128 inhibits glioma cells proliferation by targeting transcription factor E2F3a. J Mol Med (Berl) 87:43–51

Zhao S, Deng Y, Liu Y, Chen X, Yang G, Mu Y et al (2013a) MicroRNA-153 is tumor suppressive in glioblastoma stem cells. Mol Biol Rep 40:2789–2798

Zhao WH, Wu SQ, Zhang YD (2013b) Downregulation of miR-124 promotes the growth and invasiveness of glioblastoma cells involving upregulation of PPP1R13L. Int J Mol Med 32:101–107

Zheng X, Chopp M, Lu Y, Buller B, Jiang F (2013) MiR-15b and miR-152 reduce glioma cell invasion and angiogenesis via NRP-2 and MMP-3. Cancer Lett 329:146–154

Zhou J, Xu T, Yan Y, Qin R, Wang H, Zhang X et al (2013) MicroRNA-326 functions as a tumor suppressor in glioma by targeting the Nin one binding protein (NOB1). PLoS One 8:e68469

Zhou X, Ren Y, Moore L, Mei M, You Y, Xu P et al (2010a) Downregulation of miR-21 inhibits EGFR pathway and suppresses the growth of human glioblastoma cells independent of PTEN status. Lab Invest 90:144–155

Zhou X, Zhang J, Jia Q, Ren Y, Wang Y, Shi L et al (2010b) Reduction of miR-21 induces glioma cell apoptosis via activating caspase 9 and 3. Oncol Rep 24:195–201

Autophagy in Glioma Cells

5

Iwona Anna Ciechomska, Aleksandra Ellert-Miklaszewska, and Bozena Kaminska

Abstract

Autophagy is an intracellular lysosomal degradation process controlled by a number of highly conserved autophagy-related genes (*ATG*) and tightly regulated by multiple mechanisms. This process could be induced by various cellular stress-mediated signaling pathways involved in nutrient signaling, growth factor status, energy sensing, hypoxia, oxidative and ER stress, or pathogen infection. The input from these multiple upstream signal transduction pathways is integrated by the serine/threonine protein kinase TOR (Target of Rapamycin). Although autophagy is thought to be predominantly a cell-survival mechanism, some evidence points towards a role in cell death. In tumor cells, the role of autophagy may depend on the type of tumor, the stage of tumorigenesis, and the nature and extent of the insult. Recently, autophagy has received increasing scientific attention as a promising therapeutic target. Malignant gliomas are common primary tumors of the central nervous system, characterized by aggressive cell proliferation, diffuse infiltration, and resistance to cell death. New therapies against this devastating and invariably fatal disease are needed. This chapter aims at summarizing the recent data in the field of paradoxical role of the autophagy in cancer, in particularly in glioma cells.

Keywords

Cell death • Cell survival • Signaling • Lysosomes • Autophagy • Glioma • Glioblastoma • Therapy

I.A. Ciechomska (✉) • A. Ellert-Miklaszewska • B. Kaminska
Laboratory of Molecular Neurobiology, Neurobiology Center, Nencki Institute of Experimental Biology, 3 Pasteur Str., 02-093 Warsaw, Poland
e-mail: jcjech@nencki.gov.pl

A. Sedo and R. Mentlein (eds.), *Glioma Cell Biology*,
DOI 10.1007/978-3-7091-1431-5_5, © Springer-Verlag Wien 2014

Abbreviations

AMBRA1	Autophagy/beclin-1 regulator 1
ATG	AuTophaGy genes
AV	Autophagic vesicles
Bcl-2	B-cell lymphoma 2
BH3	Bcl-2 homology domain 3
Bif-1	Bax-interacting factor-1
CMA	Chaperone-mediated autophagy
CQ	Chloroquine
CsA	Cyclosporine A
DRAM1	Damage-regulated modulator of autophagy
EGFR	Epidermal growth factor receptor
ERK	Extracellular signal-regulated kinase
GBM	*Glioblastoma multiforme*
GSC	Glioma stem cells
HCQ	Hydroxychloroquine
Hsc70	Heat shock cognate 70
LAMP-2A	Lysosome-associated membrane protein type 2A
mTOR	Mammalian target of rapamycin
PDGF	Platelet-derived growth factor
PDGFR	PDGF receptor
PTEN	Phosphatase and tensin homolog deleted from chromosome 10
ROS	Reactive oxygen species
SGLT1	Sodium/glucose co-transporter 1
TMZ	Temozolomide
TP53	Tumor protein p53
TRB3	Pseudo-kinase tribbles homologue 3
UVRAG	UV irradiation resistance-associated tumor suppressor gene
VEGF	Vascular endothelial cell growth factor
Δ9-THC	(-)-*trans*-Δ9-tetrahydrocannabinol

5.1 Introduction

Malignant gliomas, the most common primary brain tumors, account for more than 40 % of all central nervous system (CNS) neoplasms and are highly resistant to available therapeutic approaches, including neurosurgical resection, radiation, and chemotherapy (Hosli et al. 1998; Galanis and Buckner 2000; Kleihues et al. 2002). Despite current efforts to develop more effective clinical treatments against gliomas, there has been only a minimal improvement in the median survival time for patients with glioblastoma (GBM, *glioblastoma multiforme*), the most aggressive glioma, from an average of 10–14 months after diagnosis in the last 5 years

(Van Meir et al. 2010). Multiple molecular mechanisms involved in the drug resistance of human glioblastoma cells have been described, including frequent genomic alterations (*EGFR*, *PTEN*, and *TP53*), activation of signaling cascades that prevents apoptosis, and defects in apoptotic machinery (Iwadate et al. 1996; Haas-Kogan et al. 1998; Roth and Weller 1999; Mohri et al. 2000; Benjamin et al. 2003; Kaufmann and Vaux 2003). Poor patient prognosis, the lack of effective therapeutics, and the inherent resistance of cancer cells to pro-apoptotic drugs accentuate the need to develop new therapies against this devastating and invariably fatal disease. Recently autophagy, an intracellular lysosomal degradation process, has received increasing scientific attention as a promising therapeutic concept.

The term "autophagy" comes from the Greek words "phagy" meaning eat and "auto" meaning self. This word was invented in 1963 by Christian de Duve, the discoverer of lysosomes, who initiated the first experiments that provided clear biochemical proof of the involvement of lysosomes in this process (Klionsky 2008). Lysosomes contain the acidic hydrolases responsible for degradation of materials. Based on the way how a cargo is delivered to the lysosome or vacuole, autophagy is divided into three main types: chaperone-mediated autophagy (CMA), microautophagy, and macroautophagy (Klionsky 2005) (Fig. 5.1).

CMA is selective for specific cytosolic proteins that contain a pentapeptide motif (KFERQ). This motif is recognized by the chaperone heat shock cognate 70 (Hsc70), which transfers protein substrates to the lysosomal membrane, where they bind to the lysosome-associated membrane protein type 2A (LAMP-2A). LAMP-2A organizes a multimeric complex that mediates translocation of the substrate protein across the membrane into the lysosomal lumen for degradation. CMA is a process that has been characterized in higher eukaryotes but not in yeast. Microautophagy refers to the sequestration of cytosolic components directly by lysosomes through invaginations in their limiting membrane. The function of this process in higher eukaryotes is not known, whereas microautophagy-like processes in fungi are involved in selective organelle degradation. The focus of this article, however, is mammalian macroautophagy.

Macroautophagy (referred to here as autophagy) is an intracellular process, in which cytoplasmic components, including long-lived or malfunctioning proteins and obsolete or damaged organelles, are sequestered within double-membraned vesicles called autophagosomes. The autophagosome is formed by expansion of the phagophore (a cup-shaped isolation membrane, autophagosome precursor). Subsequently the complete autophagosome matures by fusing with an endosome and/or a lysosome to form a single-membraned autolysosome, in which the contents are digested by hydrolases and recycled for biosynthesis in the cell (Levine and Klionsky 2004; Xie and Klionsky 2007). Sequestration of cytosolic components can be either nonspecific (the engulfment of bulk cytoplasm) or selective, targeting specific cargoes, such as organelles (mitochondria, endoplasmic reticulum) or invasive microbes. Autophagy is important for a wide range of physiological processes. Autophagy occurs at a low basal level in most cells to perform homeostatic functions (cytoplasmic and organelle turnover) but is also a critical mechanism for the adaptation of cells to stress, and it is induced by numerous stimuli,

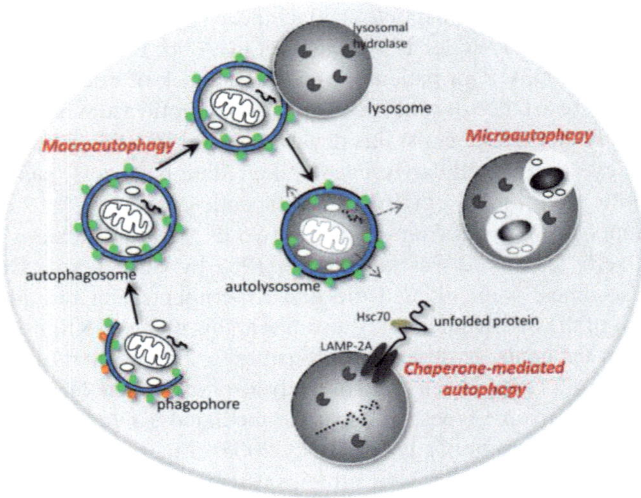

Fig. 5.1 Different types of autophagy. Microautophagy refers to the sequestration of cytosolic components directly by lysosomes through invaginations in their limiting membrane. Macroautophagy is an intracellular process, in which cytoplasmic components are sequestered within double-membraned vesicles called autophagosomes. The autophagosome is formed by expansion of the phagophore (a cup-shaped isolation membrane). Theses autophagosomes ultimately fuse with lysosomes to form autolysosomes, where contents are degraded by lysosomal hydrolases. Chaperone-mediated autophagy involves direct translocation of unfolded substrate proteins across the lysosome membrane through the action of a cytosolic and lysosomal chaperone Hsc70 and the integral membrane receptor LAMP-2A (lysosome-associated membrane protein type 2A)

including nutrient deprivation, hypoxia, hormone stimulation, and DNA damage (Klionsky and Emr 2000; He and Klionsky 2009).

Although autophagy is thought to be predominantly a cell-survival mechanism, some evidence points towards a role in cell death. In some context autophagy as distinct from type I programmed cell death (apoptosis) is called as a type II programmed cell death or autophagic cell death. We need to be aware that this term simply describes rather cell death with autophagy than cell death by autophagy (Kroemer and Levine 2008). Perhaps the most fundamental point is that either too little or too much autophagy can be deleterious, that is exemplified by its dual role in cytoprotection and cell death.

Autophagy has been implicated in several human diseases, including inflammatory diseases, neurodegeneration, aging, and cancer (Kondo and Kondo 2006; Levine and Kroemer 2008; Mizushima et al. 2008). For tumor cells, autophagy is a double-edged sword. On the one hand autophagy can be tumor suppressive through the elimination of oncogenic protein substrates, toxic unfolded proteins, and damaged organelles. But on the other hand it can be tumor promoting in established cancers through autophagy-mediated intracellular recycling that provides substrates for metabolism and that maintains the functional pool of mitochondria (White 2012). This chapter will summarize our current knowledge about the basic mechanisms of autophagy and role of autophagy in tumorigenesis, which is tissue- and genetic context-dependent.

5.2 Molecular Machinery of Autophagy

Autophagy is a tightly regulated process controlled by a number of highly conserved autophagy-related genes (*ATG*, for AuTophaGy genes). *ATGs* have been originally identified in yeasts, but many of their homologues have been discovered and characterized in mammalian cells (Mizushima et al 1998; Suzuki and Ohsumi 2007; Xie and Klionsky 2007). The regulation of autophagy by Atg proteins is associated with the autophagosome formation. The Atg proteins control autophagosome initiation, nucleation, elongation, and/or lysosome fusion step (Fig. 5.2).

Autophagy initiation requires the formation of the mammalian ortholog of yeast Atg1/unc-51-like kinase (ULK) complex consisting of ULK1 (and/or possible ULK2), mAtg13, and scaffold protein FIP200 (an ortholog of yeast Atg17). This complex integrates stress signals from the serine-threonine kinase mTOR (mammalian target of rapamycin) (Jung et al. 2009; Mizushima 2010). In mammalian cells, there are two known mTOR complexes, mTORC1 and mTORC2. mTORC1 is the predominant form associated with autophagy. Under nutrient-rich conditions, mTORC1 is associated with ULK1/2 complex and phosphorylates ULK1/2 and mAtg13. Upon inactivation of mTORC1 by nutrient starvation, mTORC1 quickly dissociates that results in the partial dephosphorylation of mAtg13 and ULK1/2; this activates ULK1/2 to phosphorylate FIP200 and to induce autophagy (Hosokawa et al. 2009; Jung et al. 2009). mTOR activity could be inhibited by

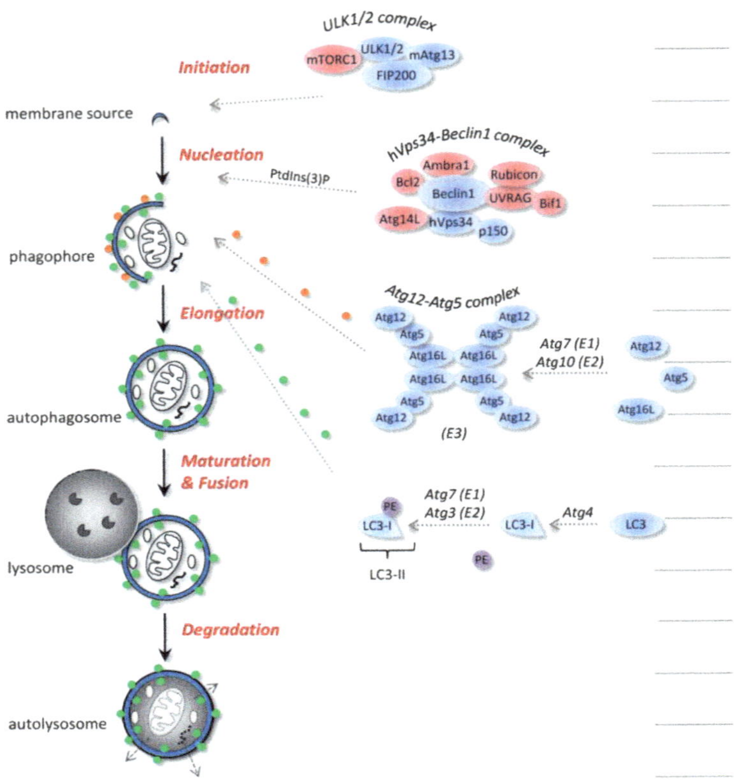

Fig. 5.2 Autophagy cascade. Autophagy is negatively regulated by mTORC1 via a multiprotein complex, including ULK1/2, Atg13, and FIP200. Autophagy is initiated by the formation of the phagophore (at the nucleation step). The nucleation phase is controlled by Beclin1/hVps34/p150-containing complexes (core complex). The Atg14L (Atg14L/Beclin1/hVps34/p150) and UVRAG (UVRAG/Beclin1/hVps34/p150) complexes function positively in autophagosome formation. The Rubicon complex (Rubicon/UVRAG/Beclin1/hVps34/p150) negatively regulates autophagy. Ambra1 and Bif-1 are essential for the induction of autophagy, whereas Bcl-2 inhibits autophagy. During elongation and expansion of the phagophore membrane, Atg12–Atg5 conjugates interact noncovalently with Atg16L, and this large multimeric complex act as an E3-like enzyme, determining the sites of Atg8/LC3 lipidation. LC3-I is conjugated to phosphatidylethanolamine in a reaction requiring Atg7 and Atg3. The lipidated form of LC3-I (LC3-II) is attached to both faces of the phagophore membrane. At the end of the elongation step, a portion of the cytosol is sequestered into a double-membrane vesicle, termed the autophagosome. A complete double-

drug such as rapamycin that results in autophagy in normal and tumor cells, including malignant glioma cells (Takeuchi et al. 2005; Iwamaru et al. 2007).

Autophagosome formation starts at a phagophore. The source of isolating membrane is unclear, although there is evidence that it may be derived from the endoplasmic reticulum (ER), *trans*-Golgi network, as well as mitochondria and plasma membrane (Axe et al. 2008; Nishida et al. 2009; Hailey et al. 2010; Ravikumar et al. 2010). One possibility is that the origin of the phagophore membrane depends upon the nature of the stress that induces autophagy. Nucleation and elongation of the pre-autophagosomal membrane are controlled by class-III phosphoinositide 3-kinase (PI3KC3)/hVps34, which is essential for generating phosphatidylinositol (3)-phosphate (PtdIns(3)P). hVps34 is a part of a large macromolecular complex that contains the proteins Beclin1 (a homologue of Atg6) and p150 (a homologue of Vps15) (Itakura et al. 2008). In mammals, there are at least three types of class-III PI3K complexes that are involved in autophagosome formation. For example, hVps34/Beclin1 complexed with Atg14L or UVRAG (UV irradiation resistance-associated tumor suppressor gene) can activate autophagy, whereas hVps34/Beclin1 complexed with Rubucon (RUN domain and cysteine-rich domain containing, Beclin1-interacting protein) protein negatively regulates autophagosome maturation and inhibits autophagy (Liang et al. 2006; Itakura et al. 2008; Matsunaga et al. 2009). Ambra1 and Bif-1 (Bax-interacting factor 1) are essential for the induction of autophagy through direct interaction with Beclin1 and UVRAG, respectively (Fimia et al. 2007; Takahashi et al. 2007) whereas Bcl-2 binds to Beclin1 and disrupts the Beclin1-associated hVps34 complex, thereby inhibiting autophagy (Pattingre et al. 2005). Chemical inhibition of hVps34 by 3-methyladenine or wortmanin blocks the formation of new autophagosomes.

The elongation and expansion of phagophore membrane are mediated by two ubiquitin-like (Atg12–Atg5 and Atg8/LC3) conjugation systems (Yang and Klionsky 2010), which regulate each other. Atg12 is conjugated to Atg5 in a reaction that requires Atg7 and Atg10 (E1 and E2-like enzymes, respectively) (Mizushima et al. 2001). The Atg12–Atg5 conjugate then interacts non-covalently with Atg16L, which oligomerizes to form a large multimeric complex (Ohsumi and Mizushima 2004). Atg12–Atg5 conjugates are localized on phagophore and dissociate upon completion of autophagosome formation. The second ubiquitination-like reaction involves the conjugation of microtubule-associated protein 1 light chain 3 (MAP1-LC3, also known as Atg8 and LC3) to the phosphatidylethanolamine (PE) (Tanida et al. 2004a, b). Firstly, LC3 is cleaved at its C terminus by Atg4 protease to generate the cytosolic LC3-I with a C-terminal glycine residue. LC3-I then conjugates to PE by the formation of an amide bond between the amino group of PE and the C-terminal glycine of LC3-I. This reaction

Fig. 5.2 (continued) membrane autophagosome then fuses with the lysosome into the autolysosome, and the cargo-containing membrane compartment is lysed and degraded by acid hydrolases

requires the E1 protein Atg7 and the E2 protein Atg3. Recruitment of LC3 to the pre-autophagosomal membrane requires the Atg12–Atg5–Atg16 complex, which acts as an E3-like enzyme (Mizushima et al. 2001; Hanada et al. 2007). Conversely, the LC3 conjugation system seems to be required for Atg12–Atg5–Atg16 complex formation (Sou et al. 2008). Lipidation of LC3 converts the soluble LC3-I into the autophagosome-associated form LC3-II, which is attached to both the outer and inner membranes of autophagosomes. After the formation of the autolysosome, intra-autophagosomal components, including also intra-autophagosomal LC3-II, are rapidly degraded by lysosomal hydrolases (Kabeya et al. 2000). LC3-II levels correlate with the number of autophagic vacuoles, which can be assessed by scoring LC3-positive cytoplasmic vesicles (Kabeya et al. 2000).

Delivery and degradation of autophagosome represent another control point. Recruitment of a cargo from the cytoplasm requires autophagy receptors such as p62, NBR1 (neighbor of BRCA1 gene 1), or NIX (BH3-only member of the Bcl-2 family). These receptors interact with the LC3 that modifies target proteins for delivery to autophagosome (Pankiv et al. 2007; Schweers et al. 2007; Amaravadi et al. 2011). A retrieval pathway is required for the disassembly of Atg protein complex from maturated autophagosomes (mediated byAtg2, Atg9, and Atg18) (Tanaka et al. 2000; Jager et al. 2004; Yang and Klionsky 2010; Amaravadi et al. 2011). Finally, autophagosomes fuse with lysosomes and the contents of two vesicles are mixed (Jahreiss et al. 2008). Autophagosomes move along microtubules in a dynein-dependent manner to lysosomes. Drugs that disrupt microtubules or inhibit dynein function abolish autophagosome fusion with lysosomes that results in accumulation of autophagic vacuoles. The small GTPase Rab7, LAMP-2 and SNARE proteins participate in the final maturation of late autophagic vacuoles, as well as in recognition and fusion of the membranes (Tanaka et al. 2000; Atlashkin et al. 2003; Gutierrez et al. 2004; Jager et al. 2004). Lysosomes are acidic organelles, with their digesting hydrolases dependent on low pH. Consequently, agents such as bafilomycin A1 or chloroquine derivatives, which disrupt the vacuolar H^+ ATPase responsible for acidifying lysosomes, block autophagy in its final step, resulting in accumulation of autophagic vacuoles (Poole and Ohkuma 1981; Gonzalez-Polo et al. 2005; Marceau et al. 2009).

5.3 Dual Role of Autophagy in Cancer

Based on numerous scientific data, it is commonly accepted that autophagy plays a key role in cancer. There are, however, some controversies regarding the character of its function. Some studies have suggested a tumor suppressive role for autophagy, while other studies have shown the opposite effect, with autophagy supporting cancer cell proliferation. It has been suggested that bi-polarity between the anti-tumorigenic and pro-tumorigenic role of autophagy may be tumor type and stage dependent. Autophagy clearly suppresses the initiation and development of cancer through the elimination of oncogenic protein substrates, toxic unfolded proteins, and damaged organelles. On the other hand, it can be tumor promoting in established cancers, due to autophagy-mediated intracellular recycling that provides substrates for metabolism and maintains the functional pool of mitochondria (White 2012).

5.3.1 Tumor Suppression by Autophagy

The first evidence of tumor suppressive autophagy comes from studies that showed reduced autophagy in cancer tissues compared with their normal counterparts (Kisen et al. 1993). The anti-tumorigenic role of autophagy was later supported by identification of **Beclin1** as both an autophagy mediator and tumor suppressor (Liang et al. 1999). While Beclin1−/− mice died early during embryogenesis (E7.5–E8.5), Beclin1+/− mice with a partial autophagy defect developed spontaneous tumors with advancing age, indicating that Beclin1 is a haploinsufficient tumor suppressor gene (Qu et al. 2003; Yue et al. 2003). Moreover, excessive stimulation of autophagy due to Beclin1 overexpression can inhibit tumor development in vivo (Liang et al. 1999). Allelic loss of *BECN1* has been reported in various human malignancies, including breast, ovarian, and prostate cancer (Qu et al. 2003). Recent findings showed that Beclin1 mRNA and protein level were reduced in high-grade brain tumors (including gliomas, meningiomas, and medulloblastomas) as compared with low-grade tumors (Miracco et al. 2007; Huang et al. 2010). Particularly cytoplasmic expression of Beclin1 (which mirrors its autophagic functionality) was decreased in high-grade glial neoplasms. The cytoplasmic expression of Beclin1 in glioma tissues was positively associated with apoptosis but negatively with cell proliferation (Pirtoli et al. 2009). Additionally, high-grade gliomas have lower protein level of autophagy-specific marker, LC3B-II, than other astrocytic tumors (Huang et al. 2010). Importantly, these differences in gene expression and protein levels have been correlated with clinicopathological characteristics of tumors and disease outcomes, including overall survival in cancer patients. High cytoplasmic expression of Beclin1 was associated with better survival, high Karnofsky classification values, and the accomplishment of an optimal postoperative therapy (Pirtoli et al. 2009; Huang et al. 2010). Moreover, high level of LC3B expression was associated with an improved outcome for patients with poorer performance scores (Karnofsky Performance Scale), whereas for patients with normal performance, survival was better for patients with low staining than with high staining of LC3B (Aoki et al. 2008). The results support the relationship between Beclin1 or LC3-II and the clinicopathologic characteristics of tumors and show significant correlations between Beclin1 and LC3-II expression and the WHO histopathologic grade of astrocytic tumors. These findings support the hypothesis that autophagy might have a suppressive role in the progression of astrocytic tumors and that Beclin1 and LC3-II might represent a biomarker that, when present in sufficient quantities, identifies subsets of patients with a less aggressive form of the disease (Huang et al. 2010).

It is worth to notice that apart from extensively studied Beclin1 and LC3-II also other autophagy-related proteins emerged to have a tumor suppressor activity. Proteins UVRAG, Bif-1, and Ambra1 have recently been shown to interact with Beclin1 and act as activators of autophagy. **UVRAG**, associated with the Beclin1/hVps34 complex, promoted hVps34 enzymatic activity and autophagy, while suppressing the proliferation of human colon cancer cells (Liang et al. 2006). Frameshift mutations in *UVRAG* have been observed in colon and gastric carcinomas (Ionov et al. 2004; Kim et al. 2008b). Similar to UVRAG, **Bif-1** enhances the binding between Beclin1 and hVps34, resulting in increased autophagy. Bif-1 knockout mice develop spontaneous tumors at the significantly higher rate compared to wild-type mice (Takahashi et al. 2007). Expression of Bif-1 is markedly reduced in human malignancies (Lee et al. 2006; Coppola et al. 2008a, b) suggesting that Bif-1 is a tumor suppressor candidate. **Ambra1** is required for Beclin1/hVps34 complex stability and activity. Ambra1 is mainly expressed in the brain, where it plays an essential role during development of the nervous system (Fimia et al. 2007). Ambra1 functional deficiency in mouse embryos leads to severe neural tube defects associated with autophagy impairment, accumulation of ubiquitinated proteins, unbalanced cell proliferation, and excessive apoptotic cell death.

The ability of Beclin1, Ambra1, UVRAG, and Bif-1 to inhibit cell proliferation and to promote non-apoptotic cell death may also contribute to their tumor suppressive effects. However, these proteins are components of the Beclin1/hVps34 complex, which not only controls autophagy but also other processes such as endosome biogenesis and sorting (Lindmo and Stenmark 2006). Moreover, Beclin1 and Bif-1 interact with apoptotic regulators Bcl-2 and Bax, respectively (Liang et al. 1998; Cuddeback et al. 2001; Takahashi et al. 2005). These findings suggest that Beclin1, UVRAG, and Bif-1 might function as tumor suppressors by mechanisms that are independent of autophagy. Beclin1 is not the only *ATG* gene reported to be a tumor suppressor gene. **Atg4** is a cysteine protease that cleaves LC3-I before it binds to a phospholipid in the autophagosomal membrane. Marino et al. (Marino et al. 2007) found that *Atg4C* knockout mice showed reduced autophagy and increased susceptibility to the development of fibrosarcomas induced by chemical carcinogenesis. Additionally, a recent study identified high incidence of frameshift mutations in *ATG2B*, *ATG5*, *ATG9B*, and *ATG12* genes in human gastric and colorectal cancers with microsatellite instability (Kang et al. 2009). Interestingly, both the *LC3* (*ATG8*) and *ATG7* genes are located in a chromosomic region that is known to be frequently deleted in cancer (Jin 2006). All these data support the tumor suppressive function of autophagy. The inhibition of tumorigenesis by autophagy is a common effect, and not limited to one unique autophagic molecule. Further studies are still required to firmly establish the role of the core autophagy machinery in tumor suppression, especially in brain tumors.

Reinforcing the hypothesis of autophagy as a tumor suppressor mechanism, it is worthy to note that signaling proteins over-activated in cancer such as class I PI3K, Akt/PKB, and mTOR suppress autophagy, whereas products of tumor suppressor genes, such as PTEN, TP53, TSC1/2 and DAPK, stimulate autophagy (Gozuacik and Kimchi 2004; Feng et al. 2005). Molecular studies of human glioblastomas have identified important genetic events, such as (1) dysregulation of growth factor signaling via amplification and mutational activation of receptor tyrosine kinase genes; (2) activation of the phosphatidylinositol-3-OH kinase (PI(3)K) pathway; and (3) inactivation of the p53 and retinoblastoma tumor suppressor pathways (Cancer Genome Atlas Research 2008). This indirect evidence suggests that autophagy may be highly relevant to pathobiology of gliomas.

The most common genetic aberration associated with malignant gliomas are amplification or mutational activation of the epidermal growth factor receptor (**EGFR**) with a frequency of about 50 % (Furnari et al. 2007). In human cancer cells, the loss of expression of EGFR resulted in autophagy activation, which was triggered by a decreased intracellular glucose level following downregulation of sodium/glucose co-transporter 1 (SGLT1) (Weihua et al. 2008). Malignant gliomas highly co-express platelet-derived growth factor (PDGF) and its receptor (Hermanson et al. 1992; Guha et al. 1995), which indicates the presence of an autocrine loop. Therefore, disruption of PDGF ligand/receptor complex represents a promising strategy for the treatment of malignant gliomas. Takeuchi et al. (Takeuchi et al. 2004) demonstrated that the inhibition of PDGF signaling by a neutralizing anti-PDGF antibody induced autophagy but not apoptosis in malignant glioma cells. Besides, amplification of the gene encoding EGFR and constitutive activity of the PDGF autocrine loop lead to augmented stimulation of the downstream PI3K/Akt/mTOR cascade in tumors. The **PI3K/Akt/mTOR** signaling pathway is activated in more than 88 % of GBM (Venkatesh et al. 2012). **PTEN** (tumor-suppressor phosphatase and tensin homologue), an inhibitor of the PI3K/Akt pathway, is frequently mutated or inactivated in glioblastomas (Krex et al. 2007; Cancer Genome Atlas Research 2008; Parsons et al. 2008; Zheng et al. 2008). Activating mutations in *PIK3CA*, coding for the **PI3K** catalytic subunit p110α, described in *glioblastoma multiforme* (Cancer Genome Atlas Research 2008; Parsons et al. 2008), generally result in Akt signaling pathway activation and tumor progression. Activation of the Akt signaling pathway switches on survival pathways, which leads to the inhibition of apoptosis and activates **mTOR** protein, which antagonizes autophagy. Therefore, activation of Akt pathway and mTOR signaling and loss of PTEN tumor suppressor may contribute to malignant transformation by the simultaneous inhibition of apoptosis and autophagy. Interestingly, a constitutively active form of Akt suppressed the rapamycin- and curcumin-induced autophagy and cytotoxicity in malignant glioma cells (Takeuchi et al. 2005; Aoki et al. 2007). Recent studies have shown that Akt inhibition suppresses the growth of tumors established from PTEN-null human cancer cells (including glioma cells). Akt knockdown or inactivation did not induce significant apoptosis but rather markedly increased autophagy (Degtyarev et al. 2008). mTORC1 negatively regulates autophagy. The inhibition of mTOR activity, which is indicated by the dephosphorylation of its substrate p70S6K, was correlated with autophagy induction in glioma cells (Ito et al. 2006; Iwamaru et al. 2007).

The tumor suppressor gene ***TP53*** is mutated in over 50 % of all human cancers. The *TP53* dysfunction is the frequent event occurring in gliomas (Zupanska and Kaminska 2002). The p53 protein can modulate autophagy depending on its cellular localization. Nuclear p53 has been reported to increase autophagy through the transcriptional upregulation of DRAM1 (damage-regulated modulator of autophagy) and Sestrin2 (Crighton et al. 2006; Maiuri et al. 2009). Interestingly, cytoplasmic p53 inhibits autophagy in a transactivation-independent manner in several tumor cells and loss of p53 results in enhanced autophagy and survival in response to hypoxia and nutrient deprivation (Tasdemir et al. 2008). However, recent studies have shown that overexpression of WT p53 (mainly localized in the nucleus) and mutant p53 (MT, mainly localized in the cytosol) in the human glioblastoma cells induced autophagic cell death. Adding 3MA blocked the WT and MT p53-induced autophagy. They also demonstrated that the p53-induced autophagy pathway was independent from the p53-induced apoptosis pathway (Sakamoto et al. 2011). In conclusion, autophagy genes are frequently mono-allelically deleted, silenced, or mutated in human tumors. On the other hand protooncogene products suppress autophagy, whereas products of tumor suppressor genes stimulate autophagy. These data support a notion of tumor suppressor functions of autophagy. Although the detailed mechanism has not been fully elucidated, several hypotheses have been recently proposed.

5.3.2 Mechanisms of Tumor Suppression by Autophagy

Several mechanisms of tumor suppression mediated by autophagy have been lately described. Accumulation of tumor-promoting DNA alterations due to exposure to oxidative stress is an important event in oncogenesis. Some evidence suggests that autophagy **limits genome instability**. In response to stress, autophagy-defective cells exhibit increased numbers of chromosomal gains or losses, that may accelerate DNA mutations (Degenhardt et al. 2006; Karantza-Wadsworth et al. 2007). Contribution of autophagy to suppression of chromosomal instability may be explained by the involvement of this process in disposal of damaged mitochondria and peroxisomes. These organelles produce reactive oxygen species (ROS) as part of their functions in oxidative metabolism, and thereby damaged mitochondria and peroxisomes represent a potential source of DNA damaging compounds. White and colleagues have found that in response to stress, autophagy-defective tumor cells preferentially accumulated p62, damaged mitochondria, ROS, and protein aggregates, which might cause DNA damage. Moreover, suppressing ROS or p62 accumulation prevented damage resulting from autophagy defects indicating that failure to regulate p62 caused oxidative stress. Importantly, sustained p62 expression was sufficient to alter NF-kappaB regulation and gene expression and to promote tumorigenesis. Thus, defective autophagy is a mechanism for p62 up-regulation commonly observed in human tumors that contributes directly to tumorigenesis likely by perturbing signaling function of p62 (Mathew et al. 2007, 2009; Mathew and White 2011).

Upon oncogene activation, a highly inflammatory microenvironment is often created that favors tumor development because of elevated oxidative stress (Mantovani et al. 2008). Indeed, combined inhibition of apoptosis and autophagy promotes necrotic cell death, a process that was associated with inflammation (Degenhardt et al. 2006). This suggests that autophagy **prevents necrosis-mediated inflammation**. Chronic inflammation is known to contribute to liver cancer (Sakurai et al. 2008; Sun and Karin 2008) and activation of inflammation is also observed with allelic loss of *Becn1* in the liver (Mathew et al. 2009) and in tumor allografts with autophagy defects (Degenhardt et al. 2006). Generally, chronic tumor inflammation favors infiltration of pro-tumorigenic immune cells (e.g., macrophages) (DeNardo et al. 2008, 2009). Thus, tumor promotion conferred by the defects in autophagy may be linked to both: increased risk and incidence of mutagenesis and the formation of pro-tumorigenic inflammatory environment (White 2012).

Alternatively, autophagy may facilitate **oncogene-induced senescence** to limit tumorigenesis in some settings (Young et al. 2009). Senescence is believed to be a tumor suppressive mechanism attributed to the induction of permanent cell-cycle exit. By facilitating senescence, autophagy can limit propagation of oncogenic mutations, thereby suppressing tumorigenesis. The detailed mechanism of autophagy-mediated senescence is still unclear, and its significance to tumor suppression remains to be determined.

Other mechanisms for autophagy-mediated tumor suppression include autophagy-regulated immunosurveillance and autophagy-inhibited angiogenesis. Autophagy was found to play an important role in **modulating adaptive immunity**, which has implications in preventing tumorigenesis by increasing tumor antigen presentation. Autophagy has been shown to facilitate both MHC class I and class II antigen presentation (Dengjel et al. 2005; English et al. 2009). More importantly, a recent study showed that autophagy is essential for cross-presentation of melanoma-cell antigens by dendritic cells in vitro and in vivo, with autophagosome as the antigen carrier (Li et al. 2008). This is the first direct evidence supporting the potential of autophagy to induce immunosurveillance against tumor cells. It has also been proposed that autophagy can inhibit cancer development by **blocking angiogenesis**. Neuropilin 1, a positive regulator of vascular endothelial cell growth factor (VEGF) signaling, is degraded by autophagy under metabolic stress (Bae et al. 2008). Inhibition of angiogenesis and tumor growth was also observed in vitro and in vivo after the treatment with mTOR inhibitors, which are known to induce autophagy (Shinohara et al. 2005; Kim et al. 2008a). Collectively, these results support the contention that autophagy plays a tumor-suppressing role in cancer, and suggest that reduced autophagic activity may constitute a hallmark of cancer (Hanahan and Weinberg 2011).

5.3.3 Tumor Promotion by Autophagy

Although autophagy can suppress tumor development, it clearly plays a role in promoting the survival of established cancers within the tumor microenvironment. The tumor environment is characterized by reduced levels of nutrients and oxygen. This suggests that a cancer cell, in addition to a high level of glycolysis, also increases **autophagy for energy production**. Accordingly, autophagosomes are found in vivo in insufficiently vascularized, metabolically stressed regions of tumors (Degenhardt et al. 2006). These data strongly suggest that in agreement with findings from cell culture experiments, autophagy is an essential mechanism for providing nutrients in tumor tissue (Boya et al. 2005; Lum et al. 2005). The central regions of solid tumors, including gliomas, are frequently hypoxic as a result of the tumor outgrowing existing vasculature. In the hypoxia core of tumors, autophagy is induced as a survival mechanism by the tumor cells (Ray et al. 2000). The mechanisms by which hypoxia induces autophagy need clarification, but the finding that BNIP3 (Bcl-2/E1B 19 kDa-interacting protein 3), is HIF-1α (a hypoxia-inducible factor-1α) downstream target gene, brings one possible explanation (Tracy et al. 2007). The mechanism involving BNIP3 was observed also in glioma cells, although authors suggest that prolonged hypoxia induces autophagic cell death in apoptosis-competent cells (Azad et al. 2008). In contrary to others they suggest that hypoxia-induced autophagy promotes tumor cell survival and adaptation to anti-angiogenic treatment in glioblastoma through upregulated BNIP3 expression (Hu et al. 2012). Hypoxia-associated growth and increased BNIP3 expression could be prevented by addition of the autophagy inhibitor chloroquine (CQ). In vivo targeting of the essential autophagy gene ATG7 also disrupted tumor growth when combined with bevacizumab (VEGF-neutralizing antibody) treatment. In conclusion, autophagy inhibitors may help prevent resistance to anti-angiogenic therapy used in the clinic.

A number of recent studies showed that active **RAS** requires autophagy to maintain its oncogenic function that again supports a positive role of autophagy in tumorigenesis (Guo et al. 2011; Kim et al. 2011; Lock et al. 2011). Autophagy promoted cancer cell survival and proliferation through its role in aerobic glycolysis and the tricarboxylic acid cycle in oncogenic Ras-transformed cells. Importantly, autophagy was required to maintain the pool of functional mitochondria necessary to support growth of Ras-driven tumors. These data suggest that tumor-promoting role of autophagy is twofold: providing energy substrates during periods of nutrient limitation and **preserving organelle function required for cell growth** (Guo et al. 2011). Whether other tumors (including malignant gliomas) are addicted to autophagy remains to be investigated.

Furthermore, autophagy may promote tumor cell survival in the process of dissemination and metastasis, during which detachment of cells from extracellular matrix may cause cell death, namely **anoikis**. In this regard, autophagy has been reported to promote epithelial cell survival during anoikis, including detached cells harboring anti-apoptotic lesions (Fung et al. 2008). Additionally, autophagy seems to be important for glioma **cell invasion**. Knockdown of *Atg12*, and the consequent inhibition of the autophagic machinery, decreases the invasiveness of glioma cells in an organotypic model (Macintosh et al. 2012). These data indicate that autophagy may play a critical role in malignant transition that is also central to the initiation of metastasis. Additionally, genetic or pharmacological inhibition of autophagy was shown to enhance the cytotoxicity of cancer chemotherapy agents and to promote tumor regression. This indirect evidence, presented further, also suggest the autophagy promote tumor cell survival. Summarizing, the role of autophagy in cancer is unquestionable, although it can influence both positively and negatively on cancer cell survival; autophagy probably functions to prevent cancer initially, but once the tumor develops, the cancer cells utilize autophagy for their own cytoprotection.

5.4 Therapeutic Potential of Autophagy in Gliomas

Autophagy has become one of the most attractive targets for designing novel anticancer treatments. Most chemotherapeutics induce cellular damage that triggers autophagy, but the impact of autophagy on tumor cell death or survival is unclear. Autophagy protects cells from death by multiple mechanisms, including improved maintenance of bioenergetic homeostasis, recycling of misfolded and aggregate prone proteins, and removal of damaged organelles such as uncoupled mitochondria (Levine and Kroemer 2008; Mizushima et al. 2008). Thanks to the availability of genetic tools to disable the autophagic machinery, it has become clear over recent years that autophagy usually constitutes a futile attempt of dying cells to adapt to lethal stress rather than a mechanism to execute a cell death program. Therefore it is postulated that "autophagic cell death" is (mostly) a misnomer and describes a reality in which cell die *with* autophagy but not *by* autophagy (Kroemer and Levine 2008; Kroemer et al. 2009; Shen et al. 2012).

5.4.1 Autophagy Inducers and Inhibitors in Glioma Treatment

Initially, autophagosomes were identified in dying cells, especially in cells where apoptotic machinery was inhibited or in cells with apoptosis-defective background suggesting induction of cell death that depends on the autophagy genes (Shimizu et al. 2004; Yu et al. 2004). The tumor cells have inherent defects in apoptotic pathways and cannot undergo cell death characterized by classic signs of apoptosis. In such case, non-apoptotic cell death occurs via alternative death pathways, including necrosis and possibly autophagy (Berry and Baehrecke 2007). Due to many genetic alterations malignant gliomas are resistant to radiation or/and chemotherapy. Recently autophagy emerged as a promising therapeutic concept in glioma treatment (Lefranc et al. 2007) as autophagy rather than apoptosis accompanies cytotoxicity induced in glioma cells by conventional anti-cancer drugs, irradiation, and other compounds (Kanzawa et al. 2003, 2004; Daido et al. 2004; Takeuchi et al. 2005; Aoki et al. 2007; Salazar et al. 2009).

mTOR is a central coordinator of cell growth that is involved in both protein translation and autophagy. **Rapamycin** is a macrolide antibiotic that inhibits the mTOR complex by binding to FK-binding protein 12. Although initially approved as an immunosuppressant, rapamycin and its derivatives are currently being used in clinical trials for the treatment of various malignancies, including malignant gliomas. Inactivation of mTORC1 by rapamycin induced autophagy but not apoptosis and inhibited the proliferation of rapamycin-sensitive glioma cell lines (Takeuchi et al. 2005). The inhibition of autophagy by small interfering RNA (siRNA) directed against the autophagy-related gene *beclin1* attenuated the cytotoxicity of rapamycin in rapamycin-sensitive glioma cells, indicating that autophagy is required for antitumor effects rather than for a protective response (Iwamaru et al. 2007). Exogenous expression of an mTOR mutant interfering with its kinase activity markedly enhanced the incidence of rapamycin-induced autophagy. Moreover, silencing of mTOR with siRNA increased the sensitivity of malignant glioma cells to rapamycin by stimulating autophagy regardless of the PTEN status of the tumor cells (Iwamaru et al. 2007). Interestingly, a PI3K inhibitor, LY294002, and an Akt inhibitor, UCN-01 (7-hydroxystaurosporine), synergistically sensitized rapamycin-sensitive and rapamycin-resistant cells to rapamycin and thus stimulated the induction of autophagy (Takeuchi et al. 2005). These results indicate that rapamycin exerts its antitumor effect on malignant glioma cells by inducing autophagy and suggest that in malignant glioma cells silencing or inhibiting mTOR kinase activity or a disruption of the PI3K/Akt signaling pathway could greatly enhance the effectiveness of rapamycin. It could be explained by the fact that in mammalian cells, there are two known mTOR complexes: mTORC1 and mTORC2. mTORC2, which is less sensitive to rapamycin, phosphorylates the Ser473 of Akt/PKB, thereby contributing to the activation of this important cell-survival kinase (Sarbassov et al. 2006). In addition, preclinical studies with individual inhibitors of PI3K and mTOR have shown only modest efficacy in gliomas. Phase II studies with temsirolimus (CCI-779), an allosteric analog of rapamycin, did not provide any survival benefit in recurrent GBM patients (Chang et al. 2005; Galanis et al. 2005). However, a dual inhibitor of PI3K and mTOR, PI-103 had potent anti-proliferative activity in gliomas in preclinical models by eliminating activation of Akt often observed with mTOR inhibitors (Fan et al. 2010).

Gliomas contain subpopulation of cells called glioma stem cells (GSCs). GSCs are defined as undifferentiated cells responsible for tumor initiation and resistance to therapy. GSCs forced to differentiate lose their properties and gain sensitivity to chemotherapy (Singh et al. 2004; Bao et al. 2006). These cells may be a critical therapeutic target. Interestingly, recent studies showed that rapamycin induced differentiation of GSCs by activating autophagy (Zhao et al. 2010; Zhuang et al. 2011). Downregulation of stem/progenitor cell markers (CD133, nestin, and neurosphere formation) and upregulation of differentiation markers (GFAP, Olig2, β-tubulin III) were shown in rapamycin-treated cells. These results may contribute to designing novel therapeutic strategies, aiming at selective elimination of GSCs or induction of their differentiation.

Arsenic trioxide (As_2O_3) induced autophagy followed by non-apoptotic cell death in malignant glioma models (Kanzawa et al. 2003, 2005). The cell death was accompanied by accumulation of autophagy-specific marker, LC3, and damage of mitochondrial membrane integrity, but not by caspase activation. Interestingly, BNIP3, but not its homologue BNIP3-like (BNIP3L), played a pivotal role in autophagic cell death induced by arsenic trioxide treatment (Kanzawa et al. 2005). Other study showed that the mechanisms of arsenic trioxide-induced autophagy could be mediated by the inhibition of PI3K/Akt and the activation of MAPK signaling pathways in human glioma cells (Chiu et al. 2011). Furthermore, survivin played a pivotal role in arsenic trioxide-induced autophagy. This study suggests that arsenic trioxide treatment or survivin inhibition could be a novel therapeutic strategy in malignant gliomas. Phase I and II clinical trials with arsenic trioxide for patients with gliomas, advanced solid tumors, metastatic liver cancer, and pediatric solid tumors were initiated (http://www.clinicaltrials.gov). BNIP3 plays also an important role in non-apoptotic cell death induced by **C_2-ceramide** in malignant glioma cells. The cell death was accompanied by features characteristic of autophagy: presence of numerous autophagic vacuoles in the cytoplasm, development of the acidic vesicular organelles, autophagosome membrane association of LC3, and a marked increase in expression levels of two forms of LC3 protein (LC3-I and LC3-II) (Daido et al. 2004). **Curcumin**, an active compound from turmeric spice, efficiently inhibited growth of malignant gliomas in vitro and in vivo by inducing autophagic cell death. Further investigation of the signal pathways of curcumin-induced autophagy revealed simultaneous inhibition of the Akt/mTOR/p70S6K pathway and stimulation of the ERK1/2 pathway (Aoki et al. 2007).

Anti-apoptotic Bcl-2 family members suppress both apoptosis and autophagy and are implicated in therapy resistance of malignant gliomas. Beclin1, classified recently as a BH3-only protein, was first discovered as a Bcl-2-interacting protein (Liang et al. 1998). The relationship between Beclin1 and Bcl2/Bcl-XL (anti-apoptotic proteins) is complex but Beclin1 cannot neutralize the anti-apoptotic function of Bcl-2 (Ciechomska et al. 2009). In contrast, Bcl-2 and Bcl-XL reduce the pro-autophagic activity of Beclin1 (Pattingre et al. 2005; Maiuri et al. 2007). The pharmacological **BH3 mimetics** (small-molecule inhibitors of pro-survival Bcl-2 proteins binding to their respective hydrophobic BH3 grooves) are capable of activating both apoptosis and autophagy (Maiuri et al. 2007; Lessene et al. 2008). Recently, it was shown that (−)-gossypol, a BH3 mimetic, induces mitochondrial dysfunction in the absence of effector caspase activation to trigger an autophagic glioma cell death. Knockdown of Beclin1 and Atg5 strongly diminished the extent of cell death induced by (−)-gossypol, indicating autophagy involvement. Moreover, BH3 mimetics induced autophagic cell death in anoxia-resistant malignant glioma cells, by inhibition of both Bcl-2 and Bcl-xL. Recently it was shown that in addition to mitochondrial pathway apoptosis, endoplasmic reticulum (ER) stress-associated apoptosis was also induced by a novel BH3 mimetic S1. Moreover, S1 can induce autophagy in human glioma cells, which may occur through ER stress and disruption of the association of Bcl-2 and Beclin1. Inhibition of autophagy by the autophagic inhibitors 3-methyladenine (3-MA) and CQ increased S1-induced apoptosis (Zhong et al. 2012). In conclusion, autophagy plays an important role in BH3 mimetic-induced glioma cell death.

Increasing number of data indicate that autophagy can be potently induced by ER stress pathways. Recently, a detailed analysis of human astrocytoma cell lines and a primary culture of human glioma cells has indicated that **Δ⁹-THC** [(-)-*trans*-Δ⁹-tetrahydrocannabinol] treatment led to formation of autophagosomes in tumor cells (Salazar et al. 2009). Autophagy was preceded by stimulation of ER stress, mTORC1 inhibition, and followed by apoptosis in cannabinoid-treated human and mouse cancer cells. Administration of Δ⁹-THC to mice bearing tumors derived from human astrocytoma cells resulted in increased ER-stress-related protein, TRB3 (pseudo-kinase tribbles homologue 3) expression, inhibition of mTOR signaling pathway, appearance of autophagy markers, and caspase-3 activation. These findings indicate that cannabinoid promotes the autophagy-mediated cell death through stimulation of ER stress in human glioma cells (Salazar et al. 2009). We observed similar changes using synthetic cannabinoids, which induced autophagy in various human glioma cell lines. However, blocking of autophagy did not prevent cell death executed by apoptosis (our unpublished observation). Signaling mechanisms linking ER stress to autophagy vary depending on specific stress conditions and organisms. Recently, we have shown that **cyclosporine A** (CsA, an immunophilin/calcineurin inhibitor) induced glioma cell death with some apoptotic features but also accompanied by the appearance of numerous cytoplasmic vacuoles, immunostained for ER and autophagy markers (Ciechomska et al. 2013). The induction of ER stress in glioma cells by CsA was evidenced by unfolded protein response activation (phosphorylation of PERK, accumulation of IRE1α) and accumulation of ER stress-associated proteins (BIP and CHOP). Using pharmacological or genetic blockers of ER stress, we showed that ER stress precedes CsA-induced autophagy. Suppression of autophagy by silencing of essential autophagy genes augmented apoptotic effects of CsA. In summary, we showed involvement of ER stress/autophagy pathways into CsA-induced programmed cell death of malignant glioma cells and presented evidence that autophagy had cytoprotective rather than cytotoxic role in that process. Likewise, neuro-steroid, 5-androstene 3β, 17α diol (**17α-AED**), induced autophagy through PERK/eIF2α signaling in human malignant glioma cells (Jia et al. 2010).

Autophagy has been proposed as a basic mechanism of cell death after **adenoviral infection**. Conditionally replicating adenoviruses induce tumor-specific autophagic cell death through a process that involves the suppression of the mTOR signaling pathway (Ito et al. 2006). In agreement with this observation, Fueyo group reported the anti-glioma effect of the tumor-selective oncolytic adenovirus Delta-24-RGD. Adenoviral infection elicited autophagic cell death in gliomas, as indicated by accumulation of critical proteins in the autophagy pathway, Atg5 and LC3-II, in the host cells. In addition, Delta-24-RGD in combination with everolimus (RAD001) induced enhanced anti-glioma effect via autophagic cell death in vitro and in vivo (Alonso et al. 2008). Delta-24-RGD was an efficacious therapeutic agent against brain tumor stem cells (BTSCs) (Jiang et al. 2007).

Treatment of xenografts derived from brain tumor stem cells with Delta-24-RGD statistically significantly improved the survival of glioma-bearing mice. Analyses of treated tumors showed that Atg5 expression co-localized with adenoviral proteins. These results show for the first time that brain tumor stem cells are susceptible to adenovirus-mediated cell death via autophagy in vitro and in vivo. Interestingly, oncolytic adenovirus Delta-24-RGD is now being tested in a phase I clinical trial for patients with malignant gliomas (Alonso et al. 2012).

The current standard of treatment for GBMs is surgical resection followed by radiotherapy and chemotherapy. It was shown that ionizing **radiation** (IR) induced autophagy, but not apoptosis, in human malignant glioma cells (Ito et al. 2005). Changes in the LC3 expression and autophagosomes formation were observed. Furthermore, gammaH2AX foci, that show the extent of DNA double-strand breaks, were more pronounced and prolonged in the cells treated with IR and autophagy inhibitors than in those cells treated with IR only. Pharmacological autophagy inhibitors sensitized glioma cells to IR (Ito et al. 2005). Similarly, other groups have shown that induction of autophagy contributes to the radioresistance of glioma stem cells (CD133-positive cells). Bafilomycin A1 and silencing of *Atg5* and *Beclin1* sensitized the CD133+ cells to γ-radiation and significantly decreased the viability of the irradiated cells and their ability to form neurospheres (Lomonaco et al. 2009). These results suggest that autophagy inhibitors may be used for radiosensitization of malignant glioma cells.

Temozolomide (TMZ) is a commonly used, oral, methylating agent with promising antitumor efficacy for malignant gliomas. Indeed, the addition of TMZ to radiotherapy resulted in a longer median survival time of glioblastoma patients (Stupp et al. 2005). TMZ achieves its cytotoxic effect mainly by methylating the O^6 position of guanine in DNA. This DNA adduct can be removed by the DNA repair protein O^6-methylguanine-DNA-methyltransferase (MGMT), which is expressed in varied levels in GBMs. Consequently, TMZ displays its highest efficacy against tumors lacking MGMT expression due to a methylated MGMT promoter (Hegi et al. 2005). It has been demonstrated that TMZ induces autophagy in malignant glioma cells (Kanzawa et al. 2004). Interestingly, depending at which step autophagy was suppressed by an inhibitor, two opposite outcomes could occur following TMZ treatment. When autophagy was prevented at an early stage by 3-MA, a type III phosphatidylinositol 3-phosphate kinase inhibitor, not only the characteristic pattern of LC3 localization but also the antitumor effect of TMZ was suppressed. On the other hand, bafilomycin A1, a specific inhibitor of vacuolar type H+-ATPase, that prevents autophagy at a late stage, sensitized tumor cells to TMZ by inducing apoptosis through activation of caspase-3 with mitochondrial membrane permeabilization. These data suggest that inhibition of autophagy at a late stage could increase the therapeutic efficacy of TMZ for malignant glioma cells. Interestingly, BH3 mimetic [(−)-gossypol] and a stable knockdown of mTOR were capable of enhancing cell death triggered by TMZ in malignant glioma cells. In contrast, knockdown of *Beclin1* and *Atg5* protected malignant glioma cells from autophagic cell death triggered by TMZ, indicating that autophagy indeed played an essential death-promoting role in this type of cell death (Voss et al. 2010).

Autophagy can enhance cell survival in nutrient-deprived cells by increasing adenosine triphosphate (ATP) production. Recently, it was shown that DNA damaging agents (such as TMZ and etoposide) induced an autophagy-associated ATP surge that protected cells and may contribute to drug resistance of malignant glioma cells (Katayama et al. 2007). The drug-induced ATP surge could be blocked by pre-incubation with the autophagy inhibitor 3-MA, an siRNA targeting *Beclin1*, or the mitochondrial ATP synthase inhibitor oligomycin. Inhibition of autophagy-induced ATP production increased non-apoptotic cell death associated with micronucleation. Recent data showed that TMZ induces autophagy through mitochondrial damage- and ER stress-dependent mechanisms in glioma cells (Lin et al. 2012). Other groups proposed that autophagy after TMZ treatment serves as a survival mechanism, stimulating cells to undergo senescence and inhibiting apoptosis. If autophagy is abrogated, glioma cells become sensitized to TMZ-induced DNA damage and undergo apoptosis, while cellular senescence is blocked (Knizhnik et al. 2013). Concluding, the impact of TMZ-induced autophagy on malignant glioma is currently unsolved. The use of different inhibitors or the selection of different knockdown targets results in conflicting results.

Preclinical models showed that some tyrosine kinase inhibitors stimulate autophagy. **Imatinib**, an inhibitor of BRC-ABL, KIT, and PGDFR, induced autophagy and non-apoptotic cell death in human malignant glioma cells in vitro. Inhibition of imatinib-induced autophagy by 3-MA or *Atg5* siRNA attenuates imatinib-induced cytotoxicity whereas inhibition of autophagy by bafilomycin A1 or RTA 203 enhances imatinib-induced cytotoxicity through increasing apoptosis following mitochondrial damage (Shingu et al. 2009). Therefore, combination of imatinib and an autolysosome inhibitor may be a potent therapeutic modality for patients with malignant gliomas. These findings also suggest that inadequate suppression of autophagy may reduce the efficacy of anticancer therapy and that appropriate modulation of autophagy is necessary for sensitizing tumor cells to anticancer therapy.

Recent studies have shown that **Akt inhibition** did not induce significant apoptosis but rather markedly increased autophagy of PTEN-null glioma cells (Degtyarev et al. 2008). Further treatment with the lysosomotropic agent chloroquine (CQ) caused accumulation of abnormal autophagolysosomes and reactive oxygen species, leading to accelerated cell death in vitro and complete tumor remission in vivo. Cell death was also promoted when Akt inhibition was combined with bafilomycin A1 or with cathepsin inhibition (Degtyarev et al. 2008). Of note, the dual PI3K-mTOR inhibitor PI-103 induced autophagy in glioma cells that are resistant to therapy. Inhibitors of autophagosome maturation cooperated with PI-103 to induce apoptosis through the mitochondrial pathway, indicating that the cellular self-digestion process of autophagy acted as a survival signal in this setting. Moreover, the PI3K-mTOR inhibitor NVP-BEZ235, which is currently in clinical trials, synergized with the lysosomotropic inhibitor of autophagy; CQ induces apoptosis in glioma xenografts in vivo. Concluding, effective cell death in preclinical models of PTEN-deficient gliomas (in vitro and in vivo) requires blockade of three targets: Akt, mTOR, and autophagy (Fan et al. 2010). Interestingly, triple therapy of γ-irradiation with CQ and selected PI3K/Akt pathway inhibitors was strongly cytotoxic for highly radioresistant stem-like glioma cells and CQ alone, at slightly higher doses, strongly promoted γ-irradiation-induced cell death in GSCs. Cell death observed in double and triple combination with CQ occurred through apoptosis triggered by inhibition of late autophagy (Firat et al. 2012). Similarly, inhibition of autophagy induced a marked increase in the death-inducing activity of erlotinib (tyrosine kinase inhibitor), confirming that EGFR targeted therapy must be associated with autophagy inhibition to achieve an antitumoral effect (Eimer et al. 2011).

Concluding, several chemotherapeutic agents have been shown to trigger autophagic cell death in glioma cells in vitro, whereas the contribution of autophagy to their antitumor effects in vivo still awaits experimental evidence. In most cases, moreover, it has not been proven that these drugs induced death via the autophagy pathway. The data rather suggest that activation of autophagy and cell death coincide and are tightly correlated. Furthermore, for many of these drugs it is hypothesized that combining them with autophagy inhibitors may improve efficacy. These results suggest that blocking lysosomal degradation can be detrimental to cancer cell survival when autophagy is activated, providing rationale for a new therapeutic approach to enhancing the anti-cancer efficacy of different drugs.

5.4.2 Clinical Trials

Multiple studies have shown that genetic knockdown of Atgs or pharmacological inhibition of autophagy can enhance antitumor efficacy of chemotherapeutic agents in preclinical models (Chen et al. 2010; Yang et al. 2011). Among pharmacological blockers, only chloroquine (CQ) and its derivative hydroxychloroquine (HCQ) which cross the blood–brain barrier have been evaluated in humans. Both drugs are commonly used as antimalarial drugs and in autoimmune disorders. They block acidification of lysosomes and thus prevent lysosome-dependent degradation of autophagosomes (Poole and Ohkuma 1981; Glaumann and Ahlberg 1987). It is noteworthy that these autophagy inhibitors are tested in clinical trials.

A phase III trial (a randomized, double-blind trial) in GBM patients treated with radiation and carmustine with or without daily CQ found a median overall survival of 24 and 11 months in CQ- and placebo-treated patients, respectively (Sotelo et al. 2006). The limitation of this study was a small number of patients (30 patients) and the difference in survival was not statistically significant. Nevertheless, the study established the safety of adding low dose CQ to DNA damaging therapy. In summary, CQ may improve mid-term survival when given in addition to conventional therapy for GBM. These results call for larger, more definitive studies of CQ as adjuvant therapy for glioblastoma is needed (Sotelo et al. 2006). This would be a cost-effective strategy for patients, who cannot afford temozolamide. In addition, it may benefit patients on chemoradiation who are receiving temozolamide in view of the consequent chemosensitization.

HCQ is preferred to CQ in humans because of its more favorable side-effects profile (Ruiz-Irastorza et al. 2010). However, pharmacokinetics of HCQ (characterized by a long delay in the onset of action) and the low potency of the drug (micromolar concentrations are required to inhibit autophagy) remain key issues that may limit its efficacy in human studies (Carmichael et al. 2003). Nevertheless, a phase I–II trial for the use of HCQ in combination with radiation (RT) and TMZ in GBM patients is currently ongoing (http://clinicaltrials.gov/ct2/show/NCT00486603). This trial is studying the side effects and best dose of HCQ in addition to the standard regimen and to see how well they work in treating patients with newly diagnosed GBM. The study aims to correlate the average change in autophagic vesicles (AV) from baseline with genotype, toxicity, and clinical outcomes and to correlate the presence of *TP53* and *PTEN* genes and *BECN1* with toxicity and clinical outcomes. A recent preliminary report from phase I trial of HCQ with TMZ and RT for glioblastoma patients showed that the maximum tolerated dose of HCQ was 600 mg. Quantitative electron microscopy (EM) and immunoblotting against the autophagy marker LC3 were used to assess therapy-associated changes in AV in peripheral blood mononuclear cells (PBMC). In that trial, investigators observed a dose-dependent inhibition of autophagy, as indicated by changes in AV in PBMC. Dose-dependent myelosuppression suggested a mechanistic interaction between HCQ and TMZ/RT. Analyses of PK-PD interactions and a phase II study are underway (Rosenfeld et al. 2010).

Currently, more than 20 trials for the use of autophagy inhibitor (HCQ) in cancer therapy have been carried out in the last few years, and many of them showed evidence of antitumor activity. Whether HCQ will be an effective autophagy inhibitor in human tumors, how patients who would benefit will be identified and their tumor assessed, and how to determine the best drugs to combine with HCQ have yet to be resolved. It must be remembered, however, that CQ and HCQ can act also by other mechanisms independently of autophagy inhibition. Nevertheless, CQ and HCQ have the potential to open new frontiers in the treatment of glial neoplasms, due to their unique features of a well-studied side-effects profile. Besides, they are inexpensive and easily available (Munshi 2009).

Concluding Remarks

Autophagy is involved in nearly all stages of tumorigenesis and tumor progression. On the one hand autophagy suppresses the initiation and development of cancer, but on the other hand, it can be tumor promoting in established cancers. Therefore, it is of great clinical interest to better understand the molecular mechanisms underlying autophagy. A key issue is to find "autophagic switch" which might mediate a transition from inhibition of autophagy in early-stage carcinogenesis to the induction of autophagy that contributes to cancer progression.

Autophagy is a dynamic process that is difficult to measure and quantify. In vitro, electron microscopy is predominantly used as the gold standard to demonstrate autophagic cells by observing autophagic vacuoles in the cytoplasm, but the fact that it is static is a real limitation of this method. Plasmids coding

GFP-LC3 are very useful in autophagy detection; because GFP-LC3 localizes on autophagosomes, autophagic cells can be identified by their characteristic GFP-LC3 dots under a fluorescence microscope. However, the major weakness of using GFP-LC3 vector is the obligatory requirement for the gene transfection. The vector is not applicable for surgical samples, although transgenic animals expressing GFP-LC3 are helpful for observing the role of autophagy in normal development (Aoki et al. 2008). The monitoring and interpretation of autophagy dynamics in clinical samples, such as tumor biopsies and PBMC are still technically challenging. However, some data suggest that Beclin1, LC3-II, and changes in autophagic vesicles in PBMC might represent biomarkers to distinguish grade of astrocytic tumors, and to measure autophagy modulation during treatment in patients with gliomas (Pirtoli et al. 2009; Huang et al. 2010; Rosenfeld et al. 2010).

Many anti-cancer treatments including novel targeted therapies stimulate autophagy, which can lead not only to increased cytotoxicity but also to therapeutic resistance. Autophagy induction has been implicated in GBM in response to radiotherapy, temozolomide, as well as targeted therapies (Kanzawa et al. 2003, 2004; Katayama et al. 2007; Shingu et al. 2009; Fan et al. 2010; Geng et al. 2010). It is important to clarify the ability of anti-cancer treatment to induce autophagy and the role of autophagy, if induced, in the response of tumor cells to that agent, in order to increase efficacy of the treatment by manipulating the autophagic process (Shingu et al. 2009). Additionally, the genetic composition of cancer cells may play a deterministic role in controlling the functional outcome of autophagy and personalized cancer therapy should be taken into consideration to optimize the therapeutic effect in patients. In this respect, differences in expression of autophagy regulators across GBM molecular subtypes could affect the response to autophagy inducing chemotherapeutics. Moreover, some studies suggest that autophagy may regulate progression in GBM (Miracco et al. 2007; Huang et al. 2010), but a systematic analysis of the autophagy status in these tumors is missing.

Abundant preclinical evidence indicates that stress-induced autophagy in tumor cells is predominantly cytoprotective and that inhibition of autophagy can enhance tumor cell death by diverse anti-cancer therapies. Although several drugs can inhibit autophagy, most of these drugs lack specificity and antitumor activity. Only CQ and HCQ are the most widely tested in preclinical models and multiple ongoing phase I and II clinical trials. These drugs, alone or in combination with cytotoxic chemotherapy or targeted agents, are evaluated mostly in patients with solid tumors. Currently, HCQ in addition to the standard regimen is being explored in the treatment of GBM (http://clinicaltrials.gov/ct2/show/NCT00486603). However, more potent and specific inhibitors of autophagy are clearly needed. High-throughput screening of chemical libraries to identify small-molecule inhibitors of autophagy is ongoing. Nevertheless, these data establish autophagy as a novel therapeutic target whose modulation presents new opportunities for cancer treatment. Identifying target molecules that regulate autophagic capacity is a potential strategy for the development of novel astrocytic tumor-directed therapies.

Acknowledgments This work was supported by a grant 2011/03/B/NZ3/04453 from The National Science Centre Poland (IAC).

References

Alonso MM, Jiang H, Gomez-Manzano C, Fueyo J (2012) Targeting brain tumor stem cells with oncolytic adenoviruses. Methods Mol Biol 797:111–125

Alonso MM, Jiang H, Yokoyama T, Xu J, Bekele NB, Lang FF, Kondo S, Gomez-Manzano C, Fueyo J (2008) Delta-24-RGD in combination with RAD001 induces enhanced anti-glioma effect via autophagic cell death. Mol Ther 16(3):487–493

Amaravadi RK, Lippincott-Schwartz J, Yin XM, Weiss WA, Takebe N, Timmer W, DiPaola RS, Lotze MT, White E (2011) Principles and current strategies for targeting autophagy for cancer treatment. Clin Cancer Res 17(4):654–666

Aoki H, Kondo Y, Aldape K, Yamamoto A, Iwado E, Yokoyama T, Hollingsworth EF, Kobayashi R, Hess K, Shinojima N, Shingu T, Tamada Y, Zhang L, Conrad C, Bogler O, Mills G, Sawaya R, Kondo S (2008) Monitoring autophagy in glioblastoma with antibody against isoform B of human microtubule-associated protein 1 light chain 3. Autophagy 4 (4):467–475

Aoki H, Takada Y, Kondo S, Sawaya R, Aggarwal BB, Kondo Y (2007) Evidence that curcumin suppresses the growth of malignant gliomas in vitro and in vivo through induction of autophagy: role of Akt and extracellular signal-regulated kinase signaling pathways. Mol Pharmacol 72(1):29–39

Atlashkin V, Kreykenbohm V, Eskelinen EL, Wenzel D, Fayyazi A, Fischer von Mollard G (2003) Deletion of the SNARE vti1b in mice results in the loss of a single SNARE partner, syntaxin 8. Mol Cell Biol 23(15):5198–5207

Axe EL, Walker SA, Manifava M, Chandra P, Roderick HL, Habermann A, Griffiths G, Ktistakis NT (2008) Autophagosome formation from membrane compartments enriched in phosphatidylinositol 3-phosphate and dynamically connected to the endoplasmic reticulum. J Cell Biol 182(4):685–701

Azad MB, Chen Y, Henson ES, Cizeau J, McMillan-Ward E, Israels SJ, Gibson SB (2008) Hypoxia induces autophagic cell death in apoptosis-competent cells through a mechanism involving BNIP3. Autophagy 4(2):195–204

Bae D, Lu S, Taglienti CA, Mercurio AM (2008) Metabolic stress induces the lysosomal degradation of neuropilin-1 but not neuropilin-2. J Biol Chem 283(42):28074–28080

Bao S, Wu Q, McLendon RE, Hao Y, Shi Q, Hjelmeland AB, Dewhirst MW, Bigner DD, Rich JN (2006) Glioma stem cells promote radioresistance by preferential activation of the DNA damage response. Nature 444(7120):756–760

Benjamin R, Capparella J, Brown A (2003) Classification of glioblastoma multiforme in adults by molecular genetics. Cancer J 9(2):82–90

Berry DL, Baehrecke EH (2007) Growth arrest and autophagy are required for salivary gland cell degradation in Drosophila. Cell 131(6):1137–1148

Boya P, Gonzalez-Polo RA, Casares N, Perfettini JL, Dessen P, Larochette N, Metivier D, Meley D, Souquere S, Yoshimori T, Pierron G, Codogno P, Kroemer G (2005) Inhibition of macroautophagy triggers apoptosis. Mol Cell Biol 25(3):1025–1040

N. Cancer Genome Atlas Research (2008) Comprehensive genomic characterization defines human glioblastoma genes and core pathways. Nature 455(7216):1061–1068

Carmichael SJ, Charles B, Tett SE (2003) Population pharmacokinetics of hydroxychloroquine in patients with rheumatoid arthritis. Ther Drug Monit 25(6):671–681

Chang SM, Wen P, Cloughesy T, Greenberg H, Schiff D, Conrad C, Fink K, Robins HI, De Angelis L, Raizer J, Hess K, Aldape K, Lamborn KR, Kuhn J, Dancey J, Prados MD, C. North American Brain Tumor, I. the National Cancer (2005) Phase II study of CCI-779 in patients with recurrent glioblastoma multiforme. Invest New Drugs 23(4):357–361

Chen S, Rehman SK, Zhang W, Wen A, Yao L, Zhang J (2010) Autophagy is a therapeutic target in anticancer drug resistance. Biochim Biophys Acta 1806(2):220–229

Chiu HW, Ho YS, Wang YJ (2011) Arsenic trioxide induces autophagy and apoptosis in human glioma cells in vitro and in vivo through downregulation of survivin. J Mol Med (Berl) 89 (9):927–941

Ciechomska IA, Gabrusiewicz K, Szczepankiewicz AA, Kaminska B (2013) Endoplasmic reticulum stress triggers autophagy in malignant glioma cells undergoing cyclosporine A-induced cell death. Oncogene 32(12):1518–1529

Ciechomska IA, Goemans GC, Skepper JN, Tolkovsky AM (2009) Bcl-2 complexed with Beclin-1 maintains full anti-apoptotic function. Oncogene 28(21):2128–2141

Coppola D, Khalil F, Eschrich SA, Boulware D, Yeatman T, Wang HG (2008a) Down-regulation of Bax-interacting factor-1 in colorectal adenocarcinoma. Cancer 113(10):2665–2670

Coppola D, Oliveri C, Sayegh Z, Boulware D, Takahashi Y, Pow-Sang J, Djeu JY, Wang HG (2008b) Bax-interacting factor-1 expression in prostate cancer. Clin Genitourin Cancer 6 (2):117–121

Crighton D, Wilkinson S, O'Prey J, Syed N, Smith P, Harrison PR, Gasco M, Garrone O, Crook T, Ryan KM (2006) DRAM, a p53-induced modulator of autophagy, is critical for apoptosis. Cell 126(1):121–134

Cuddeback SM, Yamaguchi H, Komatsu K, Miyashita T, Yamada M, Wu C, Singh S, Wang HG (2001) Molecular cloning and characterization of Bif-1. A novel Src homology 3 domain-containing protein that associates with Bax. J Biol Chem 276(23):20559–20565

Daido S, Kanzawa T, Yamamoto A, Takeuchi H, Kondo Y, Kondo S (2004) Pivotal role of the cell death factor BNIP3 in ceramide-induced autophagic cell death in malignant glioma cells. Cancer Res 64(12):4286–4293

Degenhardt K, Mathew R, Beaudoin B, Bray K, Anderson D, Chen G, Mukherjee C, Shi Y, Gelinas C, Fan Y, Nelson DA, Jin S, White E (2006) Autophagy promotes tumor cell survival and restricts necrosis, inflammation, and tumorigenesis. Cancer Cell 10(1):51–64

Degtyarev M, De Maziere A, Orr C, Lin J, Lee BB, Tien JY, Prior WW, van Dijk S, Wu H, Gray DC, Davis DP, Stern HM, Murray LJ, Hoeflich KP, Klumperman J, Friedman LS, Lin K (2008) Akt inhibition promotes autophagy and sensitizes PTEN-null tumors to lysosomotropic agents. J Cell Biol 183(1):101–116

DeNardo DG, Barreto JB, Andreu P, Vasquez L, Tawfik D, Kolhatkar N, Coussens LM (2009) CD4(+) T cells regulate pulmonary metastasis of mammary carcinomas by enhancing protumor properties of macrophages. Cancer Cell 16(2):91–102

DeNardo DG, Johansson M, Coussens LM (2008) Immune cells as mediators of solid tumor metastasis. Cancer Metastasis Rev 27(1):11–18

Dengjel J, Schoor O, Fischer R, Reich M, Kraus M, Muller M, Kreymborg K, Altenberend F, Brandenburg J, Kalbacher H, Brock R, Driessen C, Rammensee HG, Stevanovic S (2005) Autophagy promotes MHC class II presentation of peptides from intracellular source proteins. Proc Natl Acad Sci U S A 102(22):7922–7927

Eimer S, Belaud-Rotureau MA, Airiau K, Jeanneteau M, Laharanne E, Veron N, Vital A, Loiseau H, Merlio JP, Belloc F (2011) Autophagy inhibition cooperates with erlotinib to induce glioblastoma cell death. Cancer Biol Ther 11(12):1017–1027

English L, Chemali M, Duron J, Rondeau C, Laplante A, Gingras D, Alexander D, Leib D, Norbury C, Lippe R, Desjardins M (2009) Autophagy enhances the presentation of endogenous viral antigens on MHC class I molecules during HSV-1 infection. Nat Immunol 10(5):480–487

Fan QW, Cheng C, Hackett C, Feldman M, Houseman BT, Nicolaides T, Haas-Kogan D, James CD, Oakes SA, Debnath J, Shokat KM, Weiss WA (2010) Akt and autophagy cooperate to promote survival of drug-resistant glioma. Sci Signal 3(147):ra81

Feng Z, Zhang H, Levine AJ, Jin S (2005) The coordinate regulation of the p53 and mTOR pathways in cells. Proc Natl Acad Sci U S A 102(23):8204–8209

Fimia GM, Stoykova A, Romagnoli A, Giunta L, Di Bartolomeo S, Nardacci R, Corazzari M, Fuoco C, Ucar A, Schwartz P, Gruss P, Piacentini M, Chowdhury K, Cecconi F (2007) Ambra1 regulates autophagy and development of the nervous system. Nature 447(7148):1121–1125

Firat E, Weyerbrock A, Gaedicke S, Grosu AL, Niedermann G (2012) Chloroquine or chloroquine-PI3K/Akt pathway inhibitor combinations strongly promote gamma-irradiation-induced cell death in primary stem-like glioma cells. PLoS One 7(10):e47357

Fung C, Lock R, Gao S, Salas E, Debnath J (2008) Induction of autophagy during extracellular matrix detachment promotes cell survival. Mol Biol Cell 19(3):797–806

Furnari FB, Fenton T, Bachoo RM, Mukasa A, Stommel JM, Stegh A, Hahn WC, Ligon KL, Louis DN, Brennan C, Chin L, DePinho RA, Cavenee WK (2007) Malignant astrocytic glioma: genetics, biology, and paths to treatment. Genes Dev 21(21):2683–2710

Galanis E, Buckner JC (2000) Chemotherapy of brain tumors. Curr Opin Neurol 13(6):619–625

Galanis E, Buckner JC, Maurer MJ, Kreisberg JI, Ballman K, Boni J, Peralba JM, Jenkins RB, Dakhil SR, Morton RF, Jaeckle KA, Scheithauer BW, Dancey J, Hidalgo M, Walsh DJ, G. North Central Cancer Treatment (2005) Phase II trial of temsirolimus (CCI-779) in recurrent glioblastoma multiforme: a North Central Cancer Treatment Group Study. J Clin Oncol 23(23):5294–5304

Geng Y, Kohli L, Klocke BJ, Roth KA (2010) Chloroquine-induced autophagic vacuole accumulation and cell death in glioma cells is p53 independent. Neuro Oncol 12(5):473–481

Glaumann H, Ahlberg J (1987) Comparison of different autophagic vacuoles with regard to ultrastructure, enzymatic composition, and degradation capacity–formation of crinosomes. Exp Mol Pathol 47(3):346–362

Gonzalez-Polo RA, Boya P, Pauleau AL, Jalil A, Larochette N, Souquere S, Eskelinen EL, Pierron G, Saftig P, Kroemer G (2005) The apoptosis/autophagy paradox: autophagic vacuolization before apoptotic death. J Cell Sci 118(Pt 14):3091–3102

Gozuacik D, Kimchi A (2004) Autophagy as a cell death and tumor suppressor mechanism. Oncogene 23(16):2891–2906

Guha A, Dashner K, Black PM, Wagner JA, Stiles CD (1995) Expression of PDGF and PDGF receptors in human astrocytoma operation specimens supports the existence of an autocrine loop. Int J Cancer 60(2):168–173

Guo JY, Chen HY, Mathew R, Fan J, Strohecker AM, Karsli-Uzunbas G, Kamphorst JJ, Chen G, Lemons JM, Karantza V, Coller HA, Dipaola RS, Gelinas C, Rabinowitz JD, White E (2011) Activated Ras requires autophagy to maintain oxidative metabolism and tumorigenesis. Genes Dev 25(5):460–470

Gutierrez MG, Munafo DB, Beron W, Colombo MI (2004) Rab7 is required for the normal progression of the autophagic pathway in mammalian cells. J Cell Sci 117(Pt 13):2687–2697

Haas-Kogan D, Shalev N, Wong M, Mills G, Yount G, Stokoe D (1998) Protein kinase B (PKB/Akt) activity is elevated in glioblastoma cells due to mutation of the tumor suppressor PTEN/MMAC. Curr Biol 8(21):1195–1198

Hailey DW, Rambold AS, Satpute-Krishnan P, Mitra K, Sougrat R, Kim PK, Lippincott-Schwartz J (2010) Mitochondria supply membranes for autophagosome biogenesis during starvation. Cell 141(4):656–667

Hanada T, Noda NN, Satomi Y, Ichimura Y, Fujioka Y, Takao T, Inagaki F, Ohsumi Y (2007) The Atg12-Atg5 conjugate has a novel E3-like activity for protein lipidation in autophagy. J Biol Chem 282(52):37298–37302

Hanahan D, Weinberg RA (2011) Hallmarks of cancer: the next generation. Cell 144(5):646–674

He C, Klionsky DJ (2009) Regulation mechanisms and signaling pathways of autophagy. Annu Rev Genet 43:67–93

Hegi ME, Diserens AC, Gorlia T, Hamou MF, de Tribolet N, Weller M, Kros JM, Hainfellner JA, Mason W, Mariani L, Bromberg JE, Hau P, Mirimanoff RO, Cairncross JG, Janzer RC, Stupp

R (2005) MGMT gene silencing and benefit from temozolomide in glioblastoma. N Engl J Med 352(10):997–1003

Hermanson M, Funa K, Hartman M, Claesson-Welsh L, Heldin CH, Westermark B, Nister M (1992) Platelet-derived growth factor and its receptors in human glioma tissue: expression of messenger RNA and protein suggests the presence of autocrine and paracrine loops. Cancer Res 52(11):3213–3219

Hosli P, Sappino AP, de Tribolet N, Dietrich PY (1998) Malignant glioma: should chemotherapy be overthrown by experimental treatments? Ann Oncol 9(6):589–600

Hosokawa N, Hara T, Kaizuka T, Kishi C, Takamura A, Miura Y, Iemura S, Natsume T, Takehana K, Yamada N, Guan JL, Oshiro N, Mizushima N (2009) Nutrient-dependent mTORC1 association with the ULK1-Atg13-FIP200 complex required for autophagy. Mol Biol Cell 20(7):1981–1991

Hu YL, DeLay M, Jahangiri A, Molinaro AM, Rose SD, Carbonell WS, Aghi MK (2012) Hypoxia-induced autophagy promotes tumor cell survival and adaptation to antiangiogenic treatment in glioblastoma. Cancer Res 72(7):1773–1783

Huang X, Bai HM, Chen L, Li B, Lu YC (2010) Reduced expression of LC3B-II and Beclin 1 in glioblastoma multiforme indicates a down-regulated autophagic capacity that relates to the progression of astrocytic tumors. J Clin Neurosci 17(12):1515–1519

Ionov Y, Nowak N, Perucho M, Markowitz S, Cowell JK (2004) Manipulation of nonsense mediated decay identifies gene mutations in colon cancer Cells with microsatellite instability. Oncogene 23(3):639–645

Itakura E, Kishi C, Inoue K, Mizushima N (2008) Beclin 1 forms two distinct phosphatidylinositol 3-kinase complexes with mammalian Atg14 and UVRAG. Mol Biol Cell 19(12):5360–5372

Ito H, Aoki H, Kuhnel F, Kondo Y, Kubicka S, Wirth T, Iwado E, Iwamaru A, Fujiwara K, Hess KR, Lang FF, Sawaya R, Kondo S (2006) Autophagic cell death of malignant glioma cells induced by a conditionally replicating adenovirus. J Natl Cancer Inst 98(9):625–636

Ito H, Daido S, Kanzawa T, Kondo S, Kondo Y (2005) Radiation-induced autophagy is associated with LC3 and its inhibition sensitizes malignant glioma cells. Int J Oncol 26(5):1401–1410

Iwadate Y, Fujimoto S, Tagawa M, Namba H, Sueyoshi K, Hirose M, Sakiyama S (1996) Association of p53 gene mutation with decreased chemosensitivity in human malignant gliomas. Int J Cancer 69(3):236–240

Iwamaru A, Kondo Y, Iwado E, Aoki H, Fujiwara K, Yokoyama T, Mills GB, Kondo S (2007) Silencing mammalian target of rapamycin signaling by small interfering RNA enhances rapamycin-induced autophagy in malignant glioma cells. Oncogene 26(13):1840–1851

Jager S, Bucci C, Tanida I, Ueno T, Kominami E, Saftig P, Eskelinen EL (2004) Role for Rab7 in maturation of late autophagic vacuoles. J Cell Sci 117(Pt 20):4837–4848

Jahreiss L, Menzies FM, Rubinsztein DC (2008) The itinerary of autophagosomes: from peripheral formation to kiss-and-run fusion with lysosomes. Traffic 9(4):574–587

Jia W, Loria RM, Park MA, Yacoub A, Dent P, Graf MR (2010) The neuro-steroid, 5-androstene 3beta,17alpha diol; induces endoplasmic reticulum stress and autophagy through PERK/eIF2alpha signaling in malignant glioma cells and transformed fibroblasts. Int J Biochem Cell Biol 42(12):2019–2029

Jiang H, Gomez-Manzano C, Aoki H, Alonso MM, Kondo S, McCormick F, Xu J, Kondo Y, Bekele BN, Colman H, Lang FF, Fueyo J (2007) Examination of the therapeutic potential of Delta-24-RGD in brain tumor stem cells: role of autophagic cell death. J Natl Cancer Inst 99 (18):1410–1414

Jin S (2006) Autophagy, mitochondrial quality control, and oncogenesis. Autophagy 2(2):80–84

Jung CH, Jun CB, Ro SH, Kim YM, Otto NM, Cao J, Kundu M, Kim DH (2009) ULK-Atg13-FIP200 complexes mediate mTOR signaling to the autophagy machinery. Mol Biol Cell 20 (7):1992–2003

Kabeya Y, Mizushima N, Ueno T, Yamamoto A, Kirisako T, Noda T, Kominami E, Ohsumi Y, Yoshimori T (2000) LC3, a mammalian homologue of yeast Apg8p, is localized in autophagosome membranes after processing. EMBO J 19(21):5720–5728

Kang MR, Kim MS, Oh JE, Kim YR, Song SY, Kim SS, Ahn CH, Yoo NJ, Lee SH (2009) Frameshift mutations of autophagy-related genes ATG2B, ATG5, ATG9B and ATG12 in gastric and colorectal cancers with microsatellite instability. J Pathol 217(5):702–706

Kanzawa T, Germano IM, Komata T, Ito H, Kondo Y, Kondo S (2004) Role of autophagy in temozolomide-induced cytotoxicity for malignant glioma cells. Cell Death Differ 11(4):448–457

Kanzawa T, Kondo Y, Ito H, Kondo S, Germano I (2003) Induction of autophagic cell death in malignant glioma cells by arsenic trioxide. Cancer Res 63(9):2103–2108

Kanzawa T, Zhang L, Xiao L, Germano IM, Kondo Y, Kondo S (2005) Arsenic trioxide induces autophagic cell death in malignant glioma cells by upregulation of mitochondrial cell death protein BNIP3. Oncogene 24(6):980–991

Karantza-Wadsworth V, Patel S, Kravchuk O, Chen G, Mathew R, Jin S, White E (2007) Autophagy mitigates metabolic stress and genome damage in mammary tumorigenesis. Genes Dev 21(13):1621–1635

Katayama M, Kawaguchi T, Berger MS, Pieper RO (2007) DNA damaging agent-induced autophagy produces a cytoprotective adenosine triphosphate surge in malignant glioma cells. Cell Death Differ 14(3):548–558

Kaufmann SH, Vaux DL (2003) Alterations in the apoptotic machinery and their potential role in anticancer drug resistance. Oncogene 22(47):7414–7430

Kim KW, Hwang M, Moretti L, Jaboin JJ, Cha YI, Lu B (2008a) Autophagy upregulation by inhibitors of caspase-3 and mTOR enhances radiotherapy in a mouse model of lung cancer. Autophagy 4(5):659–668

Kim MJ, Woo SJ, Yoon CH, Lee JS, An S, Choi YH, Hwang SG, Yoon G, Lee SJ (2011) Involvement of autophagy in oncogenic K-Ras-induced malignant cell transformation. J Biol Chem 286(15):12924–12932

Kim MS, Jeong EG, Ahn CH, Kim SS, Lee SH, Yoo NJ (2008b) Frameshift mutation of UVRAG an autophagy-related gene, in gastric carcinomas with microsatellite instability. Hum Pathol 39 (7):1059–1063

Kisen GO, Tessitore L, Costelli P, Gordon PB, Schwarze PE, Baccino FM, Seglen PO (1993) Reduced autophagic activity in primary rat hepatocellular carcinoma and ascites hepatoma cells. Carcinogenesis 14(12):2501–2505

Kleihues P, Louis DN, Scheithauer BW, Rorke LB, Reifenberger G, Burger PC, Cavenee WK (2002) The WHO classification of tumors of the nervous system. J Neuropathol Exp Neurol 61 (3):215–225, discussion 226–219

Klionsky DJ (2005) The molecular machinery of autophagy: unanswered questions. J Cell Sci 118 (Pt 1):7–18

Klionsky DJ (2008) Autophagy revisited: a conversation with Christian de Duve. Autophagy 4 (6):740–743

Klionsky DJ, Emr SD (2000) Autophagy as a regulated pathway of cellular degradation. Science 290(5497):1717–1721

Knizhnik AV, Roos WP, Nikolova T, Quiros S, Tomaszowski KH, Christmann M, Kaina B (2013) Survival and death strategies in glioma cells: autophagy, senescence and apoptosis triggered by a single type of temozolomide-induced DNA damage. PLoS One 8(1):e55665

Kondo Y, Kondo S (2006) Autophagy and cancer therapy. Autophagy 2(2):85–90

Krex D, Klink B, Hartmann C, von Deimling A, Pietsch T, Simon M, Sabel M, Steinbach JP, Heese O, Reifenberger G, Weller M, Schackert G, German Glioma N (2007) Long-term survival with glioblastoma multiforme. Brain 130(Pt 10):2596–2606

Kroemer G, Galluzzi L, Vandenabeele P, Abrams J, Alnemri ES, Baehrecke EH, Blagosklonny MV, El-Deiry WS, Golstein P, Green DR, Hengartner M, Knight RA, Kumar S, Lipton SA, Malorni W, Nunez G, Peter ME, Tschopp J, Yuan J, Piacentini M, Zhivotovsky B, Melino G, D. Nomenclature Committee on Cell (2009) Classification of cell death: recommendations of the Nomenclature Committee on Cell Death 2009. Cell Death Differ 16(1):3–11

Kroemer G, Levine B (2008) Autophagic cell death: the story of a misnomer. Nat Rev Mol Cell Biol 9(12):1004–1010

Lee JW, Jeong EG, Soung YH, Nam SW, Lee JY, Yoo NJ, Lee SH (2006) Decreased expression of tumour suppressor Bax-interacting factor-1 (Bif-1), a Bax activator, in gastric carcinomas. Pathology 38(4):312–315

Lefranc F, Facchini V, Kiss R (2007) Proautophagic drugs: a novel means to combat apoptosis-resistant cancers, with a special emphasis on glioblastomas. Oncologist 12(12):1395–1403

Lessene G, Czabotar PE, Colman PM (2008) BCL-2 family antagonists for cancer therapy. Nat Rev Drug Discov 7(12):989–1000

Levine B, Klionsky DJ (2004) Development by self-digestion: molecular mechanisms and biological functions of autophagy. Dev Cell 6(4):463–477

Levine B, Kroemer G (2008) Autophagy in the pathogenesis of disease. Cell 132(1):27–42

Li Y, Wang LX, Yang G, Hao F, Urba WJ, Hu HM (2008) Efficient cross-presentation depends on autophagy in tumor cells. Cancer Res 68(17):6889–6895

Liang C, Feng P, Ku B, Dotan I, Canaani D, Oh BH, Jung JU (2006) Autophagic and tumour suppressor activity of a novel Beclin1-binding protein UVRAG. Nat Cell Biol 8(7):688–699

Liang XH, Jackson S, Seaman M, Brown K, Kempkes B, Hibshoosh H, Levine B (1999) Induction of autophagy and inhibition of tumorigenesis by beclin 1. Nature 402(6762):672–676

Liang XH, Kleeman LK, Jiang HH, Gordon G, Goldman JE, Berry G, Herman B, Levine B (1998) Protection against fatal Sindbis virus encephalitis by beclin, a novel Bcl-2-interacting protein. J Virol 72(11):8586–8596

Lin CJ, Lee CC, Shih YL, Lin CH, Wang SH, Chen TH, Shih CM (2012) Inhibition of mitochondria- and endoplasmic reticulum stress-mediated autophagy augments temozolomide-induced apoptosis in glioma cells. PLoS One 7(6):e38706

Lindmo K, Stenmark H (2006) Regulation of membrane traffic by phosphoinositide 3-kinases. J Cell Sci 119(Pt 4):605–614

Lock R, Roy S, Kenific CM, Su JS, Salas E, Ronen SM, Debnath J (2011) Autophagy facilitates glycolysis during Ras-mediated oncogenic transformation. Mol Biol Cell 22(2):165–178

Lomonaco SL, Finniss S, Xiang C, Decarvalho A, Umansky F, Kalkanis SN, Mikkelsen T, Brodie C (2009) The induction of autophagy by gamma-radiation contributes to the radioresistance of glioma stem cells. Int J Cancer 125(3):717–722

Lum JJ, Bauer DE, Kong M, Harris MH, Li C, Lindsten T, Thompson CB (2005) Growth factor regulation of autophagy and cell survival in the absence of apoptosis. Cell 120(2):237–248

Macintosh RL, Timpson P, Thorburn J, Anderson KI, Thorburn A, Ryan KM (2012) Inhibition of autophagy impairs tumor cell invasion in an organotypic model. Cell Cycle 11(10):2022–2029

Maiuri MC, Le Toumelin G, Criollo A, Rain JC, Gautier F, Juin P, Tasdemir E, Pierron G, Troulinaki K, Tavernarakis N, Hickman JA, Geneste O, Kroemer G (2007) Functional and physical interaction between Bcl-X(L) and a BH3-like domain in Beclin-1. EMBO J 26(10):2527–2539

Maiuri MC, Malik SA, Morselli E, Kepp O, Criollo A, Mouchel PL, Carnuccio R, Kroemer G (2009) Stimulation of autophagy by the p53 target gene Sestrin2. Cell Cycle 8(10):1571–1576

Mantovani A, Allavena P, Sica A, Balkwill F (2008) Cancer-related inflammation. Nature 454(7203):436–444

Marceau F, Bawolak MT, Bouthillier J, Morissette G (2009) Vacuolar ATPase-mediated cellular concentration and retention of quinacrine: a model for the distribution of lipophilic cationic drugs to autophagic vacuoles. Drug Metab Dispos 37(12):2271–2274

Marino G, Salvador-Montoliu N, Fueyo A, Knecht E, Mizushima N, Lopez-Otin C (2007) Tissue-specific autophagy alterations and increased tumorigenesis in mice deficient in Atg4C/autophagin-3. J Biol Chem 282(25):18573–18583

Mathew R, Karp CM, Beaudoin B, Vuong N, Chen G, Chen HY, Bray K, Reddy A, Bhanot G, Gelinas C, Dipaola RS, Karantza-Wadsworth V, White E (2009) Autophagy suppresses tumorigenesis through elimination of p62. Cell 137(6):1062–1075

Mathew R, Kongara S, Beaudoin B, Karp CM, Bray K, Degenhardt K, Chen G, Jin S, White E (2007) Autophagy suppresses tumor progression by limiting chromosomal instability. Genes Dev 21(11):1367–1381

Mathew R, White E (2011) Autophagy in tumorigenesis and energy metabolism: friend by day, foe by night. Curr Opin Genet Dev 21(1):113–119

Matsunaga K, Saitoh T, Tabata K, Omori H, Satoh T, Kurotori N, Maejima I, Shirahama-Noda K, Ichimura T, Isobe T, Akira S, Noda T, Yoshimori T (2009) Two Beclin 1-binding proteins, Atg14L and Rubicon, reciprocally regulate autophagy at different stages. Nat Cell Biol 11 (4):385–396

Miracco C, Cosci E, Oliveri G, Luzi P, Pacenti L, Monciatti I, Mannucci S, De Nisi MC, Toscano M, Malagnino V, Falzarano SM, Pirtoli L, Tosi P (2007) Protein and mRNA expression of autophagy gene Beclin 1 in human brain tumours. Int J Oncol 30(2):429–436

Mizushima N (2010) The role of the Atg1/ULK1 complex in autophagy regulation. Curr Opin Cell Biol 22(2):132–139

Mizushima N, Levine B, Cuervo AM, Klionsky DJ (2008) Autophagy fights disease through cellular self-digestion. Nature 451(7182):1069–1075

Mizushima N, Noda T, Yoshimori T, Tanaka Y, Ishii T, George MD, Klionsky DJ, Ohsumi M, Ohsumi Y (1998) A protein conjugation system essential for autophagy. Nature 395 (6700):395–398

Mizushima N, Yamamoto A, Hatano M, Kobayashi Y, Kabeya Y, Suzuki K, Tokuhisa T, Ohsumi Y, Yoshimori T (2001) Dissection of autophagosome formation using Apg5-deficient mouse embryonic stem cells. J Cell Biol 152(4):657–668

Mohri M, Nitta H, Yamashita J (2000) Expression of multidrug resistance-associated protein (MRP) in human gliomas. J Neurooncol 49(2):105–115

Munshi A (2009) Chloroquine in glioblastoma–new horizons for an old drug. Cancer 115 (11):2380–2383

Nishida Y, Arakawa S, Fujitani K, Yamaguchi H, Mizuta T, Kanaseki T, Komatsu M, Otsu K, Tsujimoto Y, Shimizu S (2009) Discovery of Atg5/Atg7-independent alternative macroautophagy. Nature 461(7264):654–658

Ohsumi Y, Mizushima N (2004) Two ubiquitin-like conjugation systems essential for autophagy. Semin Cell Dev Biol 15(2):231–236

Pankiv S, Clausen TH, Lamark T, Brech A, Bruun JA, Outzen H, Overvatn A, Bjorkoy G, Johansen T (2007) p62/SQSTM1 binds directly to Atg8/LC3 to facilitate degradation of ubiquitinated protein aggregates by autophagy. J Biol Chem 282(33):24131–24145

Parsons DW, Jones S, Zhang X, Lin JC, Leary RJ, Angenendt P, Mankoo P, Carter H, Siu IM, Gallia GL, Olivi A, McLendon R, Rasheed BA, Keir S, Nikolskaya T, Nikolsky Y, Busam DA, Tekleab H, Diaz LA Jr, Hartigan J, Smith DR, Strausberg RL, Marie SK, Shinjo SM, Yan H, Riggins GJ, Bigner DD, Karchin R, Papadopoulos N, Parmigiani G, Vogelstein B, Velculescu VE, Kinzler KW (2008) An integrated genomic analysis of human glioblastoma multiforme. Science 321(5897):1807–1812

Pattingre S, Tassa A, Qu X, Garuti R, Liang XH, Mizushima N, Packer M, Schneider MD, Levine B (2005) Bcl-2 antiapoptotic proteins inhibit Beclin 1-dependent autophagy. Cell 122(6):927–939

Pirtoli L, Cevenini G, Tini P, Vannini M, Oliveri G, Marsili S, Mourmouras V, Rubino G, Miracco C (2009) The prognostic role of Beclin 1 protein expression in high-grade gliomas. Autophagy 5(7):930–936

Poole B, Ohkuma S (1981) Effect of weak bases on the intralysosomal pH in mouse peritoneal macrophages. J Cell Biol 90(3):665–669

Qu X, Yu J, Bhagat G, Furuya N, Hibshoosh H, Troxel A, Rosen J, Eskelinen EL, Mizushima N, Ohsumi Y, Cattoretti G, Levine B (2003) Promotion of tumorigenesis by heterozygous disruption of the beclin 1 autophagy gene. J Clin Invest 112(12):1809–1820

Ravikumar B, Moreau K, Rubinsztein DC (2010) Plasma membrane helps autophagosomes grow. Autophagy 6(8):1184–1186

Ray R, Chen G, Vande Velde C, Cizeau J, Park JH, Reed JC, Gietz RD, Greenberg AH (2000) BNIP3 heterodimerizes with Bcl-2/Bcl-X(L) and induces cell death independent of a Bcl-2 homology 3 (BH3) domain at both mitochondrial and nonmitochondrial sites. J Biol Chem 275 (2):1439–1448

Rosenfeld MR, Grossman SA, Brem S, Mikkelsen T, Wang D, Piao S, Davis LE, O'Dwyer PJ, Amaravadi RK (2010) Pharmacokinetic analysis and pharmacodynamic evidence of autophagy inhibition in patients with newly diagnosed glioblastoma treated on a phase I trial of hydroxychloroquine in combination with adjuvant temozolomide and radiation (ABTC 0603). J Clin Oncol 28(15s):abstr 3086

Roth W, Weller M (1999) Chemotherapy and immunotherapy of malignant glioma: molecular mechanisms and clinical perspectives. Cell Mol Life Sci 56(5–6):481–506

Ruiz-Irastorza G, Ramos-Casals M, Brito-Zeron P, Khamashta MA (2010) Clinical efficacy and side effects of antimalarials in systemic lupus erythematosus: a systematic review. Ann Rheum Dis 69(1):20–28

Sakamoto Y, Kato S, Takahashi M, Okada Y, Yasuda K, Watanabe G, Imai H, Sato A, Ishioka C (2011) Contribution of autophagic cell death to p53-dependent cell death in human glioblastoma cell line SF126. Cancer Sci 102(4):799–807

Sakurai T, He G, Matsuzawa A, Yu GY, Maeda S, Hardiman G, Karin M (2008) Hepatocyte necrosis induced by oxidative stress and IL-1 alpha release mediate carcinogen-induced compensatory proliferation and liver tumorigenesis. Cancer Cell 14(2):156–165

Salazar M, Carracedo A, Salanueva IJ, Hernandez-Tiedra S, Lorente M, Egia A, Vazquez P, Blazquez C, Torres S, Garcia S, Nowak J, Fimia GM, Piacentini M, Cecconi F, Pandolfi PP, Gonzalez-Feria L, Iovanna JL, Guzman M, Boya P, Velasco G (2009) Cannabinoid action induces autophagy-mediated cell death through stimulation of ER stress in human glioma cells. J Clin Invest 119(5):1359–1372

Sarbassov DD, Ali SM, Sengupta S, Sheen JH, Hsu PP, Bagley AF, Markhard AL, Sabatini DM (2006) Prolonged rapamycin treatment inhibits mTORC2 assembly and Akt/PKB. Mol Cell 22 (2):159–168

Schweers RL, Zhang J, Randall MS, Loyd MR, Li W, Dorsey FC, Kundu M, Opferman JT, Cleveland JL, Miller JL, Ney PA (2007) NIX is required for programmed mitochondrial clearance during reticulocyte maturation. Proc Natl Acad Sci U S A 104(49):19500–19505

Shen S, Kepp O, Kroemer G (2012) The end of autophagic cell death? Autophagy 8(1):1–3

Shimizu S, Kanaseki T, Mizushima N, Mizuta T, Arakawa-Kobayashi S, Thompson CB, Tsujimoto Y (2004) Role of Bcl-2 family proteins in a non-apoptotic programmed cell death dependent on autophagy genes. Nat Cell Biol 6(12):1221–1228

Shingu T, Fujiwara K, Bogler O, Akiyama Y, Moritake K, Shinojima N, Tamada Y, Yokoyama T, Kondo S (2009) Inhibition of autophagy at a late stage enhances imatinib-induced cytotoxicity in human malignant glioma cells. Int J Cancer 124(5):1060–1071

Shinohara ET, Cao C, Niermann K, Mu Y, Zeng F, Hallahan DE, Lu B (2005) Enhanced radiation damage of tumor vasculature by mTOR inhibitors. Oncogene 24(35):5414–5422

Singh SK, Clarke ID, Hide T, Dirks PB (2004) Cancer stem cells in nervous system tumors. Oncogene 23(43):7267–7273

Sotelo J, Briceno E, Lopez-Gonzalez MA (2006) Adding chloroquine to conventional treatment for glioblastoma multiforme: a randomized, double-blind, placebo-controlled trial. Ann Intern Med 144(5):337–343

Sou YS, Waguri S, Iwata J, Ueno T, Fujimura T, Hara T, Sawada N, Yamada A, Mizushima N, Uchiyama Y, Kominami E, Tanaka K, Komatsu M (2008) The Atg8 conjugation system is indispensable for proper development of autophagic isolation membranes in mice. Mol Biol Cell 19(11):4762–4775

Stupp R, Mason WP, van den Bent MJ, Weller M, Fisher B, Taphoorn MJ, Belanger K, Brandes AA, Marosi C, Bogdahn U, Curschmann J, Janzer RC, Ludwin SK, Gorlia T, Allgeier A, Lacombe D, Cairncross JG, Eisenhauer E, Mirimanoff RO, R. European Organisation for, T. Treatment of Cancer Brain, G. Radiotherapy, G. National Cancer Institute of Canada

Clinical Trials (2005) Radiotherapy plus concomitant and adjuvant temozolomide for glioblastoma. N Engl J Med 352(10):987–996
Sun B, Karin M (2008) NF-kappaB signaling, liver disease and hepatoprotective agents. Oncogene 27(48):6228–6244
Suzuki K, Ohsumi Y (2007) Molecular machinery of autophagosome formation in yeast, Saccharomyces cerevisiae. FEBS Lett 581(11):2156–2161
Takahashi Y, Coppola D, Matsushita N, Cualing HD, Sun M, Sato Y, Liang C, Jung JU, Cheng JQ, Mule JJ, Pledger WJ, Wang HG (2007) Bif-1 interacts with Beclin 1 through UVRAG and regulates autophagy and tumorigenesis. Nat Cell Biol 9(10):1142–1151
Takahashi Y, Karbowski M, Yamaguchi H, Kazi A, Wu J, Sebti SM, Youle RJ, Wang HG (2005) Loss of Bif-1 suppresses Bax/Bak conformational change and mitochondrial apoptosis. Mol Cell Biol 25(21):9369–9382
Takeuchi H, Kanzawa T, Kondo Y, Kondo S (2004) Inhibition of platelet-derived growth factor signalling induces autophagy in malignant glioma cells. Br J Cancer 90(5):1069–1075
Takeuchi H, Kondo Y, Fujiwara K, Kanzawa T, Aoki H, Mills GB, Kondo S (2005) Synergistic augmentation of rapamycin-induced autophagy in malignant glioma cells by phosphatidylinositol 3-kinase/protein kinase B inhibitors. Cancer Res 65(8):3336–3346
Tanaka Y, Guhde G, Suter A, Eskelinen EL, Hartmann D, Lullmann-Rauch R, Janssen PM, Blanz J, von Figura K, Saftig P (2000) Accumulation of autophagic vacuoles and cardiomyopathy in LAMP-2-deficient mice. Nature 406(6798):902–906
Tanida I, Ueno T, Kominami E (2004a) Human light chain 3/MAP1LC3B is cleaved at its carboxyl-terminal Met121 to expose Gly120 for lipidation and targeting to autophagosomal membranes. J Biol Chem 279(46):47704–47710
Tanida I, Ueno T, Kominami E (2004b) LC3 conjugation system in mammalian autophagy. Int J Biochem Cell Biol 36(12):2503–2518
Tasdemir E, Maiuri MC, Galluzzi L, Vitale I, Djavaheri-Mergny M, D'Amelio M, Criollo A, Morselli E, Zhu C, Harper F, Nannmark U, Samara C, Pinton P, Vicencio JM, Carnuccio R, Moll UM, Madeo F, Paterlini-Brechot P, Rizzuto R, Szabadkai G, Pierron G, Blomgren K, Tavernarakis N, Codogno P, Cecconi F, Kroemer G (2008) Regulation of autophagy by cytoplasmic p53. Nat Cell Biol 10(6):676–687
Tracy K, Dibling BC, Spike BT, Knabb JR, Schumacker P, Macleod KF (2007) BNIP3 is an RB/E2F target gene required for hypoxia-induced autophagy. Mol Cell Biol 27(17):6229–6242
Van Meir EG, Hadjipanayis CG, Norden AD, Shu HK, Wen PY, Olson JJ (2010) Exciting new advances in neuro-oncology: the avenue to a cure for malignant glioma. CA Cancer J Clin 60 (3):166–193
Venkatesh HS, Chaumeil MM, Ward CS, Haas-Kogan DA, James CD, Ronen SM (2012) Reduced phosphocholine and hyperpolarized lactate provide magnetic resonance biomarkers of PI3K/Akt/mTOR inhibition in glioblastoma. Neuro Oncol 14(3):315–325
Voss V, Senft C, Lang V, Ronellenfitsch MW, Steinbach JP, Seifert V, Kogel D (2010) The pan-Bcl-2 inhibitor (-)-gossypol triggers autophagic cell death in malignant glioma. Mol Cancer Res 8(7):1002–1016
Weihua Z, Tsan R, Huang WC, Wu Q, Chiu CH, Fidler IJ, Hung MC (2008) Survival of cancer cells is maintained by EGFR independent of its kinase activity. Cancer Cell 13(5):385–393
White E (2012) Deconvoluting the context-dependent role for autophagy in cancer. Nat Rev Cancer 12(6):401–410
Xie Z, Klionsky DJ (2007) Autophagosome formation: core machinery and adaptations. Nat Cell Biol 9(10):1102–1109
Yang Z, Klionsky DJ (2010) Mammalian autophagy: core molecular machinery and signaling regulation. Curr Opin Cell Biol 22(2):124–131
Yang ZJ, Chee CE, Huang S, Sinicrope FA (2011) The role of autophagy in cancer: therapeutic implications. Mol Cancer Ther 10(9):1533–1541

Young AR, Narita M, Ferreira M, Kirschner K, Sadaie M, Darot JF, Tavare S, Arakawa S, Shimizu S, Watt FM, Narita M (2009) Autophagy mediates the mitotic senescence transition. Genes Dev 23(7):798–803

Yu L, Alva A, Su H, Dutt P, Freundt E, Welsh S, Baehrecke EH, Lenardo MJ (2004) Regulation of an ATG7-beclin 1 program of autophagic cell death by caspase-8. Science 304(5676):1500–1502

Yue Z, Jin S, Yang C, Levine AJ, Heintz N (2003) Beclin 1, an autophagy gene essential for early embryonic development, is a haploinsufficient tumor suppressor. Proc Natl Acad Sci U S A 100(25):15077–15082

Zhao Y, Huang Q, Yang J, Lou M, Wang A, Dong J, Qin Z, Zhang T (2010) Autophagy impairment inhibits differentiation of glioma stem/progenitor cells. Brain Res 1313:250–258

Zheng H, Ying H, Yan H, Kimmelman AC, Hiller DJ, Chen AJ, Perry SR, Tonon G, Chu GC, Ding Z, Stommel JM, Dunn KL, Wiedemeyer R, You MJ, Brennan C, Wang YA, Ligon KL, Wong WH, Chin L, DePinho RA (2008) p53 and Pten control neural and glioma stem/progenitor cell renewal and differentiation. Nature 455(7216):1129–1133

Zhong JT, Xu Y, Yi HW, Su J, Yu HM, Xiang XY, Li XN, Zhang ZC, Sun LK (2012) The BH3 mimetic S1 induces autophagy through ER stress and disruption of Bcl-2/Beclin 1 interaction in human glioma U251 cells. Cancer Lett 323(2):180–187

Zhuang W, Li B, Long L, Chen L, Huang Q, Liang Z (2011) Induction of autophagy promotes differentiation of glioma-initiating cells and their radiosensitivity. Int J Cancer 129(11):2720–2731

Zupanska A, Kaminska B (2002) The diversity of p53 mutations among human brain tumors and their functional consequences. Neurochem Int 40(7):637–645

Part II

Partnership: Glioma Cell Partners and the Tumor Microenvironment

Para- and Autocrine Mediators in the Glioma Microenvironment

6

Kirsten Hattermann and Rolf Mentlein

Abstract

Unregulated growth, apoptosis-resistance, invasion, strongly increased angiogenesis, and immunosuppression are hallmarks of gliomas. Drivers of these tumor-promoting processes are several auto- and paracrine factors, mostly bioactive peptides. Among them, distinct neuropeptides, chemokines, growth factors, and cytokines play a prominent role. Since receptors for neuropeptides are often overexpressed on glioma cells, derivatives are increasingly used for imaging/diagnosis and radiotherapy, whereas clinical significance of agonists or antagonists remains questionable. Chemokines and their receptors significantly contribute to tumor growth and progression by exerting not only chemotactic but also proliferative and anti-apoptotic effects. Next to these small peptides, a multitude of classical growth factors and their receptors with tyrosine kinase activity are driving players of glioma cell progression. Growth factor receptors on glioma cells are partly overexpressed or constitutively activated through mutations, and therefore appear prominent targets for anti-glioma therapies. Several cytokines are heavily produced by glioma cells and exhibit not only immune-modulatory functions, but favor glioma cell proliferation, maintenance for the stem cell character of glioma stem(-like) cells, and may attract stroma cells. Apart from peptides/proteins, some lipid and also nutrient factors are additional drivers of glioma progression. Thus, factors produced by glioma cells (or glioma stem-like cells) exert autocrine as well as paracrine actions on endothelial (angiogenesis factors), microglial, and other cells. Vice versa, blood-borne factors or those produced by the glioma microenvironment drive tumor progression in complex ways. Despite its complexity, this network of auto- and paracrine mediators provides excellent targets for glioma diagnosis, imaging, and therapies.

K. Hattermann • R. Mentlein (✉)
Department of Anatomy, University of Kiel, Olshausenstraße 40, 24098 Kiel, Germany
e-mail: rment@anat.uni-kiel.de

A. Sedo and R. Mentlein (eds.), *Glioma Cell Biology*,
DOI 10.1007/978-3-7091-1431-5_6, © Springer-Verlag Wien 2014

Keywords
Angiogenesis • Apoptosis • Cancer • Chemotaxis • Imaging • Invasion • Proliferation • Therapy

Abbreviations

7TM receptor	Seven transmembrane domain receptor
ADAM	A disintegrin and metalloproteinase
bFGF	Basic fibroblast growth factor
EGF	Epidermal growth factor
FCS	Fetal calf serum
GDNF	Glial cell-derived neurotrophic factor
GSC	Glioma stem(-like) cell
HGF/SF	Hepatocyte growth factor/scatter factor
IGF	Insulin-like growth factor
IL	Interleukin
MAPK	Mitogen-activated protein kinase
MMP	Matrix metalloproteinase
PDGF	Platelet-derived growth factor
PG	Prostaglandin
PTN	Pleiotrophin, heparin-binding brain mitogen, heparin-binding growth factor 8
RTK	Receptor tyrosine kinase
Stat	Signal transducer and activator of transcription, signal transduction and transcription
TAM	Tumor-associated macrophage
TGF	Transforming growth factor
TIL	Tumor-infiltrating lymphocyte
VEGF	Vascular endothelial cell growth factor

6.1 The Glioma Microenvironment

Glioma tissue does not only contain cancer cells but also a considerable part of nonmalignant cells. These so-termed tumor stroma cells are regarded as constitutive parts of the tumor contributing to its growth, invasion, dissemination, and metastasis. Whereas fibroblasts, tumor-associated macrophages (TAMs), tumor-infiltrating lymphocytes (TILs), and endothelial cells create a complex cellular milieu in peripheral tumors, the glioma microenvironment consists of a simpler cellular architecture. Namely, apart from heterogeneous (as the term *glioblastoma multiforme* implies) tumor cells, tumor-associated microglia/macrophages and endothelial cells dominate (Fig. 6.1). Other types of stroma cells include (rarer)

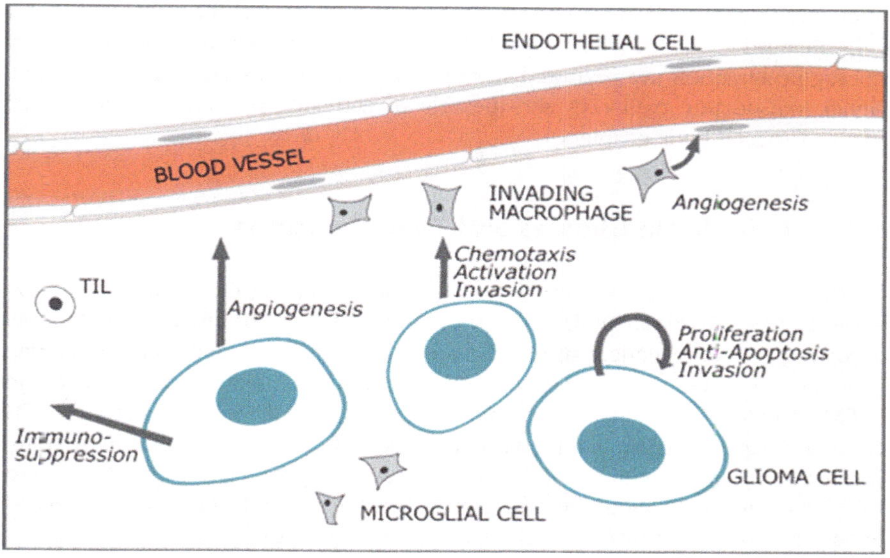

Fig. 6.1 Simplified cellular architecture of gliomas and action of auto- and paracrine factors. Gliomas are highly vascularized tumors that consist of the tumor cells, endothelial cells, perivascular macrophages, and microglial cells (=tumor-associated macrophages) apart from few other cell types (e.g., pericytes, smooth muscle cells; not drawn; lymphocytes = TILs). Factors secreted from these cells create a micro-milieu that favors tumor growth, invasion, angiogenesis, or immunosuppression. In gliomas, the normally tight blood–brain barrier may be (partially) opened

T-lymphocytes, neural precursor cells, pericytes, and vascular smooth muscle cells in the vascular niche, and astrocytes as well as meningeal cells/fibroblasts when the tumor expands (Charles et al. 2011).

Factors secreted (or exposed) from glioma cells can target glioma or glioma stem (-like) cells (GSC) to support tumor cell growth, or endothelial cells to induce angiogenesis, or attract and regulate other stroma cells like microglial cells. Vice versa factors from tumor stroma cells or blood-borne substances can support growth or invasion of glioma cells. Therefore, para- and autocrine factors in gliomas have been and are still in focus of glioma research to understand the fast growth, the unlimited invasive character of this type of tumor, its high vascularization, and its robust immunosuppression.

6.2 Types of Para- and Autocrine Factors and Their Receptors

Most factors influencing glioma cells are peptides; however, also substances like prostaglandins, sphingosine 1-phosphate, and also nutrients play a certain role. After giving an overview on the different classes of peptides and other factors, their types of receptors and modes of actions, the different functions in the glioma microenvironment are explained. Focus will be laid on factors acting on glioma cells themselves, since other aspects like angiogenesis, immunosuppression, or microglia attraction are reviewed in other chapters of this book. Also, (altered) signal transduction pathways and the role of activated transcription factors are focus of special book chapters.

6.3 Peptides in Gliomas and Their Receptors

Peptides are the largest group of bioactive factors acting on tumors. Their size varies from small oligopeptides to large proteins. They can be grouped into families according to their chemical structure, their type of receptor, and mode of actions. Neuropeptides and chemokines target seven transmembrane domain receptors (7TM receptors), growth factors bind to dimeric transmembrane receptors with tyrosine-kinase domains, and the various classes of cytokines to multimeric transmembrane receptors with mostly serine kinase activity/activation (Fig. 6.2). Here, we give an overview about the different classes, their receptors and mode of action under the aspect of their role in glioma diagnosis, progression, and spreading.

6.3.1 Neuropeptides

Neuropeptides and neuroendocrine or peripheral circulating peptide hormones are small (usually 3- to 40-residue) peptides that are typically produced by neurons or endocrine cells. They are synthesized from large precursors through proteolytic processing which is accomplished by the signal peptidase complex and subsequently by pro-protein (pro-hormone) convertases (PCs) cleaving after paired basic amino acids. Further modifications of neuropeptides involve liberation of C-terminal basic residues (Arg, Lys) by carboxypeptidase E or D, then C-terminal amidation (-NH_2) by oxidation of a glycine by peptidyl-glycine-α-amidating monooxygenase (PAM), or N-terminal cyclization of glutamine to pyroglutamic acid through pyroglutaminyl cyclase. Other posttranslational processing events may include acetylation, sulfation, phosphorylation, glycosylation, *n*-octanoylation (rare, of serine residues), and additional proteolytic cleavages. These processes occur in the *trans*-Golgi network and in secretory granules through acidification of these organelles. Neuropeptides are stored in secretory granules and released into the extracellular space/synapse upon stimulation of neurons/endocrine cells (involving Ca^{2+} influx).

As many classical neurotransmitters, neuropeptides bind to seven transmembrane domain receptors (7TM receptors; or G protein-coupled receptors) and typically generate second messengers like cAMP, cGMP (for natriuretic peptides), inositol trisphosphate (IP_3; or inositol 1,4,5-trisphosphate, $InsP_3$), diacylglycerol (DAG), and Ca^{2+}. These second messengers control phosphorylation of several intracellular targets including also transcription factors. Detailed information about neuropeptide receptors on astroglial and glioma cells, the second messengers

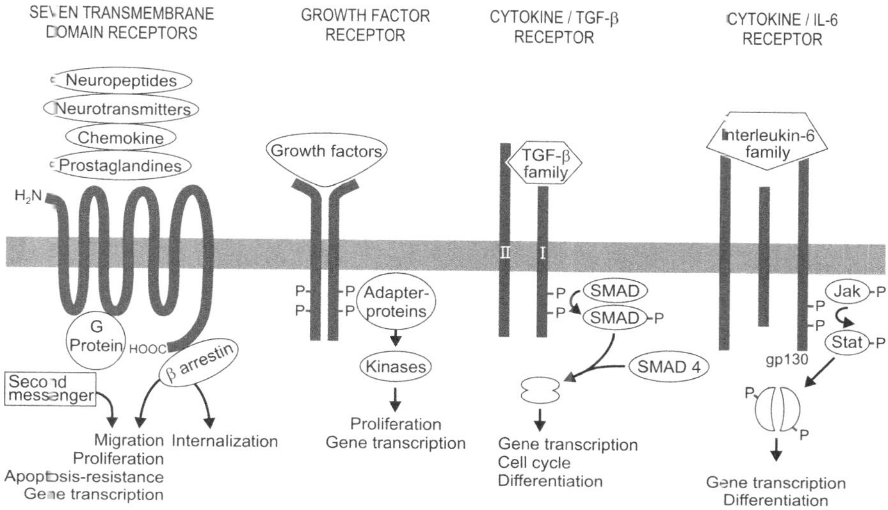

Fig. 6.2 Principal receptor types on glioma cells. Seven transmembrane domain receptors, 7TM receptors, or G protein-coupled receptors bind and are activated by some classical neurotransmitters, neuropeptides, chemokines, or prostaglandins and usually act via second messengers; however, alternative mechanisms also exists, some via β-arrestins that serves also in ligand-induced internalization of these receptors. Peptide growth factors typically bind to transmembrane receptors that dimerize upon binding; thereby, they are auto-phosphorylated by their intracellular kinase domains at tyrosine residues that then bind adapter proteins which initiate a series of phosphorylation of kinases (commonly referred as mitogen-activated protein (MAP) kinases) ending in the activation (phosphorylation) of transcription factors. Cytokine receptors are usually composed of two or three transmembrane units, here in the case of TGF-β and IL-6 receptors of binding and transducer domains (type I or gp130); upon ligand binding they are usually phosphorylated at serine/threonine residues, and then activate (phosphorylate) further signal transducer molecules (here: SMADs and Stats, see text) that form or activate transcription factors

involved, and further cellular effects can be found elsewhere (Krisch and Mentlein 1994).

In the brain, neuropeptides are far more stable than classical neurotransmitters. Namely, their half-life ranges instead of (milli)-seconds up to minutes. They are inactivated by mainly cell-surface-exposed peptidases, endo-, and exopeptidases that are partly specialized to such short peptides, but rarely to a certain neuropeptide itself (Mentlein 2004). Prominent examples of "neuropeptidases" are neprilysin (EC 3.4.24.11, CD10, neutral endopeptidase, enkephalinase), aminopeptidase N (EC 3.4.11.2, CD13), aminopeptidase A (EC 3.4.11.7), dipeptidyl peptidase IV (EC 3.4.14.5, CD26, DPP IV), angiotensin-converting enzyme (EC 3.4.15.1, CD143, ACE), or carboxypeptidase M (EC 3.4.17.12). They are exposed on the surface of microglial cells, endothelial cells, or neurons, but lesser on astrocytes or glioma cells (Lucius and Mentlein 1991; Lucius et al. 1995).

Neuropeptides or similar acting circulating peptide hormones can reach glioma cells from neurons or from the circulation through the (partial) opening of the blood–brain barrier; however, few (angiotensin, cholecystokinin) seem to be also produced by glioma cells themselves (Juillerat-Jeanneret et al. 2004; Oikonomou et al. 2008).

Contrasting the limited accessibility, surprisingly many neuropeptide/peptide hormone receptors have been identified in glioma cells in vitro and in situ, e.g , for substance P, vasoactive intestinal peptide (VIP), neuromedin B (NMB; analogue to amphibian bombesin), cholecystokinin (CCK), enkephalins, somatostatin, corticotrophin-releasing hormone (CRH), bradykinin, angiotenisin II, endothelins, natriuretic peptides (Montana and Sontheimer 2011; Arrieta et al. 2008; Feindt et al. 1995). This multitude of receptors may result from the function of neuropeptides in regulating normal astrocyte activity and metabolism (Krisch and Mentlein 1994).

In contrast, the functional importance of neuropeptide/peptide hormone receptors for glioma progression is not fully understood; they could involve growth regulating and migratory functions. However, neuropeptide receptors serve as tools for diagnosis, and some could become important drug targets (Fig. 6.3).

Some radiolabeled neuropeptides/peptide hormones are already clinically or experimentally used for the molecular imaging/diagnosis of gliomas in vivo, e.g., radioactive-labeled somatostatin or is stable derivative octreotide, neuromedin/bombesin (Luyken et al. 1994; Strauss et al. 2012). For radiotherapy and combined diagnosis, neuropeptides can also be labeled with appropriate nucleotides, e.g., with ^{90}Yttrium- or ^{213}Bismuth. Thus, radiolabeled derivatives of substance P or somatostatin/octreotide are already clinically applied to image and destroy tumors (Cordier et al. 2010; Merlo et al. 1999). Alternatively to radio-toxicity, conjugates of neuropeptides with cytotoxic drugs (e.g., doxorubicin) can be used to destroy gliomas; namely, upon receptor-mediated internalization they kill their target cells. Examples for this approach are especially conjugates with somatostatin/derivatives (Pozsgai et al. 2010) because this neuropeptide exhibits already anti-proliferative effects on its own (Feindt et al. 1995).

In few cases neuropeptides/peptide hormones appear to be directly involved in glioma pathology; therefore, agonists or antagonists could become valuable therapeutics. For example, corticotrophin-releasing hormone (CRH) was even more effective than either dexamethasone or temozolomide to reduce glioma edema in humanized rodent models (Moroz et al. 2011). Bradykinin (BK) significantly enhances glioma cell migration/invasion and guides glioma cells toward blood vessels in acute rat brain slices, a model of the frequent perivascular invasion of gliomas (Montana and Sontheimer 2011). Since somatostatin/analogues act anti-proliferative and inhibitory on the production of growth and angiogenesis factors (Feindt et al. 1995; Mentlein et al. 2001), they have been evaluated for glioma therapy. However, despite promising efficacy in preclinical models (Pinski et al. 1994), this and other approaches with neuropeptide agonists or antagonists did not reach clinical trials.

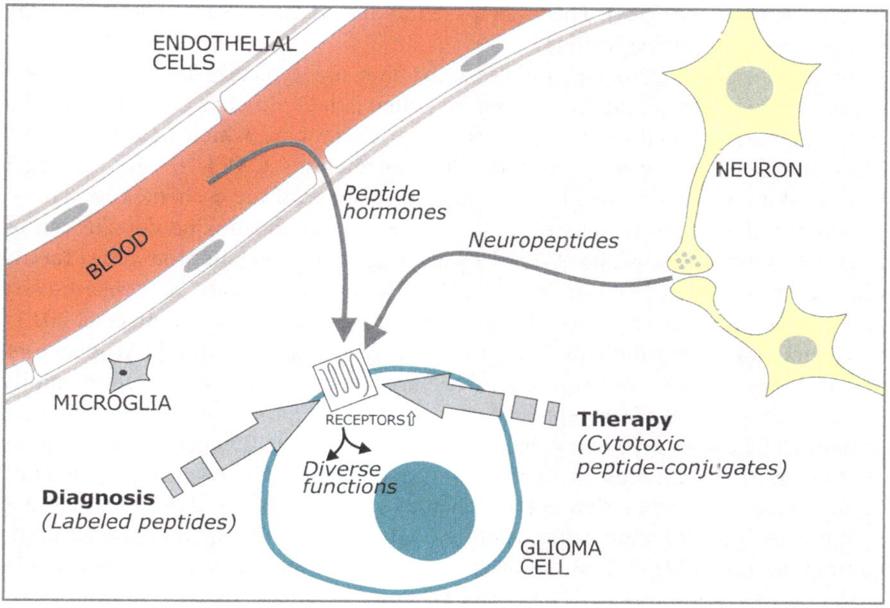

Fig. 6.3 Role of neuropeptides and peptide hormones in gliomas. Peptide hormones reach gliomas from the blood via the partially opened blood–brain barrier, whereas neuropeptides are released from neurons. Neuropeptide/peptide hormone receptors are often overexpressed in glioma cells and exert multiple functions. Due to overexpression, they can serve for in vivo diagnosis/imaging of gliomas using (radio)-labeled derivatives or for therapy applying cytotoxic peptide conjugates (or for both purposes: theragnostics)

Taken together, though neuropeptides/peptide hormones reach gliomas mainly from surrounding neurons and the circulation, neuropeptide receptors are frequently and often highly expressed on glioma cells. In particular receptors for substance P (tachykinins), somatostatin, neuromedins (bombesin-like peptides), or bradykinin are found. Though the physiological significance of neuropeptide receptors remains not always clear, they provide excellent targets for the detection and ligand-directed destruction of glioma cells.

6.3.2 Chemokines

Chemokines are small (8–10 kDa), mostly secreted peptides that were initially described as chemotactic cytokines according to their ability to attract (immune) cells. The family is divided in four subgroups based on the pattern of two conserved N-terminal cysteine residues that build disulfide bonds to two further cysteine residues accounting for the characteristic three-dimensional chemokine structure. These two cysteine residues can be either directly neighbored (CC- or β-chemokines) or separated by one (CXC- or α-chemokines) or three amino acids

(CX3C- or δ-chemokines). The C- or γ-chemokine XCL1 = lymphotactin has only one N-terminal cysteine residue.

In humans, about 50 chemokines exist that are typically produced as pre-peptides. The signal sequences (about 20 amino acids) are cleaved by the cellular signal peptidase yielding the active (mature) chemokine which then is secreted from the cell. As exceptions, the alpha-chemokine CXCL16 and CX3CL1 (=fractalkine = neurotactin), the only known member of the δ-chemokine family, are produced as transmembrane forms from which the chemokine domain can be released by proteases of the ADAM (a disintegrin and metalloproteinase) family (Gough et al. 2004; Garton et al. 2001). Chemokines can be constitutively expressed or inducible by cytokines and other growth factors (Mentlein et al. 2013).

Chemokines exert their biological effects by interacting with 7TM receptors. Shortly after the first descriptions of chemokine receptors, this receptor family came in focus as some receptors can act as co-receptors for the HIV (human immunodeficiency virus) entry (Feng et al. 1996). To date 23 chemokine receptors are known in human; thus, most receptors have more than one specific chemokine ligand. Vice versa, some (but not all) chemokines may also bind to more than one receptor. As chemokine receptors activate mostly $G_{i\alpha}$ proteins, their signal transduction often involves inhibition of adenylyl cyclase and activation of phospholipase C with generation of IP_3/Ca^{2+} and DAG as second messengers. Furthermore, several kinases of the MAPK (mitogen-activated protein kinases) and Akt (also known as protein kinase B, PKB) pathways are directly or indirectly phosphorylated and thereby activated. These multiple signal transduction effects are further complicated by heterodimerization of receptors and other downstream interactions [e.g., via β-arrestins that serve also in ligand-induced internalization of these receptors (Rajagopal et al. 2010)].

Chemokine activities are regulated by postsecretory processing through cell-surface or extracellular proteases (Mortier et al. 2011). This processing is mainly accomplished by the endoproteases matrix metalloproteinases (MMPs), cathepsins B, D, and L, plasmin/plasmin activators (urokinase), as well as exopeptidases like dipeptidyl peptidase IV (DPPIV)/CD26 or carboxypeptidase N. The chemokine activity can be eliminated by the degradation of the N-terminus (e.g., by dipeptidyl peptidases), but these N-terminal truncated chemokines can nevertheless bind to receptors through other domains and then act as antagonists (Ludwig et al. 2002; Mentlein 2004). In case of the CXCL16 and CX3CL1/fractalkine (that are highly expressed by glioma cells) liberation of the (active) chemokine domain from the membrane-bound form can also be regulated by "shedding" through proteases of the ADAM (a disintegrin and metalloproteinase) family (Abel et al. 2004; Ludwig et al. 2005). Apart from proteolytic modifications, receptor available levels of at least some chemokines can be regulated by capture and release by cell-surface or extracellular glycosaminoglycans (Allen et al. 2007; Rot 2010).

A distinct panel of chemokines is expressed in the developing and adult central nervous system, either constitutively or tightly regulated, by neurons, glial, endothelial, and meningeal cells as well as precursors. Especially the CXCL12-CXCR4-

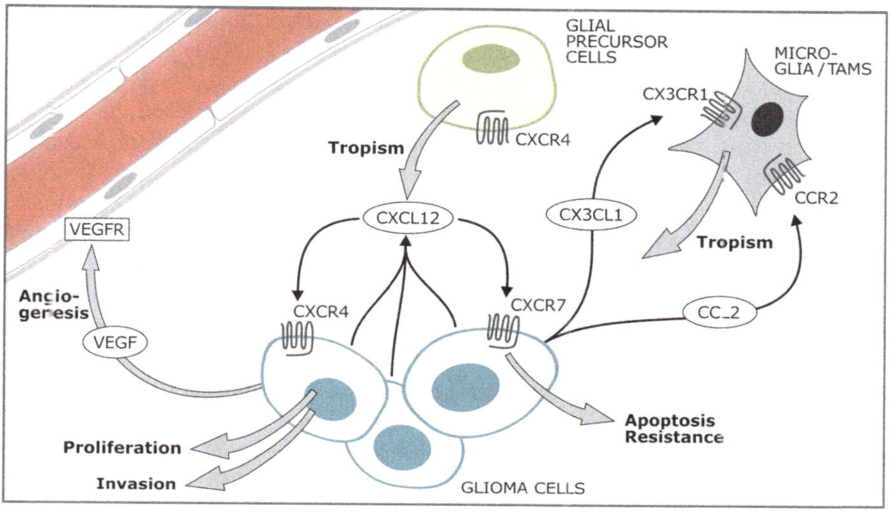

Fig. 6.4 Roles of chemokines in the glioma microenvironment. Chemokines like CXCL12/SDF-1, CCL2, or CX3CL1/fractalkine are produced by glioma (and partly stroma) cells and exert autocrine effects on gliomas cells by enhancing proliferation, apoptosis resistance, production of growth/angiogenesis factors, and/or invasion. Furthermore, they attract and activate glial precursor/stem cells and tumor stroma cells and thereby considerably contribute to the dissemination and growth of the tumor, see text

axis is known to play a decisive role in migration of stem cells and lamination, whereas several others (e.g., CCL2, CXCL8 = IL8, CX3CL1) contribute to pain and inflammation processes (Miller et al. 2008; Knerlich-Lukoschus et al. 2011; Reiss et al. 2002; Ransohoff 2009).

Accordingly, expression and functional role of the above-mentioned "CNS chemokines" is also well investigated in gliomas, where they significantly contribute to tumor growth and progression (Fig. 6.4). Here, we list the most important and/or best described chemokine/receptor axes.

CXCL12–CXCR4–CXCR7 CXCL12 = SDF-1 (stromal cell-derived factor-1) is a secreted 67 (α-splice form)/72 (β-splice form) residue peptide of the CXC family initially described as a bone marrow stromal cell-secreted factor retaining pre-B lymphocytes and supporting their maturation (Nagasawa et al. 1994). CXCL12 can bind to its long known receptor CXCR4 (gene locus 2q21) and to CXCR7 (gene locus 2q37) that was discovered a decade later and also can bind CXCL11 (=ITAC, interferon-inducible T cell alpha chemoattractant) with 10-fold lower affinity. Activation of CXCR4 by CXCL12 yields lower cAMP levels by inhibition of the adenylylcyclase by the $G_{i\alpha}$ subunit and mobilization of intracellular calcium via IP_3. In contrast, CXCR7 lacks the DRYLAIVHA motif essential for G protein recruitment and therefore fails to mobilize calcium. Nevertheless, CXCR7 can mediate signaling (e.g., activation of MAP kinases) upon binding of its specific ligands via β-arrestin, a protein that is also employed in the desensitization and internalization of activated G protein-coupled receptors (Rajagopal et al. 2010; Hattermann and Mentlein 2013; Gupta et al. 1998).

In gliomas as well as in normal brain, CXCL12 is widely produced. Its receptor CXCR4 is expressed by glioma cells with stem/progenitor cell properties that are predominantly localized in areas of necrosis, angiogenesis, and invasion (Hattermann et al. 2010; Ehtesham et al. 2009; Rempel et al. 2000). According to the character and localization of this CXCR4 expressing subpopulation of glioma cells, CXCL12 can promote their proliferation (Ehtesham et al. 2006, 2009), migration (Sciaccaluga et al. 2010), invasion by regulation of MMPs (Zhang et al. 2005), and angiogenesis by induction of VEGF (Yang et al. 2005) (Fig. 6.4). Additionally, high CXCL12 levels seem to mediate tropism of (healthy) CXCR4 expressing glial precursor cells toward gliomas (Ehtesham and Thompson 2012) that may contribute to glioma biology. In contrast, CXCR7 is expressed by more differentiated cells within gliomas and mediates resistance from apoptosis, but not proliferation or migration (Hattermann et al. 2010, 2012).

Thus, one ligand—CXCL12—can promote different processes of glioma progression in different subpopulations of glioma cells. This fine-tuning of multiple chemokine effects by differential expression of receptors is a long known feature in immune responses. It might be a promising target for the therapy of gliomas (and other tumors), especially as specific antagonists for CXCR4 and CXCR7 are available.

CX3CL1–CX3CR1 CX3CL1 (=fractalkine) is a 373 amino acid transmembrane protein. It is composed of an extracellular chemokine domain, a mucin-like stalk region, a transmembrane domain, and a short intracellular tail. The chemokine domain can be released by proteolytic cleavage through proteases of the ADAM family, in particular ADAM10 and ADAM17 (see above).

In the periphery, CX3CL1 is expressed by endothelial and vascular smooth muscle cells; it can be upregulated by inflammatory cytokines. The receptor CX3CR1 is expressed by monocytes, dendritic cells, NK cells, and subsets of T lymphocytes. Either in its transmembrane form or as liberated chemokine CX3CL1 can bind to CX3CR1. As a transmembrane ligand, it can accomplish (G protein-independent) adhesion contacts (e.g., leukocytes to endothelium), while the soluble chemokine induces chemotaxis of leukocytes (Haskell et al. 1999; Harrison et al. 1998).

In the brain, CX3CL1 is produced by neurons and astrocytes, whereas the receptor CX3CR1 is mainly expressed on microglial cells (Ludwig and Mentlein 2008). While early investigations show microglia-dependent neurotoxicity mediated by CX3CL1/CX3CR1 (Zujovic et al. 2000), more recent reports hint to neuroprotective effects of CX3CL1, especially in inflammation models with CX3CR1-deficient rodents (Re and Przedborski 2006). Not surprising, CX3CL1 is expressed by human gliomas and attracts glioma-derived CX3CR1$^+$ microglia/macrophages that can account for up to 30 % of glioma samples (Held-Feindt et al. 2010). A patient study additionally revealed that a polymorphism of the

CX3CR1 gene was associated with reduced microglia infiltration and prolonged mean survival times of glioma patients (Rodero et al. 2008).

A murine glioblastoma model in contrast does not substantiate these findings; no differences could be observed between normal and CX3CR1-deficient mice (Liu et al. 2008). This may well be due to species distinctions and compensation in rodents but may also result from regulations and microenvironmental cofactors of the ("chronic") cancerogenesis that cannot be entirely mimicked in a model. However, beside from mediating chemotaxis, CX3CL1 induces the expression of matrix metalloproteinase (MMP2, -9, and -14) in microglia/macrophages isolated from human glioma samples (Held-Feindt et al. 2010) which not only facilitate the migration of microglial cells/macrophages toward the tumor but also enable the spreading of glioma cells to degraded extracellular matrix areas.

CCL2–CCR2 CCL2 (MCP-1 = monocyte chemoattractant protein-1) was firstly isolated from glioma cell supernatants and proved to be a chemoattracting protein for monocytes in vitro (Yoshimura et al. 1989). Like many CC chemokines its gene is located on Chromosome 17, and the mature chemokine is about 13 kDa in size. It can bind to CCR2 eliciting calcium mobilization and chemotaxis (Yamagami et al. 1994). In the CNS, CCL2 is mainly produced by astrocytes and microglia, and plays an important role in brain development and in many nonneoplastic brain pathologies, e.g., multiple sclerosis, ischemic and traumatic injuries, and neuropathic pain (Semple et al. 2010).

In gliomas, CCL2 expression correlates with infiltration of microglia/macrophages (Leung et al. 1997). CCL2 can be induced by inflammatory cytokines (Desbaillets et al. 1994), and overexpression of CCL2 in experimental gliomas results in increased microglial infiltration that produce such factors (Platten et al. 2003). Apart from the recruitment of monocytic cells, CCL2 also seems to be a trophic factor for adult neural stem cells expressing CCR2 (Widera et al. 2004). According to elevated CCL2 expression levels, this attraction of neural stem/progenitor cells may occur for example upon ischemic injuries to promote repair processes (Liu et al. 2007) or can also be mediated by CCL2 expression of glioma cells where the function of these NSCs is not solved yet (Magge et al. 2009).

Especially the observation of Platten et al. (2003) that the aggressiveness of glioma growth is increased by CCL2 has rendered the CCL2-CCR2-axis an interesting target for glioma (and other cancer) therapies. First approaches with fusion proteins and neutralizing antibodies seem promising, at least for combination therapies (Rafei et al. 2011; Zhu et al. 2011).

Taken together, glioma cell-secreted chemokines may act as autocrine factors to enhance proliferation, invasion, and resistance from apoptosis, or as paracrine factors to mediate angiogenesis and chemotaxis and activation of stem/glial precursor cells and microglia/macrophages.

6.3.3 Growth and Angiogenesis Factors

Classical growth factors play a driving role in development, growth, and survival of differentiated cells as well as for tumor cell growth and apoptosis resistance. Structurally, growth factors are mostly larger proteins, often glycosylated, and sometimes forming homo- or hetero-(di)mers. Their precursors possess a signal sequence that after cleavage directs them to the Golgi apparatus where they might be glycosylated; subsequently they are constitutively secreted from the cells. Certain growth factors, e.g., those of the epidermal growth factor (EGF)-family (transforming growth factor-α, TGF-α; heparin-binding EGF-like growth factor, HB-EGF; amphiregulin, AREG), are like transmembrane cytokines and chemokines synthesized in a membrane-bound pro-form from which they are released by proteolytic cleavage through proteases of the ADAM family, in particular by ADAM 10 and 17 (Mentlein et al. 2012). Besides being cleaved from the cell surface, some growth factors (e.g., hepatocyte growth factor/scatter factor HGF/SF) require extracellular proteolytic maturation to gain full activity (Abounader and Laterra 2005). Furthermore, growth factors can be associated to the extracellular matrix (e.g., basic fibroblast growth factors, bFGF or FGF2; pleiotrophin, PTN; splice forms of vascular endothelial growth factor, VEGF-A_{165}) from where they may also be liberated by proteases.

Binding of growth factors to their receptors typically initiates dimerization of transmembrane chains (in case of insulin or insulin-like growth factors, IGFs, disulphide-linked dimers of two chains are already formed), followed by autophosphorylation of tyrosine residues by their intrinsic intracellular protein-tyrosine kinase activity, thus being termed receptor tyrosine kinase (RTK) receptors (Fig. 6.2). In glioma cells such RTK receptors (and/or the ligands) can be overexpressed or constitutively activated. Subsequently to autophosphorylation, adapter proteins (with SH2, Src Homology 2, domains) are bound, and several signal transduction cascades are initiated, principally of the MAPK and Akt pathways, which results in transcription of several genes, including transcription factors, as well as in cell proliferation.

Apart from modulated biosynthesis (and partly proteolytic processing), growth factor activities are regulated sometimes by binding to extracellular matrix components, by degradation through MMPs, and also by internalization/lysosomal degradation.

Glioma cells produce a variety of growth factors that either target glioma cells in an autocrine way, or target tumor stroma cells as paracrine growth/attraction factors, or endothelial cells as angiogenesis factors (Hoelzinger et al. 2007; Li and Graeber 2012; Nakada et al. 2013). The paracrine functions to different cell types are topics of other chapters of this book (Chaps. 7 and 8). Sometimes, these glioma and stroma cell-targeting functions are executed by a single factor. Here we list representative autocrine growth factor systems that enhance tumor cell growth, motility, and invasion.

The **EGF-EGFR**-signaling axis is one of the most important drivers of glioma cell proliferation. Members of the EGF family of ligands are EGF, heparin-binding

EGF-like growth factor (HB-EGF), transforming growth factor-α (TGF-α), neuregulins, amphiregulin, betacellulin, and others. They share one common structural motif, namely the EGF domain, which consists of six conserved cysteine residues $\underline{C}X_7\underline{C}X_{4\text{-}5}\underline{C}X_{10\text{-}13}\underline{C}X\underline{C}X_8GXR\underline{C}$ that form three intramolecular disulfide bonds. Most ligands are synthesized in membrane-associated pro-forms from which they are liberated by proteolytic cleavage through members of the ADAM-family (Mentlein et al. 2012). Soluble EGF-ligands bind to the ErbB family of receptors (EGFR = ErbB-1, HER2/c-neu = ErbB-2, Her 3 = ErbB-3, Her 4 = ErbB-4; HER = human epithelial receptor, ErbB derived from Erythroblastic Leukemia Viral Oncogene Homolog), a subfamily of the above described RTK receptor type. In tumors, ErbB receptors are often upregulated or are mutated which results in permanent activation of the intracellular tyrosine kinase activity. Therefore, ErbB receptors play an essential role in the progression of gliomas as in many other types of cancer.

Peptides of the EGF-family like EGF, TGF-α, and others are produced by glioma and glioma stroma cells; sometimes they are also overexpressed as compared to nontumor sites (Tang et al. 1997; Mishima et al. 1998). Apart from intra-tumoral production, they are synthesized by surrounding neurons and astrocytes, in particular by reactive astrocytes of the adjacent brain parenchyma (Hoelzinger et al. 2007). Thereby, auto- or paracrine loops are formed that support glioma growth considerably. Even more important than increased ligand production is the overexpression of receptors, their mutations, and an aberrant signal transduction—features that are often found also in other tumor types.

EGFR/ErbB-1 gene amplification is the most common genetic alteration in high-grade glioma (Fleming et al. 1992). Overall, this is found in 36–40 % of glioblastomas (Hatanpaa et al. 2010). About 50 % of EGFR-amplified tumors also harbor a constitutively active mutant form of the receptor termed ΔEGFR or EGFRvIII (Inda et al. 2010). This mutant is generated from a deletion of exons 2–7 of the EGFR gene resulting in an in-frame deletion of 267 amino acids from the extracellular domain of the receptor (Hatanpaa et al. 2010). EGFRvIII is unable to bind its ligand and signals constitutively. Noteworthy, EGFRvIII is usually co-expressed together with the wild-type (wt) receptor. In some glioblastomas, almost all of the tumor cells show EGFR amplification, but in others the proportion of malignant cells with EGFR amplification can vary from 10 % to 60 % in different areas of the same tumor (Hatanpaa et al. 2010). Apart from EGFR, also other members of the ErbB family, e.g., HER2, are overexpressed in part of the glioma patients (about 10–15 % in glioblastoma; (Potti et al. 2004; Koka et al. 2003), but HER2-overexpression is much more frequent in common glioma cell lines (Liu et al. 2004).

EGFR and *HER2* amplifications as well as mutations greatly enhance the EGF-mediated autocrine loop and signaling which should result in increased tumor growth and glioma cell motility. Somewhat surprising therefore, glioblastoma patients with EGFRvIII expression have no shorter survival times (Aldape et al. 2004), and also ErbB receptor family overexpression per se seems not to drastically influence patients' lifespan (Potti et al. 2004).

Nevertheless, EGFR and HER2 overexpression and mutations render ErbB receptors very attractive targets for glioma-diagnosis/imaging and in particular for glioma growth-inhibiting therapies. Several radiopharmaceuticals for PET and SPECT as well as fluorescent probes for optical imaging have been developed (Hatanpaa et al. 2010) which, however, have not entered the routine clinical practice. For glioma therapy, humanized monoclonal antibodies and especially EGFR-kinase inhibitors (e.g., Erlotinib, Gefitinib, Lapatinib) appear to be promising. However, recent clinical experiences with EGFR kinase inhibitors have been disappointing, mainly because resistance arises (Mukasa et al. 2010; Day and Waziri 2012). This resistance of gliomas and other tumors to EGFR-directed therapies may result from mutations desensitizing the kinase activity or from the activation of alternate oncogenic pathways, including also other Erb- and further growth factor receptors (Camp et al. 2005). For example, upon kinase-targeted therapy of glioblastomas other receptor tyrosine kinases such as PDGFR, IGF1-R, and c-Met are upregulated and compensate for decreased signaling by EGFRvIII or EGFR overexpression (Stommel et al. 2007).

The **PDGF** (platelet-derived growth factor) family comprises disulfide-linked dimeric proteins consisting of four homodimeric proteins, PDGF-AA, PDGF-BB, PDGF-CC, and PDGF-DD, and one heterodimeric protein, PDGF-AB. These ligands are biologically active only in their dimeric forms (about 30 kDa). Furthermore, their binding to proteins in plasma or in the extracellular matrix locally concentrates them and thereby enhances their potency. PDGFs play an important role in developmental differentiation of cells (including oligodendrocytes), in cell proliferation, in cell migration, and in angiogenesis. The finding that the *v-sis* oncogene of simian sarcoma virus (SSV) is a retroviral homolog of the *PDGF B-chain* gene was an early paradigm of oncogenes driving malignant transformation of tumor cells (Heldin and Westermark 1984). PDGFs signal through the protein tyrosine kinase receptors, PDGF receptor-α (PDGFR1) and -β (PDGFR2), each encoded by a different gene. Signal transduction is initiated by homo- or hetero-dimerization of these receptor chains, depending on the growth factor bound. The subsequent signaling cascade is similar to other RTKs.

PDGFs and their receptors are often overexpressed in gliomas, but with less frequency (11 %) as compared to EGFRs; however, this still renders them as important amplified RTK genes especially in some types of gliomas (Fleming et al. 1992; Nister et al. 1988; Cancer Genome Atlas Research Network 2008). Also *PDGFR* mutations have been described in gliomas, but they occur less frequently than *EGFR* mutations (Ozawa et al. 2010; Alentorn et al. 2012).

Overall, the PDGF-PDGFR-system is regarded one of the most important autocrine growth factor axes in gliomas, at least for a subset of them. Next to direct tumor cell growth, PDGF acts—together with other factors—as an angiogenic factor, mostly over the PDGFR-β on endothelial cells, whereas glioma cells preferentially express PDGFR-α (Hermanson et al. 1992).

The view of PDGF as a driver of glioma growth is strongly supported by analyses by the Cancer Genome Atlas Research Network (2008). Three core pathways were identified, namely the receptor tyrosine kinase/RAS/

phosphatidylinositol 3-kinase, p53, and RB signaling pathways. Later, four distinct types of glioblastoma were defined: the classical, the mesenchymal, the neural, and the proneural subtype (Verhaak et al. 2010). The proneural subtype is characterized by aberrations in the PDGF α-receptor pathway, in addition to mutations in the *isocitrate dehydrogenase 1 (IDH1)* gene.

To address the functional role of PDGF in gliomas, transgenic mouse models were generated in which overexpression of human PDGFs in astrocytic cells of the brain under control of a human GFAP promoter fragment was achieved (Nazarenko et al. 2012). The PDGF-B-overexpressing transgenic mice did not develop brain tumors; however, after crossing into a Tp53 null background, two PDGF-B transgenic mouse lines developed brain tumors at high rates of 68 and 43 % (Hede et al. 2009). In contrast to PDGF-B, introduction of the PDGF-A (long isoform) induced accumulation of immature cells in the mouse brain characterized by expression of NG2, PDGFR-α, and Olig2, markers for oligodendrocyte progenitors which might form oligodendrogliomas (Nazarenko et al. 2011). In fact, PDGFR1 is fundamentally involved in the differentiation and expansion of early ($NG2^+$, $A_2B_5^+$, $PDGFR1^+$) and late ($NG2^+$, $O4^+$, $PDGFR1^+$) oligodendrocyte precursors.

Thus, the PDGF-PDGFR-system is somewhat similar but more restricted as EGF-EGFR-system in gliomas. Therefore, subsets of gliomas seem promising targets for PDGFR-directed diagnosis and therapy. Again, though more or less specific inhibitors of the PDGFR tyrosine kinase have been developed (e.g., Imatinib, Pazopanib) and show some promising preclinical results (Dong et al. 2012), the clinical outcome even in patients with PDGFR-overexpression is not yet convincing (Paulsson et al. 2011).

HGF/SF (hepatocyte growth factor/scatter factor) is a multifunctional growth factor and angiogenesis factor. The single-chain inactive precursor is cleaved by serine proteases into a 69-kDa β-chain and a 34-kDa β-chain that are linked by a disulfide bond (Birchmeier et al. 2003; Abounader and Laterra 2005). The α-chain contains an N-terminal hairpin domain followed by four kringle domains, and the β-chain contains a serine protease-like domain with no enzymatic activity. The only known receptor for HGF/SF is the proto-oncogene c-Met. The tyrosine kinase receptor c-Met is synthesized from a 170-kDa glycosylated precursor that is cleaved into an extracellular 50-kDa α-chain and a disulfide-linked transmembrane 140-kDa β-chain with a cytoplasmic tyrosine kinase domain. Signaling is achieved via the Ras/MAPK, PI3K/Akt, and Stat pathways which mediate various cellular effects such as cell cycle regulation, cytoskeleton alterations, and gene expression including proteases.

In gliomas, HGF/SF is expressed and secreted mainly by tumor cells, whereas c-Met receptors are found on tumor and vascular endothelial cells (Rosen et al. 1996; see also Fig. 6.5). Expression levels of ligand and receptor correlate with tumor grade (Koochekpour et al. 1997). Gene transfer of HGF/SF into glioma cells enhances their tumorigenicity, tumor growth, and tumor-associated angiogenesis (Laterra et al. 1997) whereas inhibition of endogenous HGF/SF and c-Met production substantially reduces tumor growth and prolongs survival in animal models (Abounader et al. 1999). The autocrine loop formed by ligand and receptor

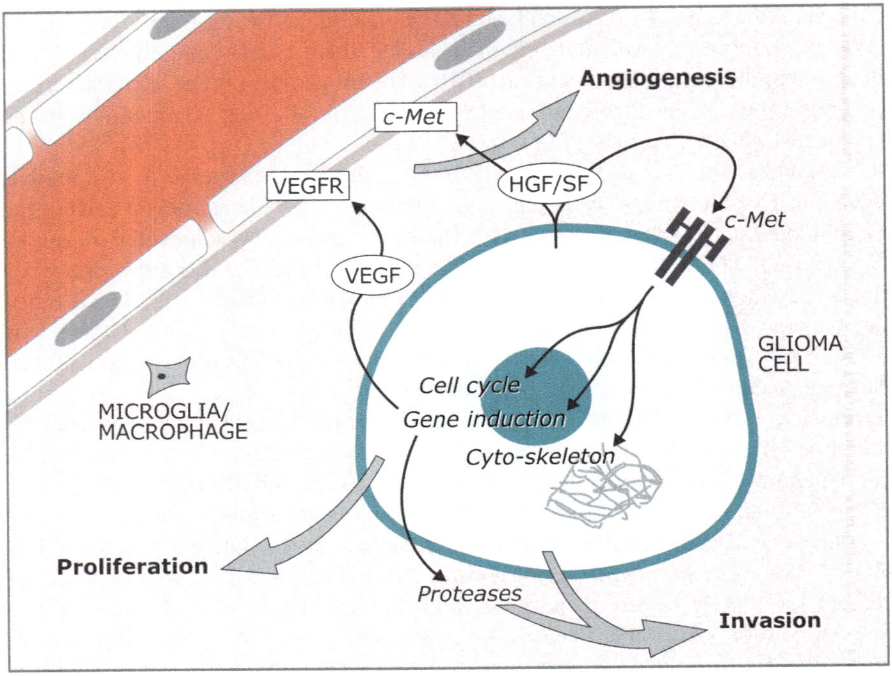

Fig. 6.5 Auto- and paracrine roles of hepatocyte growth factor/scatter factor (HGF/SF) and its receptor c-Met in gliomas. HGF/SF is synthesized prominently by glioma cells and acts as autocrine to induce cell proliferation and invasion that depends on altered motility and the secretion of proteases degrading extracellular matrix. Similar, via c-Met on endothelial cells and indirectly via induction of vascular endothelial growth factor (VEGF) and its receptors (VEGFR), the ligand induces angiogenesis and thereby increases tumor expansion

on glioma cells provides not only a mitogenic stimulation but also induces tumor cell migration and invasion.

In fact, HGF/SF has been regarded as the most potent stimulator of glioma cell migration (Brockmann et al. 2003). Furthermore, HGF/SF inhibits basal and radio/chemotherapy-induced tumor cell death and apoptosis, and as c-Met is also expressed on endothelial cells the ligand acts additionally as a powerful angiogenesis factor (Bowers et al. 2000).

Glioma cell invasion as well as angiogenesis (invasion of endothelial cells) requires extracellular matrix degradation and remodeling. This is mediated by HGF/SF-induced expression and synthesis of MMPs and urokinase. As the name scatter factor implies, dissociation and migration of tumor and endothelial cells is also induced. In addition to its direct angiogenic activities, HGF/SF acts as an indirect angiogenic factor through autocrine induction of VEGF expression and secretion in malignant gliomas (Moriyama et al. 1998). When comparing three important angiogenesis factors, HGF/SF and VEGF, but not bFGF, were highly significantly correlated with microvessel density in surgical glioma samples (Schmidt et al. 1999).

Thus, the HGF/SF–c-Met-axis appears to be an important auto- and paracrine system in gliomas. As an autocrine factor, HGF/SF does not only drive tumor cell proliferation and mediate apoptosis resistance, but also induces tumor cell dissemination and invasion. In a paracrine way, the ligand promotes directly and indirectly tumor angiogenesis. Inhibition of this axis should therefore be an excellent target for glioma therapy. NK4 is an internal fragment of HGF that binds to but does not activate the c-Met receptors, and acts as a competitive antagonist (Matsumoto and Nakamura 2003). Also, neutralizing antibodies against HGF/SF were tested (Cao et al. 2001). Furthermore, soluble c-Met receptor consisting of a recombinant protein corresponding to the entire extracellular domain of c-Met, truncated before the transmembrane domain, were used in experimental preclinical studies (Michieli et al. 2004). For clinical translations, small molecule inhibitors targeting the catalytic activity of c-Met kinase like PHA-665752 (Pfizer), SU11274 (SUGEN) or ARQ197 (Tivantinib; ArQule) seem better applicable (Puri et al. 2007). However, due to co-activation of other receptor tyrosine kinases as described above (Stommel et al. 2007), inhibitors targeting dual or multiple kinases clinically appear even more promising.

VEGF-A is the most important angiogenesis factor produced by glioma cells (Mentlein and Held-Feindt 2003). It is a glycoprotein of about 45 kDa and consists of two identical peptide chains linked by disulfide bonds. Six isoforms of human VEGF-A exit that result from alternative splicing of one mRNA and yield peptide backbones in the length of 121–206 residues. Besides from being constitutively produced in glioma cells by malignant transformation, VEGF-A expression is highly upregulated by growth factors and hypoxia. Compared to massive synthesis of VEGF-A, other factors of the VEGF-family like PlGF (placenta growth factor), VEGF-C and VEGF-D are much lesser produced in gliomas (Jenny et al. 2006). However, besides inducing angiogenesis (see Chap. 7), VEGF-A also targets tumor-associated microglial cells/macrophages (Forstreuter et al. 2002) and part of glioma cells themselves (Mentlein et al. 2004). Hereby, VEGF-A exerts also a direct tumor-promoting effects (Kerber et al. 2008) aside supporting nutrient and oxygen supply of the tumor by enhanced angiogenesis.

PTN (PleioTrophiN) is a heparin-binding growth factor produced by glioma cells, but much lesser by regular glial cells (Mentlein and Held-Feindt 2002). PTN is a 136-amino acid non-glycosylated, lysine-rich peptide (15.3 kDa) which is derived from a 168 residue precursor and then secreted. Together with midkine (MK) it forms a family of cysteine- and basic amino acid-rich peptides that are developmentally regulated and bind to extracellular, negatively charged proteoglycans and cell surface components. PTN binds and signals mainly through the anaplastic lymphoma kinase (ALK) and the receptor-type protein tyrosine phosphatase ζ/β (PTPζ). Thereby, it acts as differentiation or growth factor for various cell types (hence the name 'pleiotropin'): it has mitogenic, anti-apoptotic, transforming, angiogenic and chemotactic activities that can differ between its target cells (Mentlein 2007).

In glioma cells, both receptors are overexpressed as compared to normal brain. Targeting of each of these receptors exerts anti-proliferative and anti-apoptotic effects, and thus reduced glioma growth in vivo (Ulbricht et al. 2006; Stylianou et al. 2009). Overall, PTN and its receptors are rather "difficult" molecules with respect to their handling and signaling cascades which have certainly important roles for glioma growth and thus may represent interesting targets. In light of the tumor stem cell hypothesis (or under hypothesis of de-differentiation of glial cells during malignant transformation) this is an "embryonally" regulated autocrine system of particular importance. Thus, relation of the PTN-ALK/PTPζ to stem cell character of glioma cells and to embryonic transcription factors would be of high interest.

Taking together, a **multitude of growth factors and RTK receptors** is a characteristic feature of gliomas. As in many peripheral tumors, ligands and receptors are frequently overexpressed and form auto- and paracrine growth-promoting loops. Some receptors are mutated, frequently in those gliomas where the wild-type receptors are overexpressed. *EGFR* mutation/amplification or *PDGFR* amplification/overexpression are most commonly observed. These mutated and/or overexpressed receptors provide an interesting target for the molecular diagnosis of glioma subtypes and for in vivo imaging. Since growth factor receptors over a common general mechanism involving RTKs, these appear to be promising targets for glioma therapy.

6.3.4 Cytokines

Cytokines have been originally defined as small cell-signaling proteins in the immune system, in particular immunomodulatory molecules such as interleukins, interferons and others, but they have much broader functions in development and cell communication. We here define them as signaling proteins/peptides that act over receptors linked to serine/threonine kinase activities (Fig. 6.2), thereby listing them separately from chemokines linked to 7TM receptors and to "classical" growth factors linked to RTKs, both described above.

Transforming growth factor-β (TGF-β) is the best characterized immunosuppressive cytokine produced by glioma cells. It exists in at least three isoforms in humans: TGF-β1, TGF-β2 and TGF-β3. They are secreted from cells in latent forms which are bound to two other polypeptides, latent TGF-beta binding protein (LTBP) and latency-associated peptide (LAP). Serine proteinases such as plasmin cleave and thereby liberate TGF-β from these complexes. Thereby, TGF-β proteins dimerize to produce mature 25 kDa active molecules. TGF-β is part of a superfamily of proteins including also glial cell-derived neurotrophic factor (GDNF) described below.

TGF-β dimers bind to a type II receptor which recruits and phosphorylates a type I receptor, both transmembrane proteins (Fig. 6.2). Other members of this family use corresponding receptors. TGF-β/family members signal through the SMAD (combination of SMA = small body size, *Caenorhabditis elegans* protein and

MAD = mothers against decapentaplegic, *Drosophila* protein) pathway. The activated type I receptor recruits and thereby phosphorylates a receptor-regulated SMAD (R-SMAD), e.g. SMAD3. The R-SMAD then binds to the common SMAD (co-SMAD) and forms a heterodimeric complex with this SMAD4 (also called DPC4, deleted in pancreatic carcinoma locus 4). This R-SMAD-SMAD4 complex then enters the cell nucleus where it acts as a transcription factor for various genes.

Glioma cells (and microglial cells) produce in vivo and in vitro different amounts of TGF-β isoforms, most abundantly and considerably upregulated is TGF-β2 (Bodmer et al. 1989). Besides the ligands, also TGFRI and TGFRII are overexpressed in gliomas (Yamada et al. 1995). Also, mutations of TGFRII occur in gliomas (Izumoto et al. 1997). However, in contrast to other types of tumors, e.g. pancreatic cancer, essential components of TGFR-signal transduction like of SMAD4 have not been observed to be mutated in gliomas (Bleeker et al. 2009).

In gliomas, the TGF-β pathway has both, tumor suppressive and promoting functions, the latter being more pronounced (Joseph et al. 2013). TGF-β2 (originally described as "glioblastoma-derived T-cell suppressor factor") has been regarded as one of the most important soluble factors of immunosuppression in gliomas, e.g. by inhibiting activities of natural killer cells (NK), T cells and other immune cells (Wrann et al. 1987) and down-regulation of major histocompatibility complex (MHC) class II antigen expression on microglial cells (Black et al. 1992). Next to immunosuppression, TGF-β contributes also to tumor angiogenesis by chemotactic effects on endothelial cells and formation of extracellular matrix proteins, e.g. tenascin on which endothelial cells may adhere and spread (Mentlein and Held-Feindt 2003).

Beside these paracrine immunosuppressive and direct angiogenic effects summarized in other chapters of this book, TGF-β proteins exhibit important autocrine effects on glioma cells. Namely, they increase glioma cell proliferation by induction and crosstalk with EGFR (Held-Feindt et al. 2003). Also, TGF-β enhances the production of VEGF-A and other angiogenic factors by glioma cells (Koochekpour et al. 1996). Next to gene regulation, glioma cell migration and invasion is enhanced under TGF-β-stimulation (Merzak et al. 1994). Under the view of the cancer stem cell hypothesis, TGF-β signaling plays an important role in retention of stemness in GSCs by inducing the expression of the transcription factor Sox2, an important marker and regulator of stemness (Ikushima et al. 2009).

Considering the mostly glioma-promoting effects of TGF-β2, suppression of its production or signaling appears highly promising to treat these tumors. An antisense oligonucleotide (Trabedersen, AP 12009) that specifically blocks TGF-β2 mRNA production has been developed and applied in three phase I/II studies where Trabedersen showed some promising effects on patients' survival (Hau et al. 2011; Joseph et al. 2013). However, effectiveness of TGF-β inhibition has to be proved in further studies.

Glial cell-derived neurotrophic factor (GDNF) and other peptides of the GDNF-family like neurturin, artemin and persephin are distant members of the TGF-β superfamily. GDNF, originally isolated from the B49 rat glial cell line, is a secreted disulfide-linked homodimeric glycosylated protein of 134 amino acids (about 18–22 kDa). Unlike other factors in this superfamily, which signal through the receptor serine-threonine kinases, GDNF family ligands activate intracellular signaling cascades via the receptor tyrosine kinase RET ("rearranged during transfection"). Upon ligand binding, RET forms a complex with a member of the glycosylphosphatidylinositol (GPI)-anchored GFRα protein family, then dimerizes with another RET, followed by autophosphorylation and subsequent signal transduction. Often, TGF-β is required as a cofactor for GDNF family ligand signaling. GDNF plays an essential role in the survival of dopaminergic neurons, but also in differentiation and migration of other cell types.

As the name already implies, GDNF is produced by glial and glioma cells. Not only GDNF, but also its receptor GFRα1 is highly expressed in human gliomas (Wiesenhofer et al. 2000). In vitro, GDNF considerably enhances glioma cell migration; furthermore, production of MMPs is upregulated that favor cell invasion (Lu et al. 2010). A recent in vivo study shows that GDNF is one of the major attractors of microglia into gliomas (Ku et al. 2013). However, also other glioma cell-derived factors like GM-CSF (granulocyte macrophage colony-stimulating factor) and M-SCF (macrophage colony-stimulating factor) regulate microglia attraction and proliferation in gliomas. Overall, GDNF appears to be a further interesting candidate supporting the motility and invasion of glioma cells intro the brain.

As examples for the potential importance of other cytokines in glioma cell biology, **interleukin 6 (IL-6)** and the signaling-related cytokine **leukemia inhibitory factor (LIF)** are reviewed here. IL6 and LIF are secreted, glycosylated proteins of about 21.5–28 kDa and 58 kDa (not structurally similar). They signal through a cytokine receptor complex (Fig. 6.2) consisting of the ligand-binding chain (IL-6Rα chain, CD126; or LIFR chain), and a common signal-transducing component gp130 (CD130; also for IL-11, Oncostatin M, cardiotrophin-1 and ciliary neurotrophic factor, CNTF) that homo- (IL-6, IL-11) or hetero-dimerizes (others). Subsequently, cytoplasmic protein tyrosine kinases of the Janus kinase (JAK) family phosphorylate gp130 and latent cytoplasmic transcription factors called STATs (signal transducers and activators of transcription) are activated. These and other transcription factors activated over the alternate MAPK pathway induce multiple responses including synthesis of proteases or growth and angiogenesis factors (Brantley and Benveniste 2008). Besides by cytokines, STATs can be also activated by classical growth factors, e.g., EGF, mutated EGFR, and bFGF (Swiatek-Machado and Kaminska 2013). IL-6 acts as both, a pro-inflammatory and anti-inflammatory cytokine in the immune system and is one of the most important mediators of fever and of acute phase response. However, it is also relevant in the development and diseases of other systems. LIF (also termed cholinergic differentiation factor, CDF) affects cell growth by inhibiting differentiation and plays—beside other functions—a significant role in the development of the nervous system.

In vitro and in vivo glioma (and microglial) cells constitutively produce IL-6, LIF, and oncostatin M which can also be detected in the cerebrospinal fluid of patients (Tchirkov et al. 2001; Lilja et al. 2001; Penuelas et al. 2009). In vitro, IL-6 production is upregulated by several growth factors. Since IL-6 receptors are amplified on glioma cells in situ (Tchirkov et al. 2001), auto-/paracrine loops are formed that favor glioma cell proliferation, radiation resistance, and invasion (Saidi et al. 2009; Zhu et al. 2012). Because of its pro-inflammatory action on microglial cells/macrophages, IL-6 should also contribute to the inflammation-induced edema in glioma patients.

The importance of IL-6 for gliomas was shown in a glioblastoma mouse model where mice lacking IL-6 failed to develop tumors (Weissenberger et al. 2004). It might well be that IL-6 and other members of the IL-6-signaling family like LIF, CNTF, or oncostatin M have even a more important role in glioma initiation than in progression, since they are implicated in glial cell differentiation (Lu et al. 2012; Chen et al. 2006). In fact, STAT3 activation (or LIF stimulation) is required for maintenance of multipotency, self-renewal, and proliferation of GSCs (Sherry et al. 2009; Penuelas et al. 2009; Hattermann et al. 2010). Targeting IL-6/IL-6Rα chain expression in GSCs with the use of short hairpin RNAs (shRNAs) significantly reduced their growth in vitro and increased in vivo the survival of mice bearing intracranial human glioma xenografts (Wang et al. 2009).

Taking together, cytokines of the IL-6-signaling family not only favor glioma progression, invasion, and attenuate glioma-associated inflammation, but there is increasing evidence for an essential role in regulating GSCs. Namely, via activation of the STAT signaling pathway (that can also be induced by growth factor receptors) they maintain their stemness and proliferation.

In summary, cytokines are not only immunoregulatory factors in gliomas. As some cytokines are involved in the differentiation of neural precursor cells, they appear to also importantly influence the differentiation of glioma stem (-like) cells.

6.4 Non-peptide Factors

Since nonmalignant (astro)glial cells play an important role in nutritive functions for neurons, uptake of excess potassium ions and transmitters from neuronal activity, as well as in immune response in the nervous system, it is not surprising that also glioma cells express corresponding receptors and uptake systems. We here give some examples of receptors for lipids and for the role of amino acids in glioma cell biology whereas the roles of peptides in this context have been outlined above.

6.4.1 Prostaglandins

Prostaglandin E_2 (PGE_2) is a small lipid-soluble (cyclooxgenase-derived) derivative of arachidonic acid, a fatty acid liberated upon stimulation from cell-membrane phospholipids. PGE_2 biosynthesis is regulated by inducible membrane-associated PGE synthase cyclooxygenase-2 (COX-2) and microsomal PGE synthase (mPGES)-1. Like other arachinodyl-derivatives (eicosanoids), PGE_2 binds and signals via G protein-coupled 7TM receptors (Fig. 6.2); it acts often in an autocrine fashion raising cAMP as second messengers. Because this axis affects cell proliferation, cell adhesion, invasion, and apoptosis, PGE_2 and other eicosanoids appear to be important in the pathogenesis and progression of cancer.

As compared to normal astrocytes, glioma cells have been shown to release significant amounts of PGE_2 (and other arachinodyl derivatives like thromboxane) in vitro and in vivo (Castelli et al. 1989); however, mRNA levels of enzymes and receptors implicated in eicosanoid pathways differ significantly in different histological types of gliomas (De Armas et al. 2010). Besides tumor cells, microglial cells appear to be the most important source of PGE_2 (Li and Graeber 2012).

PGE_2 plays an important role in the generation of an immunosuppressive milieu in gliomas, e.g., by downregulation of the activity of lymphokine-activated killer cells and MHC class II expression (Li and Graeber 2012). Furthermore, this prostaglandin promotes glioma cell proliferation and cell motility in vitro (Gomes and Colquhoun 2012).

6.4.2 Endogenous Cannabinoids

Endogenous cannabinoids, such as **anandamide** (*N*-arachidonoylethanolamine; termed from Sanskrit word "ananda" meaning "delight" and amide) and **2-arachidonoylglycerol** (2-AG), are further small lipids generated from arachidonic acid that exert various functions in the nervous and immune systems. These effects are mainly mediated over two G protein-coupled 7TM receptors, called CB1 and CB2, which are found, sometimes elevated in different types of tumors, including gliomas (Ellert-Miklaszewska et al. 2013). Comparing to nontumor controls, the levels of anandamide are decreased, but those of 2-AG are increased in glioma tissues (Wu et al. 2012). Moreover, the expression levels of both cannabinoid receptors are elevated in human gliomas. Whereas CB1 is mainly found on glioma cells, CB2 is expressed on glioma-associated microglial cells/monocytes and on endothelial cells (Held-Feindt et al. 2006).

Endogenous and exogenous (plant or synthetic) cannabinoids like Δ9-tetrahydrocannabinol (Δ9-THC) from *Cannabis* species or the synthetic agonist WIN-55,212-2 exert anti-proliferative actions on glioma cells/gliomas in vitro and in vivo (Galve-Roperh et al. 2000); they may also induce apoptosis that relies on the generation of a pro-apoptotic sphingolipid ceramide (Ellert-Miklaszewska et al. 2013). However, due to the psychotropic side effects and the variable expression of CB-receptors in different patients the use of cannabionids to cure gliomas appears to be limited.

6.4.3 Sphingolipids

Sphingosine 1-phosphate (S1P) is a signaling sphingolipid that is formed from ceramides by hydrolysis through ceramidase and phosphorylation of released sphingosine by sphingosine kinases (SphK1, SphK2), enzymes ubiquitously found in the cytosol and endoplasmatic reticulum of various types of cells. S1P can be inactivated by dephosphorylation to sphingosine by sphingosine phosphatases or by irreversible degradation by sphingosine phosphate lyase. Like prostaglandins, S1P binds and signals via G protein-coupled 7TM receptors (Fig. 6.2).

In glioma cells, S1P can be produced in considerable amounts, and expression of SphK1 in *glioblastoma multiforme* tissue is upregulated and correlates with a short patient survival (Van Brocklyn et al. 2005). Stimulation of glioma cells with S1P enhances cell proliferation and invasion in vitro; these effects appear to be mediated especially via the S1P(1) receptor, but also S1P(2) and S1P(3) are involved (Young et al. 2009). Inhibition of the autocrine S1P-axis in vivo by targeting SphK1 diminished tumor growth of glioblastoma xenografts by inducing apoptosis and reducing tumor vascularization (Kapitonov et al. 2009). Thus, the S1P-system is a further lipid-derived auto- and paracrine tumor-promoting principle in gliomas.

6.4.4 Amino Acids and Derivatives

Glutamate is a major excitatory neurotransmitter in the central nervous system that is produced also by glial/glioma cells. It binds to two different types of receptors: the ionotropic- (iGluR) and metabotropic-glutamate receptor (mGluR) families (ligand-activated ion channels and G protein-coupled 7TM receptors, respectively). Receptor activation leads to the activation of multiple signaling pathways. In gliomas, glutamate enhances tumor growth, proliferation, and survival through activation of the mitogen-activated protein kinase and phosphoinositide 3-kinase/Akt pathways (Prickett and Samuels 2012). In this respect the metabotropic 7TM receptor mGluR3 plays a prominent role. Functional and pharmacological studies have shown that mGluR3 is ubiquitously expressed in glioma cells and that pharmacological blockade with an antagonist resulted in diminished glioma growth in vitro and in vivo via suppression of MAPK and Akt pathways (D'Onofrio et al. 2003; Arcella et al. 2005). However, also ionotropic glutamate receptors considerably enhance glioma growth and furthermore, glioma invasion by activation of focal adhesion kinase that regulates growth factor and integrin-stimulated cell motility (de Groot and Sontheimer 2011). Taking together, glutamate has multiple proliferative and mobilization effects on glioma cells. Clinical trials are ongoing to explore the therapeutic potential of antagonists.

Besides the amino acid transmitter glutamate, a product of another amino acid has been recently identified as a glioma-promoting agent. Namely, the **tryptophan metabolite kynureine** is an endogenous ligand of the human aryl hydrocarbon receptor (AHR) that is implicated in a variety of cellular processes including tumorigenesis and inflammation. AHR is transcription factor of the basic helix-loop-helix family which binds several exogenous ligands, mostly aromatic compounds like natural plant flavonoids, polyphenolics, indoles, and synthetic polycyclic aromatic hydrocarbons and dioxins. Kynureine is sufficiently produced in gliomas via tryptophan-2,3-dioxgenase to activate via AHR glioma cell survival and invasion in vitro and in vivo (Opitz et al. 2011). In accordance, AHR activation predicts survival of glioma patients. Also further Trp-metabolites could play a significant role in brain tumor pathogenesis (Adams et al. 2012).

In summary, small lipid mediators and amino acids/-derivatives are part of an immuno-modulatory and tumor-activation auto- and paracrine systems in gliomas. If they have essential clinical significance on their own or parts of an accompanying tumor-associated inflammation remains to be established.

Conclusion and Perspectives

Classical growth factors and RTK receptors comprise perhaps the most important auto- and paracrine tumor-promoting systems in gliomas. As in many other types of cancers, they regulate cellular proliferation, survival, differentiation, function, and motility of glioma cells. As in peripheral solid tumors, also in gliomas a multitude of growth factors and RTK receptors are found. Among them, the EGF-EGFR-, PDGF-PDGFR-, and HGF/SF-c-Met-systems appear most prominent. Overexpression and constitutive activation by receptor mutations is a further important aspect of their tumor-promoting role. Among many angiogenic factors, VEGF-A is most noteworthy. However, it cannot be excluded that other less investigated factors play a driving role, in particular in tumor invasion.

Due to their overexpression/mutations and common signal transduction mechanism, RTKs appear to be promising targets for glioma therapy. However, clinical studies with inhibitors of single RTK yielded more or less disappointing results, even in patients with a corresponding overexpression/mutation. Activation of multiple downstream signaling pathways and/or production of alternative/novel receptors have been regarded as possible mechanisms by which inhibition of a single RTK has been ineffective in gliomas. Furthermore, animal studies with single growth factor overexpression induced surprisingly low rates of gliomas. Thus, it appears that no single growth factor/receptor has an exclusive role in glioma tumorigenesis. Therefore, inhibitors of multiple kinases that are currently under investigation (Joshi et al. 2012) or combinations with other chemotherapeutics could be a better option.

Cytokines, in particular TGF-β, play an essential role in the immunosuppression of gliomas, but they also contribute like GDNF to the paracrine network in the glioma microenvironment that plays an essential role in tumor growth. As some cytokines are (together with others) important differentiation factors in development, their role in glioma stem(-like) cell differentiation has to be established.

Chemokines are perhaps beside distinct growth factors and cytokines essential drivers of glioma stem(-like) cell differentiation and of glioma cell invasion. This renders them an important future target of glioma therapy—together with other drug principles. However, preclinical studies are missing that could evaluate their importance.

The role of **neuropeptides and peptide hormones** in glioma cell biology has been neglected in the last decade, despite earlier studies revealed a remarkable overexpression of some receptors. Thus, their functional role in glioma cell biology still remains somewhat obscure. However, overexpression has turned some receptors to interesting targets of in vivo imaging and combined diagnosis and therapy in the emerging field of theragnostics.

Other, small molecules, in particular **lipids**, contribute further to the immunosuppressive and growth-promoting inflammatory auto- and paracrine networks in gliomas. As with chemokines, an evaluation of their biological and pharmaceutical role is just emerging.

Acknowledgments We thank Clemens Franke for careful drawing figures. The own experimental work cited in the text is based on long collaborations and the great support of a team of engaged coworkers. Of these, which we cannot all mention, but we would like to thank especially our ambitious technicians Judith Becker, Martina Burmester, Sonja Dahle, Dagmar Freier, Miriam Lemmer and Ursula Prange and our colleagues Prof. Dr. Dr. Janka Held-Feindt and Prof. Dr. H. Maximilian Mehdorn from the Department of Neurosurgery, University Hospital Schleswig-Holstein, 24105 Kiel, Germany. RM is also particularity indebted to his deceased colleague Prof. Dr. Brigitte Krisch, who introduced him into the fields of neuropeptides and brain morphology. The authors declare that they have no competing interests. Due to space constraints, we apologize to colleagues in the field for not citing their relevant studies.

References

Abel S, Hundhausen C, Mentlein R, Schulte A, Berkhout TA, Broadway N et al (2004) The transmembrane CXC-chemokine ligand 16 is induced by IFN-gamma and TNF-alpha and shed by the activity of the disintegrin-like metalloproteinase ADAM10. J Immunol 172 (10):6362–6372

Abounader R, Laterra J (2005) Scatter factor/hepatocyte growth factor in brain tumor growth and angiogenesis. Neuro Oncol 7(4):436–451

Abounader R, Ranganathan S, Lal B, Fielding K, Book A, Dietz H et al (1999) Reversion of human glioblastoma malignancy by U1 small nuclear RNA/ribozyme targeting of scatter factor/hepatocyte growth factor and c-met expression. J Natl Cancer Inst 91(18):1548–1556

Adams S, Braidy N, Bessesde A, Brew BJ, Grant R, Teo C et al (2012) The kynurenine pathway in brain tumor pathogenesis. Cancer Res 72(22):5649–5657

Aldape KD, Ballman K, Furth A, Buckner JC, Giannini C, Burger PC et al (2004) Immunohistochemical detection of EGFRvIII in high malignancy grade astrocytomas and evaluation of prognostic significance. J Neuropathol Exp Neurol 63(7):700–707

Alentorn A, Marie Y, Carpentier C, Boisselier B, Giry M, Labussiere M et al (2012) Prevalence, clinico-pathological value, and co-occurrence of PDGFRA abnormalities in diffuse gliomas. Neuro Oncol 14(11):1393–1403

Allen SJ, Crown SE, Handel TM (2007) Chemokine: receptor structure, interactions, and antagonism. Annu Rev Immunol 25:787–820

Arcella A, Carpinelli G, Battaglia G, D'Onofrio M, Santoro F, Ngomba RT et al (2005) Pharmacological blockade of group II metabotropic glutamate receptors reduces the growth of glioma cells in vivo. Neuro Oncol 7(3):236–245

Arrieta O, Pineda-Olvera B, Guevara-Salazar P, Hernandez-Pedro N, Morales-Espinosa D, Ceron-Lizarraga TL et al (2008) Expression of AT1 and AT2 angiotensin receptors in astrocytomas is associated with poor prognosis. Br J Cancer 99(1):160–166

Birchmeier C, Birchmeier W, Gherardi E, Vande Woude GF (2003) Met, metastasis, motility and more. Nat Rev Mol Cell Biol 4(12):915–925

Black KL, Chen K, Becker DP, Merrill JE (1992) Inflammatory leukocytes associated with increased immunosuppression by glioblastoma. J Neurosurg 77(1):120–126

Bleeker FE, Lamba S, Leenstra S, Troost D, Hulsebos T, Vandertop WP et al (2009) IDH1 mutations at residue p.R132 (IDH1(R132)) occur frequently in high-grade gliomas but not in other solid tumors. Hum Mutat 30(1):7–11

Bodmer S, Strommer K, Frei K, Siepl C, de Tribolet N, Heid I et al (1989) Immunosuppression and transforming growth factor-beta in glioblastoma. Preferential production of transforming growth factor-beta 2. J Immunol 143(10):3222–3229

Bowers DC, Fan S, Walter KA, Abounader R, Williams JA, Rosen EM et al (2000) Scatter factor/hepatocyte growth factor protects against cytotoxic death in human glioblastoma via phosphatidylinositol 3-kinase- and AKT-dependent pathways. Cancer Res 60(15):4277–4283

Brantley EC, Benveniste EN (2008) Signal transducer and activator of transcription-3: a molecular hub for signaling pathways in gliomas. Mol Cancer Res 6(5):675–684

Brockmann MA, Ulbricht U, Gruner K, Fillbrandt R, Westphal M, Lamszus K (2003) Glioblastoma and cerebral microvascular endothelial cell migration in response to tumor-associated growth factors. Neurosurgery 52(6):1391–1399, discussion 9

Camp ER, Summy J, Bauer TW, Liu W, Gallick GE, Ellis LM (2005) Molecular mechanisms of resistance to therapies targeting the epidermal growth factor receptor. Clin Cancer Res 11 (1):397–405

Cancer Genome Atlas Research Network (2008) Comprehensive genomic characterization defines human glioblastoma genes and core pathways. Nature 455(7216):1061–1068

Cao B, Su Y, Oskarsson M, Zhao P, Kort EJ, Fisher RJ et al (2001) Neutralizing monoclonal antibodies to hepatocyte growth factor/scatter factor (HGF/SF) display antitumor activity in animal models. Proc Natl Acad Sci U S A 98(13):7443–7448

Castelli MG, Chiabrando C, Fanelli R, Martelli L, Butti G, Gaetani P et al (1989) Prostaglandin and thromboxane synthesis by human intracranial tumors. Cancer Res 49(6):1505–1508

Charles NA, Holland EC, Gilbertson R, Glass R, Kettenmann H (2011) The brain tumor microenvironment. Glia 59(8):1169–1180

Chen SH, Gillespie GY, Benveniste EN (2006) Divergent effects of oncostatin M on astroglioma cells: influence on cell proliferation, invasion, and expression of matrix metalloproteinases. Glia 53(2):191–200

Cordier D, Forrer F, Bruchertseifer F, Morgenstern A, Apostolidis C, Good S et al (2010) Targeted alpha-radionuclide therapy of functionally critically located gliomas with 213Bi-DOTA-[Thi8, Met(O2)11]-substance P: a pilot trial. Eur J Nucl Med Mol Imaging 37(7):1335–1344

D'Onofrio M, Arcella A, Bruno V, Ngomba RT, Battaglia G, Lombari V et al (2003) Pharmacological blockade of mGlu2/3 metabotropic glutamate receptors reduces cell proliferation in cultured human glioma cells. J Neurochem 84(6):1288–1295

Day SE, Waziri A (2012) Clinical trials of small molecule inhibitors in high-grade glioma. Neurosurg Clin N Am 23(3):407–416

De Armas R, Durand K, Guillaudeau A, Weinbreck N, Robert S, Moreau JJ et al (2010) mRNA levels of enzymes and receptors implicated in arachidonic acid metabolism in gliomas. Clin Biochem 43(10–11):827–835

de Groot J, Sontheimer H (2011) Glutamate and the biology of gliomas. Glia 59(8):1181–1189

Desbaillets I, Tada M, de Tribolet N, Diserens AC, Hamou MF, Van Meir EG (1994) Human astrocytomas and glioblastomas express monocyte chemoattractant protein-1 (MCP-1) in vivo and in vitro. Int J Cancer 58(2):240–247

Dong Y, Han Q, Zou Y, Deng Z, Lu X, Wang X et al (2012) Long-term exposure to imatinib reduced cancer stem cell ability through induction of cell differentiation via activation of MAPK signaling in glioblastoma cells. Mol Cell Biochem 370(1–2):89–102

Ehtesham M, Mapara KY, Stevenson CB, Thompson RC (2009) CXCR4 mediates the proliferation of glioblastoma progenitor cells. Cancer Lett 274(2):305–312

Ehtesham M, Thompson RC (2012) CXCR4-expressing glial precursor cells demonstrate enhanced migratory tropism for glioma. J Cancer Ther 3(6):1086–1091

Ehtesham M, Winston JA, Kabos P, Thompson RC (2006) CXCR4 expression mediates glioma cell invasiveness. Oncogene 25(19):2801–2806

Ellert-Miklaszewska A, Ciechomska I, Kaminska B (2013) Cannabinoid signaling in glioma cells. Adv Exp Med Biol 986:209–220
Feindt J, Becker I, Blomer U, Hugo HH, Mehdorn HM, Krisch B et al (1995) Expression of somatostatin receptor subtypes in cultured astrocytes and gliomas. J Neurochem 65 (5):1997–2005
Feng Y, Broder CC, Kennedy PE, Berger EA (1996) HIV-1 entry cofactor: functional cDNA cloning of a seven-transmembrane, G protein-coupled receptor. Science 272(5263):872–877
Fleming TP, Saxena A, Clark WC, Robertson JT, Oldfield EH, Aaronson SA et al (1992) Amplification and/or overexpression of platelet-derived growth factor receptors and epidermal growth factor receptor in human glial tumors. Cancer Res 52(16):4550–4553
Forstreuter F, Lucius R, Mentlein R (2002) Vascular endothelial growth factor induces chemotaxis and proliferation of microglial cells. J Neuroimmunol 132(1–2):93–98
Galve-Roperh I, Sanchez C, Cortes ML, Gomez del Pulgar T, Izquierdo M, Guzman M (2000) Anti-tumoral action of cannabinoids: involvement of sustained ceramide accumulation and extracellular signal-regulated kinase activation. Nat Med 6(3):313–319
Garton KJ, Gough PJ, Blobel CP, Murphy G, Greaves DR, Dempsey PJ et al (2001) Tumor necrosis factor-alpha-converting enzyme (ADAM17) mediates the cleavage and shedding of fractalkine (CX3CL1). J Biol Chem 276(41):37993–38001
Gomes RN, Colquhoun A (2012) E series prostaglandins alter the proliferative, apoptotic and migratory properties of T98G human glioma cells in vitro. Lipids Health Dis 11(1):171
Gough PJ, Garton KJ, Wille PT, Rychlewski M, Dempsey PJ, Raines EW (2004) A disintegrin and metalloproteinase 10-mediated cleavage and shedding regulates the cell surface expression of CXC chemokine ligand 16. J Immunol 172(6):3678–3685
Gupta SK, Lysko PG, Pillarisetti K, Ohlstein E, Stadel JM (1998) Chemokine receptors in human endothelial cells. Functional expression of CXCR4 and its transcriptional regulation by inflammatory cytokines. J Biol Chem 273(7):4282–4287
Harrison JK, Jiang Y, Chen S, Xia Y, Maciejewski D, McNamara RK et al (1998) Role for neuronally derived fractalkine in mediating interactions between neurons and CX3CR1-expressing microglia. Proc Natl Acad Sci U S A 95(18):10896–10901
Haskell CA, Cleary MD, Charo IF (1999) Molecular uncoupling of fractalkine-mediated cell adhesion and signal transduction. Rapid flow arrest of CX3CR1-expressing cells is independent of G-protein activation. J Biol Chem 274(15):10053–10058
Hatanpaa KJ, Burma S, Zhao D, Habib AA (2010) Epidermal growth factor receptor in glioma: signal transduction, neuropathology, imaging, and radioresistance. Neoplasia 12(9):675–684
Hattermann K, Held-Feindt J, Lucius R, Muerkoster SS, Penfold ME, Schall TJ et al (2010) The chemokine receptor CXCR7 is highly expressed in human glioma cells and mediates antiapoptotic effects. Cancer Res 70(8):3299–3308
Hattermann K, Mentlein R (2013) An infernal trio: the chemokine CXCL12 and its receptors CXCR4 and CXCR7 in tumor biology. Ann Anat 195(2):103–110
Hattermann K, Mentlein R, Held-Feindt J (2012) CXCL12 mediates apoptosis resistance in rat C6 glioma cells. Oncol Rep 27(5):1348–1352
Hau P, Jachimczak P, Schlaier J, Bogdahn U (2011) TGF-beta2 signaling in high-grade gliomas. Curr Pharm Biotechnol 12(12):2150–2157
Hede SM, Hansson I, Afink GB, Eriksson A, Nazarenko I, Andrae J et al (2009) GFAP promoter driven transgenic expression of PDGFB in the mouse brain leads to glioblastoma in a Trp53 null background. Glia 57(11):1143–1153
Held-Feindt J, Dörner L, Sahan G, Mehdorn HM, Mentlein R (2006) Cannabinoid receptors in human astroglial tumors. J Neurochem 98(3):886–893
Held-Feindt J, Hattermann K, Muerkoster SS, Wedderkopp H, Knerlich-Lukoschus F, Ungefroren H et al (2010) CX3CR1 promotes recruitment of human glioma-infiltrating microglia/ macrophages (GIMs). Exp Cell Res 316(9):1553–1566

Held-Feindt J, Lutjohann B, Ungefroren H, Mehdorn HM, Mentlein R (2003) Interaction of transforming growth factor-beta (TGF-beta) and epidermal growth factor (EGF) in human glioma cells. J Neurooncol 63(2):117–127

Heldin CH, Westermark B (1984) Growth factors: mechanism of action and relation to oncogenes. Cell 37(1):9–20

Hermanson M, Funa K, Hartman M, Claesson-Welsh L, Heldin CH, Westermark B et al (1992) Platelet-derived growth factor and its receptors in human glioma tissue: expression of messenger RNA and protein suggests the presence of autocrine and paracrine loops. Cancer Res 52 (11):3213–3219

Hoelzinger DB, Demuth T, Berens ME (2007) Autocrine factors that sustain glioma invasion and paracrine biology in the brain microenvironment. J Natl Cancer Inst 99(21):1583–1593

Ikushima H, Todo T, Ino Y, Takahashi M, Miyazawa K, Miyazono K (2009) Autocrine TGF-beta signaling maintains tumorigenicity of glioma-initiating cells through Sry-related HMG-box factors. Cell Stem Cell 5(5):504–514

Inda MM, Bonavia R, Mukasa A, Narita Y, Sah DW, Vandenberg S et al (2010) Tumor heterogeneity is an active process maintained by a mutant EGFR-induced cytokine circuit in glioblastoma. Genes Dev 24(16):1731–1745

Izumoto S, Arita N, Ohnishi T, Hiraga S, Taki T, Tomita N et al (1997) Microsatellite instability and mutated type II transforming growth factor-beta receptor gene in gliomas. Cancer Lett 112 (2):251–256

Jenny B, Harrison JA, Baetens D, Tille JC, Burkhardt K, Mottaz H et al (2006) Expression and localization of VEGF-C and VEGFR-3 in glioblastomas and haemangioblastomas. J Pathol 209(1):34–43

Joseph JV, Balasubramaniyan V, Walenkamp A, Kruyt FA (2013) TGF-beta as a therapeutic target in high grade gliomas – promises and challenges. Biochem Pharmacol 85(4):478–485

Joshi AD, Loilome W, Siu IM, Tyler B, Gallia GL, Riggins GJ (2012) Evaluation of tyrosine kinase inhibitor combinations for glioblastoma therapy. PLoS One 7(10):e44372

Juillerat-Jeanneret L, Celerier J, Chapuis Bernasconi C, Nguyen G, Wostl W, Maerki HP et al (2004) Renin and angiotensinogen expression and functions in growth and apoptosis of human glioblastoma. Br J Cancer 90(5):1059–1068

Kapitonov D, Allegood JC, Mitchell C, Hait NC, Almenara JA, Adams JK et al (2009) Targeting sphingosine kinase 1 inhibits Akt signaling, induces apoptosis, and suppresses growth of human glioblastoma cells and xenografts. Cancer Res 69(17):6915–6923

Kerber M, Reiss Y, Wickersheim A, Jugold M, Kiessling F, Heil M et al (2008) Flt-1 signaling in macrophages promotes glioma growth in vivo. Cancer Res 68(18):7342–7351

Knerlich-Lukoschus F, Noack M, von der Ropp-Brenner B, Lucius R, Mehdorn HM, Held-Feindt J (2011) Spinal cord injuries induce changes in CB1 cannabinoid receptor and C-C chemokine expression in brain areas underlying circuitry of chronic pain conditions. J Neurotrauma 28 (4):619–634

Koka V, Potti A, Forseen SE, Pervez H, Fraiman GN, Koch M et al (2003) Role of Her-2/neu overexpression and clinical determinants of early mortality in glioblastoma multiforme. Am J Clin Oncol 26(4):332–335

Koochekpour S, Jeffers M, Rulong S, Taylor G, Klineberg E, Hudson EA et al (1997) Met and hepatocyte growth factor/scatter factor expression in human gliomas. Cancer Res 57 (23):5391–5398

Koochekpour S, Merzak A, Pilkington GJ (1996) Vascular endothelial growth factor production is stimulated by gangliosides and TGF-beta isoforms in human glioma cells in vitro. Cancer Lett 102(1–2):209–215

Krisch B, Mentlein R (1994) Neuropeptide receptors and astrocytes. Int Rev Cytol 148:119–169

Ku MC, Wolf SA, Respondek D, Matyash V, Pohlmann A, Waiczies S et al (2013) GDNF mediates glioblastoma-induced microglia attraction but not astrogliosis. Acta Neuropathol 125(4):609–620

Laterra J, Nam M, Rosen E, Rao JS, Lamszus K, Goldberg ID et al (1997) Scatter factor/hepatocyte growth factor gene transfer enhances glioma growth and angiogenesis in vivo. Lab Invest 76(4):565–577

Leung SY, Wong MP, Chung LP, Chan AS, Yuen ST (1997) Monocyte chemoattractant protein-1 expression and macrophage infiltration in gliomas. Acta Neuropathol 93(5):518–527

Li W, Graeber MB (2012) The molecular profile of microglia under the influence of glioma. Neuro Oncol 14(8):958–978

Lilja A, Nordborg C, Brun A, Salford LG, Aman P (2001) Expression of the IL-6 family cytokines in human brain tumors. Int J Oncol 19(3):495–499

Liu C, Luo D, Streit WJ, Harrison JK (2008) CX3CL1 and CX3CR1 in the GL261 murine model of glioma: CX3CR1 deficiency does not impact tumor growth or infiltration of microglia and lymphocytes. J Neuroimmunol 198(1–2):98–105

Liu G, Ying H, Zeng G, Wheeler CJ, Black KL, Yu JS (2004) HER-2, gp100, and MAGE-1 are expressed in human glioblastoma and recognized by cytotoxic T cells. Cancer Res 64 (14):4980–4986

Liu XS, Zhang ZG, Zhang RL, Gregg SR, Wang L, Yier T et al (2007) Chemokine ligand 2 (CCL2) induces migration and differentiation of subventricular zone cells after stroke. J Neurosci Res 85(10):2120–2125

Lu DY, Leung YM, Cheung CW, Chen YR, Wong KL (2010) Glial cell line-derived neurotrophic factor induces cell migration and matrix metalloproteinase-13 expression in glioma cells. Biochem Pharmacol 80(8):1201–1209

Lu J, Ksendzovsky A, Yang C, Mehta GU, Yong RL, Weil RJ et al (2012) CNTF receptor subunit alpha as a marker for glioma tumor-initiating cells and tumor grade: laboratory investigation. J Neurosurg 117(6):1022–1031

Lucius R, Mentlein R (1991) Degradation of the neuropeptide somatostatin by cultivated neuronal and glial cells. J Biol Chem 266(28):18907–18913

Lucius R, Sievers J, Mentlein R (1995) Enkephalin metabolism by microglial aminopeptidase N (CD13). J Neurochem 64(4):1841–1847

Ludwig A, Mentlein R (2008) Glial cross-talk by transmembrane chemokines CX3CL1 and CXCL16. J Neuroimmunol 198(1–2):92–97

Ludwig A, Schiemann F, Mentlein R, Lindner B, Brandt E (2002) Dipeptidyl peptidase IV (CD26) on T cells cleaves the CXC chemokine CXCL11 (I-TAC) and abolishes the stimulating but not the desensitizing potential of the chemokine. J Leukoc Biol 72(1):183–191

Ludwig A, Schulte A, Schnack C, Hundhausen C, Reiss K, Brodway N et al (2005) Enhanced expression and shedding of the transmembrane chemokine CXCL16 by reactive astrocytes and glioma cells. J Neurochem 93(5):1293–1303

Luyken C, Hildebrandt G, Scheidhauer K, Krisch B, Schicha H, Klug N (1994) 111Indium (DTPA-octreotide) scintigraphy in patients with cerebral gliomas. Acta Neurochir (Wien) 127(1–2):60–64

Magge SN, Malik SZ, Royo NC, Chen HI, Yu L, Snyder EY et al (2009) Role of monocyte chemoattractant protein-1 (MCP-1/CCL2) in migration of neural progenitor cells toward glial tumors. J Neurosci Res 87(7):1547–1555

Matsumoto K, Nakamura T (2003) NK4 (HGF-antagonist/angiogenesis inhibitor) in cancer biology and therapeutics. Cancer Sci 94(4):321–327

Mentlein R (2004) Cell-surface peptidases. Int Rev Cytol 235:165–213

Mentlein R (2007) Targeting pleiotropin to treat osteoarthritis. Expert Opin Ther Targets 11 (7):861–867

Mentlein R, Eichler O, Forstreuter F, Held-Feindt J (2001) Somatostatin inhibits the production of vascular endothelial growth factor in human glioma cells. Int J Cancer 92(4):545–550

Mentlein R, Forstreuter F, Mehdorn HM, Held-Feindt J (2004) Functional significance of vascular endothelial growth factor receptor expression on human glioma cells. J Neurooncol 67 (1–2):9–18

Mentlein R, Hattermann K, Held-Feindt J (2012) Lost in disruption: role of proteases in glioma invasion and progression. Biochim Biophys Acta 1825(2):178–185

Mentlein R, Hattermann K, Held-Feindt J (2013) Migration, metastasis and more: the role of chemokines in the proliferation, spreading and metastasis of tumors. In: Resende R, Ulrich H (eds) Trends in stem cell proliferation and cancer research. Springer, Vienna, pp. 339–358

Mentlein R, Held-Feindt J (2002) Pleiotrophin, an angiogenic and mitogenic growth factor, is expressed in human gliomas. J Neurochem 83(4):747–753

Mentlein R, Held-Feindt J (2003) Angiogenesis factors in gliomas: a new key to tumour therapy? Naturwissenschaften 90(9):385–394

Merlo A, Hausmann O, Wasner M, Steiner P, Otte A, Jermann E et al (1999) Locoregional regulatory peptide receptor targeting with the diffusible somatostatin analogue 90Y-labeled DOTA0-D-Phe1-Tyr3-octreotide (DOTATOC): a pilot study in human gliomas. Clin Cancer Res 5(5):1025–1033

Merzak A, McCrea S, Koocheckpour S, Pilkington GJ (1994) Control of human glioma cell growth, migration and invasion in vitro by transforming growth factor beta 1. Br J Cancer 70 (2):199–203

Michieli P, Mazzone M, Basilico C, Cavassa S, Sottile A, Naldini L et al (2004) Targeting the tumor and its microenvironment by a dual-function decoy Met receptor. Cancer Cell 6 (1):61–73

Miller RJ, Rostene W, Apartis E, Banisadr G, Biber K, Milligan ED et al (2008) Chemokine action in the nervous system. J Neurosci 28(46):11792–11795

Mishima K, Higashiyama S, Asai A, Yamaoka K, Nagashima Y, Taniguchi N et al (1998) Heparin-binding epidermal growth factor-like growth factor stimulates mitogenic signaling and is highly expressed in human malignant gliomas. Acta Neuropathol 96(4):322–328

Montana V, Sontheimer H (2011) Bradykinin promotes the chemotactic invasion of primary brain tumors. J Neurosci 31(13):4858–4867

Moriyama T, Kataoka H, Hamasuna R, Yokogami K, Uehara H, Kawano H et al (1998) Up-regulation of vascular endothelial growth factor induced by hepatocyte growth factor/scatter factor stimulation in human glioma cells. Biochem Biophys Res Commun 249(1):73–77

Moroz MA, Huang R, Kochetkov T, Shi W, Thaler H, de Stanchina E et al (2011) Comparison of corticotropin-releasing factor, dexamethasone, and temozolomide: treatment efficacy and toxicity in U87 and C6 intracranial gliomas. Clin Cancer Res 17(10):3282–3292

Mortier A, Gouwy M, Van Damme J, Proost P (2011) Effect of posttranslational processing on the in vitro and in vivo activity of chemokines. Exp Cell Res 317(5):642–654

Mukasa A, Wykosky J, Ligon KL, Chin L, Cavenee WK, Furnari F (2010) Mutant EGFR is required for maintenance of glioma growth in vivo, and its ablation leads to escape from receptor dependence. Proc Natl Acad Sci U S A 107(6):2616–2621

Nagasawa T, Kikutani H, Kishimoto T (1994) Molecular cloning and structure of a pre-B-cell growth-stimulating factor. Proc Natl Acad Sci U S A 91(6):2305–2309

Nakada M, Kita D, Teng L, Pyko IV, Watanabe T, Hayashi Y et al (2013) Receptor tyrosine kinases: principles and functions in glioma invasion. Adv Exp Med Biol 986:143–170

Nazarenko I, Hede SM, He X, Hedren A, Thompson J, Lindstrom MS et al (2012) PDGF and PDGF receptors in glioma. Ups J Med Sci 117(2):99–112

Nazarenko I, Hedren A, Sjodin H, Orrego A, Andrae J, Afink GB et al (2011) Brain abnormalities and glioma-like lesions in mice overexpressing the long isoform of PDGF-A in astrocytic cells. PLoS One 6(4):e18303

Nister M, Libermann TA, Betsholtz C, Pettersson M, Claesson-Welsh L, Heldin CH et al (1988) Expression of messenger RNAs for platelet-derived growth factor and transforming growth factor-alpha and their receptors in human malignant glioma cell lines. Cancer Res 48 (14):3910–3918

Oikonomou E, Buchfelder M, Adams EF (2008) Cholecystokinin (CCK) and CCK receptor expression by human gliomas: evidence for an autocrine/paracrine stimulatory loop. Neuropeptides 42(3):255–265

Opitz CA, Litzenburger UM, Sahm F, Ott M, Tritschler I, Trump S et al (2011) An endogenous tumour-promoting ligand of the human aryl hydrocarbon receptor. Nature 478(7368):197–203
Ozawa T, Brennan CW, Wang L, Squatrito M, Sasayama T, Nakada M et al (2010) PDGFRA gene rearrangements are frequent genetic events in PDGFRA-amplified glioblastomas. Genes Dev 24(19):2205–2218
Paulsson J, Lindh MB, Jarvius M, Puputti M, Nister M, Nupponen NN et al (2011) Prognostic but not predictive role of platelet-derived growth factor receptors in patients with recurrent glioblastoma. Int J Cancer 128(8):1981–1988
Penuelas S, Anido J, Prieto-Sanchez RM, Folch G, Barba I, Cuartas I et al (2009) TGF-beta increases glioma-initiating cell self-renewal through the induction of LIF in human glioblastoma. Cancer Cell 15(4):315–327
Pinski J, Schally AV, Halmos G, Szepeshazi K, Groot K (1994) Somatostatin analogues and bombesin/gastrin-releasing peptide antagonist RC-3095 inhibit the growth of human glioblastomas in vitro and in vivo. Cancer Res 54(22):5895–5901
Platten M, Kretz A, Naumann U, Aulwurm S, Egashira K, Isenmann S et al (2003) Monocyte chemoattractant protein-1 increases microglial infiltration and aggressiveness of gliomas. Ann Neurol 54(3):388–392
Potti A, Forseen SE, Koka VK, Pervez H, Koch M, Fraiman G et al (2004) Determination of HER-2/neu overexpression and clinical predictors of survival in a cohort of 347 patients with primary malignant brain tumors. Cancer Invest 22(4):537–544
Pozsgai E, Schally AV, Halmos G, Rick F, Bellyei S (2010) The inhibitory effect of a novel cytotoxic somatostatin analogue AN-162 on experimental glioblastoma. Horm Metab Res 42 (11):781–786
Prickett TD, Samuels Y (2012) Molecular pathways: dysregulated glutamatergic signaling pathways in cancer. Clin Cancer Res 18(16):4240–4246
Puri N, Khramtsov A, Ahmed S, Nallasura V, Hetzel JT, Jagadeeswaran R et al (2007) A selective small molecule inhibitor of c-Met, PHA665752, inhibits tumorigenicity and angiogenesis in mouse lung cancer xenografts. Cancer Res 67(8):3529–3534
Rafei M, Deng J, Boivin MN, Williams P, Matulis SM, Yuan S et al (2011) A MCP1 fusokine with CCR2-specific tumoricidal activity. Mol Cancer 10:121
Rajagopal S, Kim J, Ahn S, Craig S, Lam CM, Gerard NP et al (2010) Beta-arrestin- but not G protein-mediated signaling by the "decoy" receptor CXCR7. Proc Natl Acad Sci U S A 107 (2):628–632
Ransohoff RM (2009) Chemokines and chemokine receptors: standing at the crossroads of immunobiology and neurobiology. Immunity 31(5):711–721
Re DB, Przedborski S (2006) Fractalkine: moving from chemotaxis to neuroprotection. Nat Neurosci 9(7):859–861
Reiss K, Mentlein R, Sievers J, Hartmann D (2002) Stromal cell-derived factor 1 is secreted by meningeal cells and acts as chemotactic factor on neuronal stem cells of the cerebellar external granular layer. Neuroscience 115(1):295–305
Rempel SA, Dudas S, Ge S, Gutierrez JA (2000) Identification and localization of the cytokine SDF1 and its receptor, CXC chemokine receptor 4, to regions of necrosis and angiogenesis in human glioblastoma. Clin Cancer Res 6(1):102–111
Rodero M, Marie Y, Coudert M, Blondet E, Mokhtari K, Rousseau A et al (2008) Polymorphism in the microglial cell-mobilizing CX3CR1 gene is associated with survival in patients with glioblastoma. J Clin Oncol 26(36):5957–5964
Rosen EM, Laterra J, Joseph A, Jin L, Fuchs A, Way D et al (1996) Scatter factor expression and regulation in human glial tumors. Int J Cancer 67(2):248–255
Rot A (2010) Chemokine patterning by glycosaminoglycans and interceptors. Front Biosci 15:645–660
Saidi A, Hagedorn M, Allain N, Verpelli C, Sala C, Bello L et al (2009) Combined targeting of interleukin-6 and vascular endothelial growth factor potently inhibits glioma growth and invasiveness. Int J Cancer 125(5):1054–1064

Schmidt NO, Westphal M, Hagel C, Ergun S, Stavrou D, Rosen EM et al (1999) Levels of vascular endothelial growth factor, hepatocyte growth factor/scatter factor and basic fibroblast growth factor in human gliomas and their relation to angiogenesis. Int J Cancer 84(1):10–18

Sciaccaluga M, Fioretti B, Catacuzzeno L, Pagani F, Bertollini C, Rosito M et al (2010) CXCL12-induced glioblastoma cell migration requires intermediate conductance Ca2+-activated K+ channel activity. Am J Physiol Cell Physiol 299(1):C175–C184

Semple BD, Kossmann T, Morganti-Kossmann MC (2010) Role of chemokines in CNS health and pathology: a focus on the CCL2/CCR2 and CXCL8/CXCR2 networks. J Cereb Blood Flow Metab 30(3):459–473

Sherry MM, Reeves A, Wu JK, Cochran BH (2009) STAT3 is required for proliferation and maintenance of multipotency in glioblastoma stem cells. Stem Cells 27(10):2383–2392

Stommel JM, Kimmelman AC, Ying H, Nabioullin R, Ponugoti AH, Wiedemeyer R et al (2007) Coactivation of receptor tyrosine kinases affects the response of tumor cells to targeted therapies. Science 318(5848):287–290

Strauss LG, Koczan D, Seiz M, Tuettenberg J, Schmieder K, Pan L et al (2012) Correlation of the Ga-68-bombesin analog Ga-68-BZH3 with receptors expression in gliomas as measured by quantitative dynamic positron emission tomography (dPET) and gene arrays. Mol Imaging Biol 14(3):376–383

Stylianou DC, Auf der Maur A, Kodack DP, Henke RT, Hohn S, Toretsky JA et al (2009) Effect of single-chain antibody targeting of the ligand-binding domain in the anaplastic lymphoma kinase receptor. Oncogene 28(37):3296–3306

Swiatek-Machado K, Kaminska B (2013) STAT signaling in glioma cells. Adv Exp Med Biol 986:189–208

Tang P, Steck PA, Yung WK (1997) The autocrine loop of TGF-alpha/EGFR and brain tumors. J Neurooncol 35(3):303–314

Tchirkov A, Rolhion C, Bertrand S, Dore JF, Dubost JJ, Verrelle P (2001) IL-6 gene amplification and expression in human glioblastomas. Br J Cancer 85(4):518–522

Ulbricht U, Eckerich C, Fillbrandt R, Westphal M, Lamszus K (2006) RNA interference targeting protein tyrosine phosphatase zeta/receptor-type protein tyrosine phosphatase beta suppresses glioblastoma growth in vitro and in vivo. J Neurochem 98(5):1497–1506

Van Brocklyn JR, Jackson CA, Pearl DK, Kotur MS, Snyder PJ, Prior TW (2005) Sphingosine kinase-1 expression correlates with poor survival of patients with glioblastoma multiforme: roles of sphingosine kinase isoforms in growth of glioblastoma cell lines. J Neuropathol Exp Neurol 64(8):695–705

Verhaak RG, Hoadley KA, Purdom E, Wang V, Qi Y, Wilkerson MD et al (2010) Integrated genomic analysis identifies clinically relevant subtypes of glioblastoma characterized by abnormalities in PDGFRA, IDH1, EGFR, and NF1. Cancer Cell 17(1):98–110

Wang H, Lathia JD, Wu Q, Wang J, Li Z, Heddleston JM et al (2009) Targeting interleukin 6 signaling suppresses glioma stem cell survival and tumor growth. Stem Cells 27 (10):2393–2404

Weissenberger J, Loeffler S, Kappeler A, Kopf M, Lukes A, Afanasieva TA et al (2004) IL-6 is required for glioma development in a mouse model. Oncogene 23(19):3308–3316

Widera D, Holtkamp W, Entschladen F, Niggemann B, Zanker K, Kaltschmidt B et al (2004) MCP-1 induces migration of adult neural stem cells. Eur J Cell Biol 83(8):381–387

Wiesenhofer B, Stockhammer G, Kostron H, Maier H, Hinterhuber H, Humpel C (2000) Glial cell line-derived neurotrophic factor (GDNF) and its receptor (GFR-alpha 1) are strongly expressed in human gliomas. Acta Neuropathol 99(2):131–137

Wrann M, Bodmer S, de Martin R, Siepl C, Hofer-Warbinek R, Frei K et al (1987) T cell suppressor factor from human glioblastoma cells is a 12.5-kd protein closely related to transforming growth factor-beta. EMBO J 6(6):1633–1636

Wu X, Han L, Zhang X, Li L, Jiang C, Qiu Y et al (2012) Alteration of endocannabinoid system in human gliomas. J Neurochem 120(5):842–849

Yamada N, Kato M, Yamashita H, Nister M, Miyazono K, Heldin CH et al (1995) Enhanced expression of transforming growth factor-beta and its type-I and type-II receptors in human glioblastoma. Int J Cancer 62(4):386–392

Yamagami S, Tokuda Y, Ishii K, Tanaka H, Endo N (1994) cDNA cloning and functional expression of a human monocyte chemoattractant protein 1 receptor. Biochem Biophys Res Commun 202(2):1156–1162

Yang SX, Chen JH, Jiang XF, Wang QL, Chen ZQ, Zhao W et al (2005) Activation of chemokine receptor CXCR4 in malignant glioma cells promotes the production of vascular endothelial growth factor. Biochem Biophys Res Commun 335(2):523–528

Yoshimura T, Robinson EA, Tanaka S, Appella E, Kuratsu J, Leonard EJ (1989) Purification and amino acid analysis of two human glioma-derived monocyte chemoattractants. J Exp Med 169 (4):1449–1459

Young N, Pearl DK, Van Brocklyn JR (2009) Sphingosine-1-phosphate regulates glioblastoma cell invasiveness through the urokinase plasminogen activator system and CCN1/Cyr61. Mol Cancer Res 7(1):23–32

Zhang J, Sarkar S, Yong VW (2005) The chemokine stromal cell derived factor-1 (CXCL12) promotes glioma invasiveness through MT2-matrix metalloproteinase. Carcinogenesis 26 (12):2069–2077

Zhu VF, Yang J, Lebrun DG, Li M (2012) Understanding the role of cytokines in Glioblastoma Multiforme pathogenesis. Cancer Lett 316(2):139–150

Zhu X, Fujita M, Snyder LA, Okada H (2011) Systemic delivery of neutralizing antibody targeting CCL2 for glioma therapy. J Neurooncol 104(1):83–92

Zujovic V, Benavides J, Vige X, Carter C, Taupin V (2000) Fractalkine modulates TNF-alpha secretion and neurotoxicity induced by microglial activation. Glia 29(4):305–315

Angiogenesis in Gliomas

7

Mujeeburahim Cheerathodi and Joseph H. McCarty

Abstract

Malignant brain tumors, including *glioblastoma* (GBM), display growth, survival, and invasive properties that are coupled to blood vessels and vascular-derived factors. For example, GBM stem cells (GSCs) home to perivascular niches and invasive tumor cells commonly disperse through the brain microenvironment via extracellular matrix (ECM)-rich vascular basement membranes. Anti-vascular agents that target angiogenesis, and particularly those involving vascular endothelial cell growth factor-A (VEGF-A) and its receptors, improve progression-free survival in GBM patients. However, these benefits are often transient due to compensation by alternative angiogenic pathways. The detailed molecular mechanisms that couple GBM cells to blood vessels during tumor growth and progression as well as following anti-angiogenesis therapies are just beginning to be elucidated, with various cytokines, growth factors, and ECM proteins playing important roles. In this review we will highlight molecular pathways that link cerebral blood vessels and GBM cells during tumor growth, progression, and invasion. We will also discuss mechanisms underlying GBM-induced angiogenesis, with a particular focus placed on roles for integrin adhesion receptors and their ECM protein ligands. Therapies that target angiogenesis in GBM and other brain cancers will also be summarized.

Keywords

Brain cancer • *Glioblastoma* • Growth factors • Integrins • Extracellular matrix • Invasion • Neurovascular unit • Vascular niches

M. Cheerathodi • J.H. McCarty (✉)
Department of Neurosurgery, Unit 1004 S5.8136c, University of Texas M.D. Anderson Cancer Center, 1515 Holcombe Blvd., Houston, TX 77030, USA
e-mail: jhmccarty@mdanderson.org

A. Sedo and R. Mentlein (eds.), *Glioma Cell Biology*,
DOI 10.1007/978-3-7091-1431-5_7, © Springer-Verlag Wien 2014

Abbreviations

BBB	Blood–brain barrier
CNS	Central nervous system
EC	Endothelial cell
ECM	Extracellular matrix
EGF	Epidermal growth factor
FGF	Fibroblast growth factor
HGF	Hepatocyte growth factor
GBM	*Glioblastoma*
GSC	GBM stem cell
IL	Interleukin
MAPK	Mitogen-activated protein kinase
MMP	Matrix metalloproteinase
MRP	Multidrug resistance protein
PDGF	Platelet-derived growth factor
RTK	Receptor tyrosine kinase
TGFβ	Transforming growth factor β
TIMP	Tissue inhibitors of matrix metalloproteinase
VEGF	Vascular endothelial cell growth factor

7.1 Introduction

The formation of new blood vessels via endothelial cell (EC) proliferation and sprouting, or angiogenesis, is essential for proper development and physiology of all mammalian organs (Adams and Alitalo 2007; Potente et al. 2011). This is particularly relevant in the central nervous system (CNS)—comprised of the brain, spinal cord, and retina—where neurons and glia regulate EC behaviors via direct cell–cell contacts as well as secreted growth factors and extracellular matrix (ECM) proteins (McCarty 2009a; Zacchigna et al. 2008). Aberrant regulation of angiogenesis occurs in various diseases including brain cancers such as gliomas (Bao et al. 2006; Calabrese et al. 2007; Gilbertson and Gutmann 2007). Hallmark features of malignant gliomas include pathological neovascularization, disruption of the intratumoral blood–brain barrier (BBB), and perivascular tumor cell dispersal (Gilbertson and Rich 2007; Jain et al. 2007a; Louis 2006). Gliomas affect approximately 20,000 people within the United States each year (Holland 2001; Maher et al. 2001; Ohgaki 2005). They represent the most common type of primary brain tumors, and in their advanced stages they are one of the deadliest forms of cancer. Most high-grade gliomas are refractory to standard surgical, radiation, and chemotherapeutic interventions (Ware et al. 2003). Survival rates have changed little in the last few decades, with nearly 100 % of patients succumbing to the disease within 3 years after diagnosis. Hence, understanding the basic cellular and molecular pathways that contribute to glioma growth and invasiveness may lead to new therapeutic strategies to treat or prevent the pathogenesis of this insidious disease.

Gliomas can be subdivided into three major categories based on their histology and prognosis: astrocytomas, oligodendrogliomas, and ependymomas. Astrocytomas, or gliomas of presumptive astrocytic origin, can be further divided into four main grades (Ware et al. 2003). Grade I pilocytic astrocytomas develop mostly in young adults and are managed primarily via surgical interventions. Grade II astrocytomas are gliomas consisting of differentiated and invasive tumor cells. Grade III anaplastic astrocytoma and grade IV GBMs are poorly differentiated and highly infiltrative tumors (Louis 2006).

Multiple chromosomal abnormalities and gene expression defects have been identified in gliomas, and these alterations often correlate with histological grade and clinical prognosis (Ohgaki 2005; Phillips et al. 2006). In general, glioma initiation and progression involve gene mutations that (1) deregulate growth factor receptor tyrosine kinase signaling and (2) alter the cell cycle checkpoint machinery. Low-grade tumors often express high levels of the growth factors FGF2 and PDGF, as well as their cognate receptor tyrosine kinases (Holland 2001; Shih et al. 2004). Elevated receptor activation in turn leads to amplification of downstream signaling events, often involving Ras (Ding et al. 2001; Holland et al. 2000), and commonly correlates with loss of p53 tumor suppressor functions (Reilly et al. 2000). Disruption of the cell cycle regulatory network is linked to the progression of high-grade gliomas (Holland et al. 1998b; Uhrbom et al. 2002). For example, anaplastic astrocytomas often contain deletions of the tumor suppressors Ink4a/Arf and retinoblastoma (Rb) (Bachoo et al. 2004; Xiao et al. 2002). GBMs also commonly display amplification of EGFR signaling, which can lead to Ras hyperactivation (Holland et al. 1998a). Collectively, these alterations disrupt multiple intracellular signaling pathways that contribute to the progression of glioma from low grade to high grade (Wechsler-Reya and Scott 2001).

The genetic mutations that contribute to gliomagenesis are commonly mutated in other cancers. Thus, glioma progression is likely influenced by a combination of tumor cell-extrinsic factors (Fukumura et al. 2001; Winkler et al. 2004), as well as alterations in a distinct tumor-initiating cell of origin (Sanai et al. 2005; Shih and Holland 2004; Wechsler-Reya and Scott 2001). The exact cell type that gives rise to glioma remains uncertain. However, most neural cells in the adult brain are terminally differentiated. Thus, the tumor-initiating cell of origin for glioma is limited to those compartments that retain proliferative potential, i.e., neural stem cells, glial progenitors, and differentiated glia. Genetically engineered mouse models reveal that astrocytomas arise from presumptive neural stem cells (Alcantara Llaguno et al. 2009; Zheng et al. 2008) and/or oligodendroglial cells (Liu et al. 2011). These cells reside in various regions of the adult brain (Gilbertson and Gutmann 2007), and abnormal regulation of their proliferative and differentiative capabilities likely triggers glioma onset and progression (Aboody et al. 2000; Fomchenko and Holland 2006; Maher et al. 2001).

Most brain tumors, and particularly GBM, harbor a subpopulation of proliferative and multipotent tumor-initiating cells, or GSCs (Dirks 2006). GSCs have several similarities with nonmalignant neural stem cells, including expression of common molecular markers, for example, Nestin and CD133/Prominin-1 (Read et al. 2006). Neural stem cells and brain tumor stem cells also intimately associate with vascular basement membranes in vascular niches. Importantly, contact and communication events between brain tumor stem cells and angiogenic blood vessels positively regulate tumor growth and progression (Bao et al. 2006; Calabrese et al. 2007; Salmaggi et al. 2006). Recently, subpopulations of GBM cells have been shown to transdifferentiate to ECs and pericytes and contribute to vascular pathologies. These events are also influenced by cues within the microenvironment (Ricci-Vitiani et al. 2011; Soda et al. 2011b; Wang et al. 2011) although their pathophysiological significance remains to be determined (Rodriguez et al. 2012). The focus of this review is to highlight how glioma-derived growth factors and adhesion proteins impact angiogenesis in the brain tumor microenvironment.

7.2 Blood Vessel Pathologies in Glioma

Malignant gliomas are defined, in part, by the development of hallmark angiogenesis pathologies including florid microvascular cell proliferation leading to the formation of capillaries with glomeruloid-like tufts (Fischer et al. 2005; Jain et al. 2007b). These abnormal blood vessel morphologies are accompanied by enhanced vascular permeability due to loss of the intratumoral BBB (Jansen et al. 2004; Lopes 2003; Rong et al. 2006). Although traditionally defined as the tight junctions between ECs, the BBB is now considered just one component of a larger multicellular complex, or neurovascular unit (NVU) (Abbott et al. 2006), consisting of neurons and astrocytes, vascular ECs and pericytes, as well as various growth factors and extracellular matrix (ECM) proteins in vascular basement membranes (McCarty 2009b). Comprised mainly of EC tight junctions and multidrug resistance transporters, the BBB regulates the exchange of ions, molecules, and cells between the circulation and brain and is an impediment for drug delivery (Liebner et al. 2011; Pardridge 2002). The molecular mechanisms that control BBB development and physiology remain largely unknown, although Wnts (Daneman et al. 2009; Liebner et al. 2008; Stenman et al. 2008), G protein coupled receptors such as Gpr124 (Anderson et al. 2011; Cullen et al. 2011; Kuhnert et al. 2010), and integrin-activated TGFβs (McCarty et al. 2005; Proctor et al. 2005) play important roles as detailed below.

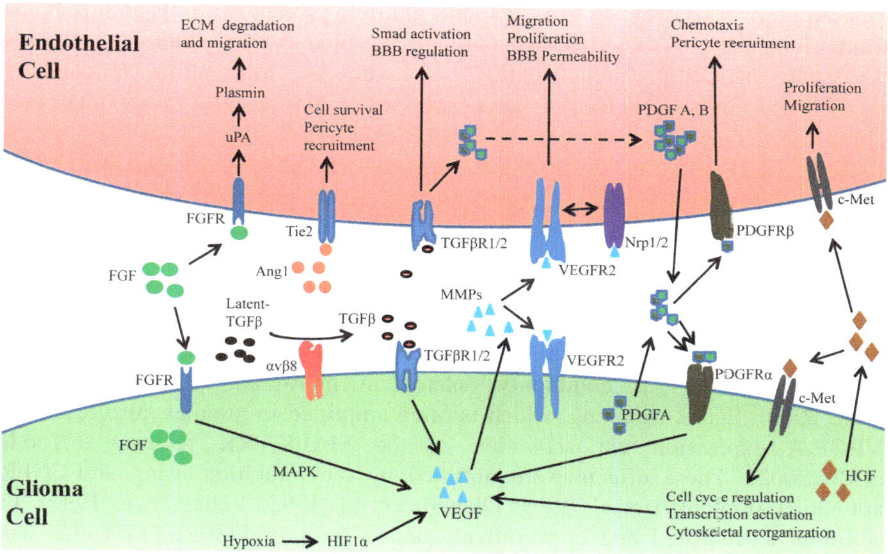

Fig. 7.1 A summary of factors involved in glioma exploitation of angiogenesis. Glioma cells dynamically communicate with blood vessels via various secreted growth factors and ECM proteins. These factors control cerebral endothelial cell and pericyte survival, proliferation, and BBB permeability

7.3 Growth Factors and Cell Adhesion Molecules in Glioma Angiogenesis

A balance between angiogenic activators and inhibitors regulates blood vessel growth, stability, and permeability (Hanahan and Folkman 1996). In glioma, this balance is disrupted by a number of pro-angiogenic and anti-angiogenic factors. Below we will discuss various growth factors and cell adhesion proteins that control angiogenesis during glioma initiation and progression (Fig. 7.1).

7.3.1 VEGF

Vascular endothelial cell growth factor-A (VEGF-A) is a critical regulator of angiogenesis during organ development as well as tumor growth and progression. VEGF-A was first discovered by Dvorak and colleagues who initially named it vascular permeability factor for its ability to enhance permeability properties of blood vessels (Dvorak 2006). Efforts by Ferrara and colleagues revealed that VEGF-A also regulates vascular EC proliferation, migration, and survival (Leung et al. 1989). Subsequent studies by a number of independent groups identified a larger gene family consisting of at least six different VEGF family members

(Jansen et al. 2004). VEGF genes express multiple protein isoforms and each can bind to distinct or shared transmembrane receptor tyrosine kinases including VEGFR1, VEGFR2, and VEGFR3 as well as Neuropilins (Bielenberg et al. 2006). These receptors are also expressed in glioma cells, suggesting autocrine VEGF signaling pathways (Ellis and Hicklin 2008).

VEGF-A expression is upregulated in glioma cells and this correlates with tumor growth and malignancy (Bulnes et al. 2012). Hypoxia-inducible factor 1a (HIF1α) is one of the major transcription factors that regulate VEGF-A gene expression (Kaelin and Ratcliffe 2008; Semenza 2003). In addition to HIF1α, other transcription factors also bind to the VEGF-A promoter to regulate gene expression. For example, p53 and VHL tumor suppressors form complexes with SP1 transcription factors and inhibit VEGF transcription (Kargiotis et al. 2006; Mukhopadhyay et al. 1997). p53 is commonly deleted in high-grade gliomas (Verhaak et al. 2010). EGFR signaling, which is often amplified in gliomas, also regulates VEGF-A expression via activation of the MAPK/ERK pathway (Woods et al. 2002). These effects were nullified by the inhibition using anti-EGFR antibodies (Goldman et al. 1993; Okamura et al. 1992; Valter et al. 1999). In addition, a truncated and constitutively active form of EGFR, EGFRvIII, has been shown to upregulate VEGF expression in glioma cells via Ras-dependent mechanisms (Feldkamp et al. 1999).

7.3.2 Notch/Delta

Cross talk between the VEGF-A and Notch pathways coordinately regulates blood vessel growth and stability (Chappell et al. 2009; Jakobsson et al. 2010). For example, VEGF-A stimulates Notch 1 expression which induces the formation of specialized endothelial "tip cells" found at the leading front of sprouting blood vessels (Hellstrom et al. 2007a). VEGFR2 signaling and tip cell formation are dampened by the anti-angiogenic Notch ligand Dll4 (Hellstrom et al. 2007b; Noguera-Troise et al. 2006). Deletion of one Dll4 allele or blockade of Notch activation with γ-secretase inhibitors induces similar phenotypes including hyperactive tip cell formation (Hellstrom et al. 2007b; Noguera-Troise et al. 2006; Siekmann and Lawson 2007). In contrast, Jag1 is a pro-angiogenic Notch ligand that counterbalances Dll4-Notch signaling and stimulates tip cell formation (Benedito et al. 2009). Exploitation of Jag1 by cancer cells has been reported; for example, epithelial carcinoma cells overexpress Jag1 and activate Notch in ECs (Zeng et al. 2005). Additionally, Jag1 in metastatic breast cancer cells mediates interactions with Notch in osteoblasts of the bone microenvironment (Sethi et al. 2011). Lastly, Notch signaling pathways are often hyperactivated in GBM (Fan et al. 2010; Hambardzumyan et al. 2008; Stockhausen et al. 2010). Inhibition of Notch activation diminishes mouse and human GSC self-renewal (Fan et al. 2010; Jeon et al. 2008) and can synergize with temozolomide to reduce glioma growth in xenograft models (Gilbert et al. 2010).

7.3.3 FGFs

Fibroblast growth factors (FGFs) are a family of structurally related proteins that regulate a wide range of developmental and pathophysiological processes (Friesel and Maciag 1995). Among the nine FGF family members, FGF-1 and FGF-2 are well characterized as angiogenic mediators and are often overexpressed in gliomas. FGF signaling has also been reported to promote VEGF expression in glioma cells (Friesel and Maciag 1995; Stefanik et al. 1991; Tsai et al. 1995). Degradation of ECM is an important step in blood vessel sprouting. FGF-2 facilitates EC migration through the ECM by upregulating urokinase-type plasminogen activator (uPA), which activates plasmin, a protease for many ECM protein components (Dunn et al. 2000).

7.3.4 PDGFs

Members of the platelet-derived growth factor (PDGF) family signal through different receptor tyrosine kinases (PDGFRs). Receptor binding activates multiple kinase cascades including PI3kinase, MAPK, JAK, SRC, and phospholipase C gamma (Fomchenko and Holland 2007). Gliomas express high levels of PDGFA and PDGFRα, but the tumor vasculature expresses low levels of PDGFRα. Instead, many glioma blood vessels express robust levels of PDGFRβ. These data suggest PDGFRα-dependent autocrine/paracrine signaling mechanisms in tumor cells and PDGFRβ-dependent paracrine signaling in ECs and pericytes (Hermanson et al. 1992). To study the effects of PDGFB in brain tumorigenesis, a mouse model was generated by overexpressing PDGFRβ in glial cells (Hermanson et al. 1992). These transgenic mice do not develop spontaneous tumors and showed normal brain development. However, when crossed to a p53−/− background mice developed tumors with pathologies similar to human GBMs including pseudopalisading necrosis, glomeruloid vessels, and BBB breakdown. Interestingly, overexpression of PDGFA in neurogenic regions of the adult mouse brain leads to premalignant gliomas via uncontrolled proliferation of neural stem and progenitor cells (Jackson et al. 2006). Interestingly, PDGFA is also a molecular marker for the classical GBM subtype (Verhaak et al. 2010). Using in vitro models, PDGFB was found to induce chemotaxis of rat brain microvascular ECs verifying the direct action of PDGFs during angiogenesis. PDGFs did not induce migratory effects on glioma cells, but were chemotactic for ECs (Brockmann et al. 2003).

7.3.5 TGFβs

The TGFβ superfamily of cytokines consists of bone morphogenetic proteins, Mullerian inhibiting substance, and activins. These proteins are involved in regulating a number of cellular processes ranging from proliferation to apoptosis (Massague et al. 2000). Members of the TGFβ family (TGFβ 1, 2, and 3) signal via canonical receptor serine/threonine kinases, TGFβR2 and TGFβR1. TGFβR2 is shared by all ligands and dimerizes with different TGFβR1s to form signaling-competent receptor complexes. Endoglin is a TGFβ co-receptor that facilitates ligand presentation to TGFβR1/TGFβR2 heterodimers (Massague and Gomis 2006). Immunohistochemical analysis of gliomas has shown upregulation of TGFβ as well as TGFβR1 and TGFβR2. TGFβ signaling via Smad transcription factors, or canonical TGFβ signaling, is also hyperactivated in many high-grade gliomas likely via uncontrolled growth and differentiation of GSCs (Bruna et al. 2007; Penuelas et al. 2009).

TGFβs often inhibit proliferation by cell cycle arrest in the G1 phase and this is mediated by regulation of INK4B expression. Interestingly, at higher TGFβ concentrations these growth inhibitory effects are negligible or in some cases potentiate glioma cell proliferation, in part owing to loss of p15 and p16 (Jen et al. 1994; Rich et al. 1999). The effects of TGFβs on angiogenesis remain controversial. In vitro studies using bovine aortic ECs treated with TGFβs showed an inhibitory effect while in vivo studies using angiogenesis system showed pro-angiogenic effects (Fajardo et al. 1996; Frater-Schroder et al. 1986). Ablation of TGFβ receptors in ECs leads to early lethality due to impaired yolk sac angiogenesis and cardiovascular development (Carvalho et al. 2007; Park et al. 2008). Additionally, TGFβ can stimulate VEGF production in glioma cells and pharmacological inhibition of TGFβR1 leads to decreased expression of VEGF and plasminogen activator inhibitor-1 (PAI-1) in gliomas (Hjelmeland et al. 2004; Koochekpour et al. 1996). PDGFA and PDGFB are downstream effectors of TGFβ in ECs while PDGFRβ expression is upregulated in vascular smooth muscle cells (Dunn et al. 2000; Helseth et al. 1988).

7.3.6 Angiopoietins

Angiopoietins (Ang1 and Ang2) play essential roles in regulating blood vessel development and stability. During embryogenesis Ang1 binds to its receptor tyrosine kinase, Tie2, and regulates stability of pericyte–EC interactions (Suri et al. 1996). Tumor cells also express Ang1, but Ang2 expression is generally limited to activated endothelium (Augustin et al. 2009). Ang2 competes with Ang1 for Tie2 binding and antagonizes Ang1 signaling (Maisonpierre et al. 1997). Hypoxia induces Ang2 expression in ECs, which disrupts Ang1–Tie2 signaling probably by acting as an antagonist to Ang1 (Holash et al. 1999). Antagonists that inhibit angiopoietin interactions with Tie receptors are currently being tested in clinical trials as anti-angiogenic agents (Peeters et al. 2013).

7.3.7 HGF

Scatter factor/hepatocyte growth factor (SF/HGF) signaling plays versatile roles in physiological and pathological processes including organogenesis and cancer

(Abounader and Laterra 2005; Birchmeier and Gherardi 1998). SF/HGF and its receptor tyrosine kinase, c-Met, are expressed by glioma cells which correlate with malignancy and vascular pathologies (Koochekpour et al. 1997; Moriyama et al. 1998; Rosen et al. 1996). Overexpression of SF/HGF caused increased tumorigenesis and tumor angiogenesis while inhibition of c-Met signaling using blocking antibodies or siRNAs suppresses tumor growth (Abounader et al. 1999, 2002; Laterra et al. 1997). In addition to glioma cells, c-Met is also expressed in tumor-associated blood vessels suggesting paracrine signals from tumor cells lead to EC growth and sprouting (Ding et al. 2003; Nakamura et al. 1995).

HGF contributes to degradation of vascular basement membranes and promotes EC migration by upregulating expression of MMPs such as MT1-MMP, MMP2, and urokinase. Another possible way SF/HGF contributes to tumor angiogenesis is by promoting proliferation through MAPK/Stat3 pathway and inhibiting apoptosis of tumor ECs (Lamszus et al. 1998; Ma et al. 2002; Wang et al. 2004). In Matrigel assays using human umbilical vein ECs, SF/HGF induces EC tube formation in a dose-dependent manner. This effect was abolished by treating with anti-HGF antibodies. In another experiment when ECs and SF/HGF secreting keratinocytes were cocultured in an in vitro system it led to the formation of EC tubes (Jiang et al. 1999; Martin et al. 1999; Wojta et al. 1999).

7.3.8 IL-6 and IL-8

Interleukins are cytokines secreted by normal and tumor cells, and in gliomas they promote proliferation and directional migration (Brat et al. 2005). Many glioma cells are capable of secreting IL-6, which can activate Sp1 and Sp3 transcription factors to induce expression of VEGF-A mRNA. IL-8 is also expressed at high levels in many glioma cells (Van Meir et al. 1990, 1992). IL-8 is a potent mediator of tumor angiogenesis via its cell surface receptors CXCR1, CXCR2, and DARC (Holmes et al. 1991; Murphy and Tiffany 1991). Glioma cells express all three receptors while DARC, but not CXCR1 and CXCR2, is expressed in microvascular ECs. CXCR1 and CXCR2 are expressed in perivascular leukocytes; hence, the angiogenic properties of IL-8 involve inflammatory responses as well. Lastly, under hypoxic conditions, IL-8 expression is upregulated via Ap-1 binding to IL-8 promoter sequences (Brat et al. 2005; Desbaillets et al. 1997, 1999).

7.3.9 TNFα

Tumor necrosis factor alpha (TNFα) is a macrophage-derived cytokine that has pleiotropic effect on cells. At low concentrations TNFα is pro-angiogenic while at high concentrations it displays anti-angiogenic activities (Fajardo et al. 1992). In high-grade gliomas, TNFα is expressed in multiple cell types including tumor cells and ECs (Maruno et al. 1997), while its receptors are expressed by ECs (Slowik et al. 1993). Angiogenic effects of TNFα are mediated indirectly by inducing expression of a number of other pro-angiogenic molecules. For example, upon TNFα treatment VEGF-A expression is upregulated in glioma cells. TNFα also upregulates expression of VEGF, IL-8, and FGFs in human microvascular ECs in vitro and blocking antibodies directed against TNFα inhibit these effects (Kargiotis et al. 2006; Ryuto et al. 1996).

7.3.10 Other Pro-angiogenic Factors

Additional growth factors and cytokines play important yet less characterized roles in angiogenesis. For example, the inducible early response gene product Cyr61/CNN1 and connective tissue growth factor CTGF/CNN2 are growth factors belonging to CNN family that induce proliferative effects on glioma cells and are downstream targets of c-Met (Goodwin et al. 2010; Jedsadayanmata et al. 1999). Expression of these proteins correlates with glioma malignancy. Tumor-associated ECs also express CTGF, suggesting pro-angiogenic roles (Pan et al. 2002; Xie et al. 2004).

The cytokine stromal cell-derived factor 1 (SDF-1/CXCL12) and its chemokine receptor CXCR4 regulate glioma cell migration and tumor cell homing to blood vessels (Rao et al. 2012). Immunohistochemical analysis revealed expression and co-localization of SDF-1 and CXCR4 in glioma cells, with an increasing intensity correlating with tumor grade. Expression of these proteins was absent in normal brain (Rempel et al. 2000). This suggests that the SDF-1/CXCR4 signaling axis may be a novel target for inhibiting glioma growth and invasion.

Various signaling effectors that control neural development also play central roles in glioma growth and angiogenesis (Eichmann et al. 2005). For example, semaphorins have important functions in controlling axonal guidance and also regulate angiogenesis. Semaphorins bind to plexin as well as Nr cell surface receptors. Nrps are co-receptors for VEGF-A in ECs and tumor cells and promote cell proliferation. Whereas VEGFR2-dependent angiogenesis results in increased vascular permeability, plexin and Nrp elicit anti-angiogenic effects upon semaphorin binding. Additionally, application of anti-Nrp inhibitory antibodies in preclinical brain tumor models results in suppression of tumor growth (Snuderl et al. 2013). Lastly, Slit-Robo interactions are important regulatory pathways in angiogenesis. Depending on its interacting receptor, Slit has opposing roles in angiogenesis. For example, when Slit binding to Robo1 leads to pro-angiogenic effects, interactions with Robo4 have anti-angiogenic outcomes (Jain et al. 2007a; Tate and Aghi 2009).

7.3.11 Anti-angiogenic Growth Factors and Cytokines

A balance between pro-angiogenic and anti-angiogenic factors, termed the angiogenic switch, controls vessel growth and stability. Alterations in this switch, for example, overexpression of pro-angiogenic factors or diminished expression of anti-angiogenic factors, promote blood vessel growth and sprouting (Hanahan and Folkman 1996). Below we detail a partial list of anti-angiogenic molecules and their likely roles in regulating glioma angiogenesis.

Angiostatin is a 38 kDa fragment of plasminogen generated by cathepsin D and MMP activities. It was the first anti-angiogenic factor to be identified in mouse models of metastatic cancer (O'Reilly et al. 1994; Tate and Aghi 2009). Angiostatin is a ligand for αvβ3 integrin and downstream signaling leads to apoptosis of ECs and tumor cells (Kirsch et al. 1998; Nishida et al. 2006; Tarui et al. 2001). An angiostatin receptor is NG2, a chondroitin sulfate proteoglycan expressed by pericytes, oligodendrocytes, and tumor cells (Stallcup and Huang 2008). NG2 can bind and sequester angiostatin and impact angiogenesis by altering the angiogenic switch (Chekenya et al. 2002; Chekenya and Pilkington 2002). Another receptor for angiostatin is ATP synthase (Rege et al. 2005). Interactions with angiostatin inhibit the enzymatic activities of ATP synthase and reduce cellular ATP production (Moser et al. 2001). Angiomotin was also identified as an angiostatin binding partner in yeast two-hybrid assays. Angiostatin functions by antagonizing the normal pro-migratory and pro-invasive functions of angiomotin (Rege et al. 2005).

Endostatin is a C-terminal fragment of type XVIII collagen, a basement membrane protein, and is another protein with anti-angiogenic properties. Endostatin induces its effects by binding to fibronectin and α5β1 and αvβ3 integrins and potentially blocking the formation of endothelial focal adhesions (O'Reilly et al. 1997; Rehn et al. 2001; Wickstrom et al. 2002).

Thrombospondins are ECM proteins that induce pro- and anti-angiogenic outcomes. In the aortic ring assay, overexpression of thrombospondins inhibits vascular cell migration and blood vessel sprouting. These effects are mediated through the CLESH domain of the cells surface receptor CD36 and type I repeats of thrombospondins-1 and -2 (Klenotic et al. 2013). Thrombospondin knockout mice also display defective wound healing and tumor-induced angiogenesis (Lawler 2000).

Tissue inhibitors of matrix metalloproteases (TIMPs) negatively regulate MMP enzymatic activities; they control EC proliferation and downregulate expression of VEGF-A. TIMPs also have pro-angiogenic properties owing to their potential to block MMP activities. For example, reduced levels of MMP-dependent expression of angiostatin and endostatin result in anti-tumorigenic and anti-angiogenic properties (Jiang et al. 2002). Lastly, pigment epithelial-derived factor (PEDF) is a member of serpin family of serine proteases that regulate neuronal differentiation and survival and are also negative regulators of angiogenesis. A specific receptor pathway, through which PEDF contributes to anti-angiogenesis, has not revealed but a key pathway involves Fas signaling (Bouck 2002; Rege et al. 2005).

7.4 Integrins in Glioma Angiogenesis

Integrins are αβ heterodimeric receptors for many ECM protein ligands that play central roles in controlling cell growth, migration, and other responses (Hynes 2002). Integrin-ECM affinity is modulated by "inside-out" signaling mechanisms (Kim et al. 2011; Vinogradova et al. 2002) involving proteins such as talins (Calderwood et al. 1999; Tadokoro et al. 2003) and kindlins (Harburger et al. 2009; Ma et al. 2008) that bind to β integrin cytoplasmic domains and induce conformational changes in extracellular regions (Shattil et al. 2010; Takagi et al. 2002; Xiong et al. 2001). ECM adhesion subsequently triggers "outside-in" signaling via adhesion protein complexes and the cytoskeleton (Harburger and Calderwood 2009; Parsons et al. 2010). In vertebrates there are 26 different integrin genes: 18 genes encoding α subunits and 8 β subunit genes. The network of integrin–ligand interactions is vast: some integrins are ligand-specific while others bind many, sharing ligands. This overlap allows for one ECM ligand to have multiple effects on a cell via adhesion to different integrins.

7.5 Integrins in GBM Cells

The brain contains a rich milieu of extracellular matrix (ECM) proteins (Thiery 2003) and abnormal regulation of cell–ECM communication is associated with gliomagenesis (Bellail et al. 2004; Gladson 1999; Shi et al. 2007a); see also Chaps. 10 and 11. For example, glioma cells like nonmalignant neural stem cells migrate through the brain parenchyma along blood vessels and white matter tracts (Sanai et al. 2005). In fact, the infiltrative nature of these tumor cells is an important determinant in the poor prognosis associated with GBM. Most metazoan cells communicate with protein components of the ECM via a family of heterodimeric cell surface receptors known as integrins (Hynes 2002). In addition to their extracellular adhesion functions, integrins also regulate intracellular signal transduction pathways that control multiple cellular responses (Giancotti and Ruoslahti 1999). In vertebrates there are 26 distinct integrin genes: 18 genes encoding α subunits, and 8 genes that encode β subunits (Hynes 2002).

Various integrins and intracellular signaling partners have been linked to the onset and/or progression of glioma (Shi et al. 2007b; Tucker 2006; Uhm et al. 1999a). For example, the fibronectin receptor α5β1 integrin is expressed in human glioma cells and inhibition of α5β1 integrin with specific small molecule antagonists retards glioma cell proliferation (Maglott et al. 2006). Additionally, human glioma cell lines express the laminin receptors α3β1 and α6β1, and these integrins regulate migration on laminin substrates (Uhm et al. 1999b).

The five members of the αv integrin subfamily primarily recognize RGD tripeptide motifs present in many shared ECM ligands, most of which are abundantly expressed in the brain microenvironment. αv integrin ECM ligands include vitronectin and fibronectin (Kalluri 2003), collagen IV (Venstrom and Reichardt 1995), and the latent associated peptide of TGFβ1 (LAP-TGFβ1) (Moses and Serra

1996). Various data link abnormal regulation of αv integrin expression and function to glioma cell growth and invasiveness. For example, αvβ1 integrin expressed in U87 glioma cells binds to the extracellular matrix protein, Ang2, leading to enhanced glioma invasiveness (Hu et al. 2006). More recently, αvβ1 integrin was found to be upregulated in glioma cells treated with anti-vascular agents, with integrin expression promoting angiogenesis and tumor cell invasion (Carbonell et al. 2013; Jahangiri et al. 2013). Human malignant gliomas display elevated levels of αvβ3 and αvβ5 integrins, suggesting that these integrins contribute to glioma cell survival and invasion (Bello et al. 2001; Treasurywala and Berens 1998). Indeed, small molecule inhibitors of αvβ3 integrin reduce glioma growth and invasiveness in vitro and in vivo (Chatterjee et al. 2000). Pieper and colleagues have shown that transformed β3−/− astrocytes form abnormally large intracranial tumors, suggesting that αvβ3 integrin may act to suppress tumor cell growth (Kanamori et al. 2004). More recent studies reveal that αvβ3 integrin exerts opposing effects, depending on whether it is expressed in tumor cells or brain microenvironment (Kanamori et al. 2006).

7.6 αvβ8 in Glioma Angiogenesis and Tumor Cell Invasiveness

The normal brain depends on αvβ8 integrin and its interactions with the ECM. The blood vessels of mice null for αv or β8 dilate, the BBB is compromised, and the mice suffer from severe CNS hemorrhage (McCarty et al. 2002), (Zhu et al. 2002). αvβ8 integrin binds to ECM-associated latent-TGFβ ligands through RGD sites and mediates release of active TGFβs. The ligands then bind the TGFβRI/II and signal through Smads and other pathways resulting in a myriad of effects. In the context of glioma, αvβ8 integrin protein levels are critically important in angiogenesis and invasiveness (Fig. 7.2). Angiogenesis is more severe in tumors with low levels of endogenous β8 integrin and overexpression of the integrin diminishes these angiogenesis pathologies. Glioma cells expressing high endogenous levels of αvβ8 integrin generate less angiogenic tumors, yet the tumors are more invasive. Invasive pathologies can be attenuated by silencing integrin gene expression using lentiviral-delivered shRNAs (Tchaicha et al. 2011). More specifically, changes are detected in cell polarity and directional migration. In scratch-wound assays cells with low levels of β8 integrin had fewer ECM contacts and displayed delayed polarization into the wound region. In contrast, β8 integrin-expressing cells formed organized actin cytoskeletal networks and polarized in a uniform direction toward the wound. αvβ8 integrin control of glioma cell polarity and directional migration is mediated, in part, via binding to RhoGDI1 leading to regulation of the Rho GTPase signaling cascade (Reyes et al. 2013). It has also recently been seen that αvβ8 is negatively regulated by mir-93, leading to gliomas with increased size and neovascularization (Fang et al. 2011).

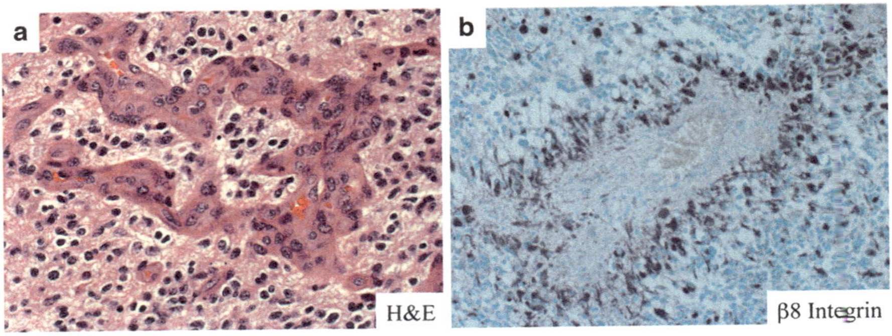

Fig. 7.2 αvβ8 integrin regulation of blood vessel pathologies in glioma. (**a**) A human GBM section stained with hematoxylin and eosin (H&E) revealing distended, glomeruloid-like blood vessels. (**b**) Human GBM section immunostained with antibodies targeting β8 integrin. Note the enrichment of β8 integrin protein expression (*brown stain*) in perivascular tumor cells

7.7 αvβ8 Integrin in Neurovascular Development and Physiology

Neural cells and vascular cells within the brain microenvironment intimately interact and communicate to form multicellular structures, or neurovascular units (Ballabh et al. 2004; Iadecola 2004; McCarty 2005; Zlokovic 2005). Proper cell–cell communication at the neurovascular unit is essential for normal CNS development, and abnormal neurovascular functions are linked to various CNS pathologies (Abbott 2002; Ballabh et al. 2004; McCarty 2005). Cerebral blood vessels are entirely compartmentalized from the surrounding neural microenvironment via a vascular basement membrane that contains a rich assortment of ECM components (Marin-Padilla 1985). Astrocyte end feet associate with the ablumenal surfaces of nearly all cerebral blood vessels via direct contacts with the vascular basement membrane (Abbott 2002). Astrocyte–blood vessel communication plays important roles in regulating molecular transport across the BBB, and also modulates rates of cerebral blood flow in response to local metabolic demands (Begley and Brightman 2003; Engelhardt 2003; Neuwelt 2004; Simard et al. 2003; Zonta et al. 2003). Astrocytes express a variety of cell surface adhesion molecules, including several integrins. At least two integrins, α6β4 and αvβ8, mediate contact and communication between perivascular neural cells and ECM components of the vascular basement membrane (Milner and Campbell 2002).

The αv and β8 integrin subunits are absolutely essential for proper neurovascular development (McCarty et al. 2002, 2005). Mouse embryos completely null for the αv integrin gene, and thus lacking all five αv integrin family members, develop CNS-specific vascular defects that include abnormal angiogenesis and intracerebral hemorrhage (Bader et al. 1998; McCarty et al. 2002, 2005). Similar integrin-dependent phenotypes are detected in the neonatal retina, which is vascularized after birth (Hirota et al. 2011).

The β8 integrin subunit pairs exclusively with αv integrin. To study αvβ8 integrin functions in the postnatal CNS Nestin-Cre transgenic mice were used to ablate αv or β8 integrin gene expression specifically in CNS neural cells. Conditional αv integrin mutants develop embryonic intracerebral hemorrhage that is grossly apparent at birth (McCarty et al. 2005). However, unlike complete αv knockouts, Nestin-Cre conditional mutants live beyond the first day of birth and survive for several months. Using a GFAP-Cre transgene, we also induced hemorrhage in the embryonic and neonatal brain after αv gene ablation (McCarty et al. 2005). Similarly, the β8 integrin gene was selectively ablated in the CNS using an identical Nestin-Cre transgene (Proctor et al. 2005). These animals also develop embryonic and neonatal intracerebral hemorrhage that is phenotypically identical to that observed in the αv integrin mutants. Deletion of the other four αv integrin-associated β subunits does not yield similar CNS vascular phenotypes (Hynes 2002). Genetic ablation of αv or β8 integrin expression in vascular ECs using the Tie2-Cre transgene did not lead to intracerebral hemorrhage or other obvious neurovascular defects (McCarty et al. 2005). These αv and β8 integrin mutant mice actually develop intestinal autoimmunity due to activities of Tie2-Cre in hematopoietic stem cells. Subsequent studies have shown that αvβ8 integrin in dendritic cells regulates latent TGFβ activation and signaling to control intestinal homeostasis.

Collectively, these molecular genetic data prove that αvβ8 integrin in CNS neural cells, particularly astroglia, regulates proper neurovascular development. Loss of αvβ8 integrin expression on CNS glia leads to defective glial-vascular cell adhesion, resulting in abnormal brain angiogenesis and intracerebral hemorrhage. αv conditional mutants also display neurological phenotypes, including sporadic seizures and a rigid gait, and mice generally do not survive beyond 8 postnatal months (McCarty et al. 2005). Similar phenotypes have been reported for the β8 integrin mutants (Proctor et al. 2005), again suggesting that the neurological defects that develop in the αv mutants are due to the specific loss of αvβ8 integrin. Additional postnatal brain deficits in β8 integrin mutant mice include impaired neuronal migration in the rostral migratory stream and widespread perivascular reactive gliosis (Mobley and McCarty 2011; Mobley et al. 2009).

αvβ8 integrin is a receptor for LAP-TGFβ1, and adhesion to an RGD peptide sequence within LAP causes activation of TGFβ signaling pathways in ECs (Cambier et al. 2005; Mu et al. 2002). Genetic ablation of TGFβ receptors in ECs leads to neurovascular phenotypes that are identical to those that develop in Nestin-Cre αv or β8 integrin mutants (Allinson et al. 2012; Arnold et al. 2012; Nguyen et al. 2011). Interestingly, TGFβ1 stimulation of vascular ECs in vitro leads to the upregulation of various ECM proteins, such as thrombospondin-1 and plasminogen activator inhibitor-1, that play established roles in regulating developmental angiogenesis and postnatal neurovascular functions (Del Zoppo 2005; Lawler 2000). Lastly, human genetic data reveal that single nucleotide polymorphisms within the TGFβ1 gene are associated with elevated risk of age-related neurovascular diseases (Kim and Lee 2006). Hyperactivation of TGFβ1-mediated signaling pathways is detected in advanced stages of glioma (Bruna et al. 2007; Rich and Bigner 2004).

Defective TGFβ activation and signaling are linked to various adult-onset CNS vascular pathologies, including Arteriovenous Malformations (Su et al. 2010), Hereditary Hemorrhagic Telangiectasia, and Pulmonary Arterial Hypertension (Orlova et al. 2011). A single nucleotide polymorphism in the TGFβ1 gene is linked to increased susceptibility to stroke (Kim and Lee 2006).

7.8 Anti-angiogenesis Therapies in Glioma

Significant progress has been made in understanding the molecular genetic events that lead to GBM initiation and progression (Furnari et al. 2007). However, only 5 % of the patients with GBM survive 5 years or more, and the medium overall survival time is about 15 months (Stupp et al. 2005; Taylor and Gerstner 2013). In addition to surgical resection, current standard-of-care treatments consist of radiation therapy and temozolomide (Stupp et al. 2005). Since gliomas are such highly vascularized neoplasms, targeting angiogenic pathways was thought to have powerful clinical benefits. Indeed, VEGF-A and VEGFRs, the main regulators of angiogenesis, as well as a number of other pro-angiogenic molecules (see above) are often overexpressed in malignant gliomas.

The US Food and Drug Administration approved the use of the anti-angiogenic antibody Bevacizumab/Avastin for the treatment of colon, lung, and breast cancers. Subsequently, in 2009 Bevacizumab was approved as a monotherapy for the treatment of gliomas (Mrugala et al. 2012). Bevacizumab is a humanized monoclonal antibody directed against VEGF-A but not other VEGF family members (Onishi et al. 2011). This antibody binds to all VEGF-A isoforms and proteolytic fragments with comparable affinities. In gliomas, Bevacizumab treatment gave promising results when combined with irinotecan. The treatments resulted in radiographic response rates of 28–40 % and a 6-month progression-free survival rate of 40–50 %. These efforts led to phase 2 clinical trials, which tested Bevacizumab as a monotherapy or in combination with irinotecan (Friedman et al. 2009; Vredenburgh et al. 2007). Combination therapies resulted in progression-free survival rates of 50.2 %, which was significantly higher than the 35 % response with Bevacizumab monotherapies. However, when compared to the medium overall survival, Bevacizumab showed promise with 9.7 months against 8.9 months in combination therapy, although progression-free survival rates were more pronounced with irinotecan. Overall survival was not significantly improved likely due to combined cytotoxic effects, leading to approval of Bevacizumab as a monotherapy for treating recurrent GBM in the United States (Kreisl et al. 2009; Taylor and Gerstner 2013). A recent publication described two patients displaying responses after receiving a combination treatment of radiation followed by temozolomide and bevacizumab, with ongoing progression-free survival of 37 and 47 months (Aguilera et al. 2013).

However, recent studies have revealed unexpected tumor cell behaviors resulting from Bevacizumab treatment. While Bevacizumab caused a reduction in tumor volumes, 30–50 % of patients developed highly infiltrative growth patterns.

Inhibition of angiogenesis results in a shift in tumor growth properties toward more infiltrative (Norden et al. 2008; Shapiro et al. 2013). Preclinical mouse models showed similar results, with inhibition of VEGF signaling causing U87 satellite lesions to form distal to the primary tumor (de Groot et al. 2010; Lucio-Eterovic et al. 2009). A separate study also yielded similar results with a medium overall survival of 8.9 months (Sahebjam et al. 2013) and other studies showed comparable data (Demirci et al. 2013; Nagane et al. 2012).

7.8.1 Cilengitide

Activation of integrin signaling in concert with growth factor receptor tyrosine kinases regulates a number of cellular processes involved in angiogenesis as well as tumor cell growth and invasion (Hood and Cheresh 2002; Kurozumi et al. 2012; Schnell et al. 2008). Cilengitide is a cyclic peptide containing an RGD sequence that binds and inhibits integrin activation and signaling (Scaringi et al. 2012). This drug is capable of antagonizing αvβ3 integrin at sub-nanomolar concentrations, and in case of α3β1 and α5β1 integrins at low nanomolar concentrations. Cilengitide also induces detachment and apoptosis in αvβ3 and αvβ5 integrin-expressing cells in culture (Taga et al. 2002). Using human xenograft models of GBM, Cilengitide suppressed tumor growth and showed anti-angiogenic and anti-tumorigenic properties (Buerkle et al. 2002; MacDonald et al. 2001; Mitjans et al. 2000; Onishi et al. 2013). These cellular outcomes are achieved through cytotoxic, anti-angiogenic, as well as anti-invasive effects (Kurozumi et al. 2012). Phase III trials in glioma are ongoing. (Eskens et al. 2003; Hariharan et al. 2007; O'Donnell et al. 2012). Early data suggest that overall survival remains modest, even though Cilengitide effectively accesses integrin targets in glioma cells and intratumoral blood vessels (Gilbert et al. 2012).

7.8.2 Sorafenib

Sorafenib is a small molecule inhibitor of VEGFRs, PDGFRs, and other kinases (Siegelin et al. 2010). In phase I clinical trials Sorafenib tested as a monotherapy or in combination with bevacizumab (Scott et al. 2010) or with radiation and temozolomide (Den et al. 2013) resulted in only modest increases in overall survival, although to date the phase II trial results have not been reported. In vitro studies have shown that sorafenib treatment of glioma cells caused a marked reduction in cell proliferation and increased apoptosis that correlated with reduced phospho-MEK and phospho-MAPK levels (Du et al. 2012). The protein kinase C δ inhibitor rottlerin has also been reported to potentiate antigrowth effects of sorafenib (Jane et al. 2006).

7.8.3 Marimastat

During blood vessel sprouting and remodeling various ECM proteins within the vascular basement membrane must be degraded. These are made possible by a class of proteins known as matrix metalloproteinases (MMPs). MMP2 (gelatinase A) and MMP9 (gelatinase B) are particularly important in glioma angiogenesis. These proteinases are secreted as proactive molecules and membrane-bound MMPs cleave and activate these proteins (Markovic et al. 2009). In comparison to normal brain and low-grade astrocytomas, GBMs overexpress many MMPs, likely leading to increased invasiveness. For example, MMP9 expression was detected at very low levels in normal brain and low-grade astrocytomas, but strong protein expression was reported in GBM (Hagemann et al. 2012). In addition, MMPs actively contribute to tumor angiogenesis by facilitating pericyte release from vascular basement membranes, releasing ECM-bound growth factors, and releasing pro-migratory ECM components helping in directed migration and in disruption of EC–cell adhesion (Rundhaug 2005). Marimastat is an MMP inhibitor that is orally administered. Activation of MMPs has proven to be essential for the tumor cell migration and angiogenesis. Various clinical trials have been conducted in different types of cancer. A phase I study identified the toxicity level of this drug with mild to severe muscle and joint pain. A phase III trial performed in different cancers, including glioma, showed only minimal improvements in overall survival (Levin et al. 2006; Steward and Thomas 2000).

7.8.4 Other Anti-vascular Agents

Thalidomide is an angiogenesis inhibitor, although the exact mechanism of action is not completely understood. Reduced expression of αvβ3 and αvβ5 integrins, as well as VEGF-A, has been reported as possible modes of action. However, clinical trials conducted using thalidomide as a monotherapy failed to improve prognosis (D'Amato et al. 1994; Fine et al. 2000; Onishi et al. 2011). Imatinib is a tyrosine kinase inhibitor of PDGFR, c-Kit, Bcr-Abl, and other targets. This compound has been shown to induce apoptosis at high concentrations. Monotherapy using imatinib was unsuccessful and did not result in clinical benefits (Morris and Abrey 2010; Radford 2002). Tenascin-C is a pro-migratory protein overexpressed in tumor ECs in GBM. Inhibition of tenascin-C using neutralizing antibodies might block angiogenesis and thereby glioma progression. However, phase II clinical trial conducted using administration of neutralizing antibodies failed to achieve survival benefits to GBM patients (Reardon et al. 2002; Zagzag et al. 1996, 2002).

7.9 Future Directions

Targeting angiogenesis was one of the more promising strategies for inhibiting tumor growth and progression, particularly in highly vascularized gliomas. Early preclinical and clinical studies yielded promising results; however, the efficacies of anti-angiogenic therapies remain in question as many reports indicate recurrence of tumors with infiltrative and drug-resistant growth properties, especially in GBM. Activation of alternative signaling pathways or compensation by pro-angiogenic molecules likely accounts for tumor recurrence and drug resistance. For example, inhibition of VEGF resulted in upregulation of placental growth factor and FGF. VEGF was also upregulated after VEGFR or EGFR inhibition, and IL8 was upregulated after HIF1α gene deletion (Carmeliet 2005). Additional mechanisms of resistance include "angiogenic mimicry" where tumor cells can transdifferentiate to ECs and contribute to blood vessel functions. Hence, once the ECs are functionally inactive, the tumor cells adapt into the function of ECs integrating into the vessel wall (El Hallani et al. 2010; Ricci-Vitiani et al. 2010; Soda et al. 2011a). This transdifferentiation is not limited to ECs, but tumor cells also give rise to mural cells such as pericytes (Cheng et al. 2013; Scully et al. 2012). Dedifferentiation of neurons and astrocytes can also contribute to gliomagenesis (Friedmann-Morvinski et al. 2012).

Another mechanism by which glioma cells acquire resistance to anti-angiogenic therapies is via enhanced invasion to distal brain regions. Mechanisms of tumor cell invasion after Bevacizumab treatment are now under intense investigation. For example, Lu et al. have shown that this invasion is mediated through c-Met activation (Lu et al. 2012) while research from our laboratory has revealed the importance of integrin αvβ8 in tumor cell invasion (Reyes et al. 2013; Tchaicha et al. 2011). A better understanding is needed for how blood vessels develop and remodel under normal and neoplastic conditions and how their regulation is altered after anti-angiogenic therapies. Combination therapies that target angiogenic effectors or both angiogenic and invasive components may lead to more effective therapies for treating gliomas.

Acknowledgments This research was supported by grant funds awarded to J.H.M. from the National Institutes of Neurological Disease and Stroke (R01NS059876 and R01NS078402) and the National Cancer Institute (P50CA127001). Due to space constraints, we apologize to colleagues in the field for not citing their relevant studies.

References

Abbott NJ (2002) Astrocyte-endothelial interactions and blood-brain barrier permeability. J Anat 200:629–638

Abbott NJ, Ronnback L, Hansson E (2006) Astrocyte-endothelial interactions at the blood-brain barrier. Nat Rev Neurosci 7:41–53

Aboody KS, Brown A, Rainov NG, Bower KA, Liu S, Yang W, Small JE, Herrlinger U, Ourednik V, Black PM et al (2000) Neural stem cells display extensive tropism for pathology in adult brain: evidence from intracranial gliomas. Proc Natl Acad Sci U S A 97:12846–12851

Abounader R, Lal B, Luddy C, Koe G, Davidson B, Rosen EM, Laterra J (2002) In vivo targeting of SF/HGF and c-met expression via U1snRNA/ribozymes inhibits glioma growth and angiogenesis and promotes apoptosis. FASEB J 16:108–110

Abounader R, Laterra J (2005) Scatter factor/hepatocyte growth factor in brain tumor growth and angiogenesis. Neuro Oncol 7:436–451

Abounader R, Ranganathan S, Lal B, Fielding K, Book A, Dietz H, Burger P, Laterra J (1999) Reversion of human glioblastoma malignancy by U1 small nuclear RNA/ribozyme targeting of scatter factor/hepatocyte growth factor and c-met expression. J Natl Cancer Inst 91:1548–1556

Adams RH, Alitalo K (2007) Molecular regulation of angiogenesis and lymphangiogenesis. Nat Rev Mol Cell Biol 8:464–478

Aguilera DG, Mazewski C, Hayes L, Jordan C, Esiashivilli N, Janns A, Macdonald TJ (2013) Prolonged survival after treatment of diffuse intrinsic pontine glioma with radiation, temozolamide, and bevacizumab: report of 2 cases. J Pediatr Hematol Oncol 35:e42–e46

Alcantara Llaguno S, Chen J, Kwon CH, Jackson EL, Li Y, Burns DK, Alvarez-Buylla A, Parada LF (2009) Malignant astrocytomas originate from neural stem/progenitor cells in a somatic tumor suppressor mouse model. Cancer Cell 15:45–56

Allinson K, Lee H, Fruttiger M, McCarty J, Arthur H (2012) Endothelial expression of TGFβ type II receptor is required to maintain vascular integrity during postnatal development of the central nervous system. PLos One 7(6):e39336

Anderson KD, Pan L, Yang XM, Hughes VC, Walls JR, Dominguez MG, Simmons MV, Burfeind P, Xue Y, Wei Y et al (2011) Angiogenic sprouting into neural tissue requires Gpr124, an orphan G protein-coupled receptor. Proc Natl Acad Sci U S A 108:2807–2812

Arnold TD, Ferrero GM, Qiu H, Phan IT, Akhurst RJ, Huang EJ, Reichardt LF (2012) Defective retinal vascular endothelial cell development as a consequence of impaired integrin alphaVbeta8-mediated activation of transforming growth factor-beta. J Neurosci 32:1197–1206

Augustin HG, Koh GY, Thurston G, Alitalo K (2009) Control of vascular morphogenesis and homeostasis through the angiopoietin-Tie system. Nat Rev Mol Cell Biol 10:165–177

Bachoo RM, Kim RS, Ligon KL, Maher EA, Brennan C, Billings N, Chan S, Li C, Rowitch DH, Wong WH et al (2004) Molecular diversity of astrocytes with implications for neurological disorders. Proc Natl Acad Sci U S A 101:8384–8389

Bader BL, Rayburn H, Crowley D, Hynes RO (1998) Extensive vasculogenesis, angiogenesis, and organogenesis precede lethality in mice lacking all alpha v integrins. Cell 95:507–519

Ballabh P, Braun A, Nedergaard M (2004) The blood-brain barrier: an overview: structure, regulation, and clinical implications. Neurobiol Dis 16:1–13

Bao S, Wu Q, Sathornsumetee S, Hao Y, Li Z, Hjelmeland AB, Shi Q, McLendon RE, Bigner DD, Rich JN (2006) Stem cell-like glioma cells promote tumor angiogenesis through vascular endothelial growth factor. Cancer Res 66:7843–7848

Begley DJ, Brightman MW (2003) Structural and functional aspects of the blood-brain barrier. Prog Drug Res 61:39–78

Bellail AC, Hunter SB, Brat DJ, Tan C, Van Meir EG (2004) Microregional extracellular matrix heterogeneity in brain modulates glioma cell invasion. Int J Biochem Cell Biol 36:1046–1069

Bello L, Francolini M, Marthyn P, Zhang J, Carroll RS, Nikas DC, Strasser JF, Villani R, Cheresh DA, Black PM (2001) Alpha(v)beta3 and alpha(v)beta5 integrin expression in glioma periphery. Neurosurgery 49:380–389, discussion 390

Benedito R, Roca C, Sorensen I, Adams S, Gossler A, Fruttiger M, Adams RH (2009) The notch ligands Dll4 and Jagged1 have opposing effects on angiogenesis. Cell 137:1124–1135

Bielenberg DR, Pettaway CA, Takashima S, Klagsbrun M (2006) Neuropilins in neoplasms: expression, regulation, and function. Exp Cell Res 312:584–593

Birchmeier C, Gherardi E (1998) Developmental roles of HGF/SF and its receptor, the c-Met tyrosine kinase. Trends Cell Biol 8:404–410

Bouck N (2002) PEDF: anti-angiogenic guardian of ocular function. Trends Mol Med 8:330–334

Brat DJ, Bellail AC, Van Meir EG (2005) The role of interleukin-8 and its receptors in gliomagenesis and tumoral angiogenesis. Neuro Oncol 7:122–133

Brockmann MA, Ulbricht U, Gruner K, Fillbrandt R, Westphal M, Lamszus K (2003) Glioblastoma and cerebral microvascular endothelial cell migration in response to tumor-associated growth factors. Neurosurgery 52:1391–1399, discussion 1399

Bruna A, Darken RS, Rojo F, Ocana A, Penuelas S, Arias A, Paris R, Tortosa A, Mora J, Baselga J et al (2007) High TGFbeta-Smad activity confers poor prognosis in glioma patients and promotes cell proliferation depending on the methylation of the PDGF-B gene. Cancer Cell 11:147–160

Buerkle MA, Pahernik SA, Sutter A, Jonczyk A, Messmer K, Dellian M (2002) Inhibition of the alpha-nu integrins with a cyclic RGD peptide impairs angiogenesis, growth and metastasis of solid tumours in vivo. Br J Cancer 86:788–795

Bulnes S, Bengoetxea H, Ortuzar N, Argandona EG, Garcia-Blanco A, Rico-Barrio I, Lafuente JV (2012) Angiogenic signalling pathways altered in gliomas: selection mechanisms for more aggressive neoplastic subpopulations with invasive phenotype. J Signal Transduct 2012:597915

Calabrese C, Poppleton H, Kocak M, Hogg TL, Fuller C, Hamner B, Oh EY, Gaber MW, Finklestein D, Allen M et al (2007) A perivascular niche for brain tumor stem cells. Cancer Cell 11:69–82

Calderwood DA, Zent R, Grant R, Rees DJ, Hynes RO, Ginsberg MH (1999) The Talin head domain binds to integrin beta subunit cytoplasmic tails and regulates integrin activation. J Biol Chem 274:28071–28074

Cambier S, Gline S, Mu D, Collins R, Araya J, Dolganov G, Einheber S, Boudreau N, Nishimura SL (2005) Integrin alpha(v)beta8-mediated activation of transforming growth factor-beta by perivascular astrocytes: an angiogenic control switch. Am J Pathol 166:1883–1894

Carbonell WS, Delay M, Jahangiri A, Park CC, Aghi MK (2013) beta1 integrin targeting potentiates antiangiogenic therapy and inhibits the growth of bevacizumab-resistant glioblastoma. Cancer Res 73:3145–3154

Carmeliet P (2005) Angiogenesis in life, disease and medicine. Nature 438:932–936

Carvalho RL, Itoh F, Goumans MJ, Lebrin F, Kato M, Takahashi S, Ema M, Itoh S, van Rooijen M, Bertolino P et al (2007) Compensatory signalling induced in the yolk sac vasculature by deletion of TGFbeta receptors in mice. J Cell Sci 120:4269–4277

Chappell JC, Taylor SM, Ferrara N, Bautch VL (2009) Local guidance of emerging vessel sprouts requires soluble Flt-1. Dev Cell 17:377–386

Chatterjee S, Matsumura A, Schradermeier J, Gillespie GY (2000) Human malignant glioma therapy using anti-alpha(v)beta3 integrin agents. J Neurooncol 46:135–144

Chekenya M, Hjelstuen M, Enger PO, Thorsen F, Jacob AL, Probst B, Haraldseth O, Pilkington G, Butt A, Levine JM et al (2002) NG2 proteoglycan promotes angiogenesis-dependent tumor growth in CNS by sequestering angiostatin. FASEB J 16:586–588

Chekenya M, Pilkington GJ (2002) NG2 precursor cells in neoplasia: functional, histogenesis and therapeutic implications for malignant brain tumours. J Neurocytol 31:507–521

Cheng L, Huang Z, Zhou W, Wu Q, Donnola S, Liu JK, Fang X, Sloan AE, Mao Y, Lathia JD et al (2013) Glioblastoma stem cells generate vascular pericytes to support vessel function and tumor growth. Cell 153:139–152

Cullen M, Elzarrad MK, Seaman S, Zudaire E, Stevens J, Yang MY, Li X, Chaudhary A, Xu L, Hilton MB et al (2011) GPR124, an orphan G protein-coupled receptor, is required for CNS-specific vascularization and establishment of the blood-brain barrier. Proc Natl Acad Sci U S A 108:5759–5764

D'Amato RJ, Loughnan MS, Flynn E, Folkman J (1994) Thalidomide is an inhibitor of angiogenesis. Proc Natl Acad Sci U S A 91:4082–4085

Daneman R, Agalliu D, Zhou L, Kuhnert F, Kuo CJ, Barres BA (2009) Wnt/beta-catenin signaling is required for CNS, but not non-CNS, angiogenesis. Proc Natl Acad Sci U S A 106:641–646

de Groot JF, Fuller G, Kumar AJ, Piao Y, Eterovic K, Ji Y, Conrad CA (2010) Tumor invasion after treatment of glioblastoma with bevacizumab: radiographic and pathologic correlation in humans and mice. Neuro Oncol 12:233–242

Del Zoppo GJ (2005) Focal cerebral ischemia and hemostasis: a PAI-1 conundrum. J Thromb Haemost 3:1376–1378

Demirci U, Tufan G, Aktas B, Balakan O, Alacacioglu A, Dane F, Engin H, Kaplan MA, Gunaydin Y, Ozdemir NY et al (2013) Bevacizumab plus irinotecan in recurrent or progressive malign glioma: a multicenter study of the Anatolian Society of Medical Oncology (ASMO). J Cancer Res Clin Oncol 139(5):829–835

Den RB, Kamrava M, Sheng Z, Werner-Wasik M, Dougherty E, Marinucchi M, Lawrence YR, Hegarty S, Hyslop T, Andrews DW et al (2013) A phase I study of the combination of sorafenib with temozolomide and radiation therapy for the treatment of primary and recurrent high-grade gliomas. Int J Radiat Oncol Biol Phys 85:321–328

Desbaillets I, Diserens AC, de Tribolet N, Hamou MF, Van Meir EG (1999) Regulation of interleukin-8 expression by reduced oxygen pressure in human glioblastoma. Oncogene 18:1447–1456

Desbaillets I, Diserens AC, Tribolet N, Hamou MF, Van Meir EG (1997) Upregulation of interleukin 8 by oxygen-deprived cells in glioblastoma suggests a role in leukocyte activation, chemotaxis, and angiogenesis. J Exp Med 186:1201–1212

Ding H, Roncari L, Shannon P, Wu X, Lau N, Karaskova J, Gutmann DH, Squire JA, Nagy A, Guha A (2001) Astrocyte-specific expression of activated p21-ras results in malignant astrocytoma formation in a transgenic mouse model of human gliomas. Cancer Res 61:3826–3836

Ding S, Merkulova-Rainon T, Han ZC, Tobelem G (2003) HGF receptor up-regulation contributes to the angiogenic phenotype of human endothelial cells and promotes angiogenesis in vitro. Blood 101:4816–4822

Dirks PB (2006) Cancer: stem cells and brain tumours. Nature 444:687–688

Du W, Zhou JR, Wang DL, Gong K, Zhang QJ (2012) Vitamin K1 enhances sorafenib-induced growth inhibition and apoptosis of human malignant glioma cells by blocking the Raf/MEK/ERK pathway. World J Surg Oncol 10:60

Dunn IF, Heese O, Black PM (2000) Growth factors in glioma angiogenesis: FGFs, PDGF, EGF, and TGFs. J Neurooncol 50:121–137

Dvorak HF (2006) Discovery of vascular permeability factor (VPF). Exp Cell Res 312:522–526

Eichmann A, Makinen T, Alitalo K (2005) Neural guidance molecules regulate vascular remodeling and vessel navigation. Genes Dev 19:1013–1021

El Hallani S, Boisselier B, Peglion F, Rousseau A, Colin C, Idbaih A, Marie Y, Mokhtari K, Thomas JL, Eichmann A et al (2010) A new alternative mechanism in glioblastoma vascularization: tubular vasculogenic mimicry. Brain 133:973–982

Ellis LM, Hicklin DJ (2008) VEGF-targeted therapy: mechanisms of anti-tumour activity. Nat Rev Cancer 8:579–591

Engelhardt B (2003) Development of the blood-brain barrier. Cell Tissue Res 314:119–129

Eskens FA, Dumez H, Hoekstra R, Perschl A, Brindley C, Bottcher S, Wynendaele W, Drevs J, Verweij J, van Oosterom AT (2003) Phase I and pharmacokinetic study of continuous twice weekly intravenous administration of Cilengitide (EMD 121974), a novel inhibitor of the integrins alphavbeta3 and alphavbeta5 in patients with advanced solid tumours. Eur J Cancer 39:917–926

Fajardo LF, Kwan HH, Kowalski J, Prionas SD, Allison AC (1992) Dual role of tumor necrosis factor-alpha in angiogenesis. Am J Pathol 140:539–544

Fajardo LF, Prionas SD, Kwan HH, Kowalski J, Allison AC (1996) Transforming growth factor beta1 induces angiogenesis in vivo with a threshold pattern. Lab Invest 74:600–608

Fan X, Khaki L, Zhu TS, Soules ME, Talsma CE, Gul N, Koh C, Zhang J, Li YM, Maciaczyk J et al (2010) NOTCH pathway blockade depletes CD133-positive glioblastoma cells and inhibits growth of tumor neurospheres and xenografts. Stem Cells 28:5–16

Fang L, Deng Z, Shatseva T, Yang J, Peng C, Du WW, Yee AJ, Ang LC, He C, Shan SW et al (2011) MicroRNA miR-93 promotes tumor growth and angiogenesis by targeting integrin-beta8. Oncogene 30:806–821

Feldkamp MM, Lau N, Rak J, Kerbel RS, Guha A (1999) Normoxic and hypoxic regulation of vascular endothelial growth factor (VEGF) by astrocytoma cells is mediated by Ras. Int J Cancer 81:118–124

Fine HA, Figg WD, Jaeckle K, Wen PY, Kyritsis AP, Loeffler JS, Levin VA, Black PM, Kaplan R, Pluda JM et al (2000) Phase II trial of the antiangiogenic agent thalidomide in patients with recurrent high-grade gliomas. J Clin Oncol 18:708–715

Fischer I, Gagner JP, Law M, Newcomb EW, Zagzag D (2005) Angiogenesis in gliomas: biology and molecular pathophysiology. Brain Pathol 15:297–310

Fomchenko EI, Holland EC (2006) Origins of brain tumors–a disease of stem cells? Nat Clin Pract Neurol 2:288–289

Fomchenko EI, Holland EC (2007) Platelet-derived growth factor-mediated gliomagenesis and brain tumor recruitment. Neurosurg Clin N Am 18:39–58, viii

Frater-Schroder M, Muller G, Birchmeier W, Bohlen P (1986) Transforming growth factor-beta inhibits endothelial cell proliferation. Biochem Biophys Res Commun 137:295–302

Friedman HS, Prados MD, Wen PY, Mikkelsen T, Schiff D, Abrey LE, Yung WK, Paleologos N, Nicholas MK, Jensen R et al (2009) Bevacizumab alone and in combination with irinotecan in recurrent glioblastoma. J Clin Oncol 27:4733–4740

Friedmann-Morvinski D, Bushong EA, Ke E, Soda Y, Marumoto T, Singer O, Ellisman MH, Verma IM (2012) Dedifferentiation of neurons and astrocytes by oncogenes can induce gliomas in mice. Science 338(6110):1080–1084

Friesel RE, Maciag T (1995) Molecular mechanisms of angiogenesis: fibroblast growth factor signal transduction. FASEB J 9:919–925

Fukumura D, Xu L, Chen Y, Gohongi T, Seed B, Jain RK (2001) Hypoxia and acidosis independently up-regulate vascular endothelial growth factor transcription in brain tumors in vivo. Cancer Res 61:6020–6024

Furnari FB, Fenton T, Bachoo RM, Mukasa A, Stommel JM, Stegh A, Hahn WC, Ligon KL, Louis DN, Brennan C et al (2007) Malignant astrocytic glioma: genetics, biology, and paths to treatment. Genes Dev 21:2683–2710

Giancotti FG, Ruoslahti E (1999) Integrin signaling. Science 285:1028–1032

Gilbert CA, Daou MC, Moser RP, Ross AH (2010) Gamma-secretase inhibitors enhance temozolomide treatment of human gliomas by inhibiting neurosphere repopulation and xenograft recurrence. Cancer Res 70:6870–6879

Gilbert MR, Kuhn J, Lamborn KR, Lieberman F, Wen PY, Mehta M, Cloughesy T, Lassman AB, Deangelis LM, Chang S et al (2012) Cilengitide in patients with recurrent glioblastoma: the results of NABTC 03-02, a phase II trial with measures of treatment delivery. J Neurooncol 106:147–153

Gilbertson RJ, Gutmann DH (2007) Tumorigenesis in the brain: location, location, location. Cancer Res 67:5579–5582

Gilbertson RJ, Rich JN (2007) Making a tumour's bed: glioblastoma stem cells and the vascular niche. Nat Rev Cancer 7:733–736

Gladson CL (1999) The extracellular matrix of gliomas: modulation of cell function. J Neuropathol Exp Neurol 58:1029–1040

Goldman CK, Kim J, Wong WL, King V, Brock T, Gillespie GY (1993) Epidermal growth factor stimulates vascular endothelial growth factor production by human malignant glioma cells: a model of glioblastoma multiforme pathophysiology. Mol Biol Cell 4:121–133

Goodwin CR, Lal B, Zhou X, Ho S, Xia S, Taeger A, Murray J, Laterra J (2010) Cyr61 mediates hepatocyte growth factor-dependent tumor cell growth, migration, and Akt activation. Cancer Res 70:2932–2941

Hagemann C, Anacker J, Ernestus RI, Vince GH (2012) A complete compilation of matrix metalloproteinase expression in human malignant gliomas. World J Clin Oncol 3:67–79

Hambardzumyan D, Becher OJ, Holland EC (2008) Cancer stem cells and survival pathways. Cell Cycle 7:1371–1378

Hanahan D, Folkman J (1996) Patterns and emerging mechanisms of the angiogenic switch during tumorigenesis. Cell 86:353–364

Harburger DS, Bouaouina M, Calderwood DA (2009) Kindlin-1 and -2 directly bind the C-terminal region of beta integrin cytoplasmic tails and exert integrin-specific activation effects. J Biol Chem 284:11485–11497

Harburger DS, Calderwood DA (2009) Integrin signalling at a glance. J Cell Sci 122:159–163

Hariharan S, Gustafson D, Holden S, McConkey D, Davis D, Morrow M, Basche M, Gore L, Zang C, O'Bryant CL et al (2007) Assessment of the biological and pharmacological effects of the alpha nu beta3 and alpha nu beta5 integrin receptor antagonist, cilengitide (EMD 121974), in patients with advanced solid tumors. Ann Oncol 18:1400–1407

Hellstrom M, Phng LK, Gerhardt H (2007a) VEGF and Notch signaling: the yin and yang of angiogenic sprouting. Cell Adh Migr 1:133–136

Hellstrom M, Phng LK, Hofmann JJ, Wallgard E, Coultas L, Lindblom P, Alva J, Nilsson AK, Karlsson L, Gaiano N et al (2007b) Dll4 signalling through Notch1 regulates formation of tip cells during angiogenesis. Nature 445:776–780

Helseth E, Unsgaard G, Dalen A, Vik R (1988) The effects of type beta transforming growth factor on proliferation and epidermal growth factor receptor expression in a human glioblastoma cell line. J Neurooncol 6:269–276

Hermanson M, Funa K, Hartman M, Claesson-Welsh L, Heldin CH, Westermark B, Nister M (1992) Platelet-derived growth factor and its receptors in human glioma tissue: expression of messenger RNA and protein suggests the presence of autocrine and paracrine loops. Cancer Res 52:3213–3219

Hirota S, Liu Q, Lee HS, Hossain MG, Lacy-Hulbert A, McCarty JH (2011) The astrocyte-expressed integrin alphavbeta8 governs blood vessel sprouting in the developing retina. Development 138:5157–5166

Hjelmeland MD, Hjelmeland AB, Sathornsumetee S, Reese ED, Herbstreith MH, Laping NJ, Friedman HS, Bigner DD, Wang XF, Rich JN (2004) SB-431542, a small molecule transforming growth factor-beta-receptor antagonist, inhibits human glioma cell line proliferation and motility. Mol Cancer Ther 3:737–745

Holash J, Maisonpierre PC, Compton D, Boland P, Alexander CR, Zagzag D, Yancopoulos GD, Wiegand SJ (1999) Vessel cooption, regression, and growth in tumors mediated by angiopoietins and VEGF. Science 284:1994–1998

Holland EC (2001) Gliomagenesis: genetic alterations and mouse models. Nat Rev Genet 2:120–129

Holland EC, Celestino J, Dai C, Schaefer L, Sawaya RE, Fuller GN (2000) Combined activation of Ras and Akt in neural progenitors induces glioblastoma formation in mice. Nat Genet 25:55–57

Holland EC, Hively WP, DePinho RA, Varmus HE (1998a) A constitutively active epidermal growth factor receptor cooperates with disruption of G1 cell-cycle arrest pathways to induce glioma-like lesions in mice. Genes Dev 12:3675–3685

Holland EC, Hively WP, Gallo V, Varmus HE (1998b) Modeling mutations in the G1 arrest pathway in human gliomas: overexpression of CDK4 but not loss of INK4a-ARF induces hyperploidy in cultured mouse astrocytes. Genes Dev 12:3644–3649

Holmes WE, Lee J, Kuang WJ, Rice GC, Wood WI (1991) Structure and functional expression of a human interleukin-8 receptor. Science 253:1278–1280

Hood JD, Cheresh DA (2002) Role of integrins in cell invasion and migration. Nat Rev Cancer 2:91–100

Hu B, Jarzynka MJ, Guo P, Imanishi Y, Schlaepfer DD, Cheng SY (2006) Angiopoietin 2 induces glioma cell invasion by stimulating matrix metalloprotease 2 expression through the alphavbeta1 integrin and focal adhesion kinase signaling pathway. Cancer Res 66:775–783

Hynes RO (2002) Integrins: bidirectional, allosteric signaling machines. Cell 110:673–687

Iadecola C (2004) Neurovascular regulation in the normal brain and in Alzheimer's disease. Nat Rev Neurosci 5:347–360

Jackson EL, Garcia-Verdugo JM, Gil-Perotin S, Roy M, Quinones-Hinojosa A, VandenBerg S, Alvarez-Buylla A (2006) PDGFR alpha-positive B cells are neural stem cells in the adult SVZ that form glioma-like growths in response to increased PDGF signaling. Neuron 51:187–199

Jahangiri A, De Lay M, Miller LM, Carbonell WS, Hu YL, Lu K, Tom MW, Paquette J, Tokuyasu TA, Tsao S et al (2013) Gene expression profile identifies tyrosine kinase c-Met as a targetable mediator of anti-angiogenic therapy resistance. Clin Cancer Res 19(7):1773–1783

Jain RK, di Tomaso E, Duda DG, Loeffler JS, Sorensen AG, Batchelor TT (2007) Angiogenesis in brain tumours. Nat Rev Neurosci 8:610–622

Jakobsson L, Franco CA, Bentley K, Collins RT, Ponsioen B, Aspalter IM, Rosewell I, Busse M, Thurston G, Medvinsky A et al (2010) Endothelial cells dynamically compete for the tip cell position during angiogenic sprouting. Nat Cell Biol 12:943–953

Jane EP, Premkumar DR, Pollack IF (2006) Coadministration of sorafenib with rottlerin potently inhibits cell proliferation and migration in human malignant glioma cells. J Pharmacol Exp Ther 319:1070–1080

Jansen M, de Witt Hamer PC, Witmer AN, Troost D, van Noorden CJ (2004) Current perspectives on antiangiogenesis strategies in the treatment of malignant gliomas. Brain Res Brain Res Rev 45:143–163

Jedsadayanmata A, Chen CC, Kireeva ML, Lau LF, Lam SC (1999) Activation-dependent adhesion of human platelets to Cyr61 and Fisp12/mouse connective tissue growth factor is mediated through integrin alpha(IIb)beta(3). J Biol Chem 274:24321–24327

Jen J, Harper JW, Bigner SH, Bigner DD, Papadopoulos N, Markowitz S, Willson JK, Kinzler KW, Vogelstein B (1994) Deletion of p16 and p15 genes in brain tumors. Cancer Res 54:6353–6358

Jeon HM, Jin X, Lee JS, Oh SY, Sohn YW, Park HJ, Joo KM, Park WY, Nam DH, DePinho RA et al (2008) Inhibitor of differentiation 4 drives brain tumor-initiating cell genesis through cyclin E and notch signaling. Genes Dev 22:2028–2033

Jiang WG, Hiscox SE, Parr C, Martin TA, Matsumoto K, Nakamura T, Mansel RE (1999) Antagonistic effect of NK4, a novel hepatocyte growth factor variant, on in vitro angiogenesis of human vascular endothelial cells. Clin Cancer Res 5:3695–3703

Jiang Y, Goldberg ID, Shi YE (2002) Complex roles of tissue inhibitors of metalloproteinases in cancer. Oncogene 21:2245–2252

Kaelin WG Jr, Ratcliffe PJ (2008) Oxygen sensing by metazoans: the central role of the HIF hydroxylase pathway. Mol Cell 30:393–402

Kalluri R (2003) Basement membranes: structure, assembly and role in tumour angiogenesis. Nat Rev Cancer 3:422–433

Kanamori M, Kawaguchi T, Berger MS, Pieper RO (2006) Intracranial microenvironment reveals independent opposing functions of host alphaVbeta3 expression on glioma growth and angiogenesis. J Biol Chem 281:37256–37264

Kanamori M, Vanden Berg SR, Bergers G, Berger MS, Pieper RO (2004) Integrin beta3 overexpression suppresses tumor growth in a human model of gliomagenesis: implications for the role of beta3 overexpression in glioblastoma multiforme. Cancer Res 64:2751–2758

Kargiotis O, Rao JS, Kyritsis AP (2006) Mechanisms of angiogenesis in gliomas. J Neurooncol 78:281–293

Kim C, Ye F, Ginsberg MH (2011) Regulation of integrin activation. Annu Rev Cell Dev Biol 27:321–345

Kim Y, Lee C (2006) The gene encoding transforming growth factor {beta}1 confers risk of ischemic stroke and vascular dementia. Stroke 37(11):2843–2845

Kirsch M, Strasser J, Allende R, Bello L, Zhang J, Black PM (1998) Angiostatin suppresses malignant glioma growth in vivo. Cancer Res 58:4654–4659
Klenotic PA, Page RC, Li W, Amick J, Misra S, Silverstein RL (2013) Molecular basis of antiangiogenic thrombospondin-1 type 1 repeat domain interactions with CD36. Arterioscler Thromb Vasc Biol 33(7):1655–1662
Koochekpour S, Jeffers M, Rulong S, Taylor G, Klineberg E, Hudson EA, Resau JH, Vande Woude GF (1997) Met and hepatocyte growth factor/scatter factor expression in human gliomas. Cancer Res 57:5391–5398
Koochekpour S, Merzak A, Pilkington GJ (1996) Vascular endothelial growth factor production is stimulated by gangliosides and TGF-beta isoforms in human glioma cells in vitro. Cancer Lett 102:209–215
Kreisl TN, Kim L, Moore K, Duic P, Royce C, Stroud I, Garren N, Mackey M, Butman JA, Camphausen K et al (2009) Phase II trial of single-agent bevacizumab followed by bevacizumab plus irinotecan at tumor progression in recurrent glioblastoma. J Clin Oncol 27:740–745
Kuhnert F, Mancuso MR, Shamloo A, Wang HT, Choksi V, Florek M, Su H, Fruttiger M, Young WL, Heilshorn SC et al (2010) Essential regulation of CNS angiogenesis by the orphan G protein-coupled receptor GPR124. Science 330:985–989
Kurozumi K, Ichikawa T, Onishi M, Fujii K, Date I (2012) Cilengitide treatment for malignant glioma: current status and future direction. Neurol Med Chir (Tokyo) 52:539–547
Lamszus K, Schmidt NO, Jin L, Laterra J, Zagzag D, Way D, Witte M, Weinand M, Goldberg ID, Westphal M et al (1998) Scatter factor promotes motility of human glioma and neuromicrovascular endothelial cells. Int J Cancer 75:19–28
Laterra J, Nam M, Rosen E, Rao JS, Lamszus K, Goldberg ID, Johnston P (1997) Scatter factor/hepatocyte growth factor gene transfer enhances glioma growth and angiogenesis in vivo. Lab Invest 76:565–577
Lawler J (2000) The functions of thrombospondin-1 and-2. Curr Opin Cell Biol 12:634–640
Leung DW, Cachianes G, Kuang WJ, Goeddel DV, Ferrara N (1989) Vascular endothelial growth factor is a secreted angiogenic mitogen. Science 246:1306–1309
Levin VA, Phuphanich S, Yung WK, Forsyth PA, Maestro RD, Perry JR, Fuller GN, Baillet M (2006) Randomized, double-blind, placebo-controlled trial of marimastat in glioblastoma multiforme patients following surgery and irradiation. J Neurooncol 78:295–302
Liebner S, Corada M, Bangsow T, Babbage J, Taddei A, Czupalla CJ, Reis M, Felici A, Wolburg H, Fruttiger M et al (2008) Wnt/beta-catenin signaling controls development of the blood-brain barrier. J Cell Biol 183:409–417
Liebner S, Czupalla CJ, Wolburg H (2011) Current concepts of blood-brain barrier development. Int J Dev Biol 55:467–476
Liu C, Sage JC, Miller MR, Verhaak RG, Hippenmeyer S, Vogel H, Foreman O, Bronson RT, Nishiyama A, Luo L et al (2011) Mosaic analysis with double markers reveals tumor cell of origin in glioma. Cell 146:209–221
Lopes MB (2003) Angiogenesis in brain tumors. Microsc Res Tech 60:225–230
Louis DN (2006) Molecular pathology of malignant gliomas. Annu Rev Pathol 1:97–117
Lu KV, Chang JP, Parachoniak CA, Pandika MM, Aghi MK, Meyronet D, Isachenko N, Fouse SD, Phillips JJ, Cheresh DA et al (2012) VEGF inhibits tumor cell invasion and mesenchymal transition through a MET/VEGFR2 complex. Cancer Cell 22:21–35
Lucio-Eterovic AK, Piao Y, de Groot JF (2009) Mediators of glioblastoma resistance and invasion during antivascular endothelial growth factor therapy. Clin Cancer Res 15:4589–4599
Ma H, Calderon TM, Fallon JT, Berman JW (2002) Hepatocyte growth factor is a survival factor for endothelial cells and is expressed in human atherosclerotic plaques. Atherosclerosis 164:79–87
Ma YQ, Qin J, Wu C, Plow EF (2008) Kindlin-2 (Mig-2): a co-activator of beta3 integrins. J Cell Biol 181:439–446

MacDonald TJ, Taga T, Shimada H, Tabrizi P, Zlokovic BV, Cheresh DA, Laug WE (2001) Preferential susceptibility of brain tumors to the antiangiogenic effects of an alpha(v) integrin antagonist. Neurosurgery 48:151–157

Maglott A, Bartik P, Cosgun S, Klotz P, Ronde P, Fuhrmann G, Takeda K, Martin S, Dontenwill M (2006) The small alpha5beta1 integrin antagonist, SJ749, reduces proliferation and clonogenicity of human astrocytoma cells. Cancer Res 66:6002–6007

Maher EA, Furnari FB, Bachoo RM, Rowitch DH, Louis DN, Cavenee WK, DePinho RA (2001) Malignant glioma: genetics and biology of a grave matter. Genes Dev 15:1311–1333

Maisonpierre PC, Suri C, Jones PF, Bartunkova S, Wiegand SJ, Radziejewski C, Compton D, McClain J, Aldrich TH, Papadopoulos N et al (1997) Angiopoietin-2, a natural antagonist for Tie2 that disrupts in vivo angiogenesis. Science 277:55–60

Marin-Padilla M (1985) Early vascularization of the embryonic cerebral cortex: Golgi and electron microscopic studies. J Comp Neurol 241:237–249

Markovic DS, Vinnakota K, Chirasani S, Synowitz M, Raguet H, Stock K, Sliwa M, Lehmann S, Kalin R, van Rooijen N et al (2009) Gliomas induce and exploit microglial MT1-MMP expression for tumor expansion. Proc Natl Acad Sci U S A 106:12530–12535

Martin TA, Harding KG, Jiang WG (1999) Regulation of angiogenesis and endothelial cell motility by matrix-bound fibroblasts. Angiogenesis 3:69–76

Maruno M, Kovach JS, Kelly PJ, Yanagihara T (1997) Distribution of endogenous tumour necrosis factor alpha in gliomas. J Clin Pathol 50:559–562

Massague J, Blain SW, Lo RS (2000) TGFbeta signaling in growth control, cancer, and heritable disorders. Cell 103:295–309

Massague J, Gomis RR (2006) The logic of TGFbeta signaling. FEBS Lett 580:2811–2820

McCarty JH (2005) Cell biology of the neurovascular unit: implications for drug delivery across the blood-brain barrier. Assay Drug Dev Technol 3:89–95

McCarty JH (2009a) Cell adhesion and signaling networks in brain neurovascular units. Curr Opin Hematol 16:209–214

McCarty JH (2009b) Integrin-mediated regulation of neurovascular development, physiology and disease. Cell Adh Migr 3:211–215

McCarty JH, Lacy-Hulbert A, Charest A, Bronson RT, Crowley D, Housman D, Savill J, Roes J, Hynes RO (2005) Selective ablation of alphav integrins in the central nervous system leads to cerebral hemorrhage, seizures, axonal degeneration and premature death. Development 132:165–176

McCarty JH, Monahan-Earley RA, Brown LF, Keller M, Gerhardt H, Rubin K, Shani M, Dvorak HF, Wolburg H, Bader BL et al (2002) Defective associations between blood vessels and brain parenchyma lead to cerebral hemorrhage in mice lacking alphav integrins. Mol Cell Biol 22:7667–7677

Milner R, Campbell IL (2002) The integrin family of cell adhesion molecules has multiple functions within the CNS. J Neurosci Res 69:286–291

Mitjans F, Meyer T, Fittschen C, Goodman S, Jonczyk A, Marshall JF, Reyes G, Piulats J (2000) In vivo therapy of malignant melanoma by means of antagonists of alphav integrins. Int J Cancer 87:716–723

Mobley AK, McCarty JH (2011) beta8 integrin is essential for neuroblast migration in the rostral migratory stream. Glia 59:1579–1587

Mobley AK, Tchaicha JH, Shin J, Hossain MG, McCarty JH (2009) {beta}8 integrin regulates neurogenesis and neurovascular homeostasis in the adult brain. J Cell Sci 122:1842–1851

Moriyama T, Kataoka H, Kawano H, Yokogami K, Nakano S, Goya T, Uchino H, Koono M, Wakisaka S (1998) Comparative analysis of expression of hepatocyte growth factor and its receptor, c-met, in gliomas, meningiomas and schwannomas in humans. Cancer Lett 124:149–155

Morris PG, Abrey LE (2010) Novel targeted agents for platelet-derived growth factor receptor and c-KIT in malignant gliomas. Target Oncol 5:193–200

Moser TL, Kenan DJ, Ashley TA, Roy JA, Goodman MD, Misra UK, Cheek DJ, Pizzo SV (2001) Endothelial cell surface F1-F0 ATP synthase is active in ATP synthesis and is inhibited by angiostatin. Proc Natl Acad Sci U S A 98:6656–6661

Moses HL, Serra R (1996) Regulation of differentiation by TGF-beta. Curr Opin Genet Dev 6:581–586

Mrugala MM, Crew LK, Fink JR, Spence AM (2012) Carboplatin and bevacizumab for recurrent malignant glioma. Oncol Lett 4:1082–1086

Mu D, Cambier S, Fjellbirkeland L, Baron JL, Munger JS, Kawakatsu H, Sheppard D, Broaddus VC, Nishimura SL (2002) The integrin alpha(v)beta8 mediates epithelial homeostasis through MT1-MMP-dependent activation of TGF-beta1. J Cell Biol 157:493–507

Mukhopadhyay D, Knebelmann B, Cohen HT, Ananth S, Sukhatme VP (1997) The von Hippel-Lindau tumor suppressor gene product interacts with Sp1 to repress vascular endothelial growth factor promoter activity. Mol Cell Biol 17:5629–5639

Murphy PM, Tiffany HL (1991) Cloning of complementary DNA encoding a functional human interleukin-8 receptor. Science 253:1280–1283

Nagane M, Nishikawa R, Narita Y, Kobayashi H, Takano S, Shinoura N, Aoki T, Sugiyama K, Kuratsu J, Muragaki Y et al (2012) Phase II study of single-agent bevacizumab in Japanese patients with recurrent malignant glioma. Jpn J Clin Oncol 42:887–895

Nakamura Y, Morishita R, Higaki J, Kida I, Aoki M, Moriguchi A, Yamada K, Hayashi S, Yo Y, Matsumoto K et al (1995) Expression of local hepatocyte growth factor system in vascular tissues. Biochem Biophys Res Commun 215:483–488

Neuwelt EA (2004) Mechanisms of disease: the blood-brain barrier. Neurosurgery 54:131–140, discussion 141–132

Nguyen HL, Lee YJ, Shin J, Lee E, Park SO, McCarty JH, Oh SP (2011) TGF-beta signaling in endothelial cells, but not neuroepithelial cells, is essential for cerebral vascular development. Lab Invest 91:1554–1563

Nishida N, Yano H, Nishida T, Kamura T, Kojiro M (2006) Angiogenesis in cancer. Vasc Health Risk Manag 2:213–219

Noguera-Troise I, Daly C, Papadopoulos NJ, Coetzee S, Boland P, Gale NW, Lin HC, Yancopoulos GD, Thurston G (2006) Blockade of Dll4 inhibits tumour growth by promoting non-productive angiogenesis. Nature 444:1032–1037

Norden AD, Young GS, Setayesh K, Muzikansky A, Klufas R, Ross GL, Ciampa AS, Ebbeling LG, Levy B, Drappatz J et al (2008) Bevacizumab for recurrent malignant gliomas: efficacy, toxicity, and patterns of recurrence. Neurology 70:779–787

O'Donnell PH, Undevia SD, Stadler WM, Karrison TM, Nicholas MK, Janisch L, Ratain MJ (2012) A phase I study of continuous infusion cilengitide in patients with solid tumors. Invest New Drugs 30:604–610

O'Reilly MS, Boehm T, Shing Y, Fukai N, Vasios G, Lane WS, Flynn E, Birkhead JR, Olsen BR, Folkman J (1997) Endostatin: an endogenous inhibitor of angiogenesis and tumor growth. Cell 88:277–285

O'Reilly MS, Holmgren L, Shing Y, Chen C, Rosenthal RA, Moses M, Lane WS, Cao Y, Sage EH, Folkman J (1994) Angiostatin: a novel angiogenesis inhibitor that mediates the suppression of metastases by a Lewis lung carcinoma. Cell 79:315–328

Ohgaki H (2005) Genetic pathways to glioblastomas. Neuropathology 25:1–7

Okamura K, Morimoto A, Hamanaka R, Ono M, Kohno K, Uchida Y, Kuwano M (1992) A model system for tumor angiogenesis: involvement of transforming growth factor-alpha in tube formation of human microvascular endothelial cells induced by esophageal cancer cells. Biochem Biophys Res Commun 186:1471–1479

Onishi M, Ichikawa T, Kurozumi K, Date I (2011) Angiogenesis and invasion in glioma. Brain Tumor Pathol 28:13–24

Onishi M, Ichikawa T, Kurozumi K, Fujii K, Yoshida K, Inoue S, Michiue H, Chiocca EA, Kaur B, Date I (2013) Bimodal anti-glioma mechanisms of cilengitide demonstrated by novel invasive glioma models. Neuropathology 33(2):162–174

Orlova VV, Liu Z, Goumans MJ, ten Dijke P (2011) Controlling angiogenesis by two unique TGF-beta type I receptor signaling pathways. Histol Histopathol 26:1219–1230

Pan LH, Beppu T, Kurose A, Yamauchi K, Sugawara A, Suzuki M, Ogawa A, Sawai T (2002) Neoplastic cells and proliferating endothelial cells express connective tissue growth factor (CTGF) in glioblastoma. Neurol Res 24:677–683

Pardridge WM (2002) Drug and gene targeting to the brain with molecular Trojan horses. Nat Rev Drug Discov 1:131–139

Park SO, Lee YJ, Seki T, Hong KH, Fliess N, Jiang Z, Park A, Wu X, Kaartinen V, Roman BL et al (2008) ALK5- and TGFBR2-independent role of ALK1 in the pathogenesis of hereditary hemorrhagic telangiectasia type 2. Blood 111:633–642

Parsons JT, Horwitz AR, Schwartz MA (2010) Cell adhesion: integrating cytoskeletal dynamics and cellular tension. Nat Rev Mol Cell Biol 11:633–643

Peeters M, Strickland AH, Lichinitser M, Suresh AV, Manikhas G, Shapiro J, Rogowski W, Huang X, Wu B, Warner D et al (2013) A randomised, double-blind, placebo-controlled phase 2 study of trebananib (AMG 386) in combination with FOLFIRI in patients with previously treated metastatic colorectal carcinoma. Br J Cancer 108:503–511

Penuelas S, Anido J, Prieto-Sanchez RM, Folch G, Barba I, Cuartas I, Garcia-Dorado D, Poca MA, Sahuquillo J, Baselga J et al (2009) TGF-beta increases glioma-initiating cell self-renewal through the induction of LIF in human glioblastoma. Cancer Cell 15:315–327

Phillips HS, Kharbanda S, Chen R, Forrest WF, Soriano RH, Wu TD, Misra A, Nigro JM, Colman H, Soroceanu L et al (2006) Molecular subclasses of high-grade glioma predict prognosis, delineate a pattern of disease progression, and resemble stages in neurogenesis. Cancer Cell 9:157–173

Potente M, Gerhardt H, Carmeliet P (2011) Basic and therapeutic aspects of angiogenesis. Cell 146:873–887

Proctor JM, Zang K, Wang D, Wang R, Reichardt LF (2005) Vascular development of the brain requires beta8 integrin expression in the neuroepithelium. J Neurosci 25:9940–9948

Radford IR (2002) Imatinib. Novartis. Curr Opin Investig Drugs 3:492–499

Rao S, Sengupta R, Choe EJ, Woerner BM, Jackson E, Sun T, Leonard J, Piwnica-Worms D, Rubin JB (2012) CXCL12 mediates trophic interactions between endothelial and tumor cells in glioblastoma. PLoS One 7:e33005

Read TA, Hegedus B, Wechsler-Reya R, Gutmann DH (2006) The neurobiology of neurooncology. Ann Neurol 60:3–11

Reardon DA, Akabani G, Coleman RE, Friedman AH, Friedman HS, Herndon JE II, Cokgor I, McLendon RE, Pegram CN, Provenzale JM et al (2002) Phase II trial of murine (131)I-labeled antitenascin monoclonal antibody 81C6 administered into surgically created resection cavities of patients with newly diagnosed malignant gliomas. J Clin Oncol 20:1389–1397

Rege TA, Fears CY, Gladson CL (2005) Endogenous inhibitors of angiogenesis in malignant gliomas: nature's antiangiogenic therapy. Neuro Oncol 7:106–121

Rehn M, Veikkola T, Kukk-Valdre E, Nakamura H, Ilmonen M, Lombardo C, Pihlajaniemi T, Alitalo K, Vuori K (2001) Interaction of endostatin with integrins implicated in angiogenesis. Proc Natl Acad Sci U S A 98:1024–1029

Reilly KM, Loisel DA, Bronson RT, McLaughlin ME, Jacks T (2000) Nf1;Trp53 mutant mice develop glioblastoma with evidence of strain-specific effects. Nat Genet 26:109–113

Rempel SA, Dudas S, Ge S, Gutierrez JA (2000) Identification and localization of the cytokine SDF1 and its receptor, CXC chemokine receptor 4, to regions of necrosis and angiogenesis in human glioblastoma. Clin Cancer Res 6:102–111

Reyes SB, Narayanan AS, Lee HS, Tchaicha JH, Aldape KD, Lang FF, Tolias KF, McCarty JH (2013) alphavbeta 8 integrin interacts with RhoGDI1 to regulate Rac1 and Cdc42 activation and drive glioblastoma cell invasion. Mol Biol Cell 24(4):474–482

Ricci-Vitiani L, Pallini R, Biffoni M, Todaro M, Invernici G, Cenci T, Maira G, Parati EA, Stassi G, Larocca LM et al (2010) Tumour vascularization via endothelial differentiation of glioblastoma stem-like cells. Nature 468:824–828

Ricci-Vitiani L, Pallini R, Biffoni M, Todaro M, Invernici G, Cenci T, Maira G, Parati EA, Stassi G, Larocca LM et al (2011) Tumour vascularization via endothelial differentiation of glioblastoma stem-like cells. Nature 468:824–828

Rich JN, Bigner DD (2004) Development of novel targeted therapies in the treatment of malignant glioma. Nat Rev Drug Discov 3:430–446

Rich JN, Zhang M, Datto MB, Bigner DD, Wang XF (1999) Transforming growth factor-beta-mediated p15(INK4B) induction and growth inhibition in astrocytes is SMAD3-dependent and a pathway prominently altered in human glioma cell lines. J Biol Chem 274:35053–35058

Rodriguez FJ, Orr BA, Ligon KL, Eberhart CG (2012) Neoplastic cells are a rare component in human glioblastoma microvasculature. Oncotarget 3:98–106

Rong Y, Durden DL, Van Meir EG, Brat DJ (2006) 'Pseudopalisading' necrosis in glioblastoma: a familiar morphologic feature that links vascular pathology, hypoxia, and angiogenesis. J Neuropathol Exp Neurol 65:529–539

Rosen EM, Laterra J, Joseph A, Jin L, Fuchs A, Way D, Witte M, Weinand M, Goldberg ID (1996) Scatter factor expression and regulation in human glial tumors. Int J Cancer 67:248–255

Rundhaug JE (2005) Matrix metalloproteinases and angiogenesis. J Cell Mol Med 9:267–285

Ryuto M, Ono M, Izumi H, Yoshida S, Weich HA, Kohno K, Kuwano M (1996) Induction of vascular endothelial growth factor by tumor necrosis factor alpha in human glioma cells. Possible roles of SP-1. J Biol Chem 271:28220–28228

Sahebjam S, Garoufalis E, Guiot MC, Muanza T, Del Maestro R, Petrecca K, Sharma R, Kavan P (2013) Bevacizumab use for recurrent high-grade glioma at McGill University Hospital. Can J Neurol Sci 40:241–246

Salmaggi A, Boiardi A, Gelati M, Russo A, Calatozzolo C, Ciusani E, Sciacca FL, Ottolina A, Parati EA, La Porta C et al (2006) Glioblastoma-derived tumorospheres identify a population of tumor stem-like cells with angiogenic potential and enhanced multidrug resistance phenotype. Glia 54:850–860

Sanai N, Alvarez-Buylla A, Berger MS (2005) Neural stem cells and the origin of gliomas. N Engl J Med 353:811–822

Scaringi C, Minniti G, Caporello P, Enrici RM (2012) Integrin inhibitor cilengitide for the treatment of glioblastoma: a brief overview of current clinical results. Anticancer Res 32:4213–4223

Schnell O, Krebs B, Wagner E, Romagna A, Beer AJ, Grau SJ, Thon N, Goetz C, Kretzschmar HA, Tonn JC et al (2008) Expression of integrin alphavbeta3 in gliomas correlates with tumor grade and is not restricted to tumor vasculature. Brain Pathol 18:378–386

Scott BJ, Quant EC, McNamara MB, Ryg PA, Batchelor TT, Wen PY (2010) Bevacizumab salvage therapy following progression in high-grade glioma patients treated with VEGF receptor tyrosine kinase inhibitors. Neuro Oncol 12:603–607

Scully S, Francescone R, Faibish M, Bentley B, Taylor SL, Oh D, Schapiro R, Moral L, Yan W, Shao R (2012) Transdifferentiation of glioblastoma stem-like cells into mural cells drives vasculogenic mimicry in glioblastomas. J Neurosci 32:12950–12960

Semenza GL (2003) Targeting HIF-1 for cancer therapy. Nat Rev Cancer 3:721–732

Sethi N, Dai X, Winter CG, Kang Y (2011) Tumor-derived JAGGED1 promotes osteolytic bone metastasis of breast cancer by engaging notch signaling in bone cells. Cancer Cell 19:192–205

Shapiro LQ, Beal K, Goenka A, Karimi S, Iwamoto FM, Yamada Y, Zhang Z, Lassman AB, Abrey LE, Gutin PH (2013) Patterns of failure after concurrent bevacizumab and hypofractionated stereotactic radiation therapy for recurrent high-grade glioma. Int J Radiat Oncol Biol Phys 85:636–642

Shattil SJ, Kim C, Ginsberg MH (2010) The final steps of integrin activation: the end game. Nat Rev Mol Cell Biol 11:288–300

Shi Q, Bao S, Song L, Wu Q, Bigner DD, Hjelmeland AB, Rich JN (2007a) Targeting SPARC expression decreases glioma cellular survival and invasion associated with reduced activities of FAK and ILK kinases. Oncogene 26:4084–4094

Shi Q, Hjelmeland AB, Keir ST, Song L, Wickman S, Jackson D, Ohmori O, Bigner DD, Friedman HS, Rich JN (2007b) A novel low-molecular weight inhibitor of focal adhesion kinase, TAE226, inhibits glioma growth. Mol Carcinog 46:488–496

Shih AH, Dai C, Hu X, Rosenblum MK, Koutcher JA, Holland EC (2004) Dose-dependent effects of platelet-derived growth factor-B on glial tumorigenesis. Cancer Res 64:4783–4789

Shih AH, Holland EC (2004) Developmental neurobiology and the origin of brain tumors. J Neurooncol 70:125–136

Siegelin MD, Raskett CM, Gilbert CA, Ross AH, Altieri DC (2010) Sorafenib exerts anti-glioma activity in vitro and in vivo. Neurosci Lett 478:165–170

Siekmann AF, Lawson ND (2007) Notch signalling limits angiogenic cell behaviour in developing zebrafish arteries. Nature 445:781–784

Simard M, Arcuino G, Takano T, Liu QS, Nedergaard M (2003) Signaling at the gliovascular interface. J Neurosci 23:9254–9262

Slowik MR, De Luca LG, Fiers W, Pober JS (1993) Tumor necrosis factor activates human endothelial cells through the p55 tumor necrosis factor receptor but the p75 receptor contributes to activation at low tumor necrosis factor concentration. Am J Pathol 143:1724–1730

Snuderl M, Batista A, Kirkpatrick ND, Ruiz de Almodovar C, Riedemann L, Walsh EC, Anolik R, Huang Y, Martin JD, Kamoun W et al (2013) Targeting placental growth factor/neuropilin 1 pathway inhibits growth and spread of medulloblastoma. Cell 152:1065–1076

Soda Y, Marumoto T, Friedmann-Morvinski D, Soda M, Liu F, Michiue H, Pastorino S, Yang M, Hoffman RM, Kesari S et al (2011) Transdifferentiation of glioblastoma cells into vascular endothelial cells. Proc Natl Acad Sci U S A 108:4274–4280

Stallcup WB, Huang FJ (2008) A role for the NG2 proteoglycan in glioma progression. Cell Adh Migr 2:192–201

Stefanik DF, Rizkalla LR, Soi A, Goldblatt SA, Rizkalla WM (1991) Acidic and basic fibroblast growth factors are present in glioblastoma multiforme. Cancer Res 51:5760–5765

Stenman JM, Rajagopal J, Carroll TJ, Ishibashi M, McMahon J, McMahon AP (2008) Canonical Wnt signaling regulates organ-specific assembly and differentiation of CNS vasculature. Science 322:1247–1250

Steward WP, Thomas AL (2000) Marimastat: the clinical development of a matrix metalloproteinase inhibitor. Expert Opin Investig Drugs 9:2913–2922

Stockhausen MT, Kristoffersen K, Poulsen HS (2010) The functional role of Notch signaling in human gliomas. Neuro Oncol 12:199–211

Stupp R, Mason WP, van den Bent MJ, Weller M, Fisher B, Taphoorn MJ, Belanger K, Brandes AA, Marosi C, Bogdahn U et al (2005) Radiotherapy plus concomitant and adjuvant temozolomide for glioblastoma. N Engl J Med 352:987–996

Su H, Kim H, Pawlikowska L, Kitamura H, Shen F, Cambier S, Markovics J, Lawton MT, Sidney S, Bollen AW et al (2010) Reduced expression of integrin alphavbeta8 is associated with brain arteriovenous malformation pathogenesis. Am J Pathol 176:1018–1027

Suri C, Jones PF, Patan S, Bartunkova S, Maisonpierre PC, Davis S, Sato TN, Yancopoulos GD (1996) Requisite role of angiopoietin-1, a ligand for the TIE2 receptor, during embryonic angiogenesis. Cell 87:1171–1180

Tadokoro S, Shattil SJ, Eto K, Tai V, Liddington RC, de Pereda JM, Ginsberg MH, Calderwood DA (2003) Talin binding to integrin beta tails: a final common step in integrin activation. Science 302:103–106

Taga T, Suzuki A, Gonzalez-Gomez I, Gilles FH, Stins M, Shimada H, Barsky L, Weinberg KI, Laug WE (2002) alpha v-Integrin antagonist EMD 121974 induces apoptosis in brain tumor cells growing on vitronectin and tenascin. Int J Cancer 98:690–697

Takagi J, Petre BM, Walz T, Springer TA (2002) Global conformational rearrangements in integrin extracellular domains in outside-in and inside-out signaling. Cell 110:599–611

Tarui T, Miles LA, Takada Y (2001) Specific interaction of angiostatin with integrin alpha(v)beta (3) in endothelial cells. J Biol Chem 276:39562–39568

Tate MC, Aghi MK (2009) Biology of angiogenesis and invasion in glioma. Neurotherapeutics 6:447–457

Taylor J, Gerstner ER (2013) Anti-angiogenic therapy in high-grade glioma (treatment and toxicity). Curr Treat Options Neurol 15(3):328–337

Tchaicha JH, Reyes SB, Shin J, Hossain MG, Lang FF, McCarty JH (2011) Glioblastoma angiogenesis and tumor cell invasiveness are differentially regulated by beta8 integrin. Cancer Res 71:6371–6381

Thiery JP (2003) Cell adhesion in development: a complex signaling network. Curr Opin Genet Dev 13:365–371

Treasurywala S, Berens ME (1998) Migration arrest in glioma cells is dependent on the alphav integrin subunit. Glia 24:236–243

Tsai JC, Goldman CK, Gillespie GY (1995) Vascular endothelial growth factor in human glioma cell lines: induced secretion by EGF, PDGF-BB, and bFGF. J Neurosurg 82:864–873

Tucker GC (2006) Integrins: molecular targets in cancer therapy. Curr Oncol Rep 8:96–103

Uhm JH, Dooley NP, Kyritsis AP, Rao JS, Gladson CL (1999a) Vitronectin, a glioma-derived extracellular matrix protein, protects tumor cells from apoptotic death. Clin Cancer Res 5:1587–1594

Uhm JH, Gladson CL, Rao JS (1999b) The role of integrins in the malignant phenotype of gliomas. Front Biosci 4:D188–D199

Uhrbom L, Dai C, Celestino JC, Rosenblum MK, Fuller GN, Holland EC (2002) Ink4a-Arf loss cooperates with KRas activation in astrocytes and neural progenitors to generate glioblastomas of various morphologies depending on activated Akt. Cancer Res 62:5551–5558

Valter MM, Wiestler OD, Pietsche T (1999) Differential control of VEGF synthesis and secretion in human glioma cells by IL-1 and EGF. Int J Dev Neurosci 17:565–577

Van Meir E, Ceska M, Effenberger F, Walz A, Grouzmann E, Desbaillets I, Frei K, Fontana A, de Tribolet N (1992) Interleukin-8 is produced in neoplastic and infectious diseases of the human central nervous system. Cancer Res 52:4297–4305

Van Meir E, Sawamura Y, Diserens AC, Hamou MF, de Tribolet N (1990) Human glioblastoma cells release interleukin 6 in vivo and in vitro. Cancer Res 50:6683–6688

Venstrom K, Reichardt L (1995) Beta 8 integrins mediate interactions of chick sensory neurons with laminin-1, collagen IV, and fibronectin. Mol Biol Cell 6:419–431

Verhaak RG, Hoadley KA, Purdom E, Wang V, Qi Y, Wilkerson MD, Miller CR, Ding L, Golub T, Mesirov JP et al (2010) Integrated genomic analysis identifies clinically relevant subtypes of glioblastoma characterized by abnormalities in PDGFRA, IDH1, EGFR, and NF1. Cancer Cell 17:98–110

Vinogradova O, Velyvis A, Velyviene A, Hu B, Haas T, Plow E, Qin J (2002) A structural mechanism of integrin alpha(IIb)beta(3) "inside-out" activation as regulated by its cytoplasmic face. Cell 110:587–597

Vredenburgh JJ, Desjardins A, Herndon JE II, Marcello J, Reardon DA, Quinn JA, Rich JN, Sathornsumetee S, Gururangan S, Sampson J et al (2007) Bevacizumab plus irinotecan in recurrent glioblastoma multiforme. J Clin Oncol 25:4722–4729

Wang R, Chadalavada K, Wilshire J, Kowalik U, Hovinga KE, Geber A, Fligelman B, Leversha M, Brennan C, Tabar V (2011) Glioblastoma stem-like cells give rise to tumour endothelium. Nature 468:829–833

Wang X, Zhou Y, Kim HP, Song R, Zarnegar R, Ryter SW, Choi AM (2004) Hepatocyte growth factor protects against hypoxia/reoxygenation-induced apoptosis in endothelial cells. J Biol Chem 279:5237–5243

Ware ML, Berger MS, Binder DK (2003) Molecular biology of glioma tumorigenesis. Histol Histopathol 18:207–216

Wechsler-Reya R, Scott MP (2001) The developmental biology of brain tumors. Annu Rev Neurosci 24:385–428

Wickstrom SA, Alitalo K, Keski-Oja J (2002) Endostatin associates with integrin alpha5beta1 and caveolin-1, and activates Src via a tyrosyl phosphatase-dependent pathway in human endothelial cells. Cancer Res 62:5580–5589

Winkler F, Kozin SV, Tong RT, Chae SS, Booth MF, Garkavtsev I, Xu L, Hicklin DJ, Fukumura D, di Tomaso E et al (2004) Kinetics of vascular normalization by VEGFR2 blockade governs brain tumor response to radiation: role of oxygenation, angiopoietin-1, and matrix metalloproteinases. Cancer Cell 6:553–563

Wojta J, Kaun C, Breuss JM, Koshelnick Y, Beckmann R, Hattey E, Mildner M, Weninger W, Nakamura T, Tschachler E et al (1999) Hepatocyte growth factor increases expression of vascular endothelial growth factor and plasminogen activator inhibitor-1 in human keratinocytes and the vascular endothelial growth factor receptor flk-1 in human endothelial cells. Lab Invest 79:427–438

Woods SA, McGlade CJ, Guha A (2002) Phosphatidylinositol 3′-kinase and MAPK/ERK kinase 1/2 differentially regulate expression of vascular endothelial growth factor in human malignant astrocytoma cells. Neuro Oncol 4:242–252

Xiao A, Wu H, Pandolfi PP, Louis DN, Van Dyke T (2002) Astrocyte inactivation of the pRb pathway predisposes mice to malignant astrocytoma development that is accelerated by PTEN mutation. Cancer Cell 1:157–168

Xie D, Yin D, Wang HJ, Liu GT, Elashoff R, Black K, Koeffler HP (2004) Levels of expression of CYR61 and CTGF are prognostic for tumor progression and survival of individuals with gliomas. Clin Cancer Res 10:2072–2081

Xiong JP, Stehle T, Diefenbach B, Zhang R, Dunker R, Scott DL, Joachimiak A, Goodman SL, Arnaout MA (2001) Crystal structure of the extracellular segment of integrin alpha Vbeta3. Science 294:339–345

Zacchigna S, Lambrechts D, Carmeliet P (2008) Neurovascular signalling defects in neurodegeneration. Nat Rev Neurosci 9:169–181

Zagzag D, Friedlander DR, Dosik J, Chikramane S, Chan W, Greco MA, Allen JC, Dorovini-Zis K, Grumet M (1996) Tenascin-C expression by angiogenic vessels in human astrocytomas and by human brain endothelial cells in vitro. Cancer Res 56:182–189

Zagzag D, Shiff B, Jallo GI, Greco MA, Blanco C, Cohen H, Hukin J, Allen JC, Friedlander DR (2002) Tenascin-C promotes microvascular cell migration and phosphorylation of focal adhesion kinase. Cancer Res 62:2660–2668

Zeng Q, Li S, Chepeha DB, Giordano TJ, Li J, Zhang H, Polverini PJ, Nor J, Kitajewski J, Wang CY (2005) Crosstalk between tumor and endothelial cells promotes tumor angiogenesis by MAPK activation of Notch signaling. Cancer Cell 8:13–23

Zheng H, Ying H, Yan H, Kimmelman AC, Hiller DJ, Chen AJ, Perry SR, Tonon G, Chu GC, Ding Z et al (2008) p53 and Pten control neural and glioma stem/progenitor cell renewal and differentiation. Nature 455:1129–1133

Zhu J, Motejlek K, Wang D, Zang K, Schmidt A, Reichardt LF (2002) beta8 integrins are required for vascular morphogenesis in mouse embryos. Development 129:2891–2903

Zlokovic BV (2005) Neurovascular mechanisms of Alzheimer's neurodegeneration. Trends Neurosci 28:202–208

Zonta M, Angulo MC, Gobbo S, Rosengarten B, Hossmann KA, Pozzan T, Carmignoto G (2003) Neuron-to-astrocyte signaling is central to the dynamic control of brain microcirculation. Nat Neurosci 6:43–50

Immune Response: Glioma-Associated Immunosuppression

8

Selby Chen and Ian F. Parney

Abstract

Gliomas such as glioblastoma have a complex relationship with the immune system. Glioblastomas have a harshly immunosuppressive microenvironment due in large part to the expression of multiple factors by tumor cells that inhibit T-cell responses. In addition, glioblastomas are heavily infiltrated with monocytic cells. These cells appear to have become immunosuppressive under the influence of the tumor and share characteristics with myeloid-derived suppressor cells. To a lesser degree, gliomas have T-cell infiltrates. Similarly, these largely appear to have adopted the immunosuppressive phenotype of regulatory T cells. Glioblastoma patients also have marked systemic immunosuppression characterized by globally reduced T-cell counts and impaired T-cell function coupled with increased circulating immunosuppressive regulatory T cells and myeloid-derived suppressor cells. The relationships between these various immunosuppressive cell populations, their impact on T cells, and their implications for immunotherapies are reviewed in this chapter.

Keywords

Glioma • Immunosuppression • Regulatory T cell • Myeloid-derived suppressor cell

S. Chen • I.F. Parney (✉)
Department of Neurological Surgery, Mayo Clinic, 200 First Street SW, Rochester, MN 55902, USA
e-mail: Parney.ian@mayo.edu

A. Sedo and R. Mentlein (eds.), *Glioma Cell Biology*,
DOI 10.1007/978-3-7091-1431-5_8, © Springer-Verlag Wien 2014

Abbreviations

APC	Antigen-presenting cell
BBB	Blood–brain barrier
CNS	Central nervous system
CSF	Colony-stimulating factor-1
EGFR	Epidermal growth factor receptor
GBM	*Glioblastoma multiforme*
HLA	Human leukocyte antigen
IFN	Interferon
IL	Interleukin
LAK	Lymphokine-activated killer cells
MDSC	Myeloid-derived suppressor cell
MHC	Major histocompatibility complex
PG	Prostaglandin
PTEN	Phosphatase and tensin homologue deleted from chromosome 10
RANTES	Regulated on activation normal T cell expressed and secreted, CCL5
STAT	Signal transducer and activator of transcription signal transduction and transcription
TGF	Transforming growth factor
TIL	Tumor-infiltrating lymphocytes
TNF	Tumor necrosis factor
T_{reg}	Regulatory T cell
VEGF	Vascular endothelial cell growth factor

8.1 Background

Glioblastoma multiforme (GBM) represents the most common and most lethal form of primary brain tumors, with a median survival of only 15 months, despite advances in surgery, chemotherapy, and radiation therapy (Stupp et al. 2005). Therefore, novel adjuvant therapies need to be developed to better combat this pathology. Immunotherapies represent one such field of significant development in recent years. However, clinical trials involving these new treatments have been met with modest success at best, largely because gliomas exhibit significant ability not only to evade the immune system, but also to actively suppress it. This chapter will provide an overview of the mechanisms of glioma-mediated immunosuppression, as well as a brief discussion of new immunologically based therapies that seek to circumvent these methods of evasion.

8.2 Basic Principles of Immunology and Neuro-immunology

The immune system is composed of two branches: innate immunity and adaptive immunity. Innate immunity deals with nonspecific responses to foreign proteins mediated by macrophages, neutrophils, and natural killer cells; it is not antigen specific but rather responds to certain molecular motifs that are commonly represented in infectious agents (Janeway and Medzhitov 2002; Medzhitov and Janeway 2000). Adaptive immunity, on the other hand, is antigen-specific and consists of both cell-mediated and humoral components (Abbas and Lichtman 2005a). Cell-mediated adaptive immunity begins with the scavenging and processing of foreign antigens by antigen-presenting cells (APC), including dendritic cells and macrophages. Antigens are presented to both cytotoxic CD8+ T cells and helper CD4+ T cells in the context of major histocompatibility complexes (MHC), with MHC class I required for CD8+ T-cell activation, and MHC class II required for CD4+ T-cell activation. In addition to antigen presentation in the context of MHC, T-cell activation requires a second costimulatory molecule, most commonly B7-1 and B7-2 on the surface of the APC's binding CD28 on the surface of the T cells (Abbas and Lichtman 2005b). Additionally, adaptive immunity has a humoral component, mediated by B cells. When B cells are activated by stimulation with an antigen and a helper T cell, it can differentiate into antibody-secreting plasma cells. Antibodies aid in the neutralization of foreign cells by assisting phagocytic cells to ingest the foreign cell (opsonization), or by activating the complement system, which can either form pores in the cellular membranes of the foreign cells and thus directly killing them or coat the foreign cells to again assist in the phagocytosis of these cells by phagocytes (Janeway et al. 2005).

The central nervous system (CNS) has long been considered an immunologically privileged environment. Multiple factors have led to this belief: the shielding of the CNS from the peripheral circulation by the blood–brain barrier (BBB), the lack of obvious connections with the lymphatic system, and the low levels of MHC molecules expressed by cells of the CNS (Lampson and Hickey 1986). However, it has now been established that the BBB is only a relative barrier to lymphocyte tracking. A small number of lymphocytes are found in normal, healthy brain, and both naïve lymphocytes and activated T cells are able to cross the BBB (Hickey and Kimura 1987; Krakowski and Owens 2000; Hickey et al. 1991; Hickey 2001). Furthermore, in pathological states, such as multiple sclerosis and various paraneoplastic syndromes, many lymphocytes also infiltrate the CNS (Sampson et al. 1996; Gordon et al. 1997; Sawamura et al. 1989). Additionally, connections between the cerebrospinal fluid and cervical lymphatics and lymphatic drainage basins within the nasal mucosa have been discovered (Cserr and Knopf 1992; Weller et al. 1996). And in the setting of injury to the CNS, microglia, the resident APCs of the CNS and descendants of the monocyte/macrophage lineage, undergo activation and upregulate both MHC and costimulatory molecules and also contribute to both CD4+ and CD8+ specific T-cell responses (Aloisi et al. 1998, 1999; Brannan and Roberts 2004). Therefore, it appears that the central nervous system may, in fact, be an immunologically distinct environment as opposed to a strictly immunologically privileged one.

8.3 The Glioma Microenvironment

There is evidence to indicate that gliomas are actually recognized by both the innate and adaptive forms of immunity (Levi-Strauss and Mallat 1987; Sandberg-Wollheim et al. 1986). The question of why there is a disconnect between immune detection and an immune response within the tumor microenvironment has been the subject of much investigation. Studies have shown that there are many mechanisms through which the glioma can subvert the immune system. One pertains directly to their expression of human leukocyte antigens (HLA), the human form of MHCs. Many glioma cells express low levels of HLA or defective HLA, leading to deficient antigen presentation to and activation of local T cells (Yang et al. 2004). One study showed that 22 of 47 glioblastoma samples displayed loss of the HLA type I antigen (Facoetti et al. 2005). Interestingly, loss of HLA type I antigen was more common among higher grade tumors, suggesting a role of immuno-evasion in glioma progression (Facoetti et al. 2005). However, HLA downregulation has not been a universal finding in glioma immunology studies (Parney et al. 2000).

Additionally, tumor cells express various cell surface markers, such as Fas ligand, galectin, and B7-H1, that participate in local immune suppression (Yang et al. 2003; Parsa et al. 2007; Hahn et al. 2004). Fas ligand (FasL, CD95L) and its receptor Fas (CD95) participate in mediating apoptosis of CD8+ T cells. Fas is expressed by glioblastoma cells (Parney et al. 2000). High levels of Fas ligand expression by glioma cells have been shown to correlate with low levels of infiltrating T lymphocytes, suggesting increased T-cell depletion via apoptosis (Ichinose et al. 2001). B7-H1 is another immunosuppressive surface molecule that is involved in T-cell apoptosis. B7-H1 is a T-cell costimulatory molecule homologue whose expression is regulated by the tumor suppressor gene PTEN. Gliomas with a loss of PTEN function have been shown to have high levels of B7-H1 expression on their cell surface and demonstrate an especially immunoresistant phenotype (Parsa et al. 2007). Additionally, expression of B7-H1 by tumor cells promotes apoptosis in T cells and stimulates the proliferation of immunosuppressive regulatory T cells (T_{reg}) (Dong et al. 2002; Francisco et al. 2010). Finally, expression of other surface molecules such as Galectin-1 by gliomas likely augments T lymphocyte death in a manner similar to glioma overexpression of FasL (Hahn et al. 2004). The effects of these immunosuppressive surface molecules are shown schematically in Fig. 8.1.

In addition to PTEN, another oncogenic pathway has recently undergone significant investigation as it pertains to glioma immunosuppression. Signal transducers and activators of transcription 3 (STAT3) have been shown to be constitutively activated in various tumors, including high-grade gliomas (Weissenberger et al. 2004). STAT3 is a potent regulator of immunologic activity; it suppresses macrophage activation in vitro and limits the host's inflammatory response

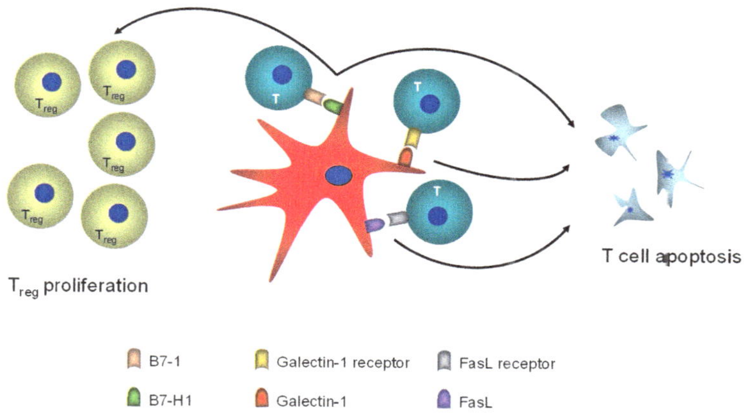

Fig. 8.1 Gliomas express multiple cell surface proteins that bind to corresponding receptors on T cells, leading either to cell death or proliferation of regulatory T cells

(O'Farrell et al. 1998; Takeda et al. 1999; Lin and Bost 2004). STAT3 blockade has led to increased expression of the pro-inflammatory cytokine interleukin-12 (IL-12) and the chemotactic cytokine RANTES (regulated on activation, normal T cell expressed and secreted, CCL5) by macrophages, as well as reversing systemic tolerance (Cheng et al. 2003). STAT3 has also been shown to effect expansion of a subpopulation of immature myeloid cells called myeloid-derived suppressor cells (MDSC) (Wu et al. 2011) that has significant immunosuppressive capacity, including inhibition of T-cell activity while causing expansion of T_{reg}'s (Serafini et al. 2004). There also appears to be a link between STAT3 inhibition and downregulation of B7-H1 expression (Matta et al. 2012), which is significant given immunosuppressive function of B7-H1. Clearly, the cellular proliferative pathways and immunosuppressive pathways mediated by various oncogenes are interconnected.

Beyond the differential expression of various membrane-bound proteins, glioma cells secrete various immunosuppressive cytokines and molecules that affect the immunologic milieu of the tumor microenvironment. Gliomas express high levels of prostaglandin E_2 (PDE_2) (Fontana et al. 1982; Sawamura et al. 1990), which inhibits interleukin-2 (IL-2) activation of lymphocytes. They also secrete transforming growth factor-β2 (TGF-β2), which suppresses IL-2-dependent T-cell proliferation (Kuppner et al. 1989; Siepl et al. 1988). In fact, TGF-β2 was initially described as glioblastoma cell-derived T-cell suppressor factor (G-TsF) because of its inhibition of cytotoxic T cells. Studies have shown that inhibition of the TGF-β2 signaling pathway in a rat glioma model significantly prolonged survival (Liau et al. 1998) and occasionally eradicated the tumor completely (Fakhrai et al. 1996). Interleukin-10 (IL-10) is also selectively secreted by invasive gliomas, although the source of IL-10 in gliomas may in fact be tumor-infiltrating monocytes/macrophages (Nitta et al. 1994). IL-10 downregulates expression of MHC-II antigens, costimulatory molecules on macrophages, and is involved in the

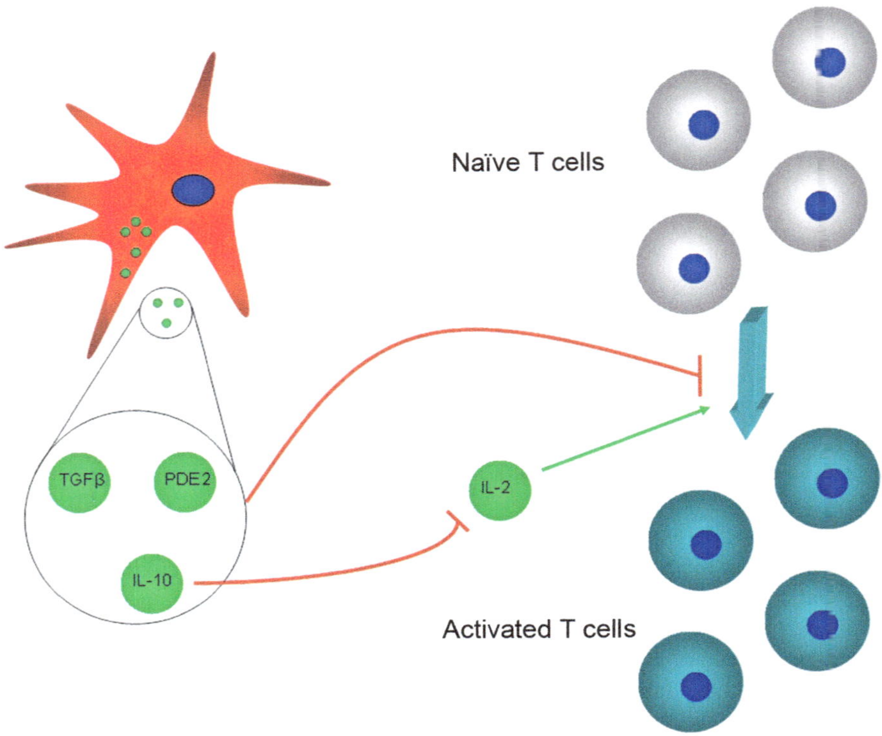

Fig. 8.2 Gliomas also secrete multiple factors that cause local immunosuppression within the tumor microenvironment. TGF-beta, PDE-2, and IL-10 all inhibit T-cell activation, while IL-10 also inhibits IL-2

JAK-STAT signaling pathway (Dix et al. 1999). It also can inhibit the synthesis of various pro-inflammatory cytokines, including IL-2, tumor necrosis factor-α (TNF-α), and interferon-γ (IFN-γ) (Dix et al. 1999). The effects of these immunosuppressive cytokines and other secreted molecules are shown schematically in Fig. 8.2.

In addition to frankly immunosuppressive cytokines, immunomodulatory cytokines are also highly secreted by gliomas. Glioma cells secrete interleukin-6 (IL-6) and colony-stimulating factor-1 (CSF-1), which shift adaptive immunity to humoral responses that may be less effective against solid tumors and/or less effective at attracting and stimulating monocytes (Parney et al. 2000; Hao et al. 2002; Bender et al. 2010). High-grade gliomas often overexpress tumor-derived vascular endothelial cell growth factor (VEGF), which primarily acts as an angiogenic factor, but also stimulates the proliferation of immunosuppressive myeloid-derived suppressor cells (Gabrilovich 2004).

The combination of cell surface proteins and secreted cytokines within the tumor microenvironment alters the profile of immune cells found within gliomas, in terms of both number and function. Studies have shown that glioblastomas are heavily infiltrated with monocytes/microglia, with these cells comprising 10–30 % of viable cells within the tumor mass (Parney et al. 2009). Data suggest that these cells likely represent invading macrophages from the periphery rather than resident microglia (Parney et al. 2009). However, microglia within the tumor microenvironment appear to have impaired antigen presentation (Flugel et al. 1999). MHC-II induction by stimulation has been shown to be significantly less in microglia and infiltrating macrophages derived from gliomas than in those obtained from normal brain samples (Schartner et al. 2005). The cytokine profile of microglia is also altered in the presence of tumor cells, with a decrease in the secretion of the pro-inflammatory cytokine TNF-α and an increase in the secretion of the inhibitory cytokine IL-10 (Kostianovsky et al. 2008). When in contact with GBM cells in vitro, monocytes also express high levels of TGF-β and B7-H1, have reduced phagocytic ability, and induce apoptosis in activated T cells (Rodrigues et al. 2010). Furthermore, Gustafson et al. have reported that immunosuppressive CD14+/HLA-DR-monocytes increase after being exposed to malignant glioma cells in culture (Gustafson et al. 2010).

Lymphocytes are also altered in the setting of gliomas. T cells, specifically CD4 + helper T cells, exhibit impaired function within the tumor microenvironment, displaying weak proliferative responses and decreased synthesis of IL-2 (Elliott et al. 1984). Additionally, tumor-infiltrating lymphocytes (TIL) exhibit a distinct phenotype. Studies have demonstrated that most CD8+ T cells found within gliomas are not activated (Hussain and Heimberger 2005). It remains unclear whether these are simply naïve lymphocytes that have passively crossed a compromised blood–brain barrier or if active T cells are rendered inactive by the host of immunosuppressive mechanisms within the tumor immunologic milieu (Han et al. 2010). Also, regulatory T cells are overrepresented in the TIL population as compared to the peripheral blood (El Andaloussi and Lesniak 2006; Fecci et al. 2006a). T_{reg}'s are a subset of T cells that play an important role in maintaining self-tolerance and preventing autoimmunity by monitoring ongoing immune activity. In oncologic states, they can serve to prevent the immune system from mounting a proper response to tumor antigens. While T_{reg}'s constitute 5–10 % of the peripheral CD4+ T-cell population (Sakaguchi 2005), they represent up to 60 % of the TIL population (Heimberger et al. 2008a; Jacobs et al. 2010). Glioblastomas have been shown to secrete CCL22 and other soluble chemokines that preferentially attract T_{reg}'s to the tumor (Jacobs et al. 2010; Crane et al. 2012). T_{reg}'s suppress the activity of effector T cells and downregulate the production of IL-2 and IFN-γ, thereby inhibiting T-cell activation and proliferation (Thornton and Shevach 1998; Camara et al. 2003). They also induce lymphocytes to secrete IL-2 (Dieckmann et al. 2002) and TGF-β (Liu et al. 2007), which further propagate the regulatory T-cell phenotype (Fecci et al. 2006b). Antibodies against T_{reg}'s have improved survival of mice harboring gliomas (Fecci et al. 2006b; El Andaloussi et al. 2006). There is an increasing amount of evidence that the proliferation of T_{reg}'s is central to the glioma's immunosuppressive capacity.

8.4 Systemic Immunosuppression

In addition to the local immunosuppressive microenvironment, gliomas also induce systemic immunosuppression in patients. It has long been known that patients bearing gliomas had impaired cell-mediated immunity, as evidenced by lymphopenia, absence of cutaneous anergy to common bacterial antigens, and an inability to mount a delayed-type hypersensitivity reaction (Brooks et al. 1972, 1974, 1976; Young et al. 1976). Peripheral blood lymphocytes from glioma patients have also been shown to have poor proliferation in response to mitogen and/or antigen stimulation in vitro (Elliott et al. 1984, 1987; Roszman et al. 1982, 1985). Some have speculated that this may be due to the circulation of immunosuppressive cytokines secreted by the tumor. However, most known immunosuppressive factors secreted by glioblastoma cells, including IL-10, TGF-β, IL-4, and IL-6, are either undetectable in glioblastoma patients' serum or not increased compared to healthy controls (Rodrigues et al. 2010; Gustafson et al. 2010).

The lymphopenia seen in glioma patients is limited primarily to the T-cell population. Peripheral blood obtained from patients with gliomas has been shown to have a significant decrease in the number of CD4+ T cells as compared to normal controls. Indeed, CD4+ counts approach those of patients with Acquired Immune Deficiency Syndrome (AIDS) (Gustafson et al. 2010). Functionally, these T cells are also impaired, including having decreased IL-2 secretion (Elliott et al. 1984) in addition to decreased expression of the IL-2 receptor (Elliott et al. 1990). Studies have also suggested the presence of T-cell signaling defects that may lead to deficient T-cell/APC contacts and hence decreased immune upregulation (Dix et al. 1999; Lowin-Kropf et al. 1998). Our lab has demonstrated that glioblastoma patients have decreased circulating immature dendritic cells, suggesting defects in antigen-presenting capacity (Zhang L et al; unpublished). The combination of decreased IL-2 receptors, poor T-cell/APC contacts, and decreased APCs likely explains the poor proliferative capacity of T cells in response to antigen stimulation.

In addition to having decreased numbers of CD4+ T cells, glioma patients have a larger proportion of their CD4+ cells represented by T_{reg}'s. In one study, patients with a glioma had 16 % of their CD4+ cells represented by T_{reg}'s, compared to 6 % in healthy volunteers (Fecci et al. 2006a). Though this number is not as high as the fraction of TILs that are T_{reg}'s, it nonetheless represents a significant increase over normal. The source of these regulatory T cells remains unclear. It is unlikely that peripheral T cells are responding to cytokines secreted by the glioma itself as these cytokines have not been detected in the serum of glioma patients. Some have speculated that this expansion of the T_{reg} population in the peripheral blood is due to reentry of T_{reg}'s from the tumor microenvironment into the circulation after T cells adopt a regulatory phenotype upon contact with glioma cells or the cytokines within the glioma immunologic milieu. However, the incidence of lymphocytes within the tumor as reported in the literature is variable, with at least some studies that show a relative paucity of tumor-infiltrating lymphocytes, making this hypothesis seem somewhat problematic (Parney et al. 2009; Hitchcock and Morris 1988; Farmer et al. 1989; Kuppner et al. 1988). However, increasing evidence seems to suggest that glioblastomas secrete soluble cytokines that attract T_{reg}'s from the periphery to the tumor microenvironment where the T_{reg}'s exhibit increased survival and proliferate, and that a portion of these regulatory T cells escape back into the circulation. Crane et al. showed that circulating T_{reg} populations after a gross total tumor resection in GBM patients decreased from 16.1 % to 5.97 %, which rebounded to 14.5 % by the time of recurrence (Crane et al. 2012). The correlation between tumor burden and peripheral T_{reg} prevalence points to the glioma itself as a possible source of circulating regulatory T cells.

In addition to T_{reg}'s, circulating myeloid-derived suppressor cells (MDSCs; CD33+/HLA-DR-/Lin-monocytes) may also play a role in systemic immunosuppression in glioma patients. Our lab has shown that MDSCs are increased in the peripheral blood of glioblastoma patients (Rodrigues et al. 2010). A similar population of immunosuppressive monocytes (CD14+/HLA-DR-) has also been reported to be increased in glioblastoma patients (Gustafson et al. 2010). MDSCs have the ability to induce apoptosis in activated T cells, to alter T-cell recognition, and to stimulate T_{reg} proliferation among naïve T cells (Serafini et al. 2004; Sinha et al. 2005; Nagaraj et al. 2007). Therefore, it is likely that MDSCs and other immunosuppressive monocytic cells play a significant role in systemic immunosuppression in malignant glioma patients. The origin of circulating MDSCs remains unclear, though given the prevalence of immunosuppressive monocytes/microglia within the tumor mass, the glioma itself seems to be a possibility. Potential cellular interactions between the tumor microenvironment and the systemic immune system are shown schematically in Fig. 8.3.

In addition to the intrinsic aspects of glioma-associated immunosuppression, one has to consider the iatrogenic factors that contribute to impaired immune function when treating patients with gliomas, including the effects of steroids and chemotherapy. Glucocorticoids given for tumor-associated edema have been shown to decrease the production of pro-inflammatory cytokines and interfere with cytokine receptor signaling (Kunicka et al. 1993). Corticosteroids may also cause sequestration of CD4+ T cells (Barshes et al. 2004; Su et al. 2004). Dexamethasone, a synthetic glucocorticoid and the most common steroid used in the treatment of high-grade gliomas, has been shown to decrease both CD4+ and CD8+ T-cell counts in the peripheral blood of steroid-treated glioblastoma patients as compared to non-treated glioblastoma patients (Gustafson et al. 2010). Temozolomide, a common chemotherapeutic agent used in the treatment of high-grade gliomas, has also been associated with lymphopenia, particularly in the CD4+ population (Brock et al. 1998; Khan et al. 2002; Heimberger et al. 2008b). The impact of common therapies on systemic immune function in glioblastoma patients is shown schematically in Fig. 8.4.

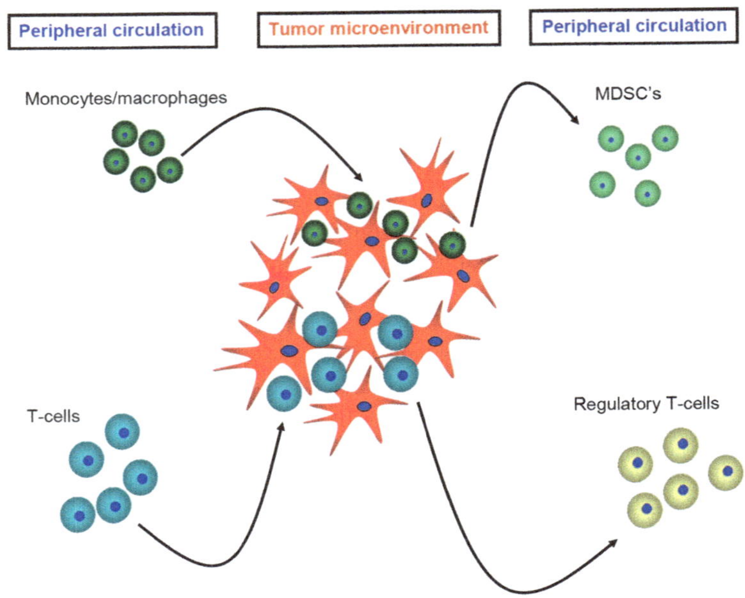

Fig. 8.3 Theoretical model regarding how the glioma causes increased populations of myeloid-derived suppressor cells and regulatory T cells within the peripheral circulation. Monocytes/microglia and T cells enter the tumor microenvironment, are converted to MDSCs and Treg's, and subsequently reenter the peripheral circulation to effect systemic immunosuppression

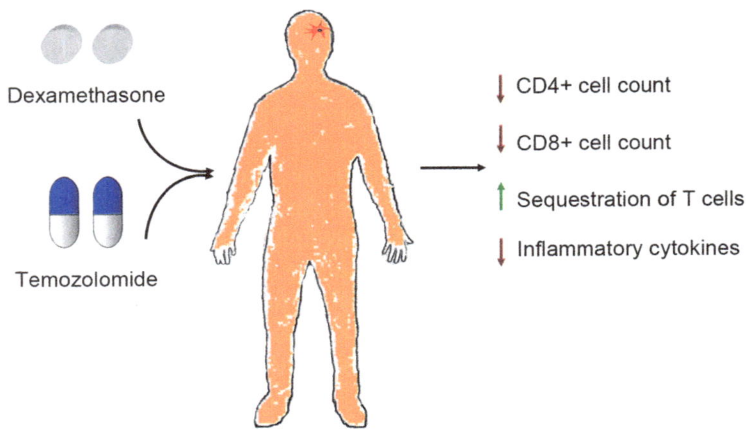

Fig. 8.4 Iatrogenic factors, such as corticosteroids and chemotherapy, may also add to systemic immunosuppression through a variety of mechanisms

8.5 Immunotherapy

The immunosuppression seen in glioma patients makes developing effective immunotherapies challenging. The usual mechanisms that immune-based therapies try to utilize are often subverted by the glioma itself. However, understanding the immunosuppressive mechanisms in place also allows us to target various immunologic pathways for augmentation and recruitment in attacking the tumor. While an in-depth review of current glioma immunotherapies is outside the scope of this chapter, a discussion of glioma-associated immunosuppression would not be complete without a brief overview of how our knowledge of glioma immunology has contributed to translational and clinical research on treatments for gliomas.

One category of immunotherapy is cytokine therapy. Given the various pro-inflammatory cytokines that are downregulated in gliomas, the thought arose that infusing glioma patients with these cytokines would induce an immune reaction that would subsequently eradicate the tumor. Intratumoral injection of IFN-α was the first such trial to be undertaken (Jereb et al. 1989; Farkkila et al. 1994). Unfortunately, no survival benefit was proven. Follow-up studies using systemic or intrathecal administration of IFN-α, IFN-γ, and/or IL-2 again showed treatment-associated toxicities but no significant survival benefit (Merchant et al. 1992; Buckner et al. 2001; Chamberlain 2002). Intratumoral injection of neural stem cells or progenitor cells virally transfected to produce IL-2, IL-4, IL-12, and IL-23 have shown promise in animal studies (Kikuchi et al. 1999; Benedetti et al. 2000; Ehtesham et al. 2002; Yuan et al. 2006). Alternate vehicles of cytokine delivery, such as liposomes containing a plasmid with the IFN-β gene, have also shown benefit in mice (Mizuno and Yoshida 1998; Natsume et al. 1999). These modalities, however, have yet to reach clinical trials.

Other therapies have centered on administering antibodies to antigens selectively expressed by glioma cells. Tenascin is an extracellular matrix protein expressed in gliomas but not normal brain (Bourdon et al. 1983). Intrathecal injection of a radioactively conjugated monoclonal antibody to tenascin showed dose-limiting toxicity (Bigner et al. 1998; Brown et al. 1996), while injection into glioblastoma resection cavities has shown improved survival in phase I and II clinical trials (Reardon et al. 2006a, b). Epidermal growth factor receptor (EGFR) is another protein overexpressed by gliomas. Trials involving radioactively conjugated antibodies to EGFR and murine monoclonal anti-EGFR antibodies have shown varying levels of success (Kalofonos et al. 1989; Faillot et al. 1996; Stragliotto et al. 1996).

Adoptive immunotherapy focuses on isolating immune effector cells from the patient, expanding that cell population ex vivo under controlled conditions, and reintroducing them to the patient in hopes of augmenting the immune response. Many studies have utilized harvested lymphocytes stimulated with IL-2 to produce lymphokine-activated killer cells (LAK) (Han et al. 2012). These studies have shown modest survival benefit, though some are complicated by dose-dependent toxicity related to IL-2-induced cerebral edema (Barba et al. 1989; Lillehei et al. 1991; Hayes et al. 1995). Use of tumor-infiltrating lymphocytes as a source of adoptive immunotherapy has largely been fruitless, likely related to the immunosuppressive microenvironment of the glioma (Han et al. 2012; Saris et al. 1992). Researchers have also investigated the use of T cells from tumor site-draining lymph nodes and injecting these into patients after expansion and activation ex vivo. This showed modest success in a small cohort of patients in a phase I trial, and likely warrants further investigation (Tsurushima et al. 1996).

Active immunotherapy involves the vaccination of patients against various tumor antigens. The National Cancer Institute has recently attempted to answer the question of which antigens are the most attractive targets for cancer immunotherapy. After analyzing 75 tumor-associated antigens based on therapeutic function, immunogenicity, oncogenicity, specificity, expression level in tumors, cancer stem cell expression, percentage of patients with positive tumors, and cellular location of expression, the NCI included the following in their highest ranked antigens: WT1, MUC1, LMP2, HPV E6/E7, HER2/Neu, EGFRviii, MAGE A3, and NY-ESO-1 (Cheever et al. 2009). While the expression of some of these antigens (e.g., EGFRviii) by glioblastomas is well documented (Gan et al. 2009), the presence or absence of others (e.g., MUC1) has not been reported at all. Therefore, early efforts at creating tumor vaccines used autologous tumor cells or tumor cell lysates given in conjunction with various cytokines (Steiner et al. 2004; Sloan et al. 2000; Ishikawa et al. 2007; Okada et al. 2007). However, because of the poor antigen-presenting capacity of glioma tumor cells and their low levels of costimulatory molecule expression, these early efforts were not met with much success (Han et al. 2012).

To increase antigen presentation to effector lymphocytes, recent studies have turned to dendritic cells (DCs) as a tool in glioma vaccines. DCs are professional APCs that highly express costimulatory molecules and effectively activate T cells. In clinical trials, autologous DCs are harvested from the patient's peripheral blood or bone marrow, primed to maturation, exposed to tumor antigen (either via whole tumor cells, tumor lysates, specific antigens, or tumor RNA), and then reintroduced into the patient (Yu et al. 2004; Yamanaka et al. 2003). The results of clinical trials have been promising, with increased lymphocyte infiltration into the tumor (Yu et al. 2001) as well as radiographic regression (Kikuchi et al. 2004). A phase I trial of EGFRviii peptide-pulsed DCs showed a median overall survival of 18.7 months from the time of vaccination in glioblastoma patients (Sampson et al. 2009), despite the fact that EGFRviii has been shown to be present in only 30 % of high-grade gliomas (Gan et al. 2009). A phase II randomized trial using a tumor lysate-pulsed DC vaccine for high-grade gliomas is ongoing (Liau et al. 2005). Dendritic cell vaccines approaches are shown schematically in Fig. 8.5.

Conclusion

Gliomas exhibit multiple mechanisms that aid in their evasion and suppression of the immune system. Through the expression of various cell surface proteins and the secretion of several immunosuppressive cytokines, they are able to subvert both the innate and adaptive forms of immunity by altering the phenotype of immune cells within their microenvironment that in turn causes systemic

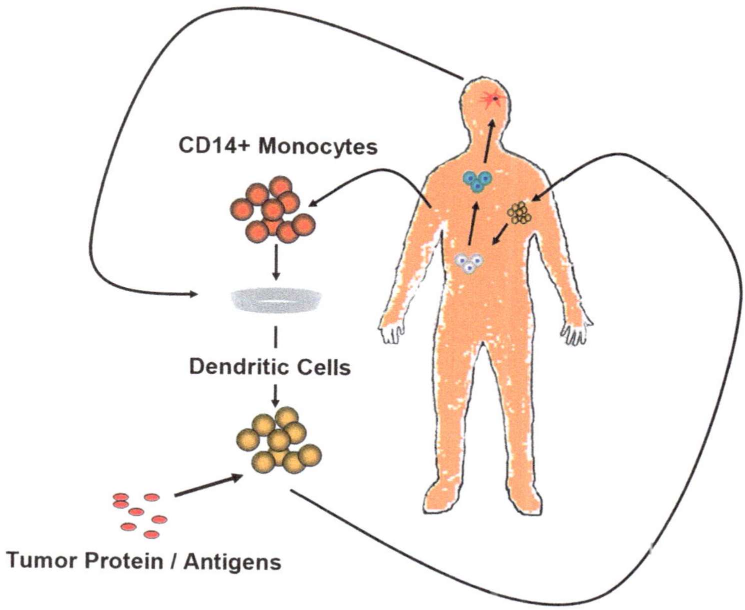

Fig. 8.5 Dendritic cell vaccine therapy. Dendritic cells are harvested from the patient's peripheral blood. The patient's glioma is resected and the DCs are pulsed with either whole tumor lysate, or specific proteins expressed by the tumor. The DCs are then reintroduced into the patient's peripheral blood, where they activate T cells, which subsequently home to the glioma and cause tumor lysis

immunosuppression. Many forms of immunotherapy have been developed to try to address these issues. The future success of immune-based therapies likely hinges on a multimodality approach; with the combination of cytokine-based therapy, adoptive immunotherapy, and active immunotherapy all playing a role. Furthermore, an individual's immunologic profile, much like their genetic profile, may dictate which therapy or cocktail of therapies is optimal for any given patient. With the ongoing findings regarding glioma immunology and tumor-associated immunosuppression, the quantity and quality of immunotherapies continue to increase. Hopefully, this will culminate in the eventual discovery of a cure for gliomas.

References

Abbas AK, Lichtman AH (2005a) General properties of immune responses. Elsevier Saunders, Philadelphia, PA

Abbas AK, Lichtman AH (2005b) Activation of T lymphocytes. Elsevier Saunders, Philadelphia, PA

Aloisi F, Ria F, Penna G, Adorini L (1998) Microglia are more efficient than astrocytes in antigen processing and in Th1 but not Th2 cell activation. J Immunol 160:4671–4680

Aloisi F, Ria F, Columba-Cabezas S, Hess H, Penna G, Adorini L (1999) Relative efficiency of microglia, astrocytes, dendritic cells and B cells in naive CD4+ T cell priming and Th1/Th2 cell restimulation. Eur J Immunol 29:2705–2714

Barba D, Saris SC, Holder C, Rosenberg SA, Oldfield EH (1989) Intratumoral LAK cell and interleukin-2 therapy of human gliomas. J Neurosurg 70:175–182

Barshes NR, Goodpastor SE, Goss JA (2004) Pharmacologic immunosuppression. Front Biosci 9:411–420

Bender AM, Collier LS, Rodriguez FJ et al (2010) Sleeping beauty-mediated somatic mutagenesis implicates CSF1 in the formation of high-grade astrocytomas. Cancer Res 70:3557–3565

Benedetti S, Pirola B, Pollo B et al (2000) Gene therapy of experimental brain tumors using neural progenitor cells. Nat Med 6:447–450

Bigner DD, Brown MT, Friedman AH et al (1998) Iodine-131-labeled antitenascin monoclonal antibody 81C6 treatment of patients with recurrent malignant gliomas: phase I trial results. J Clin Oncol 16:2202–2212

Bourdon MA, Wikstrand CJ, Furthmayr H, Matthews TJ, Bigner DD (1983) Human glioma-mesenchymal extracellular matrix antigen defined by monoclonal antibody. Cancer Res 43:2796–2805

Brannan CA, Roberts MR (2004) Resident microglia from adult mice are refractory to nitric oxide-inducing stimuli due to impaired NOS2 gene expression. Glia 48:120–131

Brock CS, Newlands ES, Wedge SR et al (1998) Phase I trial of temozolomide using an extended continuous oral schedule. Cancer Res 58:4363–4367

Brooks WH, Netsky MG, Normansell DE, Horwitz DA (1972) Depressed cell-mediated immunity in patients with primary intracranial tumors. Characterization of a humoral immunosuppressive factor. J Exp Med 136:1631–1647

Brooks WH, Caldwell HD, Mortara RH (1974) Immune responses in patients with gliomas. Surg Neurol 2:419–423

Brooks WH, Roszman TL, Rogers AS (1976) Impairment of rosette-forming T lymphocytes in patients with primary intracranial tumors. Cancer 37:1869–1873

Brown MT, Coleman RE, Friedman AH et al (1996) Intrathecal 131I-labeled antitenascin monoclonal antibody 81C6 treatment of patients with leptomeningeal neoplasms or primary brain tumor resection cavities with subarachnoid communication: phase I trial results. Clin Cancer Res 2:963–972

Buckner JC, Schomberg PJ, McGinnis WL et al (2001) A phase III study of radiation therapy plus carmustine with or without recombinant interferon-alpha in the treatment of patients with newly diagnosed high-grade glioma. Cancer 92:420–433

Camara NO, Sebille F, Lechler RI (2003) Human CD4+ CD25+ regulatory cells have marked and sustained effects on CD8+ T cell activation. Eur J Immunol 33:3473–3483

Chamberlain MC (2002) A phase II trial of intra-cerebrospinal fluid alpha interferon in the treatment of neoplastic meningitis. Cancer 94:2675–2680

Cheever MA, Allison JP, Ferris AS et al (2009) The prioritization of cancer antigens: a national cancer institute pilot project for the acceleration of translational research. Clin Cancer Res 15:5323–5337

Cheng F, Wang HW, Cuenca A et al (2003) A critical role for Stat3 signaling in immune tolerance. Immunity 19:425–436

Crane CA, Ahn BJ, Han SJ, Parsa AT (2012) Soluble factors secreted by glioblastoma cell lines facilitate recruitment, survival, and expansion of regulatory T cells: implications for immunotherapy. Neuro Oncol 14:584–595

Cserr HF, Knopf PM (1992) Cervical lymphatics, the blood-brain barrier and the immunoreactivity of the brain: a new view. Immunol Today 13:507–512

Dieckmann D, Bruett CH, Ploettner H, Lutz MB, Schuler G (2002) Human CD4(+)CD25(−) regulatory, contact-dependent T cells induce interleukin 10-producing, contact-independent type 1-like regulatory T cells [corrected]. J Exp Med 196:247–253

Dix AR, Brooks WH, Roszman TL, Morford LA (1999) Immune defects observed in patients with primary malignant brain tumors. J Neuroimmunol 100:216–232

Dong H, Strome SE, Salomao DR et al (2002) Tumor-associated B7-H1 promotes T-cell apoptosis: a potential mechanism of immune evasion. Nat Med 8:793–800

Ehtesham M, Kabos P, Kabosova A, Neuman T, Black KL, Yu JS (2002) The use of interleukin 12-secreting neural stem cells for the treatment of intracranial glioma. Cancer Res 62:5657–5663

El Andaloussi A, Lesniak MS (2006) An increase in CD4+ CD25+ FOXP3+ regulatory T cells in tumor-infiltrating lymphocytes of human glioblastoma multiforme. Neuro Oncol 8:234–243

El Andaloussi A, Han Y, Lesniak MS (2006) Prolongation of survival following depletion of CD4 + CD25+ regulatory T cells in mice with experimental brain tumors. J Neurosurg 105:430–437

Elliott LH, Brooks WH, Roszman TL (1984) Cytokinetic basis for the impaired activation of lymphocytes from patients with primary intracranial tumors. J Immunol 132:1208–1215

Elliott LH, Brooks WH, Roszman TL (1987) Activation of immunoregulatory lymphocytes obtained from patients with malignant gliomas. J Neurosurg 67:231–236

Elliott LH, Brooks WH, Roszman TL (1990) Inability of mitogen-activated lymphocytes obtained from patients with malignant primary intracranial tumors to express high affinity interleukin 2 receptors. J Clin Invest 86:80–86

Facoetti A, Nano R, Zelini P et al (2005) Human leukocyte antigen and antigen processing machinery component defects in astrocytic tumors. Clin Cancer Res 11:8304–8311

Faillot T, Magdelenat H, Mady E et al (1996) A phase I study of an anti-epidermal growth factor receptor monoclonal antibody for the treatment of malignant gliomas. Neurosurgery 39:478–483

Fakhrai H, Dorigo O, Shawler DL et al (1996) Eradication of established intracranial rat gliomas by transforming growth factor beta antisense gene therapy. Proc Natl Acad Sci U S A 93:2909–2914

Farkkila M, Jaaskelainen J, Kallio M et al (1994) Randomised, controlled study of intratumoral recombinant gamma-interferon treatment in newly diagnosed glioblastoma. Br J Cancer 70:138–141

Farmer JP, Antel JP, Freedman M, Cashman NR, Rode H, Villemure JG (1989) Characterization of lymphoid cells isolated from human gliomas. J Neurosurg 71:528–533

Fecci PE, Mitchell DA, Whitesides JF et al (2006a) Increased regulatory T-cell fraction amidst a diminished CD4 compartment explains cellular immune defects in patients with malignant glioma. Cancer Res 66:3294–3302

Fecci PE, Sweeney AE, Grossi PM et al (2006b) Systemic anti-CD25 monoclonal antibody administration safely enhances immunity in murine glioma without eliminating regulatory T cells. Clin Cancer Res 12:4294–4305

Flugel A, Labeur MS, Grasbon-Frodl EM, Kreutzberg GW, Graeber MB (1999) Microglia only weakly present glioma antigen to cytotoxic T cells. Int J Dev Neurosci 17:547–556

Fontana A, Kristensen F, Dubs R, Gemsa D, Weber E (1982) Production of prostaglandin E and an interleukin-1 like factor by cultured astrocytes and C6 glioma cells. J Immunol 129:2413–2419

Francisco LM, Sage PT, Sharpe AH (2010) The PD-1 pathway in tolerance and autoimmunity. Immunol Rev 236:219–242

Gabrilovich D (2004) Mechanisms and functional significance of tumour-induced dendritic-cell defects. Nat Rev Immunol 4:941–952

Gan HK, Kaye AH, Luwor RB (2009) The EGFRvIII variant in glioblastoma multiforme. J Clin Neurosci 16:748–754

Gordon LB, Nolan SC, Cserr HF, Knopf PM, Harling-Berg CJ (1997) Growth of P511 mastocytoma cells in BALB/c mouse brain elicits CTL response without tumor elimination: a new tumor model for regional central nervous system immunity. J Immunol 159:2399–2408

Gustafson MP, Lin Y, New KC et al (2010) Systemic immune suppression in glioblastoma: the interplay between CD14+ HLA-DRlo/neg monocytes, tumor factors, and dexamethasone. Neuro Oncol 12:631–644

Hahn HP, Pang M, He J et al (2004) Galectin-1 induces nuclear translocation of endonuclease G in caspase- and cytochrome c-independent T cell death. Cell Death Differ 11:1277–1286

Han SJ, Kaur G, Yang I, Lim M (2010) Biologic principles of immunotherapy for malignant gliomas. Neurosurg Clin N Am 21:1–16

Han SJ, Zygourakis C, Lim M, Parsa AT (2012) Immunotherapy for glioma: promises and challenges. Neurosurg Clin N Am 23:357–370

Hao C, Parney IF, Roa WH, Turner J, Petruk KC, Ramsay DA (2002) Cytokine and cytokine receptor mRNA expression in human glioblastomas: evidence of Th1, Th2 and Th3 cytokine dysregulation. Acta Neuropathol 103:171–178

Hayes RL, Koslow M, Hiesiger EM et al (1995) Improved long term survival after intracavitary interleukin-2 and lymphokine-activated killer cells for adults with recurrent malignant glioma. Cancer 76:840–852

Heimberger AB, Abou-Ghazal M, Reina-Ortiz C et al (2008a) Incidence and prognostic impact of FoxP3+ regulatory T cells in human gliomas. Clin Cancer Res 14:5166–5172

Heimberger AB, Sun W, Hussain SF et al (2008b) Immunological responses in a patient with glioblastoma multiforme treated with sequential courses of temozolomide and immunotherapy: case study. Neuro Oncol 10:98–103

Hickey WF (2001) Basic principles of immunological surveillance of the normal central nervous system. Glia 36:118–124

Hickey WF, Kimura H (1987) Graft-vs.-host disease elicits expression of class I and class II histocompatibility antigens and the presence of scattered T lymphocytes in rat central nervous system. Proc Natl Acad Sci U S A 84:2082–2086

Hickey WF, Hsu BL, Kimura H (1991) T-lymphocyte entry into the central nervous system. J Neurosci Res 28:254–260

Hitchcock ER, Morris CS (1988) Mononuclear cell infiltration in central portions of human astrocytomas. J Neurosurg 68:432–437

Hussain SF, Heimberger AB (2005) Immunotherapy for human glioma: innovative approaches and recent results. Expert Rev Anticancer Ther 5:777–790

Ichinose M, Masuoka J, Shiraishi T, Mineta T, Tabuchi K (2001) Fas ligand expression and depletion of T-cell infiltration in astrocytic tumors. Brain Tumor Pathol 18:37–42

Ishikawa E, Tsuboi K, Yamamoto T et al (2007) Clinical trial of autologous formalin-fixed tumor vaccine for glioblastoma multiforme patients. Cancer Sci 98:1226–1233

Jacobs JF, Idema AJ, Bol KF et al (2010) Prognostic significance and mechanism of Treg infiltration in human brain tumors. J Neuroimmunol 225:195–199

Janeway CA Jr, Medzhitov R (2002) Innate immune recognition. Annu Rev Immunol 20:197–216

Janeway C, Travers P, Walport M, Schlomchik M (2005) Immunobiology: the immune system in health and disease. Garland Science, New York, NY

Jereb B, Petric J, Lamovec J, Skrbec M, Soss E (1989) Intratumor application of human leukocyte interferon-alpha in patients with malignant brain tumors. Am J Clin Oncol 12:1–7

Kalofonos HP, Pawlikowska TR, Hemingway A et al (1989) Antibody guided diagnosis and therapy of brain gliomas using radiolabeled monoclonal antibodies against epidermal growth factor receptor and placental alkaline phosphatase. J Nucl Med 30:1636–1645

Khan RB, Raizer JJ, Malkin MG, Bazylewicz KA, Abrey LE (2002) A phase II study of extended low-dose temozolomide in recurrent malignant gliomas. Neuro Oncol 4:39–43

Kikuchi T, Joki T, Saitoh S et al (1999) Anti-tumor activity of interleukin-2-producing tumor cells and recombinant interleukin 12 against mouse glioma cells located in the central nervous system. Int J Cancer 80:425–430

Kikuchi T, Akasaki Y, Abe T et al (2004) Vaccination of glioma patients with fusions of dendritic and glioma cells and recombinant human interleukin 12. J Immunother 27:452–459

Kostianovsky AM, Maier LM, Anderson RC, Bruce JN, Anderson DE (2008) Astrocytic regulation of human monocytic/microglial activation. J Immunol 181:5425–5432

Krakowski ML, Owens T (2000) Naive T lymphocytes traffic to inflamed central nervous system, but require antigen recognition for activation. Eur J Immunol 30:1002–1009

Kunicka JE, Talle MA, Denhardt GH, Brown M, Prince LA, Goldstein G (1993) Immunosuppression by glucocorticoids: inhibition of production of multiple lymphokines by in vivo administration of dexamethasone. Cell Immunol 149:39–49

Kuppner MC, Hamou MF, de Tribolet N (1988) Immunohistological and functional analyses of lymphoid infiltrates in human glioblastomas. Cancer Res 48:6926–6932

Kuppner MC, Hamou MF, Sawamura Y, Bodmer S, de Tribolet N (1989) Inhibition of lymphocyte function by glioblastoma-derived transforming growth factor beta 2. J Neurosurg 71:211–217

Lampson LA, Hickey WF (1986) Monoclonal antibody analysis of MHC expression in human brain biopsies: tissue ranging from "histologically normal" to that showing different levels of glial tumor involvement. J Immunol 136:4054–4062

Levi-Strauss M, Mallat M (1987) Primary cultures of murine astrocytes produce C3 and factor B, two components of the alternative pathway of complement activation. J Immunol 139:2361–2366

Liau LM, Fakhrai H, Black KL (1998) Prolonged survival of rats with intracranial C6 gliomas by treatment with TGF-beta antisense gene. Neurol Res 20:742–747

Liau LM, Prins RM, Kiertscher SM et al (2005) Dendritic cell vaccination in glioblastoma patients induces systemic and intracranial T-cell responses modulated by the local central nervous system tumor microenvironment. Clin Cancer Res 11:5515–5525

Lillehei KO, Mitchell DH, Johnson SD, McCleary EL, Kruse CA (1991) Long-term follow-up of patients with recurrent malignant gliomas treated with adjuvant adoptive immunotherapy. Neurosurgery 28:16–23

Lin T, Bost KL (2004) STAT3 activation in macrophages following infection with Salmonella. Biochem Biophys Res Commun 321:828–834

Liu VC, Wong LY, Jang T et al (2007) Tumor evasion of the immune system by converting CD4+ CD25- T cells into CD4+ CD25+ T regulatory cells: role of tumor-derived TGF-beta. J Immunol 178:2883–2892

Lowin-Kropf B, Shapiro VS, Weiss A (1998) Cytoskeletal polarization of T cells is regulated by an immunoreceptor tyrosine-based activation motif-dependent mechanism. J Cell Biol 140:861–871

Matta BM, Raimondi G, Rosborough BR, Sumpter TL, Thomson AW (2012) IL-27 production and STAT3-dependent upregulation of B7-H1 mediate immune regulatory functions of liver plasmacytoid dendritic cells. J Immunol 188:5227–5237

Medzhitov R, Janeway C Jr (2000) Innate immunity. N Engl J Med 343:338–344

Merchant RE, McVicar DW, Merchant LH, Young HF (1992) Treatment of recurrent malignant glioma by repeated intracerebral injections of human recombinant interleukin-2 alone or in combination with systemic interferon-alpha. Results of a phase I clinical trial. J Neurooncol 12:75–83

Mizuno M, Yoshida J (1998) Effect of human interferon beta gene transfer upon human glioma, transplanted into nude mouse brain, involves induced natural killer cells. Cancer Immunol Immunother 47:227–232

Nagaraj S, Gupta K, Pisarev V et al (2007) Altered recognition of antigen is a mechanism of CD8+ T cell tolerance in cancer. Nat Med 13:828–835

Natsume A, Mizuno M, Ryuke Y, Yoshida J (1999) Antitumor effect and cellular immunity activation by murine interferon-beta gene transfer against intracerebral glioma in mouse. Gene Ther 6:1626–1633

Nitta T, Hishii M, Sato K, Okumura K (1994) Selective expression of interleukin-10 gene within glioblastoma multiforme. Brain Res 649:122–128

O'Farrell AM, Liu Y, Moore KW, Mui AL (1998) IL-10 inhibits macrophage activation and proliferation by distinct signaling mechanisms: evidence for Stat3-dependent and -independent pathways. EMBO J 17:1006–1018

Okada H, Lieberman FS, Walter KA et al (2007) Autologous glioma cell vaccine admixed with interleukin-4 gene transfected fibroblasts in the treatment of patients with malignant gliomas. J Transl Med 5:67

Parney IF, Farr-Jones MA, Chang LJ, Petruk KC (2000) Human glioma immunobiology in vitro: implications for immunogene therapy. Neurosurgery 46:1169–1177, discussion 1177–1168

Parney IF, Waldron JS, Parsa AT (2009) Flow cytometry and in vitro analysis of human glioma-associated macrophages. Laboratory investigation. J Neurosurg 110:572–582

Parsa AT, Waldron JS, Panner A et al (2007) Loss of tumor suppressor PTEN function increases B7-H1 expression and immunoresistance in glioma. Nat Med 13:84–88

Reardon DA, Akabani G, Coleman RE et al (2006a) Salvage radioimmunotherapy with murine iodine-131-labeled antitenascin monoclonal antibody 81C6 for patients with recurrent primary and metastatic malignant brain tumors: phase II study results. J Clin Oncol 24:115–122

Reardon DA, Quinn JA, Akabani G et al (2006b) Novel human IgG2b/murine chimeric antitenascin monoclonal antibody construct radiolabeled with 131I and administered into the surgically created resection cavity of patients with malignant glioma: phase I trial results. J Nucl Med 47:912–918

Rodrigues JC, Gonzalez GC, Zhang L et al (2010) Normal human monocytes exposed to glioma cells acquire myeloid-derived suppressor cell-like properties. Neuro Oncol 12:351–365

Roszman TL, Brooks WH, Elliott LH (1982) Immunobiology of primary intracranial tumors. VI. Suppressor cell function and lectin-binding lymphocyte subpopulations in patients with cerebral tumors. Cancer 50:1273–1279

Roszman TL, Brooks WH, Steele C, Elliott LH (1985) Pokeweed mitogen-induced immunoglobulin secretion by peripheral blood lymphocytes from patients with primary intracranial tumors. Characterization of T helper and B cell function. J Immunol 134:1545–1550

Sakaguchi S (2005) Naturally arising Foxp3-expressing CD25+ CD4+ regulatory T cells in immunological tolerance to self and non-self. Nat Immunol 6:345–352

Sampson JH, Archer GE, Ashley DM et al (1996) Subcutaneous vaccination with irradiated, cytokine-producing tumor cells stimulates CD8+ cell-mediated immunity against tumors located in the "immunologically privileged" central nervous system. Proc Natl Acad Sci U S A 93:10399–10404

Sampson JH, Archer GE, Mitchell DA et al (2009) An epidermal growth factor receptor variant III-targeted vaccine is safe and immunogenic in patients with glioblastoma multiforme. Mol Cancer Ther 8:2773–2779

Sandberg-Wollheim M, Zweiman B, Levinson AI, Lisak RP (1986) Humoral immune responses within the human central nervous system following systemic immunization. J Neuroimmunol 11:205–214

Saris SC, Spiess P, Lieberman DM, Lin S, Walbridge S, Oldfield EH (1992) Treatment of murine primary brain tumors with systemic interleukin-2 and tumor-infiltrating lymphocytes. J Neurosurg 76:513–519

Sawamura Y, Hosokawa M, Kuppner MC et al (1989) Antitumor activity and surface phenotypes of human glioma-infiltrating lymphocytes after in vitro expansion in the presence of interleukin 2. Cancer Res 49:1843–1849

Sawamura Y, Diserens AC, de Tribolet N (1990) In vitro prostaglandin E2 production by glioblastoma cells and its effect on interleukin-2 activation of oncolytic lymphocytes. J Neurooncol 9:125–130

Schartner JM, Hagar AR, Van Handel M, Zhang L, Nadkarni N, Badie B (2005) Impaired capacity for upregulation of MHC class II in tumor-associated microglia. Glia 51:279–285

Serafini P, De Santo C, Marigo I et al (2004) Derangement of immune responses by myeloid suppressor cells. Cancer Immunol Immunother 53:64–72

Siepl C, Bodmer S, Frei K et al (1988) The glioblastoma-derived T cell suppressor factor/transforming growth factor-beta 2 inhibits T cell growth without affecting the interaction of interleukin 2 with its receptor. Eur J Immunol 18:593–600

Sinha P, Clements VK, Ostrand-Rosenberg S (2005) Reduction of myeloid-derived suppressor cells and induction of M1 macrophages facilitate the rejection of established metastatic disease. J Immunol 174:636–645

Sloan AE, Dansey R, Zamorano L et al (2000) Adoptive immunotherapy in patients with recurrent malignant glioma: preliminary results of using autologous whole-tumor vaccine plus granulocyte-macrophage colony-stimulating factor and adoptive transfer of anti-CD3-activated lymphocytes. Neurosurg Focus 9:e9

Steiner HH, Bonsanto MM, Beckhove P et al (2004) Antitumor vaccination of patients with glioblastoma multiforme: a pilot study to assess feasibility, safety, and clinical benefit. J Clin Oncol 22:4272–4281

Stragliotto G, Vega F, Stasiecki P, Gropp P, Poisson M, Delattre JY (1996) Multiple infusions of anti-epidermal growth factor receptor (EGFR) monoclonal antibody (EMD 55,900) in patients with recurrent malignant gliomas. Eur J Cancer 32A:636–640

Stupp R, Mason WP, van den Bent MJ et al (2005) Radiotherapy plus concomitant and adjuvant temozolomide for glioblastoma. N Engl J Med 352:987–996

Su YB, Sohn S, Krown SE et al (2004) Selective CD4+ lymphopenia in melanoma patients treated with temozolomide: a toxicity with therapeutic implications. J Clin Oncol 22:610–616

Takeda K, Clausen BE, Kaisho T et al (1999) Enhanced Th1 activity and development of chronic enterocolitis in mice devoid of Stat3 in macrophages and neutrophils. Immunity 10:39–49

Thornton AM, Shevach EM (1998) CD4+ CD25+ immunoregulatory T cells suppress polyclonal T cell activation in vitro by inhibiting interleukin 2 production. J Exp Med 188:287–296

Tsurushima H, Liu SQ, Tsuboi K, Yoshii Y, Nose T, Ohno T (1996) Induction of human autologous cytotoxic T lymphocytes against minced tissues of glioblastoma multiforme. J Neurosurg 84:258–263

Weissenberger J, Loeffler S, Kappeler A et al (2004) IL-6 is required for glioma development in a mouse model. Oncogene 23:3308–3316

Weller RO, Engelhardt B, Phillips MJ (1996) Lymphocyte targeting of the central nervous system: a review of afferent and efferent CNS-immune pathways. Brain Pathol 6:275–288

Wu L, Du H, Li Y, Qu P, Yan C (2011) Signal transducer and activator of transcription 3 (Stat3C) promotes myeloid-derived suppressor cell expansion and immune suppression during lung tumorigenesis. Am J Pathol 179:2131–2141

Yamanaka R, Abe T, Yajima N et al (2003) Vaccination of recurrent glioma patients with tumour lysate-pulsed dendritic cells elicits immune responses: results of a clinical phase I/II trial. Br J Cancer 89:1172–1179

Yang BC, Lin HK, Hor WS et al (2003) Mediation of enhanced transcription of the IL-10 gene in T cells, upon contact with human glioma cells, by Fas signaling through a protein kinase A-independent pathway. J Immunol 171:3947–3954

Yang I, Kremen TJ, Giovannone AJ et al (2004) Modulation of major histocompatibility complex Class I molecules and major histocompatibility complex-bound immunogenic peptides induced by interferon-alpha and interferon-gamma treatment of human glioblastoma multiforme. J Neurosurg 100:310–319

Young HF, Sakalas R, Kaplan AM (1976) Inhibition of cell-mediated immunity in patients with brain tumors. Surg Neurol 5:19–23

Yu JS, Wheeler CJ, Zeltzer PM et al (2001) Vaccination of malignant glioma patients with peptide-pulsed dendritic cells elicits systemic cytotoxicity and intracranial T-cell infiltration. Cancer Res 61:842–847

Yu JS, Liu G, Ying H, Yong WH, Black KL, Wheeler CJ (2004) Vaccination with tumor lysate-pulsed dendritic cells elicits antigen-specific, cytotoxic T-cells in patients with malignant glioma. Cancer Res 64:4973–4979

Yuan X, Hu J, Belladonna ML, Black KL, Yu JS (2006) Interleukin-23-expressing bone marrow-derived neural stem-like cells exhibit antitumor activity against intracranial glioma. Cancer Res 66:2630–2638

Microglia in Gliomas: Friend or Foe?

9

Bozena Kaminska

Abstract

Microglia are myeloid cells residing in the central nervous system that regulate innate and adaptive immune responses in development, infections, and neuropathological disorders. Clinical and experimental data show that microglia constitute the dominant immune infiltrate in malignant gliomas. The clinical significance of the mononuclear infiltrate in gliomas remains the subject of controversy. Despite accumulation of these immune cells in both low and high grade gliomas, the antitumor immune response is defective in glioblastomas. Some evidence shows the immunosuppressive and pro-invasive action of glioblastoma-infiltrating microglia. Molecular mechanisms responsible for dual and likely opposite role of these cells are being recently unraveled. This chapter summarizes the latest findings on the heterogeneity of glioma-infiltrating microglia/macrophages, functional characterization of their phenotype, and contribution to glioma pathology. Recent attempts to determine a profile of cytokine/chemokine production and gene expression profiling in CD11b^{+} cells isolated from patients or rodent gliomas revealed their similarity to alternatively activated, the M2-type macrophages observed in other tissues. Glioma-infiltrating microglia/macrophages acquire the alternative, pro-invasive phenotype in which their phagocytic, trophic, and tissue remodeling functions are enhanced. Cell culture and organotypic brain slice culture studies demonstrated the pro-invasive activity of microglia, and their polarization into tumor supportive cells. This is supported by studies of rodent experimental gliomas which reproduce glioblastoma pathology. Genetic or pharmacological ablation of microglia/macrophages impairs glioma growth, extends survival, and in some cases restores to some extent antitumor immune responses. The potential for targeting interactions between glioma and infiltrating microglia/macrophages in

B. Kaminska (✉)
Laboratory of Molecular Neurobiology, Nencki Institute of Experimental Biology, 3 Pasteur Str., 02-093 Warsaw, Poland
e-mail: bozenakk@nencki.gov.pl

A. Sedo and R. Mentlein (eds.), *Glioma Cell Biology*,
DOI 10.1007/978-3-7091-1431-5_9, © Springer-Verlag Wien 2014

therapeutic interventions is discussed. Small molecule inhibitors of mitogen-activated protein kinase (MAPK)-signaling with immunosuppressive or anti-inflammatory properties, such as cyclosporine A and minocycline, were shown to block infiltration and activation of microglia in vitro and in organotypic brain slices. These molecules reduce infiltration of microglia/macrophages, angiogenesis, and tumor growth of experimental gliomas providing a rationale for blocking pro-invasive functions of microglia/macrophages as a new therapeutic strategy in glioblastomas.

Keywords
Glioma • Infiltrating microglia/macrophages • M1/M2 phenotype • Gene expression • Pro/anti-inflammatory cytokines • Tumor angiogenesis

Abbreviations

BM	Bone marrow
G-CM	Glioma-conditioned medium
GDNF	Glial cell-derived neurotrophic factor
GFP	Green fluorescent protein
GLUT5	Glucose transporter 5
GM-CSF	Granulocyte-macrophage colony stimulating factor
HO	Heme oxygenase
IFN	Interferon
IL	Interleukin
LPS	Lipopolysaccharide
MAPK	Mitogen-activated protein kinase (ERK1 and ERK2)
MCP-1	Monocyte chemoattractant protein-1 (CCL2)
M-CSF	Macrophage colony stimulating factor
MMP	Matrix metalloproteinase
MT1-MMP	Membrane type 1-MMP (=MMP14)
RANTES	Regulated on activation normal T cell expressed and secreted, CCL5
TGF	Transforming growth factor
TLR	Toll-like receptor
TNF	Tumor necrosis factor
Treg	Regulatory T cell
TREM2	Triggering receptor expressed on myeloid cells 2

9.1 Introduction

Microglia are an abundant portion of the central nervous system (CNS) cell population comprising 5–20 % of the total glial cell population. Microglial cells, in contrast to other CNS cell types, are of myeloid origin. The consensus phenotypic profile for the microglial cell is: $CD68^{+}$, $CD45^{low}$, $CD11b^{+}$, $CD11c^{high}$, MHC class II^{+}, CD14 (Guillemin and Brew 2004). Apart from resident microglia, extraparenchymal brain macrophages normally exist in the CNS as perivascular, choroid plexus-associated, and meningeal macrophages; therefore, the term "brain macrophage" is used to encompass all types of macrophages infiltrating the brain. Neurons control microglial function by physical contact or by releasing neurotransmitters, peptides, and/or growth factors including gamma-aminobutyric acid, glutamate, catecholamines, CD22, CCL21, fraktalkine (CX3CL1), which act on receptors present on microglia membrane. The surface molecule CD200 widely expressed on neurons, astrocytes, and oligodendrocytes via its receptor CD200R (exclusively expressed on microglia) leads to inactivation of microglia and keeps them in a resting state (Cardona et al. 2006; Kierdorf and Prinz 2013). Under physiological conditions, microglia perform tissue surveillance in the nervous system and produce neurotrophic factors supporting cell survival, regeneration, and neurogenesis (Neumann et al. 2009). These cells can respond to changes in sensory activity and can influence neuronal activity. During development and neurogenesis, microglia interactions with neurons help to shape the final patterns of neural circuits. Activated microglia can remove damaged cells as well as dysfunctional synapses, in a process termed "synaptic stripping" (Kettenmann et al. 2013).

One of the most remarkable features of microglia is their ability of morphological and functional plasticity in response to activating stimuli. In vivo imaging studies of transgenic mice, in which all microglia are fluorescently labeled after replacing the *Cx3cr1* gene with the gene encoding enhanced green fluorescent protein, have revealed that in the brain, microglia are highly dynamic, moving constantly to actively survey the brain parenchyma. Transcranial time-lapse two-photon imaging of GFP-labeled microglia demonstrates that microglial cells display a ramified morphology with microglial processes highly dynamic in the intact mouse cortex. However, under number of pathological conditions, ramified microglia will activate and engage a series of morphological alterations that leads to a hypertrophy of microglial cell body and a retraction of their ramifications (Nimmerjahn et al. 2005; Hanisch and Kettenmann 2007). Fully activated (reactive) microglia harbor an amoeboid morphology similar to an activated macrophage. Microglial processes are highly motile and respond to chemoattractant molecules released by damaged or apoptotic cells ("find-me" signals) such as fractalkine and extracellular nucleotides (ATP, UDP). An engulfment synapse is formed between microglial receptors and their ligands in the membrane of the apoptotic cell ("eat-me" signals), leading to the tethering and engulfing of the apoptotic cell in a phagosome which matures by fusing with lysosomes. The apoptotic cell is fully degraded in the phagolysosome in less than 2 h (Sierra et al. 2013).

Detection of pathogen-associated molecular patterns (PAMPS) is mediated through scavenger receptors, Toll-like receptors (TLRs) such as the CD14/TLR4 complex, or receptors of the immunoglobulin superfamily (e.g., c-type lectins). Detection of apoptotic cells, in particular exposure of phosphatidylserine in the outer leaflet of the cell membrane, is mediated directly by several receptors, including brain-specific angiogenesis inhibitor 1—BAI-1 (Armstrong and Ravichandran 2011), and by bridging molecules such as milk fat globule-epidermal growth factor (MFG-E8). Another receptor is triggering receptor expressed on myeloid cells-2 (TREM2), whose loss of function prevents microglial phagocytosis (Takahashi et al. 2007). Anionic oligosaccharides such as bacterial lipopolysaccharides and heat shock protein 60 (Hsp60) exposed in the surface of apoptotic cells have been proposed as TREM2 ligands (Stefano et al. 2009). Antibodies (IgG) and proteins of the complement system such as C3b bind to Fc receptors and complement receptor 3 (CR3) on microglia, and mediate phagocytosis (Goodridge et al. 2012). Apoptotic cell clearance is generally an immunologically silent process and a common feature of microglial phagocytosis via TREM-2 or phophatidylserine receptors is a release of anti-inflammatory cytokines such as interleukin (IL)-10 and transforming growth factor beta (TGFβ) (Ravichandran 2010).

Microglial cells rapidly respond to pathological insults, becoming activated to induce a variety of effects that may contribute to both pathogenesis, or to confer neuronal protection. Once activated, they become immune effector cells mediating both innate and adaptive responses (Graeber and Streit 2010; Yang et al. 2010). Proper antigen presentation is critical in the generation of specific, durable responses by the adaptive immune system, and requires interaction between the T cell receptor and processed antigen peptide presented on major histocompatibility complex (MHC) molecules by the antigen presenting cells (APC).

Gliomas are the most frequent primary CNS tumors. According to the WHO classification gliomas are divided in low-grade (grades I and II) and high-grade (grades III and IV) tumors. Low-grade tumors are well-differentiated, slow-growing lesions. Grade I tumors are well restricted and surgically curable, whereas grade II tumors are diffuse, infiltrating lesions with a potential for progression towards a high-grade tumor. Pilocytic astrocytomas (grade I) also differ from diffuse astrocytomas in their altered and increased expression of immune response genes (Huang et al. 2005). The WHO classification divides the malignant gliomas of adults into astrocytic tumors—the most malignant of which is termed glioblastoma—as well as oligodendrogliomas and oligoastrocytomas. Glioblastoma is considered to be one of the most difficult human malignancies to manage, due to frequent dysfunctions of tumor suppressors and oncogenes. These highly infiltrative tumors often invade into normal brain tissues preventing surgical resection. Glioblastoma develops several mechanisms that mediate tumor escape from immune surveillance and produce immunosuppressive factors which results in T-cell anergy or apoptosis (Prins and Liau 2004; Albesiano et al. 2010). The mean survival of patients with glioblastoma has only increased slightly and generally ranges between 1 and 2 years in spite of recent therapeutic advancements.

Histopathologic studies of glioma tissue have shown consistently high levels of infiltrating microglia. Though, microglia could play a role in the immune defense against gliomas, their unharmed growth and highly malignant behavior indicates that microglial immune defense mechanisms do not function properly in these tumors (Graeber et al. 2002; Hussain et al. 2006a, b). In fact, recent evidence indicates that glioma-infiltrating microglia/macrophages promote glioma growth, invasion, and facilitate immunosuppression (Charles et al. 2012). In this chapter I will review microglial origin, phenotype, and activities in gliomas, and provide a plausible explanation for their dual role in these tumors.

9.2 Microglia/Macrophage Accumulation in Gliomas

9.2.1 Evaluation of Microglia/Macrophage Content in Gliomas by Immunohistochemistry

Although the presence of leukocytes within tumors has been recognized for over 100 years, advancements in histological techniques and flow cytometry allowed estimation of the extent of tumor infiltration by myeloid cells. In one of the first studies, the authors have analyzed 47 CNS (11 glioblastomas, 9 meningiomas, 3 medulloblastomas, 12 primary neural tumors, and 12 brain metastases) for their content of macrophages. Cell suspensions were prepared by enzymatic digestion and macrophages were determined by IgGEAC rosette assay (Morantz et al. 1979). The 11 glioblastomas had a mean macrophage content of 45 % (from 8 % to 78 %), the 9 meningiomas had 44 % (from 5 % to 81 %), the 3 medulloblastomas 6 % (from 2 % to 15 %), and the metastatic tumors 24 % (range: 4 % to 70 %) (Morantz et al. 1979). Immunohistochemical studies by Morimura et al. (1990) using macrophage markers: Fc-gamma (Fcγ) and complement receptors found in average 20–30 % positive cells of all cells present in 12 gliomas. Microglial cells were recognized by a restricted panel of macrophage markers (anti-Fcγ receptors 1, 2, 3, complement receptor CR3, HLA-DR, common leukocyte antigen CD45, and the monocyte marker RM3/1). Only low numbers of T cells were detected in the tumors. In peritumoral tissue mainly dendritic, microglia-like cells were present, with decreased expression of antigens CD4, RM3/1, and Fcγ receptors in comparison to those in gliomas (Morimura et al. 1990).

Morris and Esiri (1991) used monoclonal antibodies Mac387, KP1, and the lectin RCA-1 to stain for macrophages and microglia in 27 gliomas. They found RCA-1 to be the superior tool for detecting most macrophages and microglia, and detected more macrophages and microglia in high-grade gliomas than in low-grade (Morris and Esiri 1991). A study by Wierzba-Bobrowicz et al. (1994) using lectin RCA-1 on a panel of 40 gliomas demonstrated the presence of both the ramified and amoeboid microglia in the protoplasmic and fibrillary astrocytomas, while in glioblastoma and anaplastic astrocytomas the greatest number of amoeboid microglia and very rarely ramified microglial cells were found. In the non-anaplastic tumors such as gemistocytic astrocytomas, numerous but mostly ramified microglia have been observed.

Further studies of 72 brain tumors of different types using the monocyte/macrophage lineage markers (Ki-M1P, HLA-DR, KP1, My4, My7, Ki-M1, Ki-M6, EBM 11) showed the largest number of ramified and ameboid microglia as well as macrophages among glioblastoma and anaplastic gliomas (Roggendorf et al. 1996). Fewer, predominantly amoeboid, microglia cells were found in glial tumors of low malignancy. Neuronal tumors showed only a mild increase of microglia suggesting some type of specificity for microglia–glioma interactions.

Gemistocytic astrocytomas contain unusually high numbers of microglial cells and aberrant MHC Class II expression by tumor cells correlates with loss of immune-competent microglia. Although these tumors are graded as WHO grade II astrocytomas and their proliferative potential is low, they behave aggressively. Their poor prognosis could be due to pro-invasive role of microglia (Klein and Roggendorf 2001). An immunohistochemical double-labeling study of pilocytic astrocytomas and astrocytomas WHO grade II–IV using the antibodies Ki67 (as proliferation-marker) and CD68 (as microglia marker) demonstrated that microglial cells in astrocytic brain tumors proliferate, with the highest rates of proliferating microglia especially in pilocytic astrocytomas. The proliferation indices of microglia were lowest in fibrillary astrocytoma (Klein and Roggendorf 2001).

Our unpublished studies using HLA-DP, DQ, DR immunohistochemistry on a panel of brain tumors confirmed the strong immunoreactivity reflecting microglia/macrophage accumulation in glial tumors. Positive cells were localized diffusely throughout the tumor, around vessels; we did not observe the presence of positive cells in the areas of palisading necrosis. Malignant embryonal tumors such as PNET (primitive neuroectodermal tumors) or medulloblastoma did not show HLA-DP, DQ, DR immunoreactivity (Fig. 9.1). In adult gliomas we found the significant correlation between HLA-DP, DQ, DR immunoreactivity and tumor grade. Such correlation was not observed among pediatric tumors. Interestingly, we found numerous, but mostly hypertrophic HLA-DP, DQ, DR positive cells in pilocytic astrocytomas, while most HLA-DP, DQ, DR positive cells were amoeboid in glioblastoma. A meta-analysis of five microarray datasets characterizing the expression profiles of immune-defense associated and inflammatory genes in gliomas of different grades showed the reduced expression of many immune response and TLR signaling pathway genes in glioblastomas (unpublished). It suggests a dual role of glioma-infiltrating microglia/macrophages which in low-grade gliomas may exert effective antitumor responses.

Several studies reported different accumulation of macrophages/microglial cells in tumors of different grades. The number of $CD68^+$ cells was higher in glioblastoma (*glioblastoma multiforme*, GBM) than that in grade II or III gliomas and positively correlated with vascular density (Nishie et al. 1999). Infiltrating $CD68^+$ cells were positively stained for heme oxygenase (HO)-1 that indicated HO-1 expression in infiltrating microglia/macrophages in gliomas. Deininger et al. (2000) reported that the number of macrophages/microglial cells, expressing cyclooxygenase (COX)-1, was significantly lower in oligodendrogliomas than in

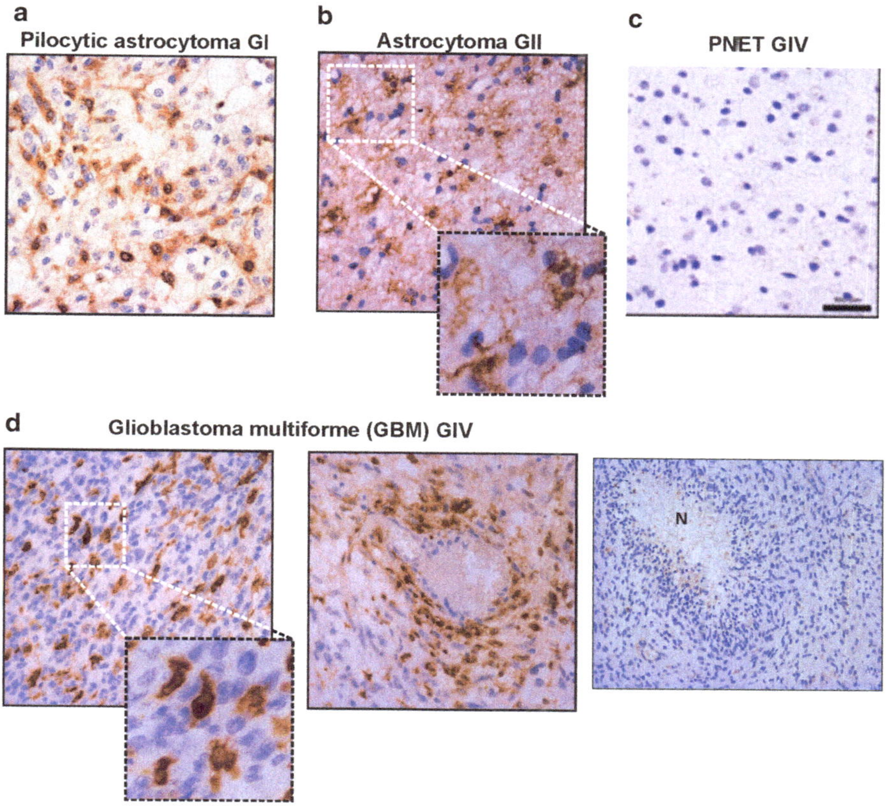

Fig. 9.1 Accumulation of microglia/macrophages in human brain tumors of different grades. The abundance of microglia/macrophages was determined immunohistochemically by evaluating HLA-DP, DQ, DR expression on paraffin-embedded sections of glioma biopsies from juvenile pilocytic astrocytoma (**a**), pleomorphic xanthoastrocytoma (**b**), primitive neuroectodermal tumor—PNET (**c**), and glioblastoma multiforme (**d**). Intense positive staining (*brown*) is present (magnification: ×100, inset: ×200) intratumorally and around vessels. Note the ramified morphology of HLA-DP, DQ, DR immunoreactive cells in anaplastic astrocytoma (**b**) and more amoeboid morphology of HLA-DP, DQ, DR immunoreactive cells in glioblastoma (**c**). No positive staining was detected in the areas of palisading necrosis (N) in glioblastomas

anaplastic oligodendrogliomas (Deininger et al. 2000). A study using another microglial marker, glucose transporter 5 (GLUT5), indicated that the number of GLUT5-positive microglia/macrophages was significantly higher in astrocytic tumors than in oligodendroglial tumors (Sasaki et al. 2004).

Infiltration experimental gliomas with microglia/macrophages have been demonstrated in rodent animal glioma models using immunohistochemical staining for Iba1 or CD68 antigens. Mouse glioma 261 (GL261) cells that carry point mutations in the K-ras and p53 genes are frequently used as a model of experimental glioblastoma. Transplanted GL261 cells develop both subcutaneous and intracranial tumors in immunocompetent C57BL/6 mice (Szatmári et al. 2006; Maes and Van Gool 2011). Microglia/macrophages accumulation was detected by Iba1 staining in experimental GL261 gliomas (Färber et al. 2008; Markovic et al. 2009). Histological detection of Iba1 antigen and flow cytometry studies show accumulation of microglia and macrophages in intracranial GL261 gliomas (Gabrusiewicz et al. 2011; Sielska et al. 2013).

The experimental rat RG-2 gliomas in rats were heavily infiltrated with microglia/macrophages, and infiltrating cells were positive for MHC Class II (Ia) antigens (Morioka et al. 1992a). Activated cells expressing CR3 complement receptor and MHC class II (Ia) antigen were found throughout the tumor and with increased density along the tumor periphery (Morioka et al. 1992b). Cells expressing the ED2 epitope (considered to be macrophages) were almost exclusively of the perivascular type and did not show distribution similar to MHC class II (Ia) expressing cells. The ED2 epitope was found sporadically on ramified microglial cells. The results showed glioma infiltration with microglia and blood mononuclear cells, and no evidence of tumor destruction (Morioka et al. 1992b).

Another glioma model is a transgenic rat expressing v-erbB (a viral, oncogenic form of the epidermal growth factor receptor) under transcriptional regulation by the S100-β promoter that develops a brain tumor. A recent study that assessed Iba1 staining for microglia/macrophages in brain tumors in transgenic rats (five malignant glioma, four anaplastic oligodendroglioma, four astrocytoma) demonstrates Iba1^{+} cells in the tumor core of malignant gliomas that were more activated than Iba1^{+} microglia of non-neoplastic brain tissue and intraparenchymal anaplastic oligodendrogliomas. Iba1 expression was positively correlated to Ki-67 expression in all the gliomas that suggests ongoing microglial proliferation. Most Iba1^{+} cells showed no or little expression of putative M2 phenotype markers, CD163 or CD204, in transgenic rats (Sasaki et al. 2013).

A new glioma model (ALTS1C1) derived from primary astrocytes transformed by SV40 large T antigen, followed by serial subcutaneous and intracranial passages in syngeneic C57BL/6J mice, was recently introduced. It recapitulates numerous neuropathological features of human high-grade gliomas and shows accumulation of microglia/macrophages in ALTS1C1 tumor-bearing brains (Wang et al. 2012). The authors noticed different expression levels of F4/80 and CD68 markers: in the primary tumor core both markers were expressed almost equally and overlapped, while in the tumor margin and the infiltration islands there were higher levels of F4/80 than CD68, and there was a subset of F4/80^{+}/CD68^{-} tumor-infiltrating cells which were rarely found in the primary tumor core (Wang et al. 2012).

Several studies using CD11b-TK mice in which a treatment with ganciclovir (GCV) results in ablation of microglia/macrophages (CD11b+, CD68+, Iba1+, CD45^{low} cells) addressed the question of a role of these cells in glioma progression. Although, a preliminary study using this model has shown that depletion of microglia/macrophages in CD11b-TK$^{mt\text{-}30}$ mice via GCV systemic injection resulted in the increased tumor size (Galarneau et al. 2007), a recent study using the same model showed that local ablation of microglia/macrophages in CD11b-TK$^{mt\text{-}30}$ mice decreased tumor size and improved survival (Zhai et al. 2011).

Studies of Neurofibromatosis-1 (NF1) genetically engineered mice that develop optic gliomas demonstrated the role of microglia in optic glioma proliferation. GCV treatment of CD11b-TK mice reduced Nf1 optic gliomas proliferation during tumor development and progression (Simmons et al. 2011). Altogether, findings from clinical and animal studies strongly suggest that malignant gliomas recruit microglia/macrophages, and induce production of tumor survival and invasion promoting factors, which in turn facilitates glioma growth and malignancy.

9.2.2 Evaluation of Microglia and Macrophage Content in Gliomas by Flow Cytometry

Flow cytometry is used to identify microglial cells within the central nervous system. A flow cytometric phenotype for ramified microglia isolated from adult CNS was defined ($CD45^{low}$ CD11b/c+) in the Lewis and Brown Norway rats and clearly distinguished these cells from all blood-derived leukocytes, the latter being $CD45^{high}$. Isolated microglia cells were mostly MHC class II positive (Sedgwick et al. 1991, 1993). Highly purified populations of $CD45^{low}CD11b/c^{+}$ microglia and $CD45^{high}CD11b/c^{+}$ macrophages have been obtained from the adult CNS (Dick et al. 1995).

Badie and Schartner (2000) determined infiltration of immune cells into experimental gliomas (intracranially implanted rat C6, 9L, and RG-2 tumor cells) based on detection of CD11b/c, CD45, and CD8a antigens by flow cytometry. The extent of microglia ($CD11b^{high}$, $CD45^{low}$), macrophage ($CD11b^{high}$, $CD45^{high}$), and lymphocyte ($CD11b^{negative}$, $CD45^{high}$) infiltration into tumors, tumor periphery, and contralateral tumor-free hemispheres was measured for each glioma type. They found that microglia account for 13–34 % of viable cells and were present in the tumors, tumor periphery, and contralateral tumor-free hemispheres. In contrast, macrophages were less prominent within the tumors and tumor periphery (4.2–12 %) and were rare in the contralateral tumor-free hemispheres (0.9–1.1 %). Among the tumor types, RG-2 gliomas had the least microglia/macrophage infiltration. The distribution pattern of lymphocytes varied among tumor models: whereas lymphocytes accounted for more than one-third of the cells in C6 and 9L tumors; they represented only 1 % of cells in RG-2 gliomas.

A recent study using immunomagnetic sorting of $CD11b^{+}$ cells from intracranial murine EGFP-GL261 gliomas followed by detection of CD45 antigens by flow cytometry, demonstrated an early accumulation of activated microglia (8 days postimplantation) followed by accumulation of peripheral macrophages in gliomas at day 15th. Quantification of glioma-infiltrating microglia/macrophages revealed a ~2.6-fold increase in the number of microglia and a ~30-fold increase in the number of macrophages 15 days after tumor implantation (Gabrusiewicz et al. 2011). The Iba1 staining of tissue sections showed the strong increase in Iba1-positive cells at day 8 after tumor implantation. At the day 15 the tumor was heavily infiltrated with amoeboid, Iba1-positive cells that accumulated within and around the implanted glioma cells. Interestingly, Iba1-positive cells in the close vicinity of the tumor

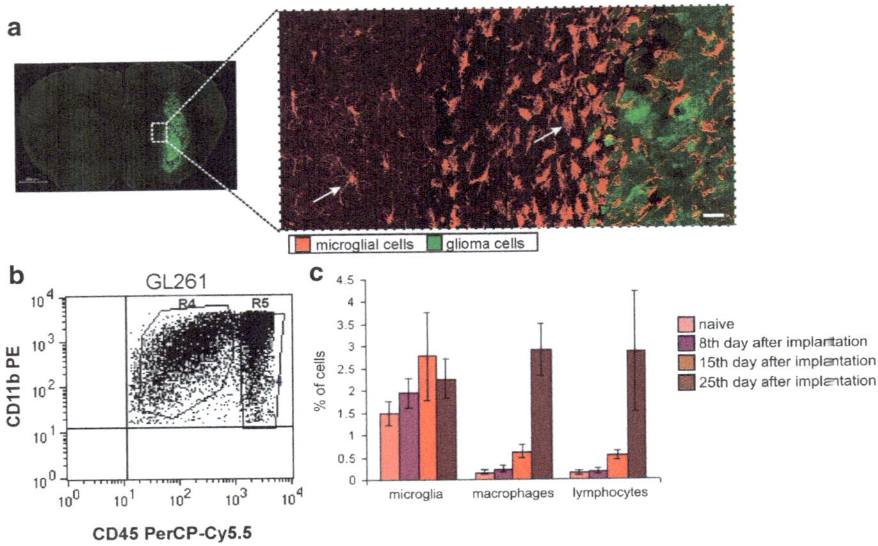

Fig. 9.2 Accumulation of activated microglia/macrophages in murine experimental gliomas. (**a**) Representative confocal images of brain sections 15 days after implantation of pEGFP-N1 GL261 cells into the striatum of C57BL/6 mice. Note the infiltration and morphological transformation of glioma-infiltrating Iba1$^+$ cells (*white arrows*). Scale bar: *left image*—1,000 μm, *right image*—20 μm. (**b**) Microglia and macrophages were separated using a magnetic-bead-conjugated anti-CD11b antibody and stained with CD45 PerCP-Cy5.5 and CD11b PE prior to FACS acquisition. Representative *dot plots* for microglia (Gate R4, CD11b$^+$/CD45^{low}) and macrophages (Gate R5, CD11b$^+$/CD45^{high}) from tumor-bearing hemispheres. (**c**) Quantification of glioma infiltrating microglia, macrophages, and T lymphocytes 8, 15, and 25 days after tumor cell implantation

were evidently more activated and amoeboid than the cells located distantly (Fig. 9.2). Ramified microglia, with thin branching processes and a small cell body, were detected in the tumor-free parenchyma (Gabrusiewicz et al. 2011).

Altogether, histological and flow cytometry studies show accumulation of microglia and macrophages in human gliomas and rodent experimental gliomas. Intratumoral microglia/macrophage density increases during glioma progression and correlates with the grade of malignancy. Neuronal or embryonal CNS tumors do not exhibit accumulation of microglia and macrophages within tumors.

9.3 Origin of Myeloid Cells Infiltrating Gliomas

To date, no single microglia-specific marker that does not also label peripheral macrophages or extraparenchymal brain macrophages has been identified; therefore the question remains whether all CD11b+ cells are CNS resident microglia or are at least in part from other sources such as bone marrow. Flow cytometry studies, based on evaluation of CD11b and CD45 antigens, show accumulation of both microglia and macrophages in human gliomas and rodent experimental gliomas.

Microglial progenitors are yolk sac-derived, invade the brain during early embryonic development, and then locally proliferate in the brain (Ginhoux et al. 2010; Schulz et al. 2012). In contrast to other yolk sac-derived macrophages, they are not replaced during the postnatal period and later life by liver- or bone marrow (BM)-derived macrophages (Hoeffel et al. 2012). Results from irradiation/bone marrow reconstitution studies indicated that microglia in the brains of chimeric mice are slowly replenished by BM-derived cells and microglial turnover could accelerate under neurological conditions. There are also interspecies differences: reactive microglia in rats originate from resident microglia, whereas they have a mixed BM-derived and resident origin in mice, depending on the severity of ischemic tissue damage (Lambertsen et al. 2011). Recent works questioned the use of irradiation/reconstitution experiments and re-evaluation of microglia origin in the CNS suggests that postnatal hematopoietic progenitors do not significantly contribute to microglia population in the adult brain in mice (Davoust et al. 2008). In vivo lineage tracing studies established that adult microglia originate from primitive hematopoietic progenitors that arise before embryonic day 8 (Ginhoux et al. 2010).

Our studies of glioma-infiltrating microglia/macrophages in chimeric mice with the reconstituted bone marrow from GFP transgenic mice showed the significant infiltration of peripheral macrophages to the experimental murine DsRed-GL261 gliomas. Using flow cytometry we found that the infiltrating macrophages ($CD11b^{+}CD45^{hi}$) consist 60 % of all GFP cells found in the tumor-bearing hemisphere (unpublished). These results suggest that myeloid cells infiltrating malignant gliomas consist of a heterogeneous population that may arise from different origin (brain or blood and bone marrow derived) and exhibits different functionalities. Even if irradiation/bone marrow reconstitution studies exaggerate the extent of contribution of peripheral or BM-derived monocytes/macrophages to a pool of glioma-infiltrating myeloid cells, together with flow cytometry studies, it shows considerable accumulation of microglia and macrophages in glioblastoma. The lack of appropriate animal models and flow cytometry data from human low-grade gliomas do not allow estimating the contribution of peripheral or BM-derived monocytes/macrophages to pathobiology of these tumors.

9.4 Defining a Phenotype of Glioma-Infiltrating Microglia/Macrophages

9.4.1 Functional and Molecular Definition of Microglia/Macrophage Phenotype in Cocultures and Animal Glioma Models

9.4.1.1 Immune Responses of Microglia/Macrophages Exposed to Glioma

Although glioblastoma is immunogenic, immune-mediated eradication does not occur efficiently, and attempts at immunotherapy directed against brain tumors have been minimally successful (Grauer et al. 2009; Rolle et al. 2012). There are numerous impairments in glioma immunity which include: low peripheral lymphocyte counts, reduced hypersensitivity reactions to recall antigens, impaired mitogen-induced blastogenic responses by peripheral blood mononuclear cells, and increased $CD8^+$ suppressor T cells (reviewed in Hussain and Heimberger 2005). Adaptive immune responses are remarkably deficient, with diminished responsiveness of peripheral T cells due to impaired transmembrane signaling through the T-cell receptor/CD3 complex (Morford et al. 1997). Soluble factors derived from glioblastoma cell lines increased chemotaxis of Treg, compared with conventional T cells, Treg infiltration, proliferation, and survival (Crane et al. 2012).

Activation of adaptive immune responses that are essential for tumor eradication/suppression is triggered by activation of innate immunity (Akira et al. 2001). Rodent studies have shown that microglia play a critical effector role in rapid responses to injury, autoimmune stimuli, and viral infections, and initiate CNS innate and adaptive immune responses through multiple TLRs (Olson and Miller 2004). The phenotype and function of microglia isolated from resected human glioma tissues by sequential Percoll density gradients have been characterized. Those studies demonstrated that glioma-infiltrating microglia/macrophages express Toll-like receptors (TLR1, 2, 3, 4) and the adaptor protein CD14, were capable of phagocytosis, however did not produce pro-inflammatory cytokines: IL-6, IFNα, and TNFα. Microglia isolated from normal brain were more efficient at tumor cytotoxicity than those derived from tumors (Hussain et al. 2006a).

Further studies revealed that despite surface MCH class II expression, glioma-infiltrating microglia/macrophages lack expression of the co-stimulatory molecules CD86, CD80, and CD40 critical for T-cell activation. Glioma patients demonstrate lack of effector/activated T cells (glioma-infiltrating CD8+ lymphocytes were CD8+CD25−) and an increased population of regulatory CD4 T cells (CD4+CD25+FOXP3+) infiltrating the tumor. Overall, glioma-infiltrating microglia/macrophages may have a few innate immune functions intact, but their capacity to secrete cytokines, to upregulate co-stimulatory molecules, and subsequently to activate antitumor effector T cells could not be sufficient to initiate immune responses (Hussain et al. 2006a).

There is no data on the innate immune phenotype and function of microglia isolated from low-grade gliomas. However, gene expression studies in pilocytic and diffusely infiltrating grade II astrocytomas shed some light on differences in immune defense responses in various tumors. Comparison of gene expression profiles in WHO grade I benign pilocytic astrocytoma, diffusely infiltrating grade II astrocytomas and oligodendrogliomas revealed that the number of immune system-related genes (such as *HLA-DRα*, *HLA-DPB1*, *HLA-DQB1*, *IgG3*, *IgGK*, *FCER1G*, *A2M*, *FCRN*, *IFI-56K*, and *DAP12*) were upregulated in all tested tumors relative to normal cerebellum. However, immune-defense genes such as *HLA-DRalpha*, *HLA-DPA1*, *HLA-DPB1*, *HLA-DQB1*, *A2M*, *TIMP1*, *TIMP2*, *CDKN1A*, and *SOCS3* were expressed at higher levels in pilocytic astrocytomas than diffuse astrocytomas and oligodendrogliomas (Huang et al. 2005). The results suggest that a pattern of immune gene expression in pilocytic astrocytomas differs from those of diffusely infiltrating low-grade gliomas and that their benign behavior may be related to upregulation of immune defense-associated genes.

The importance of ability to upregulate immune defense-associated genes and to launch effective antitumor responses by tumor infiltrating microglia/macrophages is strengthened by recent gene expression profiling data of human gliomas. Ten epidemiologic studies indicated that allergies appear to reduce glioma risk. In a meta-analysis of 12 studies evaluating allergy and risk of gliomas, a significant negative association (a protective effect) was found between allergy and glioma (Chen et al. 2011). The systematic analysis of 919 allergy- and inflammation-related genes in 142 glioblastoma tissue samples revealed downregulation of the majority of allergy- and inflammation-related genes in glioblastoma (Schwartzbaum et al. 2010). Moreover, 69 % of these genes were found negatively correlated with the expression of a glioma stem-like cell marker CD133, including allergy-related (e.g., IL-4R-α) and immunoregulatory genes (e.g., TGFβ1). A recent study evaluating gene expression microarray profiles of high grade gliomas from long-term survivors showed that increased immune gene expression and immune cell infiltration in tumors distinguished long-term from short-term survivors (Donson et al. 2012). Using immune cell-specific gene classifiers, both T cell-associated and myeloid linage-associated genes were shown to be enriched in high grade gliomas from long-term versus short-term survivors.

Taking into consideration previous results on the role of glioma initiating cells in tropism of microglia/macrophages to gliomas (Yi et al. 2011) and their polarization into immunosuppressive cells (Wu et al. 2010), we can postulate a critical role of polarized microglia/macrophages in shaping immune microenvironment of malignant human gliomas and the lack of effective antitumor defense against these tumors.

9.4.1.2 Phagocytic Properties of Microglia/Macrophages Exposed to Glioma

Microglia from human glioblastoma patients have the capability of phagocytosing latex microbeads ex vivo (Hussain et al. 2006a). Cocultures of human microglial cells (microglial CHME5 cell line) grown on collagen beads or on coverslips and placed on monolayer of C6 cells showed decreased bead phagocytosis after a transient increase (Voisin et al. 2010). Cell motility and viability were not affected in such cocultures. In contrast, a recent study shows enhancement of phagocytosis by primary rat microglial cultures exposed to rat C6 glioma cells or glioma-conditioned medium (Ellert-Miklaszewska et al. 2013). The same study shows the strong increase of microglial motility and sustained proliferation under such conditions. This discrepancy could arise from the use of an immortalized microglia CHME5 cell line which has different properties than primary microglial cultures. Studies of cyclodextrin-based nanoparticle (CDP-NP) uptake shows that CDP-NPs were preferentially taken up by BV2 and N9 microglial cells compared with GL261 glioma cells (Alizadeh et al. 2010). Fluorescent microscopy and flow cytometry analysis of intracranial GL261 gliomas demonstrated a predominant CDP-NP uptake by macrophages and microglia, within and around the tumor site (Alizadeh et al. 2010).

Tumor cells are known to evade phagocytosis by deficient "eat-me" signals on their membrane. Microglia phagocytose tumor cells induced to apoptosis with the cytotoxic agent etoposide (Chang et al. 2000). Normal human astrocytes, glioma cells, and microglia all phagocytosed apoptotic U-251 MG glioma cells; however, microglia had phagocytose fourfold more than did other cells. Binding of annexin-V to phosphatidylserine on apoptotic glioma cell membranes inhibited microglial phagocytosis by 90 % (Chang et al. 2000). Furthermore, the ability of microglia to phagocytose undamaged or rat 9L gliosarcoma cells damaged by alloreactive cytotoxic T lymphocytes (aCTL) has been demonstrated. In vitro, 5.5 ± 0.9 % of microglial cells isolated from tumor-bearing rat brains phagocytosed aCTL-damaged 9L cells. At 3 days following intracranial 9L cell infusion, 17.5 ± 0.1 % of the microglia phagocytosed CFSE-labeled aCTL-damaged 9L tumor cells within glioma-bearing brain (Kulprathipanja and Kruse 2004).

9.4.2 Characterization of Microglia/Macrophage Phenotype in Human Gliomas

Activated macrophages can be broadly classified in two main groups: classical, inflammatory macrophages (M1 type), activated by interferon gamma (IFNγ) and lipopolysaccharide (LPS), and alternatively activated macrophages (M2 type). The latter could be further subdivided depending on stimuli. In general, M1 type macrophages exhibit potent microbiocidal properties and promote strong IL-12-mediated Th1 responses, while M2 support Th2-associated effector functions. M2 polarized macrophages play a role in resolution of inflammation through high endocytic clearance capacities, trophic factor synthesis, accompanied by reduced pro-inflammatory cytokine secretion (Martinez et al. 2008; Solinas et al. 2009). Hallmarks of M2 macrophages are IL-10^{high} IL-12^{low} IL-$1ra^{high}$ IL-1 decoyR^{high} production, CCL17 and CCL22 secretion, high expression of mannose, scavenger and galactose-type receptors, poor antigen-presenting capability, enhanced phagocytic and wound-healing activities. M2 macrophages express specific changes in some metabolic pathways: arginine metabolism is oriented toward the production of ornitine and polyamine instead of citrulline and NO. Arginase-1 (Arg1) is upregulated in alternatively activated macrophages and, due to its higher affinity for arginine, competes with inducible nitric oxide synthase (iNOS), which metabolizes arginine. Tumor-associated macrophages (TAMs) resemble M2-polarized cells and play a pivotal role in tumor growth and progression. TAMs promote angiogenesis, remodeling, and repair of tissues. Activated M2 cells control the inflammatory response by downregulating M1-mediated functions (Solinas et al. 2009). Monocyte differentiation could be influenced by soluble factors in tumor microenvironment. In tumors, there is an established gradient of IL-10 that can switch monocyte differentiation toward macrophages rather than to dendritic cells (Li and Flavell 2008).

Compelling evidence suggests that glioma-associated microglia/macrophages resemble M2-polarized cells and support tumor growth and progression. The immune functions of CD11b/c$^+$CD45$^+$ glioma-infiltrating microglia/macrophages from postoperative tissue specimens of glioma patients have been reduced, as they did not produce pro-inflammatory cytokines (TNFα, interleukin 1β, or interleukin 6), and do not mediate T-cell proliferation. They express MHC class II, but they lacked expression of the co-stimulatory molecules CD86, CD80, and CD40 that are critical for T-cell activation. The presence of regulatory T cells may also contribute to the lack of effective immune activation against malignant human gliomas (Hussain et al. 2006a).

Macrophage scavenger receptors CD163 and CD204 are believed to be markers for "M2 type" macrophages. CD163 and CD204 positive cells were detected in gliomas. The ratio of CD163 and CD204 positive cells among glioma-infiltrating microglia/macrophages correlates with the histological grade of glioma (Komohara et al. 2008). Significant increases in arginase activity and G-CSF levels were observed in plasma specimens obtained from patients with glioblastoma (Raychaudhuri et al. 2011). Further, myeloid-derived suppressor cell (MDSC)-like cells have been detected in glioma patients (Rodrigues et al. 2010). Patients with glioblastoma (GBM) have increased MDSC counts (CD33$^+$HLADR$^-$) in their blood that are composed of neutrophilic (CD15$^+$; >60 %), lineage-negative (CD15$^-$CD14$^-$; 31 %), and monocytic (CD14$^+$; 6 %) subsets. T cells from patients with glioblastoma had suppressed IFN-γ production after stimulation and removal of MDSCs by pre-incubation with anti-CD33/CD15-coated beads significantly restored T cell function.

Increased tumor infiltration by MSDC has been reported in rat glioma models (Prins et al. 2002; Graf et al. 2005). Animal studies on GFAP-V(12)*HA-ras* mouse astrocytomas demonstrated that immune infiltration at the tumor site in mice is dominated by immunosuppressive cells from the early stages of tumor development, even at very early asymptomatic stages (Tran Thang et al. 2010). Phenotypical characterization of MSDC in GL261 murine glioma suggests that tumor-infiltrating MSDC are pleiotropic monocytes/macrophages that bear M1- and M2-type characteristics. Over 90 % were of the CD11b$^+$F4/80$^+$ monocyte/macrophage lineage, displayed a CD11b/c$^+$Gr-1^{low}IL-4Rα$^+$ phenotype, and suppressed the proliferation of activated splenic CD8$^+$ T cells. These MDSCs expressed both M1 and M2 activation markers: *CD206, CXCL10, IL-1β, TGF-β, and TNF-α* mRNAs and CXCL10, CD206 proteins. In addition, the cells expressed CX3CR1 and CCR2 which are the markers of an inflammatory monocyte (Umemura et al. 2008).

Soluble factors secreted by glioblastoma cells have been shown to stimulate cultured peripheral blood monocytes to differentiate into the M2-like or MSDC-like cells, characterized by increased CD163 and CD204 staining, reduced IL-12 and TNFα production (Komohara et al. 2008), and upregulation of IL-10 and TGF-β (Rodrigues et al. 2010). Peripheral blood monocytes cocultured with glioblastoma cells (U87 and U251) acquired immunosuppressive, MDSC-like features, including reduced CD14 (but not CD11b) expression, increased immunosuppressive IL-10,

TGF-β, and B7-H1 expression, decreased phagocytic ability, and increased ability to induce apoptosis in activated lymphocytes. Direct contact between monocytes and glioblastoma cells was necessary for complete induction of such phenotype (Rodrigues et al. 2010). These results support a hypothesis that normal human monocytes exposed to malignant glioma cells may adopt an MDSC-like phenotype in glioma microenvironment. It means that many immunologically active cell types are present in malignant glioma patients and may contribute to immunosuppression. Interestingly, a recent study has demonstrated immunosuppressive properties of glioma cancer stem cells which produce soluble CSF-1, TGF-β1, and macrophage inhibitory cytokine (MIC)-1, thus inducing recruitment and polarization of macrophages/microglia into immunosuppressive cells (Wu et al. 2010).

Studies on cultured microglial cells demonstrate remarkable plasticity and ability to adapt different fate or differentiation route depending on a stimulus. Rat primary microglial cultures exposed to glioma-conditioned medium (G-CM) undergo morphological alterations and become motile and highly phagocytic. Global gene expression profiling of LPS- or G-CM-stimulated microglial cultures followed by computational analysis of gene expression pattern in differentially stimulated microglia revealed activation of alternative genetic programs. Most genes characteristic for innate and inflammatory immune responses, which are commonly upregulated during classical inflammation, were not induced by G-CM stimulation. The analysis of signaling pathways activated in microglia after G-CM demonstrates defective NFκB and STAT 1 activation, resulting in failure to mount production of inflammation mediators (IL-1β, iNOS, Cox2) and polarize to the classical inflammatory phenotype. Furthermore, the increased expression/production of many M2-like factors (Arg1, MT1-MMP, IL-10, TGF-β) in G-CM activated microglial cultures could be responsible for sustained proliferation and acquisition of the glioma-associated phenotype (Ellert-Miklaszewska et al. 2013). In particular, upregulation of genes coding for CD69—a C-type lectin R family member, a negative regulator of the immune response; CD86—the co-stimulatory molecule implicated in dendritic cell maturation; CCL2 (MCP-1, monocyte chemoattractant protein-1), CCL5 (RANTES, regulated on activation, normal T cell expressed and secreted), CXCL chemokines CXCL1, CXCL2, CXCL7, and CXCL14 was detected (Ellert-Miklaszewska et al. 2013). Glioma secretome-exposed microglia express the microglial fractalkine receptor CX3CR1 that could be important in the regulation of myeloid cell trafficking and polarization. These findings are summarized in Fig. 9.3.

Recent data support the notion that microglia/macrophage accumulation in diffuse glial tumors reflects participation of these cells in supporting the invasive potential of gliomas. Microglia and macrophages can secrete various cytokines, growth factors, and enzymes, including extracellular matrix proteases which directly or indirectly may influence tumor migration/invasiveness and proliferation (Watters et al. 2005; Li and Graeber 2012).

Experimental studies using brain organotypic slices, microglia-glioma cocultures, and genetic models, in which microglial cells are ablated in the tumor, strongly support pro-invasive role of glioma-infiltrating microglia

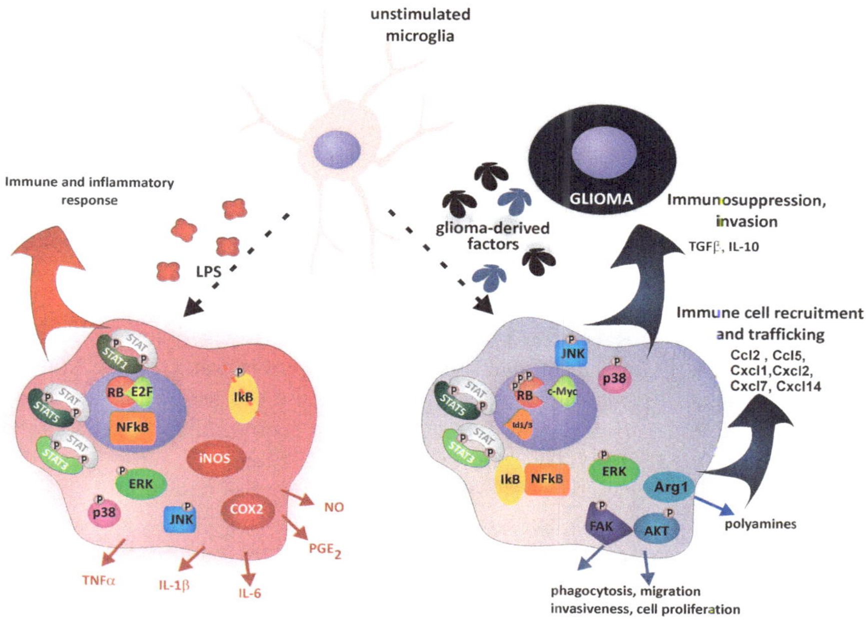

Fig. 9.3 Graphical summary of the classical, inflammatory, and pro-tumorigenic phenotypes of activated microglia. Primary rat microglial cultures exposed to lipopolysaccharide (LPS) or glioma secreted factors adapt different fates and polarize into pro-inflammatory or alternatively activated cells. Glioma-derived factors increase focal adhesion kinase and PI-3K/Akt signaling and activate ERK and p38 MAPK but not JNK signaling. In contrast to LPS stimulation, glioma secretome do not activate pro-inflammatory Stat1 and NFκB signaling in microglial cells. The genetic program induced by glioma secretome include: transcription factors c-Myc and Id1/3 (inhibitors of differentiation), and numerous coding for chemokines and cytokines regulating immune cell recruitment and trafficking. Glioma-induced activation is associated with increased TGF-β and IL-10 production, while LPS-induced cells produce inflammation mediators

(Markovic et al. 2005, 2009; Sliwa et al. 2007). The invasion of GFP-labeled GL261 glioblastoma cells in organotypic brain slices that were depleted of microglia by treatment with clodronate-filled liposomes was significantly decreased. Inoculation of exogenous microglia together with glioma cells into cultured brain slices increased the infiltrative behavior of the tumor depending on the microglia/glioma cell ratio and increased activity of metalloproteinase-2 (MMP-2). It has been shown that soluble factors released from glioma cells strongly stimulate MMP-2 activity in microglia (Markovic et al. 2009). Membrane bound MT-MMPs, in particular membrane-type MT1- and MT2-MMP, play a major role in activating MMP-2. Newly synthesized MMP-2 is secreted as an inactive pro-enzyme, which is cleaved on the cell surface by membrane-type MT1-MMP complexed with TIMP-2. Metalloproteinase MT1-MMP secreted by glioma-exposed microglia activates pro-MMP-2 in glioma cells that promotes tumor invasion, as was shown using brain slices from MT1-MMP-deficient mice and in a microglia depletion model (Markovic et al. 2009). Glioma-released factors induce the expression and activity of MT1-MMP via microglial toll-like receptors and the p38 MAPK pathway, as deletion of the toll-like receptor adapter protein MyD88 or p38 inhibition prevented MT1-MMP expression and activity in cultured microglial cells. Microglial MT1-MMP in turn activates glioma-derived pro-MMP-2 and promotes glioma expansion, as shown in an ex vivo model using MT1-MMP-deficient brain tissue and in microglia depleted mice (Markovic et al. 2009)

Activated microglia release cytokines which enhance tumor cell invasion (Wesolowska et al. 2008). Microglial cells exposed to glioblastoma cells secrete active TGF-β1, which through a paracrine loop stimulates glioblastoma invasion. Plasmid-transcribed small hairpin RNAs (shRNAs) which downregulate the TGF-β type II receptor (Tβ IIR) expression, effectively inhibited TGF-β-dependent signaling pathways and transcriptional responses in glioblastoma cells. Microglia strongly enhanced glioma invasiveness in the coculture system, but this activity was lost in glioma cells depleted of TGF-β type II receptor indicating an important role of microglia-derived TGF-β in tumor invasion. Moreover, tumorigenicity of glioblastoma cells depleted of TGF-β IIR in nude mice was reduced by 50 %.

Fas (CD95/APO-1) is a cell surface "death receptor" that mediates apoptosis upon engagement by its ligand—FasL but can also promote tumor invasion when apoptosis is compromised. Kleber and coworkers (Kleber et al. 2008) demonstrated that interaction of glioma cells with the surrounding brain tissue induces expression of FasL in both tumor and host cells. FasL modulated glioma invasion via PI-3K/MMP-dependent mechanism and neutralization of Fas activity blocked migration of glioma cells in a mouse syngenic model of intracranial glioblastoma. We demonstrated that a recombinant FasL Interfering Protein (FIP), which interferes with Fas signaling in C6 glioma cells, impaired cell motility and invasiveness of glioma cells in vitro. A blockade of Fas signaling reduced MMP-2 activity in glioma cells. Interestingly, reduction of MMP-2 activity was not due to downregulation of *mmp-2* and *mt1-mmp* expression but was dependent on modulation of *timp-2* mRNA and TIMP-2 protein levels by Fas signaling (Wisniewski et al. 2010). FasL expression is higher in many glioblastoma cell lines in comparison to non-transformed astrocytes and is upregulated in microglia exposed to glioma-conditioned medium (Wisniewski et al. 2010). Microglia accumulating within tumors account for half of the FasL expression in the murine intracranial tumors (Badie et al. 2001).

9.5 Molecules and Mechanisms Responsible for Microglia/Macrophage Accumulation in Gliomas

Signals, signaling pathways, and molecular mechanisms underlying "switch" or "reeducation" of glioma-infiltrating microglia and macrophages into pro-invasive cells are poorly known. Glioma cells produce several factors with a chemotactic activity which attract microglia, including monocyte chemotactic protein (MCP)-3 (Okada et al. 2009) and hepatocyte growth factor/scatter factor (HGF/SF), as shown by in vitro and in vivo studies (Wang et al. 2012). Macrophage proliferation, differentiation, and chemotaxis can be regulated by several factors, including macrophage colony stimulating factor (M-CSF), granulocyte-macrophage colony stimulating factor (GM-CSF), interleukin-34, and the chemokine CCL2 (Hibbs et al. 2007; Hamilton 2008; Pollard 2009). Increased M-CSF expression has been associated with a poor prognosis and increased angiogenesis in various non-brain tumors (Sapi and Kacinski 1999; Lin et al. 2002; Mroczko et al. 2007; Zhu et al. 2008; Mantovani and Sica 2010; Davies et al. 2011). Overexpression of M-CSF accelerates tumor progression and increases pulmonary metastasis in murine experimental breast cancers (Lin et al. 2001). RNAi-mediated silencing of M-CSF or its receptor c-FMS suppresses the growth of mammary tumor xenografts in mice and reduces macrophage infiltration, metalloproteinase expression, and angiogenesis (Aharinejad et al. 2004; Lin et al. 2006). Osteopetrotic ($Csf1^{op}/Csf1^{op}$) mice harbor a mutation in the *Csf1* gene that causes an M-CSF deficiency (Wiktor-Jedrzejczak et al. 1990; Yoshida et al. 1990), affects myeloid lineage development, and reduces the number of monocytes/macrophages in many tissues (Wiktor-Jedrzejczak et al. 1992; Cecchini et al. 1994). Experimental breast cancer progression was delayed and lung metastases were reduced in op/op mice (Nowicki et al. 1996; Qian et al. 2009).

We found upregulation of GM-CSF levels in glioma-bearing brains and increased expression of *Csf2*, but not *Csf1*, in murine GL261 glioma cells compared with non-transformed astrocytes (Gabrusiewicz et al. 2011). Knockdown of GM-CSF expression in GL261 glioma cells strongly reduced microglia-dependent invasion in organotypical brain slices. The number of infiltrating microglia/macrophages ($Iba1^+$ cells) and intratumoral angiogenesis were reduced in murine gliomas depleted of GM-CSF. It resulted in reduction of tumor growth and extended animal survival. M1/M2 gene profiling in sorted microglia/macrophages suggests impairment of their pro-invasive activation in GM-CSF-depleted gliomas (Sielska et al. 2013). Deficiency of M-CSF did not affect glioma growth in op/op mice in vivo and the accumulation of $Iba1^+$ cells. This suggests that glioma-derived GM-CSF is responsible for the recruitment/activation of microglia/macrophages and contributes to tumor progression. The increased expression of *CSF2* and inverse correlation with patient survival was found in human glioblastoma (Sielska et al. 2013). The elevated expression of *CSF2* in human glioblastoma and the autocrine or paracrine action of GM-CSF contributing to glioma proliferation in vitro have been recently demonstrated (Revoltella et al. 2012).

The chemokine receptor CX3CR1 and its ligand CX3CL1 (fractalkine) are known to be involved in immune responses, and influence migration of macrophages, microglia, and lymphocytes in vivo (Combadiere et al. 2007). CX3CR1 was localized on Iba1 and CD11b/c positive glioma-infiltrating microglia/macrophages (GIMs). Cultured human GIMs responded to CX3CL1-triggered activation of CX3CR1 with adhesion and migration in vitro (Held-Feindt et al. 2010). However, formation of GL261 glioma and animal survival rates were not affected in CX3CR1-deficient C57BL/6 mice. Tumor-bearing CX3CR1−/− mice had similar numbers of infiltrating microglia and CD4+, CD8+, FoxP3+, or

Ly49G2+ lymphocytes as control mice. These data indicate that CX3CR1 has little or no effects on glioma formation or the migration of microglia and lymphocytes into GL261 tumors (Liu et al. 2008).

The experimental ALTS1C1 gliomas expressed a relatively high level of CXCL12 (stromal cell-derived factor-1, SDF-1) in vitro and in vivo. Inhibition of CXCL12 production reduced tumor invasiveness and the resulting tumors had well-defined borders and were lacking infiltration tracts. Inhibition of CXCL12 production reduced tropism of microglia/macrophages (determined by confocal imaging using CD68 and F4/80 markers) toward hypoxia (Wang et al. 2012). The authors postulate that a CXCL12-concentration gradient could be the driving force for tropism microglia/macrophages toward hypoxia.

Glial cell line-derived neurotrophic factor (GDNF) is expressed in gliomas, Gl261, and human glioma cell lines, and this factor has been shown as a strong chemoattractant for cultured microglia. Knockdown of GDNF expression in mouse GL261 glioma cells by shRNA diminished accumulation of microglia in tumors, while overexpression of GDNF in fibroblasts seeded into hollow fibers implanted to the brain promoted microglia attraction. GDNF release from human or mouse glioma had a strong effect on microglia accumulation; glioma-induced astrogliosis was not affected. Injection of GDNF depleted GL261 glioma cells into mouse brains resulted in reduced tumor expansion and improved survival (Ku et al. 2013).

9.6 Targeting Glioma-Infiltrating Microglia/Macrophages as a Innovative Therapeutic Strategy

Recent studies from our laboratory demonstrated that pharmacological inhibition of infiltration and activation of brain resident microglia and peripheral macrophages effectively impairs tumor growth in mice (Sliwa et al. 2007; Gabrusiewicz et al. 2011). Cyclosporine A (CsA), a widely used immunosuppressive drug and atypical MAPK signaling inhibitor (Zawadzka et al. 2012), reduced amoeboid transformation of microglial cells, and inhibited microglia-dependent invasion of glioma cells in vitro and in organotypic brain slices. A systemically applied CsA inhibited microglia/macrophages infiltration (as determined by flow cytometry and immunofluorescence) and blocked expression/activity of proteolytic enzymes (MMP-2, MT1-MMP) and production of cytokines (IL-10 and GM-CSF) (Gabrusiewicz et al. 2011). Our data demonstrate that blockade of microglia/macrophage infiltration and inhibition of their pro-invasive behavior reduce glioma growth in mice.

Minocycline, a semisynthetic antibiotic of the tetracycline family, has emerged as a potent anti-inflammatory and microglia targeting drug, beneficial in animal models of several CNS disorders. For the reason of the good tolerance and penetration into the brain, minocycline has been clinically tested for stroke, multiple sclerosis, spinal cord injury, and some neurodegenerative diseases (Kim and Suh 2009). The treatment with minocycline reduced the expression of MT1-MMP in glioma-infiltrating microglia in vitro and in organotypic brain slices. This reduction is dependent on the presence of microglia. Glioma growth in an experimental mouse model was strongly reduced by the addition of minocycline to drinking water, compared to untreated controls. Coherently, MT1-MMP was abundantly expressed in glioma-associated microglia in controls, but was strongly attenuated in tumors of minocycline treated animals (Markovic et al. 2011). This suggests that the clinically approved antibiotic minocycline is a promising new candidate for adjuvant therapy against malignant gliomas. However, minocycline and CsA treatments have some drawbacks because while blocking tumor-associated activation of microglia, they exert immunosuppressive action and inactivate many components of the immune system. Minocycline represses MHC II expression in microglia (an event requisite for T cell reactivation) in an experimental model of autoimmune diseases via the inhibition of transcription factor CIITA expression (Nikodemova et al. 2007).

Recent studies defined more specific, intracellular signaling pathways which could be targeted. Signal transducer and activator of transcription 3 (STAT 3) plays a suppressive role in antitumor immunity. STAT 3 was activated in immortalized N9 microglial cells exposed to GL261 glioma-conditioned medium that resulted in upregulation of IL-10 and IL-6, and downregulation of IL1-β. Inhibition of STAT 3 by CPA-7 (the platinum (IV)-based anticancer drug) or specific siRNA reversed glioma-induced cytokine expression profile in N9 cells. Furthermore, inactivation of STAT 3 in intracranial GL261 gliomas by siRNA resulted in inhibition of tumor growth (Zhang et al. 2009).

A novel small molecule inhibitor of STAT3—WP1066, which can penetrate the CNS in mice, reversed tolerance in immune cells isolated from GBM patients. Specifically, WP1066 induced the expression of co-stimulatory molecules on peripheral macrophages and glioma-infiltrating microglia, stimulated the production of the immune-stimulatory cytokines interleukin (IL)-2, IL-4, IL-12, and IL-15, and induced proliferation of effector T cells from GBM patients that are refractory to CD3 stimulation (Hussain et al. 2007).

Glioblastoma is heterogeneous and (as many other tumors) contains glioma stem cells or glioma-initiating cells, a subpopulation of cells that possess the capacity for self-renewal and forming neurospheres in vitro, are capable of pluripotent differentiation, and could initiate tumors in vivo. Glioma-initiating cells have been shown to suppress adaptive immunity by affecting capacity to inhibit T-cell proliferation, triggering T-cell apoptosis, and induction of FoxP3(+) regulatory T cells (Wei et al. 2010). A recent study demonstrates that glioma cancer stem cells produce sCSF-1, TGF-β1, and MIC-1, cytokines known to recruit and polarize the macrophages/microglia to immunosuppressive phenotype. Glioma-initiating cells polarized human macrophages/microglia to an M2-like phenotype, inhibited phagocytosis, induced the secretion of the immunosuppressive cytokines IL-10 and TGF-β1, and enhanced their capacity to inhibit T-cell proliferation. The inhibition of phagocytosis and the secretion of IL-10 were reversed when the STAT 3 pathway in glioma stem cells was blocked with a pharmacological inhibitor WP1066 or specific STAT shRNA (Wei et al. 2010). However, such approaches are in infancy, these findings illustrate that interactions between glioma and brain macrophages are potential, promising targets in glioma therapy.

There are also approaches based on local TLR stimulation to induce antitumor immunity. TLR9 is overexpressed in human and murine glioma cell lines. Injection of CpG-oligonucleotides—CpG-ODN (TLR9 inducers) into C57BL/6 mice implanted with GL261 glioma prolonged the survival of mice with tumors. CpG-ODN induced TLR9 downregulation and apoptosis of GL261 cells in vitro as well as in vivo. Furthermore, the CpG stimulation enhanced the antigen presenting capacity of microglia, shifted the immune response toward $CD8^+$ T cells, and decreased the number of regulatory T cells (El Andaloussi et al. 2006). Intratumoral injection of TLR1/2 (Pam3Cys-SK4) or TLR7 (R848) agonist produced a significant survival benefit, whereas agonists of TLR3 (poly(I:C)) or TLR4 (purified LPS) alone were not effective. Studies using wild-type and TLR9(−/−) knockout mice revealed that the efficacy of CpG-ODN treatment required TLR9 expression on nontumor cells (Grauer et al. 2008).

Concluding Remarks

A simple response to a question whether microglia in gliomas are friends or foes proved to be difficult. Histopathological studies of human and rodent glioma tissues have consistently shown high levels of infiltrating microglia/macrophages. The abundance of amoeboid, activated microglia/macrophages is higher in diffusive and invasive gliomas than in benign tumors. The presence of activated microglia/macrophages does not reflect activation of immune defense response in malignant gliomas; on the contrary unharmed growth and highly malignant behavior indicates that microglial immune defense mechanisms do not function properly in these tumors. Moreover, these cells are responsible for local and generalized immunosuppression. Glioma-infiltrating microglia/macrophages acquire the alternative, pro-invasive phenotype in which phagocytic, trophic, and tissue remodeling functions of these cells are enhanced. Cell culture and organotypic brain slice culture studies demonstrated the pro-invasive activity of microglia and their polarization into tumor supportive cells. Several mechanisms underlying the pro-invasive activity of microglia have been demonstrated such as TGFβ1 and FasL production, MT1-MMP-dependent activation of MMP-2. This is well supported by the results of animal glioma studies because most of them represent malignant gliomas. Genetic or pharmacological ablation of microglia/macrophages impairs glioma growth, extends survival, and in some cases restores to some extent antitumor immune responses.

On the other hand, low-grade gliomas are characterized by the presence of activated, hypertrophic, or ramified microglia/macrophages, non-diffusive growth and effective antitumor responses. Unfortunately, due to a lack of proper cellular and animal models, there is no conclusive data if microglia/macrophages infiltrating low-grade gliomas are "friends" which have the classical, inflammatory phenotype and perform protective functions, typical for microglia under physiological conditions.

Since no single microglia-specific marker exists that does not also label peripheral macrophages or extraparenchymal brain macrophages, it is difficult to estimate proportion and contribution of different populations of myeloid cells to glioma pathobiology. Flow cytometry studies of human and rodent glioma tissues clearly demonstrated the presence of both microglia and peripheral macrophages among immune cells recruited to gliomas. It is possible that the myeloid-derived infiltrating cells are functionally distinct from the resident microglia and peripheral macrophages undergo specific "education" in glioma microenvironment. Some faint and fragmentary evidence support this idea, f.e. IL-10 is produced mainly by $CD11b^{+}CD45^{low}$ cells isolated from GL261 gliomas (Gabrusiewicz et al. 2011). Studies on the role of myeloid cell populations under neurodegenerative conditions suggest that peripheral macrophages infiltrating CNS are regulatory and can perform immune-resolving functions (London et al. 2013). As proposed by Michal Schwartz (Schwartz 2010), the blood-derived macrophages are similar to "alternatively activated" macrophages (also known as tissue repairing or M2). They are not spontaneously recruited but their infiltration to CNS could be enhanced by glioma-activated microglia. These blood-derived, immunosuppressive, wound healing macrophages should reinforce repair mechanisms but instead they help tumor to grow.

Understanding the mechanisms responsible for high invasiveness of glioblastoma cells—which is even accelerated by interactions of the brain tumor with surrounding cells—may lead to identification of specific targets for a future treatment. A better understanding of microglia–glioma interaction may provide better methods to manipulate the glioma microenvironment to allow the generation of a specific and lasting anti-glioma immunity. Therapies that will effectively target invasive glioblastoma cells may significantly improve therapeutic outcome. Thus, counteracting accumulation and activation of brain macrophages should be taken into account when considering the development of more effective therapy against malignant gliomas.

Acknowledgments I would like to thank Dr. Aleksandra Ellert-Miklaszewska and Malgorzata Sielska for reading and help with editing of the manuscript. This work was supported by a grant 2012/04/A/NZ3/00630 from The National Science Centre Poland.

References

Aharinejad S, Paulus P, Sioud M, Hofmann M, Zins K, Schäfer R, Stanley ER, Abraham D (2004) Colony-stimulating factor-1 blockade by antisense oligonucleotides and small interfering RNAs suppresses growth of human mammary tumor xenografts in mice. Cancer Res 64(15): 5378–5384

Akira S, Takeda K, Kaisho T (2001) Toll-like receptors: critical proteins linking innate and acquired immunity. Nat Immunol 2(8):675–680

Albesiano E, Han JE, Lim M (2010) Mechanisms of local immunoresistance in glioma. Neurosurg Clin N Am 21(1):17–29

Alizadeh D, Zhang L, Hwang J, Schluep T, Badie B (2010) Tumor-associated macrophages are predominant carriers of cyclodextrin-based nanoparticles into gliomas. Nanomedicine 6(2): 382–390

Armstrong A, Ravichandran KS (2011) Phosphatidylserine receptors: what is the new RAGE? EMBO Rep 12(4):287–288

Badie B, Schartner JM (2000) Flow cytometric characterization of tumor-associated macrophages in experimental gliomas. Neurosurgery 46(4):957–961, discussion 961–952

Badie B, Schartner J, Prabakaran S, Paul J, Vorpahl J (2001) Expression of Fas ligand by microglia: possible role in glioma immune evasion. J Neuroimmunol 120(1–2):19–24

Cardona AE, Pioro EP, Sasse ME, Kostenko V, Cardona SM, Dijkstra IM, Huang D, Kidd G, Dombrowski S, Dutta R, Lee JC, Cook DN, Jung S, Lira SA, Littman DR, Ransohoff RM (2006) Control of microglial neurotoxicity by the fractalkine receptor. Nat Neurosci 9(7): 917–924

Cecchini MG, Dominguez MG, Mocci S, Wetterwald A, Felix R, Fleisch H, Chisholm O, Hofstetter W, Pollard JW, Stanley ER (1994) Role of colony stimulating factor-1 in the establishment and regulation of tissue macrophages during postnatal development of the mouse. Development 120(6):1357–1372

Chang GH, Barbaro NM, Pieper RO (2000) Phosphatidylserine-dependent phagocytosis of apoptotic glioma cells by normal human microglia, astrocytes, and glioma cells. Neuro Oncol 2(3): 174–183

Charles NA, Holland EC, Gilbertson R, Glass R, Kettenmann H (2012) The brain tumor microenvironment. Glia 60(3):502–514

Chen C, Xu T, Chen J, Zhou J, Yan Y, Lu Y, Wu S (2011) Allergy and risk of glioma: a meta-analysis. Eur J Neurol 18(3):387–395

Combadiere C, Feumi C, Raoul W, Keller N, Rodero M, Pezard A, Lavalette S, Houssier M, Jonet L, Picard E, Debre P, Sirinyan M, Deterre P, Ferroukhi T, Cohen SY, Chauvaud D, Jeanny JC, Chemtob S, Behar-Cohen F, Sennlaub F (2007) CX3CR1-dependent subretinal microglia cell accumulation is associated with cardinal features of age-related macular degeneration. J Clin Invest 117(10):2920–2928

Crane CA, Ahn BJ, Han SJ, Parsa AT (2012) Soluble factors secreted by glioblastoma cell lines facilitate recruitment, survival, and expansion of regulatory T cells: implications for immunotherapy. Neuro Oncol 14(5):584–595

Davies LC, Rosas M, Smith PJ, Fraser DJ, Jones SA, Taylor PR (2011) A quantifiable proliferative burst of tissue macrophages restores homeostatic macrophage populations after acute inflammation. Eur J Immunol 41(8):2155–2164

Davoust N, Vuaillat C, Androdias G, Nataf S (2008) From bone marrow to microglia: barriers and avenues. Trends Immunol 29(5):227–234

Deininger MH, Meyermann R, Trautmann K, Morgalla M, Duffner F, Grote EH, Wickboldt J, Schluesener HJ (2000) Cyclooxygenase (COX)-1 expressing macrophages/microglial cells and COX-2 expressing astrocytes accumulate during oligodendroglioma progression. Brain Res 885(1):111–116

Dick AD, Ford AL, Forrester JV, Sedgwick JD (1995) Flow cytometric identification of a minority population of MHC class II positive cells in the normal rat retina distinct from CD45lowCD11b/c+ CD4low parenchymal microglia. Br J Ophthalmol 79(9):834–840

Donson AM, Birks DK, Schittone SA, Kleinschmidt-DeMasters BK, Sun DY, Hemenway MF, Handler MH, Waziri AE, Wang M, Foreman NK (2012) Increased immune gene expression and immune cell infiltration in high-grade astrocytoma distinguish long-term from short-term survivors. J Immunol 189(4):1920–1927

El Andaloussi A, Sonabend AM, Han Y, Lesniak MS (2006) Stimulation of TLR9 with CpG ODN enhances apoptosis of glioma and prolongs the survival of mice with experimental brain tumors. Glia 54(6):526–535

Ellert-Miklaszewska A, Dabrowski M, Lipko M, Sliwa M, Maleszewska M, Kaminska B (2013) Molecular definition of the pro-tumorigenic phenotype of glioma-activated microglia. Glia 61 (7):1178–1190

Färber K, Synowitz M, Zahn G, Vossmeyer D, Stragies R, van Rooijen N, Kettenmann H (2008) An alpha5beta1 integrin inhibitor attenuates glioma growth. Mol Cell Neurosci 39(4):579–585

Gabrusiewicz K, Ellert-Miklaszewska A, Lipko M, Sielska M, Frankowska M, Kaminska B (2011) Characteristics of the alternative phenotype of microglia/macrophages and its modulation in experimental gliomas. PLoS One 6(8):e23902

Galarneau H, Villeneuve J, Gowing G, Julien JP, Vallières L (2007) Increased glioma growth in mice depleted of macrophages. Cancer Res 67(18):8874–8881

Ginhoux F, Greter M, Leboeuf M, Nandi S, See P, Gokhan S, Mehler MF, Conway SJ, Ng LG, Stanley ER, Samokhvalov IM, Merad M (2010) Fate mapping analysis reveals that adult microglia derive from primitive macrophages. Science 330(6005):841–845

Goodridge HS, Underhill DM, Touret N (2012) Mechanisms of Fc receptor and dectin-1 activation for phagocytosis. Traffic 13(8):1062–1071

Graeber MB, Streit WJ (2010) Microglia: biology and pathology. Acta Neuropathol 119(1): 89–105

Graeber MB, Scheithauer BW, Kreutzberg GW (2002) Microglia in brain tumors. Glia 40(2): 252–259

Graf MR, Sauer JT, Merchant RE (2005) Tumor infiltration by myeloid suppressor cells in response to T cell activation in rat gliomas. J Neurooncol 73(1):29–36

Grauer OM, Molling JW, Bennink E, Toonen LW, Sutmuller RP, Nierkens S, Adema GJ (2008) TLR ligands in the local treatment of established intracerebral murine gliomas. J Immunol 181(10):6720–6729

Grauer OM, Wesseling P, Adema GJ (2009) Immunotherapy of diffuse gliomas: biological background, current status and future developments. Brain Pathol 19(4):674–693

Guillemin GJ, Brew BJ (2004) Microglia, macrophages, perivascular macrophages, and pericytes: a review of function and identification. J Leukoc Biol 75(3):388–397

Hamilton JA (2008) Colony-stimulating factors in inflammation and autoimmunity. Nat Rev Immunol 8(7):533–544

Hanisch UK, Kettenmann H (2007) Microglia: active sensor and versatile effector cells in the normal and pathologic brain. Nat Neurosci 10(11):1387–1394

Held-Feindt J, Hattermann K, Muerkoster SS, Wedderkopp H, Knerlich-Lukoschus F, Ungefroren H, Mehdorn HM, Mentlein R (2010) CX3CR1 promotes recruitment of human glioma-infiltrating microglia/macrophages (GIMs). Exp Cell Res 316(9):1553–1566

Hibbs ML, Quilici C, Kountouri N, Seymour JF, Armes JE, Burgess AW, Dunn AR (2007) Mice lacking three myeloid colony-stimulating factors (G-CSF, GM-CSF, and M-CSF) still produce macrophages and granulocytes and mount an inflammatory response in a sterile model of peritonitis. J Immunol 178(10):6435–6443

Hoeffel G, Wang Y, Greter M, See P, Teo P, Malleret B, Leboeuf M, Low D, Oller G, Almeida F, Choy SH, Grisotto M, Renia L, Conway SJ, Stanley ER, Chan JK, Ng LG, Samokhvalov IM, Merad M, Ginhoux F (2012) Adult Langerhans cells derive predominantly from embryonic fetal liver monocytes with a minor contribution of yolk sac-derived macrophages. J Exp Med 209(6):1167–1181

Huang H, Hara A, Homma T, Yonekawa Y, Ohgaki H (2005) Altered expression of immune defense genes in pilocytic astrocytomas. J Neuropathol Exp Neurol 64(10):891–901

Hussain SF, Heimberger AB (2005) Immunotherapy for human glioma: innovative approaches and recent results. Expert Rev Anticancer Ther 5(5):777–790

Hussain SF, Yang D, Suki D, Aldape K, Grimm E, Heimberger AB (2006a) The role of human glioma-infiltrating microglia/macrophages in mediating antitumor immune responses. Neuro Oncol 8(3):261–279

Hussain SF, Yang D, Suki D, Grimm E, Heimberger AB (2006b) Innate immune functions of microglia isolated from human glioma patients. J Transl Med 4:15

Hussain SF, Kong LY, Jordan J, Conrad C, Madden T, Fokt I, Priebe W, Heimberger AB (2007) A novel small molecule inhibitor of signal transducers and activators of transcription 3 reverses immune tolerance in malignant glioma patients. Cancer Res 67(20):9630–9636

Kettenmann H, Kirchhoff F, Verkhratsky A (2013) Microglia: new roles for the synaptic stripper. Neuron 77(1):10–18

Kierdorf K, Prinz M (2013) Factors regulating microglia activation. Front Cell Neurosci 7:44

Kim HS, Suh YH (2009) Minocycline and neurodegenerative diseases. Behav Brain Res 196(2): 168–179

Kleber S, Sancho-Martinez I, Wiestler B, Beisel A, Gieffers C, Hill O, Thiemann M, Mueller W, Sykora J, Kuhn A, Schreglmann N, Letellier E, Zuliani C, Klussmann S, Teodorczyk M, Gröne HJ, Ganten TM, Sültmann H, Tüttenberg J, von Deimling A, Regnier-Vigouroux A, Herold-Mende C, Martin-Villalba A (2008) Yes and PI3K bind CD95 to signal invasion of glioblastoma. Cancer Cell 13(3):235–248

Klein R, Roggendorf W (2001) Increased microglia proliferation separates pilocytic astrocytomas from diffuse astrocytomas: a double labeling study. Acta Neuropathol 101(3):245–248

Komohara Y, Ohnishi K, Kuratsu J, Takeya M (2008) Possible involvement of the M2 anti-inflammatory macrophage phenotype in growth of human gliomas. J Pathol 216(1):15–24

Ku MC, Wolf SA, Respondek D, Matyash V, Pohlmann A, Waiczies S, Waiczies H, Niendorf T, Synowitz M, Glass R, Kettenmann H (2013) GDNF mediates glioblastoma-induced microglia attraction but not astrogliosis. Acta Neuropathol 125(4):609–620

Kulprathipanja NV, Kruse CA (2004) Microglia phagocytose alloreactive CTL-damaged 9L gliosarcoma cells. J Neuroimmunol 153(1–2):76–82

Lambertsen KL, Deierborg T, Gregersen R, Clausen BH, Wirenfeldt M, Nielsen HH, Dalmau I, Diemer NH, Dagnaes-Hansen F, Johansen FF, Keating A, Finsen B (2011) Differences in origin of reactive microglia in bone marrow chimeric mouse and rat after transient global ischemia. J Neuropathol Exp Neurol 70(6):481–494

Li MO, Flavell RA (2008) Contextual regulation of inflammation: a duet by transforming growth factor-beta and interleukin-10. Immunity 28(4):468–476

Li W, Graeber MB (2012) The molecular profile of microglia under the influence of glioma. Neuro Oncol 14(8):958–978

Lin EY, Nguyen AV, Russell RG, Pollard JW (2001) Colony-stimulating factor 1 promotes progression of mammary tumors to malignancy. J Exp Med 193(6):727–740

Lin EY, Gouon-Evans V, Nguyen AV, Pollard JW (2002) The macrophage growth factor CSF-1 in mammary gland development and tumor progression. J Mammary Gland Biol Neoplasia 7(2): 147–162

Lin EY, Li JF, Gnatovskiy L, Deng Y, Zhu L, Grzesik DA, Qian H, Xue XN, Pollard JW (2006) Macrophages regulate the angiogenic switch in a mouse model of breast cancer. Cancer Res 66(23):11238–11246

Liu C, Luo D, Streit WJ, Harrison JK (2008) CX3CL1 and CX3CR1 in the GL261 murine model of glioma: CX3CR1 deficiency does not impact tumor growth or infiltration of microglia and lymphocytes. J Neuroimmunol 198(1–2):98–105

London A, Cohen M, Schwartz M (2013) Microglia and monocyte-derived macrophages: functionally distinct populations that act in concert in CNS plasticity and repair. Front Cell Neurosci 7:34

Maes W, Van Gool SW (2011) Experimental immunotherapy for malignant glioma: lessons from two decades of research in the GL261 model. Cancer Immunol Immunother 60(2):153–160

Mantovani A, Sica A (2010) Macrophages, innate immunity and cancer: balance, tolerance, and diversity. Curr Opin Immunol 22(2):231–237

Markovic DS, Glass R, Synowitz M, Rooijen N, Kettenmann H (2005) Microglia stimulate the invasiveness of glioma cells by increasing the activity of metalloprotease-2. J Neuropathol Exp Neurol 64(9):754–762

Markovic DS, Vinnakota K, Chirasani S, Synowitz M, Raguet H, Stock K, Sliwa M, Lehmann S, Kälin R, van Rooijen N, Holmbeck K, Heppner FL, Kiwit J, Matyash V, Lehnardt S, Kaminska B, Glass R, Kettenmann H (2009) Gliomas induce and exploit microglial MT1-MMP expression for tumor expansion. Proc Natl Acad Sci U S A 106(30):12530–12535

Markovic DS, Vinnakota K, van Rooijen N, Kiwit J, Synowitz M, Glass R, Kettenmann H (2011) Minocycline reduces glioma expansion and invasion by attenuating microglial MT1-MMP expression. Brain Behav Immun 25(4):624–628

Martinez FO, Sica A, Mantovani A, Locati M (2008) Macrophage activation and polarization. Front Biosci 13:453–461

Morantz RA, Wood GW, Foster M, Clark M, Gollahon K (1979) Macrophages in experimental and human brain tumors. Part 2: studies of the macrophage content of human brain tumors. J Neurosurg 50(3):305–311

Morford LA, Elliott LH, Carlson SL, Brooks WH, Roszman TL (1997) T cell receptor-mediated signaling is defective in T cells obtained from patients with primary intracranial tumors. J Immunol 159(9):4415–4425

Morimura T, Neuchrist C, Kitz K, Budka H, Scheiner O, Kraft D, Lassmann H (1990) Monocyte subpopulations in human gliomas: expression of Fc and complement receptors and correlation with tumor proliferation. Acta Neuropathol 80(3):287–294

Morioka T, Baba T, Black KL, Streit WJ (1992a) Immunophenotypic analysis of infiltrating leukocytes and microglia in an experimental rat glioma. Acta Neuropathol 83(6):590–597

Morioka T, Baba T, Black KL, Streit WJ (1992b) Inflammatory cell infiltrates vary in experimental primary and metastatic brain tumors. Neurosurgery 30(6):891–896

Morris CS, Esiri MM (1991) Immunocytochemical study of macrophages and microglial cells and extracellular matrix components in human CNS disease. 1. Gliomas. J Neurol Sci 101(1): 47–58

Mroczko B, Groblewska M, Wereszczyńska-Siemiatkowska U, Okulczyk B, Kedra B, Łaszewicz W, Dabrowski A, Szmitkowski M (2007) Serum macrophage-colony stimulating factor levels in colorectal cancer patients correlate with lymph node metastasis and poor prognosis. Clin Chim Acta 380(1–2):208–212

Neumann H, Kotter MR, Franklin RJ (2009) Debris clearance by microglia: an essential link between degeneration and regeneration. Brain 132(Pt 2):288–295

Nikodemova M, Watters JJ, Jackson SJ, Yang SK, Duncan ID (2007) Minocycline down-regulates MHC II expression in microglia and macrophages through inhibition of IRF-1 and protein kinase C (PKC)alpha/betaII. J Biol Chem 282(20):15208–15216

Nimmerjahn A, Kirchhoff F, Helmchen F (2005) Resting microglial cells are highly dynamic surveillants of brain parenchyma in vivo. Science 308(5726):1314–1318

Nishie A, Ono M, Shono T, Fukushi J, Otsubo M, Onoue H, Ito Y, Inamura T, Ikezaki K, Fukui M, Iwaki T, Kuwano M (1999) Macrophage infiltration and heme oxygenase-1 expression correlate with angiogenesis in human gliomas. Clin Cancer Res 5(5):1107–1113

Nowicki A, Szenajch J, Ostrowska G, Wojtowicz A, Wojtowicz K, Kruszewski AA, Maruszynski M, Aukerman SL, Wiktor-Jedrzejczak W (1996) Impaired tumor growth in colony-stimulating factor 1 (CSF-1)-deficient, macrophage-deficient op/op mouse: evidence for a role of CSF-1-dependent macrophages in formation of tumor stroma. Int J Cancer 65(1): 112–119

Okada M, Saio M, Kito Y, Ohe N, Yano H, Yoshimura S, Iwama T, Takami T (2009) Tumor-associated macrophage/microglia infiltration in human gliomas is correlated with MCP-3, but not MCP-1. Int J Oncol 34(6):1621–1627

Olson JK, Miller SD (2004) Microglia initiate central nervous system innate and adaptive immune responses through multiple TLRs. J Immunol 173(6):3916–3924

Pollard JW (2009) Trophic macrophages in development and disease. Nat Rev Immunol 9(4): 259–270

Prins RM, Liau LM (2004) Cellular immunity and immunotherapy of brain tumors. Front Biosci 9: 3124–3136

Prins RM, Scott GP, Merchant RE, Graf MR (2002) Irradiated tumor cell vaccine for treatment of an established glioma. II. Expansion of myeloid suppressor cells that promote tumor progression. Cancer Immunol Immunother 51(4):190–199

Qian B, Deng Y, Im JH, Muschel RJ, Zou Y, Li J, Lang RA, Pollard JW (2009) A distinct macrophage population mediates metastatic breast cancer cell extravasation, establishment and growth. PLoS One 4(8):e6562

Ravichandran KS (2010) Find-me and eat-me signals in apoptotic cell clearance: progress and conundrums. J Exp Med 207(9):1807–1817

Raychaudhuri B, Rayman P, Ireland J, Ko J, Rini B, Borden EC, Garcia J, Vogelbaum MA, Finke J (2011) Myeloid-derived suppressor cell accumulation and function in patients with newly diagnosed glioblastoma. Neuro Oncol 13(6):591–599

Revoltella RP, Menicagli M, Campani D (2012) Granulocyte-macrophage colony-stimulating factor as an autocrine survival-growth factor in human gliomas. Cytokine 57(3):347–359

Rodrigues JC, Gonzalez GC, Zhang L, Ibrahim G, Kelly JJ, Gustafson MP, Lin Y, Dietz AB, Forsyth PA, Yong VW, Parney IF (2010) Normal human monocytes exposed to glioma cells acquire myeloid-derived suppressor cell-like properties. Neuro Oncol 12(4):351–365

Roggendorf W, Strupp S, Paulus W (1996) Distribution and characterization of microglia/macrophages in human brain tumors. Acta Neuropathol 92(3):288–293

Rolle CE, Sengupta S, Lesniak MS (2012) Mechanisms of immune evasion by gliomas. Adv Exp Med Biol 746:53–76

Sapi E, Kacinski BM (1999) The role of CSF-1 in normal and neoplastic breast physiology. Proc Soc Exp Biol Med 220(1):1–8

Sasaki A, Yamaguchi H, Horikoshi Y, Tanaka G, Nakazato Y (2004) Expression of glucose transporter 5 by microglia in human gliomas. Neuropathol Appl Neurobiol 30(5):447–455

Sasaki A, Yokoo H, Tanaka Y, Homma T, Nakazato Y, Ohgaki H (2013) Characterization of microglia/macrophages in gliomas developed in S-100β-v-erbB transgenic rats. Neuropathology 33(5):505–514

Schulz C, Gomez Perdiguero E, Chorro L, Szabo-Rogers H, Cagnard N, Kierdorf K, Prinz M, Wu B, Jacobsen SE, Pollard JW, Frampton J, Liu KJ, Geissmann F (2012) A lineage of myeloid cells independent of Myb and hematopoietic stem cells. Science 336(6077):86–90

Schwartz M (2010) "Tissue-repairing" blood-derived macrophages are essential for healing of the injured spinal cord: from skin-activated macrophages to infiltrating blood-derived cells? Brain Behav Immun 24(7):1054–1057

Schwartzbaum JA, Huang K, Lawler S, Ding B, Yu J, Chiocca EA (2010) Allergy and inflammatory transcriptome is predominantly negatively correlated with CD133 expression in glioblastoma. Neuro Oncol 12(4):320–327

Sedgwick JD, Schwender S, Imrich H, Dörries R, Butcher GW, ter Meulen V (1991) Isolation and direct characterization of resident microglial cells from the normal and inflamed central nervous system. Proc Natl Acad Sci U S A 88(16):7438–7442

Sedgwick JD, Schwender S, Gregersen R, Dörries R, ter Meulen V (1993) Resident macrophages (ramified microglia) of the adult brown Norway rat central nervous system are constitutively major histocompatibility complex class II positive. J Exp Med 177(4):1145–1152

Sielska M, Przanowski P, Wylot B, Gabrusiewicz K, Maleszewska M, Kijewska M, Zawadzka M, Kucharska J, Vinnakota K, Kettenmann H, Kotulska K, Grajkowska W, Kaminska B (2013) Distinct roles of CSF family cytokines in macrophage infiltration and activation in glioma progression and injury response. J Pathol 230(3):310–321

Sierra A, Abiega O, Shahraz A, Neumann H (2013) Janus-faced microglia: beneficial and detrimental consequences of microglial phagocytosis. Front Cell Neurosci 7:6

Simmons GW, Pong WW, Emnett RJ, White CR, Gianino SM, Rodriguez FJ, Gutmann DH (2011) Neurofibromatosis-1 heterozygosity increases microglia in a spatially and temporally restricted pattern relevant to mouse optic glioma formation and growth. J Neuropathol Exp Neurol 70(1):51–62

Sliwa M, Markovic D, Gabrusiewicz K, Synowitz M, Glass R, Zawadzka M, Wesolowska A, Kettenmann H, Kaminska B (2007) The invasion promoting effect of microglia on glioblastoma cells is inhibited by cyclosporin A. Brain 130(Pt 2):476–489

Solinas G, Germano G, Mantovani A, Allavena P (2009) Tumor-associated macrophages (TAM) as major players of the cancer-related inflammation. J Leukoc Biol 86(5):1065–1073

Stefano L, Racchetti G, Bianco F, Passini N, Gupta RS, Panina Bordignon P, Meldolesi J (2009) The surface-exposed chaperone, Hsp60, is an agonist of the microglial TREM2 receptor. J Neurochem 110(1):284–294

Szatmári T, Lumniczky K, Désaknai S, Trajcevski S, Hídvégi EJ, Hamada H, Sáfrány G (2006) Detailed characterization of the mouse glioma 261 tumor model for experimental glioblastoma therapy. Cancer Sci 97(6):546–553

Takahashi K, Prinz M, Stagi M, Chechneva O, Neumann H (2007) TREM2-transduced myeloid precursors mediate nervous tissue debris clearance and facilitate recovery in an animal model of multiple sclerosis. PLoS Med 4(4):e124

Tran Thang NN, Derouazi M, Philippin G, Arcidiaco S, Di Berardino-Besson W, Masson F, Hoepner S, Riccadonna C, Burkhardt K, Guha A, Dietrich PY, Walker PR (2010) Immune infiltration of spontaneous mouse astrocytomas is dominated by immunosuppressive cells from early stages of tumor development. Cancer Res 70(12):4829–4839

Umemura N, Saio M, Suwa T, Kitoh Y, Bai J, Nonaka K, Ouyang GF, Okada M, Balazs M, Adany R, Shibata T, Takami T (2008) Tumor-infiltrating myeloid-derived suppressor cells are pleiotropic-inflamed monocytes/macrophages that bear M1- and M2-type characteristics. J Leukoc Biol 83(5):1136–1144

Voisin P, Bouchaud V, Merle M, Diolez P, Duffy L, Flint K, Franconi JM, Bouzier-Sore AK (2010) Microglia in close vicinity of glioma cells: correlation between phenotype and metabolic alterations. Front Neuroenergetics 2:131

Wang SC, Hong JH, Hsueh C, Chiang CS (2012) Tumor-secreted SDF-1 promotes glioma invasiveness and TAM tropism toward hypoxia in a murine astrocytoma model. Lab Invest 92(1):151–162

Watters JJ, Schartner JM, Badie B (2005) Microglia function in brain tumors. J Neurosci Res 81(3):447–455

Wei J, Barr J, Kong LY, Wang Y, Wu A, Sharma AK, Gumin J, Henry V, Colman H, Priebe W, Sawaya R, Lang FF, Heimberger AB (2010) Glioblastoma cancer-initiating cells inhibit T-cell proliferation and effector responses by the signal transducers and activators of transcription 3 pathway. Mol Cancer Ther 9(1):67–78

Wesolowska A, Kwiatkowska A, Slomnicki L, Dembinski M, Master A, Sliwa M, Franciszkiewicz K, Chouaib S, Kaminska B (2008) Microglia-derived TGF-beta as an important regulator of glioblastoma invasion–an inhibition of TGF-beta-dependent effects by shRNA against human TGF-beta type II receptor. Oncogene 27(7):918–930

Wierzba-Bobrowicz T, Kuchna I, Matyja E (1994) Reaction of microglial cells in human astrocytomas (preliminary report). Folia Neuropathol 32(4):251–252

Wiktor-Jedrzejczak W, Bartocci A, Ferrante AW, Ahmed-Ansari A, Sell KW, Pollard JW, Stanley ER (1990) Total absence of colony-stimulating factor 1 in the macrophage-deficient osteopetrotic (op/op) mouse. Proc Natl Acad Sci U S A 87(12):4828–4832

Wiktor-Jedrzejczak W, Ratajczak MZ, Ptasznik A, Sell KW, Ahmed-Ansari A, Ostertag W (1992) CSF-1 deficiency in the op/op mouse has differential effects on macrophage populations and differentiation stages. Exp Hematol 20(8):1004–1010

Wisniewski P, Ellert-Miklaszewska A, Kwiatkowska A, Kaminska B (2010) Non-apoptotic Fas signaling regulates invasiveness of glioma cells and modulates MMP-2 activity via NFkappaB-TIMP-2 pathway. Cell Signal 22(2):212–220

Wu A, Wei J, Kong LY, Wang Y, Priebe W, Qiao W, Sawaya R, Heimberger AB (2010) Glioma cancer stem cells induce immunosuppressive macrophages/microglia. Neuro Oncol 12(11): 1113–1125

Yang I, Han SJ, Kaur G, Crane C, Parsa AT (2010) The role of microglia in central nervous system immunity and glioma immunology. J Clin Neurosci 17(1):6–10

Yi L, Xiao H, Xu M, Ye X, Hu J, Li F, Li M, Luo C, Yu S, Bian X, Feng H (2011) Glioma-initiating cells: a predominant role in microglia/macrophages tropism to glioma. J Neuroimmunol 232(1–2):75–82

Yoshida H, Hayashi S, Kunisada T, Ogawa M, Nishikawa S, Okamura H, Sudo T, Shultz LD (1990) The murine mutation osteopetrosis is in the coding region of the macrophage colony stimulating factor gene. Nature 345(6274):442–444

Zawadzka M, Dabrowski M, Gozdz A, Szadujkis B, Sliwa M, Lipko M, Kaminska B (2012) Early steps of microglial activation are directly affected by neuroprotectant FK506 in both in vitro inflammation and in rat model of stroke. J Mol Med (Berl) 90(12):1459–1471

Zhai H, Heppner FL, Tsirka SE (2011) Microglia/macrophages promote glioma progression. Glia 59(3):472–485

Zhang L, Alizadeh D, Van Handel M, Kortylewski M, Yu H, Badie B (2009) Stat3 inhibition activates tumor macrophages and abrogates glioma growth in mice. Glia 57(13):1458–1467

Zhu XD, Zhang JB, Zhuang PY, Zhu HG, Zhang W, Xiong YQ, Wu WZ, Wang L, Tang ZY, Sun HC (2008) High expression of macrophage colony-stimulating factor in peritumoral liver tissue is associated with poor survival after curative resection of hepatocellular carcinoma. J Clin Oncol 26(16):2707–2716

Glioma and Extracellular Matrix

10

Christian Mawrin and Constanze Seidenbecher

Abstract

Invasion of glioma cells into brain tissue involves a multitude of molecular interactions of invading cells with host cells and with the surrounding extracellular matrix (ECM). The brain ECM is a complex structure composed of proteoglycans and glycoproteins of glial and neuronal origin, which is largely based on hyaluronic acid (HA) as central organizing glycosaminoglycan. It is particularly rich in different carbohydrate epitopes and undergoes a developmental shift around synaptogenesis and terminal maturation. Glioma cells do not only bind to ECM components via specific receptors which are highly expressed on their surfaces but they actively shape the ECM by secreting matrix constituents and ECM-modulating enzymes like proteases, glycosidases, or sulfatases, thus creating a permissive environment. Proteomic approaches have shown that the individual secretomes of different glioma cells can vary largely in their composition. This mutual glioma–host–ECM interaction, which also involves the microvasculature of the brain, contributes to the devastating nature of glioma tumors, because it enables the malignant cells to pave their way into the parenchyma. The complex nature of normal as well as glioma-derived ECM makes it challenging to identify entry points for a successful therapeutic treatment. First attempts targeting ECM receptors on glioma cells or technological approaches to change the biomechanical properties of brain ECM are promising but need more time.

C. Mawrin
Institute for Neuropathology, Otto von Guericke University, Magdeburg, Germany

Center for Behavioral Brain Sciences (CBBS), Magdeburg, Germany

C. Seidenbecher (✉)
AG Perisynaptic Extracellular Matrix, Leibniz Institute for Neurobiology, Brenneckestr. 6, 39118 Magdeburg, Germany

Center for Behavioral Brain Sciences (CBBS), Magdeburg, Germany
e-mail: seidenc@lin-magdeburg.de

A. Sedo and R. Mentlein (eds.), *Glioma Cell Biology*,
DOI 10.1007/978-3-7091-1431-5_10, © Springer-Verlag Wien 2014

Keywords
Glioma • Infiltration • Hyaluronic acid • Proteoglycan • Glycoprotein • SPARC (secreted protein acidic and rich in cysteine) • Integrin • Tenascin • Lectican • Brevican

Abbreviations

ADAM	A disintegrin and metalloproteinase
ADAMTS	A disintegrin and metalloproteinase with thrombospondin motifs
AKT	v-akt murine thymoma viral oncogene homolog
AMPA	2-Amino-3-(3-hydroxy-5-methyl-isoxazol-4-yl)propanoic acid
CS	Chondroitin sulfate
CSF	Cerebrospinal fluid
CTGF	Connective tissue growth factor
ECM	Extracellular matrix
EGFR	Epidermal growth factor receptor
EPHA2	Ephrin receptor A2
ERK	Extracellular signal-regulated kinase
FAK	Focal adhesion kinase
FGF2	Fibroblast growth factor
FGFR1c	FGF receptor 1c
GAG	Glycosaminoglycan
GPI	Glycosylphosphatidylinositol
HA	Hyaluronic acid
HAPLN	HA and proteoglycan link protein
HARE	Hyaluronan receptor for endocytosis
HAS	HA synthase
HS	Heparan sulfate
HSPG	HS proteoglycan
HYAL	Hyaluronidase
IGF1R	Insulin-like growth factor 1 receptor
LGI1	Leucine-rich glioma inactivated-1
LYVE-1	Lymphatic vessel endothelial hyaluronan receptor 1
MHC	Major histocompatibility complex
MMP	Matrix metalloproteinase
PDGFR	Platelet-derived growth factor receptor
PNN	Perineuronal net
RHAMM	Receptor of HA-mediated motility
RTK	Receptor tyrosine kinase
SMOC1	SPARC-related modular calcium-binding protein 1
SPAM-1	Sperm adhesion molecule 1
SPARC	Secreted protein acidic and rich in cysteine

SULF	Sulfatase
SVZ	Subventricular zone
TGF-β	Transforming growth factor-β
TIMP	Tissue inhibitors of matrix metalloproteinases
VEGF	Vascular endothelial cell growth factor
WHO	World Health Organization
YKL-40	Chitinase-3-like protein 1

10.1 The Extracellular Matrix of Brain Tissue

Invasion of glioma into brain tissue involves manifold interactions with host cells but also with the neural extracellular matrix (ECM) surrounding them. The brain contains a unique form of ECM which accounts for approx. 20 % of the volume and which is molecularly discrete from other tissues. It compartmentalizes the extracellular space into functional microdomains important for many forms of neuronal plasticity [for review see Dityatev et al. (2010)]. In contrast to other tissues it is almost devoid of collagens but instead contains a network of proteoglycans of the chondroitin sulfate and heparan sulfate subtypes as well as glycoproteins. The central backbone of brain ECM is hyaluronic acid (HA), a huge carbohydrate polymer, which is bound by link proteins and proteoglycans, thus forming a three-dimensional extracellular meshwork filling the extracellular space. HA is an unbranched polymer that is composed of repeating disaccharides of glucuronic acid and *N*-acetylglucosamine: [-β(1,4)-GlcUA-β(1,3)-GlcNAc-]. It consists typically of 2,000–25,000 disaccharides of a molecular weight of 10^3–10^4 kDa and a polymer length of 2–25 μm [summarized in Frischknecht and Seidenbecher (2008)].

An important family of brain proteoglycans is the so-called lectican family, whose members aggrecan, versican, neurocan, and brevican are characterized by a lectin-like domain in their C-terminus. Lecticans bind to HA via their N-terminus and to tenascins or plasma membrane glycoproteins or glycolipids via their C-terminus and thereby link cell surfaces to the ECM (Fig. 10.1). A specialized formation of brain ECM is found in close contact with the surface of distinct neurons, particularly parvalbumin-positive interneurons. This specialization is called perineuronal net (PNN) and contributes to the establishment of the proper extracellular ion and pH milieu to optimize neuronal transmission and control plasticity [for review see Wang and Fawcett (2012)].

The ECM meshwork undergoes substantial reorganization during brain maturation. While in juvenile CNS a rather flexible ECM based on tenascin-C and neurocan is permissive for developmental structural plasticity in the mature brain the ECM meshwork is more close-mashed and less permissive and flexible. This reorganization coincides with the closure of sensitive periods and restricts structural plasticity [summarized in Gundelfinger et al. (2010)].

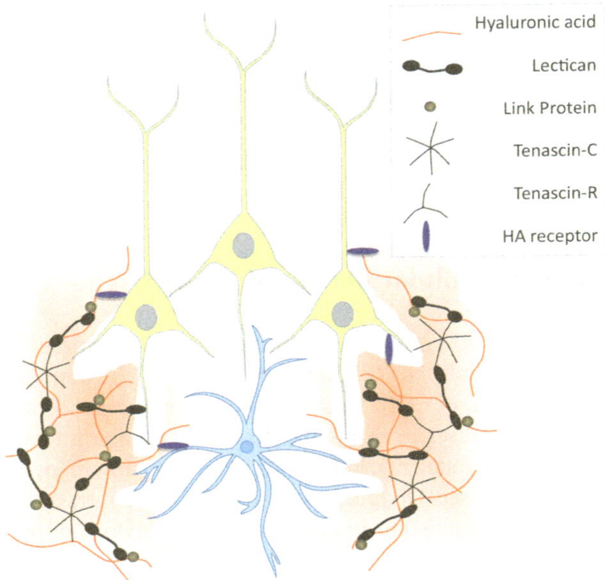

Fig. 10.1 Schematic representation of the brain ECM. Important components of the hyaluronic acid-based ECM, which are secreted by neurons and astrocytes, are shown

Primary brain tumors show unique patterns of invasion and infiltration (Fig. 10.2) which suggest that they are molecularly adapted to the extracellular microenvironment produced by the brain ECM [for a comprehensive review see Bellail et al. (2004)].

10.2 Glioma-Specific ECM and Modification of Brain ECM via Tumor-Derived Mechanisms

10.2.1 Glioma and Hyaluronic Acid

Gliomas are found in HA-rich environments in the brain. Figure 10.3 summarizes most significant changes in the HA-based matrix under neoplastic conditions.

A variety of cell lines like U-87 MG (glioblastoma-astrocytoma), U-373 MG (glioblastoma), and Hs 683 (glioma) but not U-138 MG (glioblastoma) had been shown to bind to HA (Asher and Bignami 1992). HA is interconnected with proteoglycans and link proteins to form the three-dimensional structure, and in contrast to most other ECM components the *link proteins* HAPLN1, HAPLN2, and HAPLN4 are clearly reduced in the tumor site, possibly reflecting a reduction of neurons secreting link proteins. Functionally this lack of link proteins may lead to a

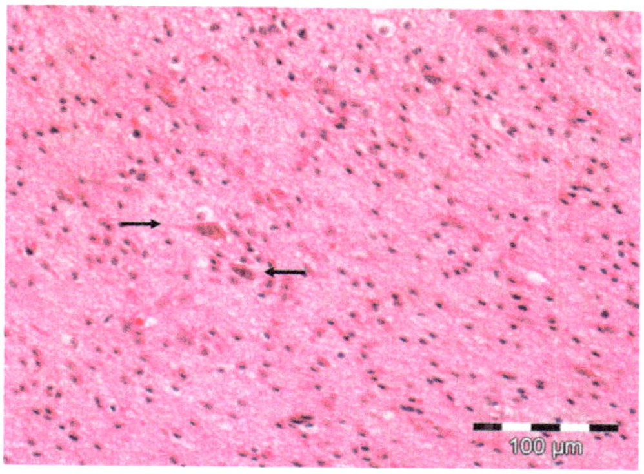

Fig. 10.2 Human brain tissue with diffuse infiltration of astrocytic tumor cells in a patient diagnosed with glioblastoma. Invasion occurs in HA-rich environments. Note the entrapped neurons (*arrows*) surrounded by astrocytic tumor cells

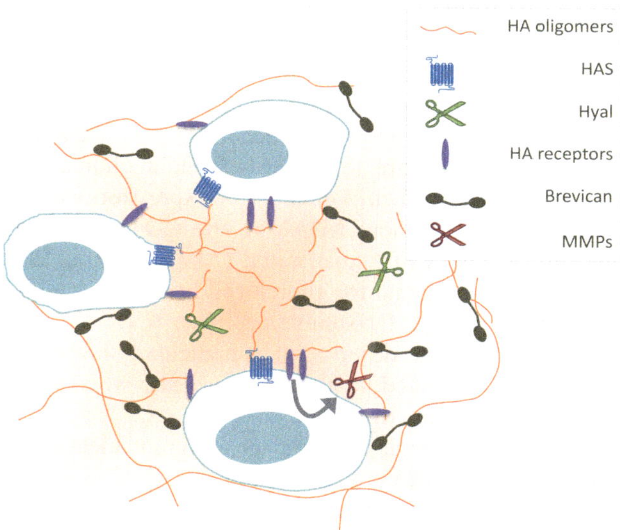

Fig. 10.3 Glioma cells sculpture their extracellular environment. Schematic drawing illustrating changes in the HA-based ECM in glioblastoma. Increased expression of hyaluronidases (Hyal), HA synthases (HAS), HA receptors like CD44 or RHAMM, and lectican-type proteoglycans like brevican results in a highly permissive matrix rich in short HA oligomers, which is more flexible due to a lack in link proteins. HA-induced secretion of matrix metalloproteinases (MMPs) from glioma cells results in further disintegration of the host tissue (for details see text)

disintegration of the matrix, because proteoglycans and HA, which are overexpressed in most gliomas, are less associated and free to transiently bind other partners in the tumor-remodeled ECM (Sim et al. 2009). But glioma cells like U-373 and G-26 also synthesize HA and secrete the majority (80–90 %) into the extracellular space (Wiranowska et al. 2010), thus creating their own permissive environment.

10.2.1.1 Hyaluronic Acid Synthases

There are three different HA synthases (HAS) encoded in the human genome: HAS1, 2, and 3. Of these, the glioma cell line U-118 MG expresses only the HAS2 and HAS3 genes (Jacobson et al. 2000). Overexpression of HAS2 in glioblastoma cell lines leads to augmented HA production and promotes tumor cell growth in vivo, but only if the cells express hyaluronidases as well (Enegd et al. 2002).

HA production was observed to be mechanistically linked to the function of the HA receptor CD44, as a function-blocking anti-CD44 antibody reduced HA secretion. Along this line, application of HA oligomers, which are known competitors of CD44 binding to large HA polymers, augmented HA release (Wiranowska et al. 2010). Since small HA oligomers are preferentially found in neoplastic tissue this can be interpreted as an ECM-based molecular mechanism of tumor growth.

10.2.1.2 Hyaluronic Acid Receptors

Glioma cells indeed express large amounts of HA receptors to establish contacts to HA and to invade into brain tissue. HA receptors *CD44* and *RHAMM* (Hyaluronan-mediated motility receptor) are ubiquitously found in glioma cell lines as well as in glioma tumor samples. In fact, a gradient of expression was reported with high-grade gliomas expressing more RHAMM and CD44 than low-grade lesions or non-neoplastic specimens of human brain (Akiyama et al. 2001). This receptor interaction seems to be crucial for glioma cell line proliferation since this process can be disturbed with RHAMM soluble peptides.

RHAMM expression was observed to be particularly high at the invasive edge of human gliomas as compared to the core regions, whereas CD44 seems more pronounced inside the core (Kim et al. 2011).

Another HA receptor with high impact on tumor metastasis is *LYVE-1* (lymphatic vessel endothelial hyaluronan receptor 1), which is expressed on lymphatic vessels, e.g., in the craniofacial region (Furukawa et al. 2008). It has a CD44-like domain and serves as endocytic receptor for HA mediating uptake and clearance of extracellular HA.

Macrophages infiltrating the brain under conditions of tumor growth express the scavenger receptor *stabilin-2* (also termed HARE: hyaluronan receptor for endocytosis). The stabilins are fasciclin-like HA receptors, which do not only bind HA but also other extracellular ligands like heparins, low-density lipoprotein, or advanced glycation end products. Clearance of these ligands occurs via clathrin-mediated endocytosis. HA binding to HARE stimulates ERK1/2 activation (Kyosseva et al. 2008) and NF-κB-activated gene expression. HARE obviously senses a rather narrow size range of HA degradation products ranging from approx. 40–400 kDa (Pandey et al. 2013; Pandey and Weigel 2014). Pandey et al. suggest a model in which optimal length HA fragments bind multiple HARE proteins, thus enabling interactions of their cytoplasmic domains that stimulate intracellular signaling. Accordingly, HARE is supposed to constitute a novel signaling system, which could monitor the homeostasis of tissue biomatrix turnover throughout the body—a process which is severely affected under neoplastic conditions.

Finally, extracellular HA can also be bound by the *GPI-anchored isoform of brevican*, which is found to be upregulated in human glioma approx. sevenfold, as compared to normal tissue (Gary et al. 2000). GPI-brevican is a smaller isoform lacking the C-terminal globular domains, which results from alternative splicing of the brevican gene. The GPI-anchored isoform acts as a receptor for HA on astroglial and glioma cells.

10.2.1.3 Hyaluronic Acid-Cleaving Enzymes

Glioma cells developed a molecular toolkit to invade the brain. This toolkit includes HA-degrading enzymes termed *hyaluronidases* (originally known as "spreading factor").

As Liu et al. showed, angiogenesis is preferentially induced by hyaluronidase-positive tumor cells and blocked by hyaluronidase inhibitors. Tumor cells thus use hyaluronidase to facilitate invasion into HA-rich host tissue. Notably, HA fragments can further promote tumor establishment by inducing angiogenesis (Liu et al. 1996).

However, hyaluronidase activity is known to be higher in brain metastases of carcinomas than in glioblastoma. In general, hyaluronidase activity is associated with more aggressive cancer cells and is directly or indirectly involved in the expression of brain metastasis phenotypes (Delpech et al. 2002).

Several malignant glioma cell lines (U87MG, U251MG, U343MG-A, and U373MG) were tested for hyaluronidase expression by Jin et al. (2009) who reported that those glioma cells expressing this enzymatic activity were more invasive than deficient ones.

A study which analyzed the expression patterns of three hyaluronidases HYAL1, 2, and 3 in glioma cell lines came to the conclusion that HYAL1 is the most prominent hyaluronidase in glioma (Junker et al. 2003). The mRNA encoding this enzyme contains a retained intron within the 5′ untranslated region. The authors made the interesting observation that the hyaluronidase activity and thus the invasive and angiogenic potential seem to be correlated with the effectiveness of splicing out this intron. The genes encoding hyaluronidases HYAL4 and SPAM-1 are both located on human chromosome 7d and were reported to be amplified in Chinese patients with low-grade, but mostly in high-grade glioma due to copy number variations (Li et al. 2013).

In contrast to the above-described tumor growth-promoting role of hyaluronidases an opposite concept was developed, which ascribes a tumor suppressor potential to hyaluronidases [summarized in Stern (2005)]. This concept was based on genetic investigations of the human gene locus 3p21.3, which harbors the HYAL1, 2, and 3 genes and which was found to be deleted in some carcinomas including glia-derived tumors (Nauman and Czernicki 2002). Although the tumor suppressor activity is not associated with the HYAL gene loci but with neighboring loci this concept was further pursued, perhaps based on the assumption that an enzyme degrades HA, which is known to promote tumor metastasis, should therefore be rather suppressive. This controversy can be resolved by taking into account the concentration of hyaluronidases. While endogenous amounts are tumor-promoting exogenous administration of high concentrations exceeding the physiologically occurring by more than 100 times leads to apoptosis or can drive a tumor into necrosis (Jacobson et al. 2002; Lokeshwar et al. 2005). Therefore, hyaluronidase infusion may be considered as an anti-tumor drug if applied at very high concentrations. However in the brain this treatment may have severe side effects since hyaluronidase is able to strongly affect synaptic plasticity and function, e.g., via increasing the lateral mobility of AMPA-type glutamate receptors and their augmented exchange in synapses, which as a consequence leads to altered short-term plasticity (Frischknecht et al. 2009). Furthermore, hyaluronidase was shown to reduce dendritic calcium currents and thus use-dependent synaptic plasticity (Kochlamazashvili et al. 2010).

Taken together, HA is an important factor for growth and invasion of glioma cells. However, there seems to be an optimal HA concentration range which is established by a fine-tuned interplay between HA synthase and hyaluronidase activities and HA receptor expression. At very high HA concentrations, e.g., under conditions of lacking hyaluronidase activity or experimentally overexpressed HAS, the interaction of glioma cells with the ECM may lose its transient character which is necessary for migration and spreading. Therefore, growth arrest or apoptosis are possible consequences, thus opening a window for therapeutic intervention.

10.2.2 Glioma and Matrix-Degrading Proteases

HA-induced invasion of glioma cells into healthy brain tissue is mediated by two molecular events occurring in parallel: the secretion of matrix metalloproteinases (MMP) and the augmentation of cell migration. Release of MMPs from glioma cells can be triggered by HA-induced focal adhesion kinase (FAK) activation, which in turn elicits ERK activation and intracellular NF-kappaB translocation (Park et al. 2008).

MMP-2 and -9 play a pivotal role in brain tumors. Whereas MMP2 is found at the invasive edge of human gliomas more than in the core, MMP9 appears more concentrated and active in the core (Kim et al. 2011). Also, *MMP-19* is massively upregulated in glioblastoma via proinflammatory cytokines (Lettau et al. 2010).

Another class of extracellularly active proteases are ADAMTS family members like *ADAMTS4* and *ADAMTS5* (aggrecanases 1 and 2; ADAMTS = a disintegrin and metalloproteinase with thrombospondin motifs), which are also upregulated in human glioblastoma by cytokines like TGFβ or interleukin-1β (Held-Feindt et al. 2006). Among their substrates is brevican, an important glioma-related proteoglycan (see below).

MMPs cleave a multitude of matrix components like proteoglycans or glycoproteins but also cell adhesion molecules or receptors. In normal brain their activity is tightly controlled in a coordinated and complex manner, e.g., by activating proteases and by endogeneous tissue inhibitors of matrix metalloproteinases (*TIMPs*); however, under neoplastic conditions this balance is disturbed leading to overshooting proteolytic activity as prerequisite for malignant invasion. Interestingly, the balance is under epigenetic control, as shown for the MMP-2/TIMP-2 axis in highly migratory glioblastoma cells (Chernov et al. 2009).

For a detailed description of glioma-associated proteolytic enzymes, see Chap. 12 in this book.

10.2.3 Integrins as ECM Receptors

Most important receptors for ECM protein components are integrins. They constitute a large family of dimeric transmembrane receptors, which are composed of α and β subunits and initiate a whole plethora of intracellular signaling cascades via coupling to adapter proteins. Regulated molecular connection to the actin-based cytoskeleton ensures ECM-guided mechanocoupling. Ligands are structural proteins like collagens, tenascins, fibronectin, or laminin, but also extracellular signaling molecules like thrombospondins and vitronectin. The ligand specificity and binding strength is determined by the individual combination of α and β chains. Their tissue-integrating nature makes them prime candidates for tumor growth regulation.

In fact, many integrins are upregulated in neoplastic brain tissue [for a review see Gritsenko et al. (2012)]. For example, the integrin $\alpha v\beta 8$ is described as a major driver of cell invasion in *glioblastoma multiforme* (Reyes et al. 2013). The integrins $\alpha v\beta 3$ and $\alpha v\beta 5$ are also found enriched on glioma cells as well as on the surface of the vasculature (Bello et al. 2001). One of the ligands of αvβ3 is the cell adhesion molecule *Thy-1/CD90*, which is expressed on vascular and lymphatic endothelial cells but also on neurons [summarized in Herrera-Molina et al. (2013)]. Neuronal Thy-1 is clustered by the integrin, thus leading to growth arrest of neurites and retraction of neuronal processes, at least in juvenile cells (Herrera-Molina et al. 2012). Thy-1/CD90 is not only a potential prognostic marker for high-grade gliomas but also a marker for cancer stem cells inside gliomas. It resides within the endothelial niche and may also play a critical role in the generation of tumor vasculatures via differentiation into endothelial cells (He et al. 2012).

10.2.4 ECM Proteoglycans

Proteoglycans are large multidomain ECM molecules carrying glycosaminoglycan (GAG) side chains like heparan sulfate (HS), chondroitin sulfate (CS), or keratan/dermatan sulfate. Due to their negatively charged molecular nature and their binding characteristics they are key factors in the regulation of cell–microenvironment interactions and the extracellular milieu. Lectican-type CS proteoglycans act as a barrier for structural reorganization and thus tumor cell infiltration.

The brain-specific lectican *brevican* is a major component of the CNS extracellular matrix. It is produced by neurons as well as astrocytes and together with versican isoform V2, aggrecan, tenascin-R, and link proteins it constitutes the HA-based ECM meshwork of the mature brain. Brevican (earlier termed BEHAB for Brain-enriched Hyaluronic Acid Binding Protein) is found throughout the brain and associates also with specialized ECM structures like the perisynaptic ECM, the PNN, or the ECM at the axon initial segment [for review, see Frischknecht and Seidenbecher (2012)]. Brevican is expressed in glioblastoma specimens (Fig. 10.4). It appears associated with glioma cell adhesion, motility, and tumor growth, and is also related to glioma cell differentiation. Therefore it may be a marker for the degree of malignance (Jaworski et al. 1996; Lu et al. 2012). Due to differential glycosylation glioma-specific brevican isoforms are generated, which lack most carbohydrates normally present on the proteoglycan but instead contain increased amounts of sialic acid. These oversialated isoforms are indeed present in most high-grade gliomas and absent from low-grade tumors (Viapiano et al. 2005). Brevican enhances the expression of the ECM protein fibronectin at the surface of glioma cells, and after proteolytic cleavage by ADAMTS4 brevican binds to fibronectin, which leads to upregulation of integrins thus promoting glioma cell motility (Hu et al. 2008).

Besides brevican also versican and *neurocan* as well as the glycoprotein tenascin-C were found to be upregulated in astrocytoma and to correlate principally with the invasive phenotype of low-grade astrocytoma (Varga et al. 2012). This is an interesting finding, because in contrast to brevican neurocan and tenascin-C are typical components of the juvenile ECM, indicating the loss of brain maturation-specific characteristic features of the glioma-surrounding parenchyma.

Versican occurs in the brain in several isoforms. Full-length versican (V0) and the shorter isoform V1 (lacking the first GAG attachment region) are major isoforms during late stages of embryonic development (Landolt et al. 1995), and the V2 splice variant, which lacks the second GAG attachment domain, is prominently present in the mature brain (Schmalfeldt et al. 2000). In high-grade gliomas versican V0 and V1 (but not the V2 variant) are upregulated by TGF-β2, which contributes to the malignant phenotype and the induction of cell migration (Arslan et al. 2007). On the other hand, when the mature versican isoform V2 was transfected into the glioblastoma cell line U87 the developing tumors harbored a very elaborate vasculature. Mechanistically, fibronectin was upregulated by V2 and mediated V2 function in brain tumor angiogenesis (Yang and Yee 2013).

Under neoplastic conditions not only the expression level or alternative gene splicing of proteoglycans can be modified but also their posttranslational modification. In particular the sulfation patterns of GAG chains in HS proteoglycans are dramatically changed. The 6O-sulfation, which is known to be critical for ligand interactions and activation of receptor tyrosine kinase (RTK) pathways, can be cleaved by extracellular *sulfatases* like SULF1 and SULF2. SULF2 expression is enhanced in glioblastoma and confers a growth advantage to those cells (Phillips et al. 2012). From a mechanistic perspective, the enhanced removal of 6O-sulfates from the HS proteoglycans leads to stronger activation of RTKs like PDGFR,

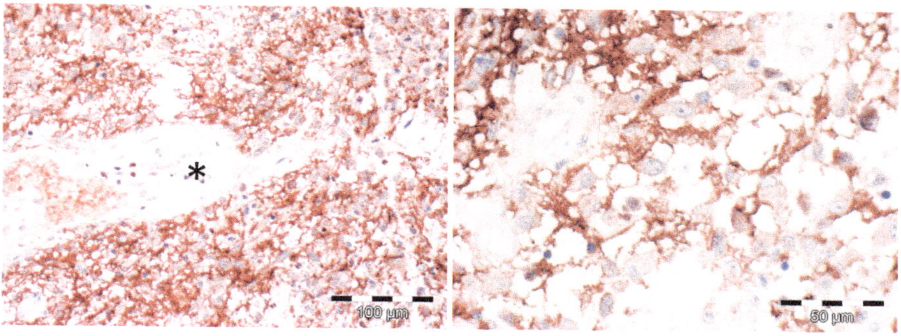

Fig. 10.4 Brevican immunoexpression in a glioblastoma. The proteoglycan is massively upregulated at cell surfaces and in the extracellular space, where it is incorporated into netlike supramolecular structures, but absent from blood vessels (*asterisk*). Brevican was detected with a polyclonal antibody recognizing all isoforms

IGF1R, or EPHA2 via enhanced bioavailability of trophic factors and finally to increased proliferation and tumor cell invasion. Thus, SULF2-positive glioma can be considered as particularly growth factor dependent.

But the picture of sulfation influence on gliomas is more complex: in LN229 glioblastoma cells high levels of *keratan sulfate Gal-6-sulfotransferase* were reported which lead to the expression of highly sulfated keratan sulfates on proteoglycans like aggrecan (Hayatsu et al. 2008). This upregulation of sulfation was also observed in anaplastic astrocytomas and glioblastomas, but hardly in low-grade astrocytoma, pointing to a role as determinant of malignancy (Kato et al. 2008).

The membrane-associated proteoglycan *NG2* belongs to those most upregulated in gliomas, in both the classical and proneural subtypes. NG2 is a characteristic component of oligodendrocyte precursor cells and of a special type of astrocytes. It is implicated in gliomagenesis and predicts poor survival and even resistance of the tumor to ionizing radiation (Svendsen et al. 2011) and to chemotherapy (Chekenya et al. 2008). In glial progenitor cells NG2 co-localizes with the RTKs PDGFRA and EGFR and is implicated in both pathways. Since NG2 is also expressed on pericytes lining the microvessels it is suspected to contribute to recruitment of progenitor cells into the tumor [for review see Stallcup and Huang (2008)].

Syndecan-1 is a transmembrane HS proteoglycan, which binds to thrombospondins (Naganuma et al. 2004). It is upregulated in glioblastoma multiforme, particularly in the mesenchymal type, and this is interpreted as a consequence of NF-κB activation [reviewed in Wade et al. (2013)]. Heparanase enhances shedding of syndecan-1. After shedding the N-terminal extracellular domain gets released from the cell surface and associates with cytokines. Syndecan-1 is discussed in the literature as a potential prognosis predictor of glioma (Xu et al. 2012).

The GPI-anchored HS proteoglycan *glypican-1* is overexpressed in glioma vessel endothelial cells and promotes angiogenesis via forming a ternary complex with FGF2 and the FGFR1c (Qiao et al. 2003). Furthermore, glypican-1 is also expressed by glioma cells and increases their response to FGF-2 (Su et al. 2006).

An exciting new cell biological concept about the function of HS proteoglycans in cancer was put forward only recently (Christianson et al. 2013). HS proteoglycans can function as internalizing receptors of cancer cell-derived extracellular vesicles, which have exosome-like characteristics. These vesicles mediate communication among glioma cells and with the vasculature (Kucharzewska et al. 2013). Exosomes internalized in the U87-MG cell line contain HS proteoglycans of the syndecan and glypican type, and this uptake can be blocked by free HS, but not CS chains, indicating a specific role of heparan sulfates, particularly the 2-O and N-sulfation types, in these processes. This is a new facet to the emerging concept of HSPGs as key receptors of macromolecular cargo (Christianson et al. 2013).

For an up-to-date comprehensive review on alterations in brain proteoglycans in glioma, see (Wade et al. 2013).

10.2.5 ECM Glycoproteins

The brain ECM is a rich source of different glycoproteins which normally serve processes of developmental and activity-induced plasticity of neural cells. In neoplastic tissue a multitude of these molecules was found to be implicated in the multifaceted interplay between invading cells, host cells, and vasculature.

A very prominent glycoprotein with clear association with glioma is the Secreted Protein Acidic and Rich in Cysteine (*SPARC*), which is dramatically upregulated in the microenvironment of the tumor and involved in ECM remodeling, cell proliferation, and migration. The multifunctional scavenger receptor *stabilin-1* has the potential to inhibit tumor growth via binding and internalizing SPARC. Decreased SPARC expression results in decreased tumor invasiveness and enhanced survival. Stabilin-1-expressing macrophages appear transiently in the tumor environment, but disappear as the tumor progresses (David et al. 2012). Therefore, stabilin-1 promises considerable potential for an immune system-based treatment of glioblastoma.

SPARC expression is downregulated by *microRNA-145*, which targets connective tissue growth factor (CTGF), an upstream signaling molecule of SPARC expression. This cascade leads to reduced migration and invasion of glioma cells. Thus, miR-145 is another attractive SPARC-related therapeutic target for anti-invasive treatment of astrocytic tumors (Lee et al. 2013).

Hevin is an ECM glycoprotein with some similarity to SPARC, but in contrast to SPARC it is not present in malignant brain tissue. However, after cleavage of hevin by MMP-3 SPARC-like fragments are produced which associate with neovascularization in high-grade gliomas, like SPARC does, and so the hevin fragments could influence the angiogenic activity of SPARC (Weaver et al. 2011).

The ECM glycoprotein *tenascin-C* is expressed by human glioma in vivo and shows a strong association with tumor blood vessels (Brosicke et al. 2013). It interacts with the SPARC-related modular calcium-binding protein 1 (SMOC1), which can counteract the chemo-attractive effect of tenascin-C on glioma cells (Brellier et al. 2011).

YKL-40 is a member of the chitinase-like glycoprotein family, which lacks enzymatic activity, and which is also associated with brain tumors. It increases VEGF and promotes angiogenesis via coordinating matrix receptors syndecan-1 and integrin αvβ5 and thus stimulating the FAK-ERK signaling cascade (Francescone et al. 2011).

Another important molecular event in malignant glioblastoma cell migration is the increase in expression of *osteopontin* triggered by AKT activation. Osteopontin possesses inflammatory activity, which can be modulated by proteolytic cleavage via thrombin and plasma carboxypeptidase B2. While full-length osteopontin is generally elevated in cerebrospinal fluid (CSF) samples from cancer patients compared with noncancer patients, the CSF levels of cleaved osteopontin are specifically increased in gliomas compared with systemic cancer and noncancer patients. Obviously, osteopontin and its cleaved forms promote motility of glioma cells and confer resistance to apoptosis (Park et al. 2008; Yamaguchi et al. 2013).

Malignant glioma cells produce the multifunctional matrix glycoprotein *thrombospondin-1*, which contributes to cell motility via binding to integrin receptors αvβ3 and α3β1 and to the HS proteoglycan syndecan-1 (Naganuma et al. 2004). An in vitro and in vivo study in malignant glioma provided evidence that thrombospondin-1 and TGF-beta expression directly correlates with malignancy (Kawataki et al. 2000).

Whereas most glycoproteins appear to be upregulated in glioma the expression of the human gene *Leucine-Rich Glioma Inactivated-1* (LGI1) is downregulated, making it an interesting candidate tumor suppressor gene. This downregulation correlates with malignant progression in astrocytic gliomas (Besleaga et al. 2003). LGI1 binds as secreted molecule to membrane-bound proteases ADAM-11, -22, and -23 (Leonardi et al. 2011) and suppresses the expression of secreted matrix metalloproteases MMP1 and MMP3 (Kunapuli et al. 2004). Beyond its involvement in glioma LGI1 is also a candidate molecule for epilepsy.

10.3 Infiltration Mechanisms of Gliomas

Compared to other human neoplasms, with an incidence of 6.5/100,000 (Ostrom et al. 2013) gliomas represent only a small group of tumors. However, the most intriguing feature of the vast majority of glial neoplasms is the highly infiltrative nature of tumor growth, defining gliomas as a very special group of neoplasms.

Although their mechanistic link to the modification of the ECM was discovered much later the infiltrative nature has been recognized for a long time. Detailed investigations by HJ Scherer in the 1940s described the infiltration pathways along myelinated nerve fibers, the basement membrane of blood vessels, and the subependymal growth of gliomas (Scherer 1940). Perineuronal accumulation of tumor cells is also frequently encountered, summarized by Scherer as the formation of "secondary structures" by diffuse glial neoplasms (Scherer 1938). This means that growth and infiltration of glioma cells depend on preexisting tissue structures. In contrast, "primary structure" means the formation of certain specific structures by the tumor cells itself. In biopsy samples, the presence of secondary structures might be regarded as indicators for diffuse tumor growth.

Glioma cells can also spread along cerebrospinal fluid pathways or meninges. In contrast, extraneuronal true "metastasis" of malignant gliomas is extremely rare. There are a few reports in the literature describing metastatic glioblastoma preferentially to the lung, soft tissue, and bones, and it has been proposed that these metastases might represent tumor subclones with different molecular features (Park et al. 2000).

The infiltrative nature of diffuse gliomas is the main reason for the limited resectability of these tumors by neurosurgery (Fig. 10.2). Historical data have shown that even resection of a whole hemisphere does not affect the tumor growth and overall survival significantly (Bell 1949). However, the introduction of modern techniques to detect infiltrative tumor cells due to a characteristic fluorescence has resulted in more optimized tumor resection approaches and increased overall survival (Stummer et al. 2006). It has been demonstrated that gross total resection of glioblastoma results in superior prognosis compared to partial resection or biopsy-only approaches (Noorbakhsh et al. 2014). However, despite all technical advances in modern neurosurgery, control of infiltrative tumor cells remains a challenge in the treatment of gliomas.

From a pathology point of view, various glial neoplasms are summarized under the term "diffuse gliomas". The main groups are tumors with an astrocytic or oligodendral line of differentiation, or a mixture of both (Louis et al. 2007). Depending on neuropathological criteria such as mitotic activity, nuclear and cellular pleomorphism, vascular proliferations, and the presence of necrosis these gliomas are designated as WHO (World Health Organization) grade II, III, or IV, respectively. Although patients suffering from a diffuse glioma WHO grade II may survive for several years, sooner or later these tumors undergo tumor progression to higher-grade forms, leading to death of the patient. The most malignant astrocytic neoplasm is glioblastoma WHO grade IV, unfortunately being also the most frequent subtype among diffuse gliomas with 3.19/100,000 (Ostrom et al. 2013). Current survival rates for patients diagnosed with glioblastoma are still poor with 1-year survival of 35 % and 5-year survival of less than 5 % (Omuro and DeAngelis 2013). The prognosis of glioblastoma patients depends on a variety of different factors, including age, tumor location, certain molecular features, extent of tumor resection, and the treatment scheme given. With regard to the infiltrative properties of glioblastoma, some cases display a highly characteristic diffuse growth from one hemisphere to the other crossing the corpus callosum, called "butterfly glioma" (Dziurzynski et al. 2012). This growth pattern is associated with reduced survival (Ramakrishna et al. 2010). In addition, other modes of spread may influence prognosis substantially. It has been demonstrated that involvement of the subventricular zone (SVZ) results in significantly reduced survival in glioblastoma (Sonoda et al. 2013). Occasionally, gliomas may arise at different, widely separated sites within the brain either synchronous (at initial presentation) or metachronous, leading to the designation of multicentric gliomas (Salvati et al. 2003; Patil et al. 2012a). The frequency of these tumors is estimated with about 2 %. In contrast, more frequently so-called "multifocal" gliomas can be seen, characterized by a pattern of dissemination by either CSF, by commissural structures, or by satellite formation. The prognosis of patients suffering from multifocal glioblastoma is even worse compared to solitary tumors (Patil et al. 2012b).

The histological hallmark of glioblastoma is the presence of necroses and microvascular proliferations. Both features indicate highly proliferative tumor parts. However, in brain areas clearly infiltrated by glioma cells it was demonstrated that the blood vessel density is in the range of normal cerebral tissue, suggesting that tumor-driven neoangiogenesis is absent. This suggests that the glioma cells co-opt and "abuse" preexisting supply structures, a behavior reminiscent of "guerrilla warriors" (Claes et al. 2007).

An attractive concept to explain tumor initiation, recurrence, and invasion in glioblastoma is related to the demonstration of glioblastoma stem cells (Yan et al. 2013). Among other characteristic features, these stem cells overexpress the integrin α3, which supports invasiveness (Nakada et al. 2013). On the other hand, preexisting structures do support stem cell migration, for instance astrocytes (Rath et al. 2013).

The most extreme example of diffuse glioma is called *gliomatosis cerebri*, a tumor involving at least three cerebral lobes but sometimes also infiltrating deep brain structures, the brain stem and spinal cord (Mawrin et al. 2005). Interestingly, molecular genetic studies have shown a common origin of widespread diffuse gliomas, suggesting that these tumors are clonal (Berkman et al. 1992; Mawrin 2005).

10.4 Therapeutic Potential

So far, ECM-based therapeutic approaches are targeted to ECM receptors, to ECM-modifying enzymes, or to changes in the biomechanical properties of the host ECM.

Proteomic screens and secretome analyses have contributed valuable data to shed light on the ability of glioma cells to "shape their own surrounding" by secreting pro-migratory and pro-angiogenic molecules. Two studies provide comparative analyses of the secretomes of different glioblastoma cell lines [U87MG, LN229, and HNGC2 in Polisetty et al. (2011) and LN18, T98, U118, and U87 in Formolo et al. (2011)]. These studies unravel that besides a common set of proteins secreted by all of them each cell line has its own characteristic secretomic signature, which emphasizes the diversity of glioma and the difficulty to find a common treatment. The cell line U87 for example, which has a particularly high invasive potential, secretes more proteases of the ADAM and cathepsin families and more YKL-40. These findings argue for a proteome-driven individual treatment strategy, which takes into account the complex interplay of ECM molecules.

The HA receptor RHAMM, which is upregulated in gliomas, is highly immunogenic and was used to develop a vaccination strategy. In a mouse glioma model injection of dendritic cells transfected with a modified RHAMM mRNA, which encodes RHAMM with a sorting signal for the late endosomal/lysosomal compartment, where MHC II comes into contact with RHAMM, yielded higher numbers of activated T lymphocytes and longer survival (Amano et al. 2007). Soluble peptides directed at the HA-binding domain of RHAMM inhibited glioma cell line proliferation in a dose-dependent manner and reduced cell migration both on and off HA-based ECM. Interestingly, this approach seems to be specific for RHAMM, as anti-CD44 antibodies did not inhibit the migration of human glioma cells in those experiments (Akiyama et al. 2001). As mentioned above, approaches targeting the immune system are most promising. A good candidate is the scavenger receptor stabilin-1, which binds to SPARC and has the potential to remove and degrade it from the surrounding extracellular space (David et al. 2012).

Another promising therapeutic target for control of SPARC levels is the pathway comprising micro-RNA miR-145 and the growth factor CTGF, which directly regulates SPARC expression (Lee et al. 2013).

Over the last years much effort was spent on targeting αvβ3 integrin as important ECM receptor. As an antiangiogenic agent in glioblastoma cilengitide was developed which is a cyclized Arg-Gly-Glu(RGD)-containing pentapeptide and blocks αv-mediated cellular functions very efficiently in preclinical studies (Scaringi et al. 2012; Kurozumi et al. 2012), but a final proof of anti-tumor action in a large clinical patient study is yet to come.

A different approach is pursued by the field of mechanobiology. Here, not the individual molecules of the ECM are in the focus but the general biomechanical properties of the matrix. There is accumulating evidence that mechanical and other biophysical signals from the ECM can powerfully influence cell proliferation, differentiation, death, and motility. Also in the growth and spread of malignant brain tumors mechanical signals conferred by the ECM influence motility, mechanics, and/or proliferation in manifold ways. Thus, tissue engineering with designed biomatrices could be an additional strategy to treat gliomas in the future (Kumar 2009).

Acknowledgment In PubMed more than 300 publications on the role of ECM in malignant gliomas are listed. Due to very limited space we could cite only a small number of these publications, but we wish to express our gratitude to all scientists who contributed to our current knowledge about the neurochemistry, molecular biology, and pathophysiological significance of brain tumor-surrounding ECM. We thank S. Hartmann, Dept. of Neuropathology, for the histological and immunohistochemical stainings.

The authors' work was supported by the COST action ECMNet, the DFG SFB779 (TPA03 to C.I.S.), the Leibniz Graduate School Synaptogenetics, and the DFG GRK 1167.

References

Akiyama Y, Jung S, Salhia B, Lee S, Hubbard S, Taylor M, Mainprize T, Akaishi K, van Furth W, Rutka JT (2001) Hyaluronate receptors mediating glioma cell migration and proliferation. J Neurooncol 53:115–127

Amano T, Kajiwara K, Yoshikawa K, Morioka J, Nomura S, Fujisawa H, Kato S, Fujii M, Fukui M, Hinoda Y, Suzuki M (2007) Antitumor effects of vaccination with dendritic cells transfected with modified receptor for hyaluronan-mediated motility mRNA in a mouse glioma model. J Neurosurg 106:638–645

Arslan F, Bosserhoff AK, Nickl-Jockschat T, Doerfelt A, Bogdahn U, Hau P (2007) The role of versican isoforms V0/V1 in glioma migration mediated by transforming growth factor-beta2. Br J Cancer 96:1560–1568

Asher R, Bignami A (1992) Hyaluronate binding and CD44 expression in human glioblastoma cells and astrocytes. Exp Cell Res 203:80–90

Bell LJ (1949) Cerebral hemispherectomy: report of a case ten years after operation. J Neurosurg 6:285–293

Bellail AC, Hunter SB, Brat DJ, Tan C, Van Meir EG (2004) Microregional extracellular matrix heterogeneity in brain modulates glioma cell invasion. Int J Biochem Cell Biol 36:1046–1069

Bello L, Francolini M, Marthyn P, Zhang J, Carroll RS, Nikas DC, Strasser JF, Villani R, Cheresh DA, Black PM (2001) Alpha(v)beta3 and alpha(v)beta5 integrin expression in glioma periphery. Neurosurgery 49:380–389, discussion 390

Berkman RA, Clark WC, Saxena A, Robertson JT, Oldfield EH, Ali IU (1992) Clonal composition of glioblastoma multiforme. J Neurosurg 77:432–437

Besleaga R, Montesinos-Rongen M, Perez-Tur J, Siebert R, Deckert M (2003) Expression of the LGI1 gene product in astrocytic gliomas: downregulation with malignant progression. Virchows Arch 443:561–564

Brellier F, Ruggiero S, Zwolanek D, Martina E, Hess D, Brown-Luedi M, Hartmann U, Koch M, Merlo A, Lino M, Chiquet-Ehrismann R (2011) SMOC1 is a tenascin-C interacting protein over-expressed in brain tumors. Matrix Biol 30:225–233

Brosicke N, van Landeghem FK, Scheffler B, Faissner A (2013) Tenascin-C is expressed by human glioma in vivo and shows a strong association with tumor blood vessels. Cell Tissue Res 354:409–430

Chekenya M, Krakstad C, Svendsen A, Netland IA, Staalesen V, Tysnes BB, Selheim F, Wang J, Sakariassen PO, Sandal T, Lonning PE, Flatmark T, Enger PO, Bjerkvig R, Sioud M, Stallcup WB (2008) The progenitor cell marker NG2/MPG promotes chemoresistance by activation of integrin-dependent PI3K/Akt signaling. Oncogene 27:5182–5194

Chernov AV, Sounni NE, Remacle AG, Strongin AY (2009) Epigenetic control of the invasion-promoting MT1-MMP/MMP-2/TIMP-2 axis in cancer cells. J Biol Chem 284:12727–12734

Christianson HC, Svensson KJ, van Kuppevelt TH, Li JP, Belting M (2013) Cancer cell exosomes depend on cell-surface heparan sulfate proteoglycans for their internalization and functional activity. Proc Natl Acad Sci U S A 110:17380–17385

Claes A, Idema AJ, Wesseling P (2007) Diffuse glioma growth: a guerilla war. Acta Neuropathol 114:443–458

David C, Nance JP, Hubbard J, Hsu M, Binder D, Wilson EH (2012) Stabilin-1 expression in tumor associated macrophages. Brain Res 1481:71–78

Delpech B, Laquerriere A, Maingonnat C, Bertrand P, Freger P (2002) Hyaluronidase is more elevated in human brain metastases than in primary brain tumours. Anticancer Res 22:2423–2427

Dityatev A, Seidenbecher CI, Schachner M (2010) Compartmentalization from the outside: the extracellular matrix and functional microdomains in the brain. Trends Neurosci 33:503–512

Dziurzynski K, Blas-Boria D, Suki D, Cahill DP, Prabhu SS, Puduvalli V, Levine N (2012) Butterfly glioblastomas: a retrospective review and qualitative assessment of outcomes. J Neurooncol 109:555–563

Enegd B, King JA, Stylli S, Paradiso L, Kaye AH, Novak U (2002) Overexpression of hyaluronan synthase-2 reduces the tumorigenic potential of glioma cells lacking hyaluronidase activity. Neurosurgery 50:1311–1318

Formolo CA, Williams R, Gordish-Dressman H, MacDonald TJ, Lee NH, Hathout Y (2011) Secretome signature of invasive glioblastoma multiforme. J Proteome Res 10:3149–3159

Francescone RA, Scully S, Faibish M, Taylor SL, Oh D, Moral L, Yan W, Bentley B, Shao R (2011) Role of YKL-40 in the angiogenesis, radioresistance, and progression of glioblastoma. J Biol Chem 286:15332–15343

Frischknecht R, Seidenbecher CI (2008) The crosstalk of hyaluronan-based extracellular matrix and synapses. Neuron Glia Biol 4:249–257

Frischknecht R, Seidenbecher CI (2012) Brevican: a key proteoglycan in the perisynaptic extracellular matrix of the brain. Int J Biochem Cell Biol 44:1051–1054

Frischknecht R, Heine M, Perrais D, Seidenbecher CI, Choquet D, Gundelfinger ED (2009) Brain extracellular matrix affects AMPA receptor lateral mobility and short-term synaptic plasticity. Nat Neurosci 12:897–904

Furukawa M, Shimoda H, Kajiwara T, Kato S, Yanagisawa S (2008) Topographic study on nerve-associated lymphatic vessels in the murine craniofacial region by immunohistochemistry and electron microscopy. Biomed Res 29:289–296

Gary SC, Zerillo CA, Chiang VL, Gaw JU, Gray G, Hockfield S (2000) cDNA cloning, chromosomal localization, and expression analysis of human BEHAB/brevican, a brain specific proteoglycan regulated during cortical development and in glioma. Gene 256:139–147

Gritsenko PG, Ilina O, Friedl P (2012) Interstitial guidance of cancer invasion. J Pathol 226:185–199

Gundelfinger ED, Frischknecht R, Choquet D, Heine M (2010) Converting juvenile into adult plasticity: a role for the brain's extracellular matrix. Eur J Neurosci 31:2156–2165

Hayatsu N, Ogasawara S, Kaneko MK, Kato Y, Narimatsu H (2008) Expression of highly sulfated keratan sulfate synthesized in human glioblastoma cells. Biochem Biophys Res Commun 368:217–222

He J, Liu Y, Zhu T, Zhu J, Dimeco F, Vescovi AL, Heth JA, Muraszko KM, Fan X, Lubman DM (2012) CD90 is identified as a candidate marker for cancer stem cells in primary high-grade gliomas using tissue microarrays. Mol Cell Proteomics 11(M111):010744

Held-Feindt J, Paredes EB, Blomer U, Seidenbecher C, Stark AM, Mehdorn HM, Mentlein R (2006) Matrix-degrading proteases ADAMTS4 and ADAMTS5 (disintegrins and metalloproteinases with thrombospondin motifs 4 and 5) are expressed in human glioblastomas. Int J Cancer 118:55–61

Herrera-Molina R, Frischknecht R, Maldonado H, Seidenbecher CI, Gundelfinger ED, Hetz C, Aylwin Mde L, Schneider P, Quest AF, Leyton L (2012) Astrocytic alphaVbeta3 integrin inhibits neurite outgrowth and promotes retraction of neuronal processes by clustering Thy-1. PLoS One 7:e34295

Herrera-Molina R, Valdivia A, Kong M, Alvarez A, Cardenas A, Quest AF, Leyton L (2013) Thy-1-interacting molecules and cellular signaling in cis and trans. Int Rev Cell Mol Biol 305:163–216

Hu B, Kong LL, Matthews RT, Viapiano MS (2008) The proteoglycan brevican binds to fibronectin after proteolytic cleavage and promotes glioma cell motility. J Biol Chem 283:24848–24859

Jacobson A, Brinck J, Briskin MJ, Spicer AP, Heldin P (2000) Expression of human hyaluronan synthases in response to external stimuli. Biochem J 348(Pt 1):29–35

Jacobson A, Rahmanian M, Rubin K, Heldin P (2002) Expression of hyaluronan synthase 2 or hyaluronidase 1 differentially affect the growth rate of transplantable colon carcinoma cell tumors. Int J Cancer 102:212–219

Jaworski DM, Kelly GM, Piepmeier JM, Hockfield S (1996) BEHAB (brain enriched hyaluronan binding) is expressed in surgical samples of glioma and in intracranial grafts of invasive glioma cell lines. Cancer Res 56:2293–2298

Jin SG, Jeong YI, Jung S, Ryu HH, Jin YH, Kim IY (2009) The effect of hyaluronic Acid on the invasiveness of malignant glioma cells: comparison of invasion potential at hyaluronic acid hydrogel and matrigel. J Korean Neurosurg Soc 46:472–478

Junker N, Latini S, Petersen LN, Kristjansen PE (2003) Expression and regulation patterns of hyaluronidases in small cell lung cancer and glioma lines. Oncol Rep 10:609–616

Kato Y, Hayatsu N, Kaneko MK, Ogasawara S, Hamano T, Takahashi S, Nishikawa R, Matsutani M, Mishima K, Narimatsu H (2008) Increased expression of highly sulfated keratan sulfate synthesized in malignant astrocytic tumors. Biochem Biophys Res Commun 369:1041–1046

Kawataki T, Naganuma H, Sasaki A, Yoshikawa H, Tasaka K, Nukui H (2000) Correlation of thrombospondin-1 and transforming growth factor-beta expression with malignancy of glioma. Neuropathology 20:161–169

Kim CS, Jung S, Jung TY, Jang WY, Sun HS, Ryu HH (2011) Characterization of invading glioma cells using molecular analysis of leading-edge tissue. J Korean Neurosurg Soc 50:157–165

Kochlamazashvili G, Henneberger C, Bukalo O, Dvoretskova E, Senkov O, Lievens PM, Westenbroek R, Engel AK, Catterall WA, Rusakov DA, Schachner M, Dityatev A (2010) The extracellular matrix molecule hyaluronic acid regulates hippocampal synaptic plasticity by modulating postsynaptic L-type Ca(2+) channels. Neuron 67:116–128

Kucharzewska P, Christianson HC, Welch JE, Svensson KJ, Fredlund E, Ringner M, Morgelin M, Bourseau-Guilmain E, Bengzon J, Belting M (2013) Exosomes reflect the hypoxic status of glioma cells and mediate hypoxia-dependent activation of vascular cells during tumor development. Proc Natl Acad Sci U S A 110:7312–7317

Kumar S (2009) Cell-matrix mechanobiology: applications to brain tumors and design of tissue engineering scaffolds. In: Annual international conference of the IEEE engineering in medicine and biology society, pp 3350–3352

Kunapuli P, Kasyapa CS, Hawthorn L, Cowell JK (2004) LGI1, a putative tumor metastasis suppressor gene, controls in vitro invasiveness and expression of matrix metalloproteinases in glioma cells through the ERK1/2 pathway. J Biol Chem 279:23151–23157

Kurozumi K, Ichikawa T, Onishi M, Fujii K, Date I (2012) Cilengitide treatment for malignant glioma: current status and future direction. Neurol Med Chir 52:539–547

Kyosseva SV, Harris EN, Weigel PH (2008) The hyaluronan receptor for endocytosis mediates hyaluronan-dependent signal transduction via extracellular signal-regulated kinases. J Biol Chem 283:15047–15055

Landolt RM, Vaughan L, Winterhalter KH, Zimmermann DR (1995) Versican is selectively expressed in embryonic tissues that act as barriers to neural crest cell migration and axon outgrowth. Development 121:2303–2312

Lee HK, Bier A, Cazacu S, Finniss S, Xiang C, Twito H, Poisson LM, Mikkelsen T, Slavin S, Jacoby E, Yalon M, Toren A, Rempel SA, Brodie C (2013) MicroRNA-145 is downregulated in glial tumors and regulates glioma cell migration by targeting connective tissue growth factor. PLoS One 8:e54652

Leonardi E, Andreazza S, Vanin S, Busolin G, Nobile C, Tosatto SC (2011) A computational model of the LGI1 protein suggests a common binding site for ADAM proteins. PLoS One 6: e18142

Lettau I, Hattermann K, Held-Feindt J, Brauer R, Sedlacek R, Mentlein R (2010) Matrix metalloproteinase-19 is highly expressed in astroglial tumors and promotes invasion of glioma cells. J Neuropathol Exp Neurol 69:215–223

Li Y, Wang D, Wang L, Yu J, Du D, Chen Y, Gao P, Wang DM, Yu J, Zhang F, Fu S (2013) Distinct genomic aberrations between low-grade and high-grade gliomas of Chinese patients. PLoS One 8:e57168

Liu D, Pearlman E, Diaconu E, Guo K, Mori H, Haqqi T, Markowitz S, Willson J, Sy MS (1996) Expression of hyaluronidase by tumor cells induces angiogenesis in vivo. Proc Natl Acad Sci U S A 93:7832–7837

Lokeshwar VB, Cerwinka WH, Isoyama T, Lokeshwar BL (2005) HYAL1 hyaluronidase in prostate cancer: a tumor promoter and suppressor. Cancer Res 65:7782–7789

Louis DN, Ohgaki H, Wiestler OD, Cavenee WK, Burger PC, Jouvet A, Scheithauer BW, Kleihues P (2007) The 2007 WHO classification of tumours of the central nervous system. Acta Neuropathol 114:97–109

Lu R, Wu C, Guo L, Liu Y, Mo W, Wang H, Ding J, Wong ET, Yu M (2012) The role of brevican in glioma: promoting tumor cell motility in vitro and in vivo. BMC Cancer 12:607

Mawrin C (2005) Molecular genetic alterations in gliomatosis cerebri: what can we learn about the origin and course of the disease? Acta Neuropathol 110:527–536

Mawrin C, Schneider T, Firsching R, Wiedemann FR, Dietzmann K, Bornemann A, Roessner BF, Sellhaus B, von Deimling A (2005) Assessment of tumor cell invasion factors in gliomatosis cerebri. J Neurooncol 73:109–115

Naganuma H, Satoh E, Asahara T, Amagasaki K, Watanabe A, Satoh H, Kuroda K, Zhang L, Nukui H (2004) Quantification of thrombospondin-1 secretion and expression of alphavbeta3 and alpha3beta1 integrins and syndecan-1 as cell-surface receptors for thrombospondin-1 in malignant glioma cells. J Neurooncol 70:309–317

Nakada M, Nambu E, Furuyama N, Yoshida Y, Takino T, Hayashi Y, Sato H, Sai Y, Tsuji T, Miyamoto KI, Hirao A, Hamada JI (2013) Integrin alpha3 is overexpressed in glioma stem-like cells and promotes invasion. Br J Cancer 108:2516–2524

Nauman P, Czernicki Z (2002) The molecular basis of glia-derived tumours of the brain. Part I. Neurol Neurochir Pol 36:553–564

Noorbakhsh A, Tang JA, Marcus LP, McCutcheon B, Gonda DD, Schallhorn CS, Talamini MA, Chang DC, Carter BS, Chen CC (2014) Gross-total resection outcomes in an elderly population with glioblastoma: a SEER-based analysis. J Neurosurg 120:31–39

Omuro A, DeAngelis LM (2013) Glioblastoma and other malignant gliomas: a clinical review. JAMA 310:1842–1850

Ostrom QT, Gittleman H, Farah P, Ondracek A, Chen Y, Wolinsky Y, Stroup NE, Kruchko C, Barnholtz-Sloan JS (2013) CBTRUS statistical report: Primary brain and central nervous system tumors diagnosed in the United States in 2006-2010. Neuro Oncol 15(Suppl 2):ii1–ii56

Pandey MS, Weigel PH (2014) HARE-mediated endocytosis of hyaluronan, heparin, dermatan sulfate and acetylated LDL, but not chondroitin sulfate types A, C, D or E activates NF-kB-regulated gene expression. J Biol Chem 289(3):1756–1767

Pandey MS, Baggenstoss BA, Washburn J, Harris EN, Weigel PH (2013) The hyaluronan receptor for endocytosis (HARE) activates NF-kappaB-mediated gene expression in response to 40-400-kDa, but not smaller or larger, hyaluronans. J Biol Chem 288:14068–14079

Park CC, Hartmann C, Folkerth R, Loeffler JS, Wen PY, Fine HA, Black PM, Shafman T, Louis DN (2000) Systemic metastasis in glioblastoma may represent the emergence of neoplastic subclones. J Neuropathol Exp Neurol 59:1044–1050

Park JB, Kwak HJ, Lee SH (2008) Role of hyaluronan in glioma invasion. Cell Adh Migr 2:202–207

Patil CG, Eboli P, Hu J (2012a) Management of multifocal and multicentric gliomas. Neurosurg Clin N Am 23:343–350, x

Patil CG, Yi A, Elramsisy A, Hu J, Mukherjee D, Irvin DK, Yu JS, Bannykh SI, Black KL, Nuno M (2012b) Prognosis of patients with multifocal glioblastoma: a case-control study. J Neurosurg 117:705–711

Phillips JJ, Huillard E, Robinson AE, Ward A, Lum DH, Polley MY, Rosen SD, Rowitch DH, Werb Z (2012) Heparan sulfate sulfatase SULF2 regulates PDGFRalpha signaling and growth in human and mouse malignant glioma. J Clin Invest 122:911–922

Polisetty RV, Gupta MK, Nair SC, Ramamoorthy K, Tiwary S, Shiras A, Chandak GR, Sirdeshmukh R (2011) Glioblastoma cell secretome: analysis of three glioblastoma cell lines reveal 148 non-redundant proteins. J Proteomics 74:1918–1925

Qiao D, Meyer K, Mundhenke C, Drew SA, Friedl A (2003) Heparan sulfate proteoglycans as regulators of fibroblast growth factor-2 signaling in brain endothelial cells. Specific role for glypican-1 in glioma angiogenesis. J Biol Chem 278:16045–16053

Ramakrishna R, Barber J, Kennedy G, Rizvi A, Goodkin R, Winn RH, Ojemann GA, Berger MS, Spence AM, Rostomily RC (2010) Imaging features of invasion and preoperative and postoperative tumor burden in previously untreated glioblastoma: correlation with survival. Surg Neurol Int 1:40

Rath BH, Fair JM, Jamal M, Camphausen K, Tofilon PJ (2013) Astrocytes enhance the invasion potential of glioblastoma stem-like cells. PLoS One 8:e54752

Reyes SB, Narayanan AS, Lee HS, Tchaicha JH, Aldape KD, Lang FF, Tolias KF, McCarty JH (2013) alphavbeta8 integrin interacts with RhoGDI1 to regulate Rac1 and Cdc42 activation and drive glioblastoma cell invasion. Mol Biol Cell 24:474–482

Salvati M, Caroli E, Orlando ER, Frati A, Artizzu S, Ferrante L (2003) Multicentric glioma: our experience in 25 patients and critical review of the literature. Neurosurg Rev 26:275–279

Scaringi C, Minniti G, Caporello P, Enrici RM (2012) Integrin inhibitor cilengitide for the treatment of glioblastoma: a brief overview of current clinical results. Anticancer Res 32:4213–4223

Scherer HJ (1938) Structural development in gliomas. Am J Cancer 34:333–351

Scherer HJ (1940) A critical review: the pathology of cerebral gliomas. J Neurol Psychiatry 3:147–177

Schmalfeldt M, Bandtlow CE, Dours-Zimmermann MT, Winterhalter KH, Zimmermann DR (2000) Brain derived versican V2 is a potent inhibitor of axonal growth. J Cell Sci 113 (Pt 5):807–816

Sim H, Hu B, Viapiano MS (2009) Reduced expression of the hyaluronan and proteoglycan link proteins in malignant gliomas. J Biol Chem 284:26547–26556

Sonoda Y, Saito R, Kanamori M, Kumabe T, Uenohara H, Tominaga T (2013) The association of subventricular zone involvement at recurrence with survival after repeat surgery in patients with recurrent glioblastoma. Neurol Med Chir (Tokyo) 2013:27

Stallcup WB, Huang FJ (2008) A role for the NG2 proteoglycan in glioma progression. Cell Adh Migr 2:192–201

Stern R (2005) Hyaluronan metabolism: a major paradox in cancer biology. Pathol Biol 53:372–382

Stummer W, Pichlmeier U, Meinel T, Wiestler OD, Zanella F, Reulen HJ (2006) Fluorescence-guided surgery with 5-aminolevulinic acid for resection of malignant glioma: a randomised controlled multicentre phase III trial. Lancet Oncol 7:392–401

Su G, Meyer K, Nandini CD, Qiao D, Salamat S, Friedl A (2006) Glypican-1 is frequently overexpressed in human gliomas and enhances FGF-2 signaling in glioma cells. Am J Pathol 168:2014–2026

Svendsen A, Verhoeff JJ, Immervoll H, Brogger JC, Kmiecik J, Poli A, Netland IA, Prestegarden L, Planaguma J, Torsvik A, Kjersem AB, Sakariassen PO, Heggdal JI, Van Furth WR, Bjerkvig R, Lund-Johansen M, Enger PO, Felsberg J, Brons NH, Tronstad KJ, Waha A, Chekenya M (2011) Expression of the progenitor marker NG2/CSPG4 predicts poor survival and resistance to ionising radiation in glioblastoma. Acta Neuropathol 122:495–510

Varga I, Hutoczki G, Szemcsak CD, Zahuczky G, Toth J, Adamecz Z, Kenyeres A, Bognar L, Hanzely Z, Klekner A (2012) Brevican, neurocan, tenascin-C and versican are mainly responsible for the invasiveness of low-grade astrocytoma. Pathol Oncol Res 18:413–420

Viapiano MS, Bi WL, Piepmeier J, Hockfield S, Matthews RT (2005) Novel tumor-specific isoforms of BEHAB/brevican identified in human malignant gliomas. Cancer Res 65:6726–6733

Wade A, Robinson AE, Engler JR, Petritsch C, James CD, Phillips JJ (2013) Proteoglycans and their roles in brain cancer. FEBS J 280:2399–2417

Wang D, Fawcett J (2012) The perineuronal net and the control of CNS plasticity. Cell Tissue Res 349:147–160

Weaver M, Workman G, Schultz CR, Lemke N, Rempel SA, Sage EH (2011) Proteolysis of the matricellular protein hevin by matrix metalloproteinase-3 produces a SPARC-like fragment (SLF) associated with neovasculature in a murine glioma model. J Cell Biochem 112:3093–3102

Wiranowska M, Ladd S, Moscinski LC, Hill B, Haller E, Mikecz K, Plaas A (2010) Modulation of hyaluronan production by CD44 positive glioma cells. Int J Cancer 127:532–542

Xu Y, Yuan J, Zhang Z, Lin L, Xu S (2012) Syndecan-1 expression in human glioma is correlated with advanced tumor progression and poor prognosis. Mol Biol Rep 39:8979–8985

Yamaguchi Y, Shao Z, Sharif S, Du XY, Myles T, Merchant M, Harsh G, Glantz M, Recht L, Morser J, Leung LL (2013) Thrombin-cleaved fragments of osteopontin are overexpressed in malignant glial tumors and provide a molecular niche with survival advantage. J Biol Chem 288:3097–3111

Yan K, Yang K, Rich JN (2013) The evolving landscape of glioblastoma stem cells. Curr Opin Neurol 26:701–707

Yang W, Yee AJ (2013) Versican V2 isoform enhances angiogenesis by regulating endothelial cell activities and fibronectin expression. FEBS Lett 587:185–192

Glioblastomas and the Special Role of Adhesion Molecules in Their Invasion

11

Vivaldo Moura-Neto*, Loraine Campanati*, Diana Matias, Claudia Maria Pereira, Catarina Freitas, Juliana M. Coelho-Aguiar, Tania Cristina Leite de Sampaio e Spohr, Ana Lucia Tavares-Gomes, Diego Pinheiro-Aguiar, Suzana Assad Kahn, Joana Silva-Balça, Bruno Pontes, Isabel Porto-Carreiro, Jane Faria, Rodrigo Alves Portela Martins, Silvia Lima-Costa, Maria de Fátima Dias-Costa, Maria Celeste Lopes, and Flavia Regina Souza Lima

Abstract

Among the cells that constitute the central nervous system (CNS), astrocytes are the most abundant cells in the brain. They originate from the neural tube, differently from other glial cells of the astrocyte-like, enteric glia, which originate from the neural crest. Cerebral astrocytes interact with other brain cells such

*These authors contributed equally to this work.

V. Moura-Neto (✉)
Instituto de Ciências Biomédicas, Universidade Federal do Rio de Janeiro, Rio de Janeiro, RJ, Brazil

Instituto Estadual do Cérebro Paulo Niemeyer, Rio de Janeiro, Brazil
e-mail: vivaldo@icb.ufrj.br; vivaldomouraneto@globo.com

L. Campanati • D. Matias • C.M. Pereira • C. Freitas • J.M. Coelho-Aguiar • T.C. Leite de Sampaio e Spohr • D. Pinheiro-Aguiar • S.A. Kahn • B. Pontes • J. Faria • R.A.P. Martins • F.R. Souza Lima
Instituto de Ciências Biomédicas, Universidade Federal do Rio de Janeiro, Rio de Janeiro, RJ, Brazil

A.L. Tavares-Gomes
Departamento de Neurobiologia, Universidade Federal Fluminense, Niterói, RJ, Brazil

J. Silva-Balça
Instituto de Ciências Biomédicas, Universidade Federal do Rio de Janeiro, Rio de Janeiro, RJ, Brazil

Universidade de Coimbra, Coimbra, Portugal

I. Porto-Carreiro
Instituto de Bioquímica Médica, Programa de Oncobiologia, Universidade Federeal do Rio de Janeiro, Rio de Janeiro, RJ, Brazil

S. Lima-Costa • M.d.F. Dias-Costa
Instituto de Ciências da Saúde, Universidade Federal da Bahia, Salvador, BA, Brazil

M.C. Lopes
Universidade de Coimbra, Coimbra, Portugal

A. Sedo and R. Mentlein (eds.), *Glioma Cell Biology*,
DOI 10.1007/978-3-7091-1431-5_11, © Springer-Verlag Wien 2014

as neurons, mediated by cellular contact involving extracellular matrix (ECM) elements and the integrin of astrocytes. Interactions can also occur through exchange of soluble molecules, cytokines, and growth factors, delivered by astrocytes and the interacting cells. Of the growth factors secreted by astrocytes, members of the transforming growth factor (TGF) beta family, especially TGF-β1, are highly important. TGF-β1 acts in brain development, in adult brain homeostasis, and also in orchestrating the brain's response to injury and/or aging. These cell–cell interactions depend on the cellular membrane plasticity expressed by biophysical forces of interaction. In parallel to the properties of normal astrocytes, a growing astrocytoma develops new malignant behavior and interactions with parenchyma that differ from those of normal astrocytes. In this chapter, we analyze important points that characterize tumor progression and discuss new therapeutic approaches to treat these tumors.

Keywords
Extracellular matrix • Adhesion molecules • Growth factors • Invasion

Abbreviations

CNS	Central nervous system
CSC	Cancer stem-like cells
CTGF	Connective tissue growth factor
DPP IV	Dipeptidylpeptidase IV
EGF	Epidermal growth factor
EV	Extracellular vesicles
FAK	Focal adhesion kinase
FAP	Fibroblast activation protein seprase
GBM	*Glioblastoma multiforme*
GFAP	Glial fibrillary acidic protein
GSC	Glioma stem(-like) cell
JNK	c-jun-N-terminal kinase
MAPK	Mitogen-activated protein kinase (ERK1 and ERK2)
MMP	Matrix metalloproteinase
NCAM	Neural cell adhesion molecule
PI3K	Phosphatidylinositide-3-kinase
TGF	Transforming growth factor
TGFR	TGF receptor
TIMP	Tissue inhibitor of MMPs
VEGF	Vascular endothelial growth factor

11.1 Introduction: Astrocytes, Their Special Properties, and Their Interactions with Brain Extracellular Matrix

11.1.1 Glial Cells in a Healthy System

Among all the cells that constitute the central nervous system (CNS), astrocytes are certainly the most abundant cells in the brain. They were first described in 1846 by Rudolf Virchow (1821–1902), as "brain glue," and were assumed to have only a passive supporting role that provides an inert scaffold necessary for neuronal distribution and interactions. The glial family comprises basically oligodendrocytes, microglia, and astrocytes, which were described comprehensively by Ramon y Cajal (1852–1934). Astrocytes themselves include a range of cellular subtypes, which differ in morphology, development, metabolism, and cellular physiology (Oberheim et al. 2012). In the past, astrocytes were thought to play only a secondary role in brain physiology, but emerging research suggests that astrocytes have an active role in brain function and information processing, not only during development but also in adulthood (Allen and Barres 2009). Glial cells are evolutionarily conserved and are present in organisms from the simplest invertebrates to humans; furthermore, the proportion of glial cells to total neural cells is correlated with an animal's size. As animals have evolved, glial cells have not only become more diverse and specialized, but are also essential for neuronal function and survival (Allen and Barres 2009). Astrocytes in the human cerebral cortex are much more complex than those in other mammals and are also involved in information processing (Allen and Barres 2009; Diniz et al. 2012).

Astrocytes are crucial for potassium homeostasis, neurotransmitter uptake, synapse formation, regulation of the blood–brain barrier, and the development of the nervous system, playing important roles in neurogenesis, myelination, neuronal migration, proliferation, differentiation, and even neuronal signaling (Diniz et al. 2012; Anton et al. 1997; Barres 2008; de Sampaio e Spohr et al. 2002; Froes et al. 1999; Gomes et al. 1999a; Lim and Alvarez-Buylla 1999). Indeed, astrocytes are crucial at each step of neural development, as they function as neural stem cells and generate neurons, stimulate neurite outgrowth, guide axon projections, promote synapse formation, and maintain neuronal survival (Diniz et al. 2012; Kriegstein and Alvarez-Buylla 2009; Spohr et al. 2008, 2011, 2012; Banker 1980; Garcia-Abreu et al. 1995; Mallat et al. 1986; Moura Neto et al. 1985, 1986). Interestingly, astrocytes and neurons are able to couple in vitro by Gap junctions in a transient phenomenon that is inversely correlated with cellular maturation (Froes et al. 1999). This cell-to-cell connection by Gap junctions via juxtaposed connexin hemi-channels greatly increases the complexity of the network formed by astrocytes and neurons (Giaume 2010). Our group has shown that astrocytes are able to modulate neuronal survival and neuritogenesis via cellular prion protein secretion (Lima et al. 2007).

However, not only astrocytes are able to influence neuronal dynamics. Neuron–glia interaction is a "two-way street," as neurons are also important for astrocyte differentiation. Our group has demonstrated that neurons are able to induce astrocyte differentiation by secreting transforming growth factor-β1 (TGF-β1), inducing the expression of the glial fibrillary acidic protein GFAP gene promoter (de Sampaio e Spohr et al. 2002; Gomes et al. 1999b). Although both neurons and astrocytes synthesize and secrete TGF-β1, the addition of neurons to astrocyte monolayers greatly increased TGF-β1 synthesis and secretion by astrocytes (de Sampaio e Spohr et al. 2002; Sousa Vde et al. 2004). We subsequently demonstrated that neuronal glutamate activates the GFAP gene promoter of cerebral cortical astrocytes, through induction of the TGF-β1 signaling pathway (Romao et al. 2008).

Our group also found that cerebellar astrocytes can mediate the effects of triiodthyronie (T3) hormone on neuron proliferation and differentiation, by secreting EGF, and that this event is mediated by epidermal growth factor (EGF) modulation of laminin and fibronectin expression by astrocytes through the mitogen-activated protein kinase (MAPK) and phosphatidylinositide-3-kinase (PI3K) pathways (Gomes et al. 1999c; Martinez and Gomes 2002, 2005). Astrocytes now appear to be an important homeostatic cellular element that controls the brain environment by regulating the volume and composition of extracellular space, which is important for neuronal trophic support (Kettenmann and Verkhratsky 2008). Interactions with other brain cells are mediated by cellular contact involving extracellular matrix (ECM) elements and integrin of astrocytes and other cell types. Interactions can also occur through exchange of soluble molecules, cytokines, and growth factors, delivered by astrocytes and the interacting cells.

11.1.2 Growth Factors Involved in Cell Motility and Migration

Among all the growth factors that astrocytes secrete, members of the TGF beta family are of great importance, especially TGF-β1. TGF-β1 is important during brain development, in adult brain homeostasis, and also in orchestrating the brain's response to injury and/or aging (Bae et al. 2011). TGF-β1 is an injury-related cytokine that has several effects on astrocytes, including inhibition of proliferation and motility control by acting on ECM (Baghdassarian et al. 1993; Flanders et al. 1993; Gagelin et al. 1995; Krieglstein et al. 1998; Laping et al. 1994; Wyss-Coray et al. 1995). This factor similarly affects glioblastoma proliferation and motility. Therefore, TGF-β1 is important for neuron–glia interaction, and is involved in several processes, specifically in the regulation of neuronal survival, migration, astrocyte development, differentiation, and cerebral gene expression (de Sampaio e Spohr et al. 2002; Bottner et al. 2000; Brionne et al. 2003; Gomes and Sousa 2005; Krieglstein et al. 2002; Lesne et al. 2002). As TGF-β1 has an important role in ECM production, this cytokine may regulate processes such as cell adhesion, migration, and proliferation (Bottner et al. 2000; Brionne et al. 2003; Abreu et al. 2002; Unsicker and Strelau 2000; Javelaud and Mauviel 2005; Massague and Gomis 2006; Moustakas and Heldin 2005; Zhang et al. 2009).

TGF-β acts via canonical signaling, which is mediated by two serine threonine kinase receptors, transforming growth factor-beta receptors I (TGFRI) and II (TGFRII), which activate the SMAD 2/3 and SMAD 4 transcription factors (SMAD, combination of SMA = small body size, *Caenorhabditis elegans* protein and MAD = mothers against decapentaplegic, *Drosophila* protein; compare Chaps. 3 and 6). Recently, a TGF-β-induced noncanonical signaling pathway was described, which is able to activate the Ras GTPase, mitogen-activated protein kinase (MAPK), p38, c-jun-N-terminal kinase (JNK), and phosphatidylinositol-3 kinase (PI3K) signaling cascades. Thus, this cytokine controls apoptosis, proliferation, ECM protein synthesis, cell differentiation, and also especially cell migration, in several extracellular contexts, either independently or as a modulator of SMAD activity (Javelaud and Mauviel 2005; Massague and Gomis 2006; Moustakas and Heldin 2005; Zhang et al. 2009) in normal astrocytes, and may also affect glioblastomas.

Astrocytes, besides being the major source of growth factor production, are also the major source of ECM components such as laminin, fibronectin, and proteoglycans, both in vivo and in vitro (Spohr et al. 2011, 2012; Garcia-Abreu et al. 2000). It has been demonstrated that the pattern of laminin protein expression on the astrocyte surface provides directional cues for neurite outgrowth, since the ability of astrocytes to induce neurite outgrowth is dependent on the region where the astrocytes originate. Astrocytes derived from different sections of the embryonic mouse midbrain show different capabilities to support neurite outgrowth (Spohr et al. 2011, 2012; Garcia-Abreu et al. 1995; Martinez and Gomes 2002; Mendes et al. 2003). Interestingly, during neuron–glial interactions, the ECM has a pivotal role. The punctate pattern of the glial laminin expression was not permissive for neurite outgrowth, although the fibrillary pattern was (Spohr et al. 2011, 2012; Garcia-Abreu et al. 1995, 2000; Martinez and Gomes 2002; Mendes et al. 2003). However, astrocytes lose their permissivity to induce neurite outgrowth with age, and embryonic astrocytes are more permissive than adult astrocytes (Bahr et al. 1995). In addition, this astrocyte–neuron interaction is bidirectional, as neurons modulate GFAP expression in astrocytes by inducing the synthesis of this protein (Froes et al. 1999; Gomes et al. 1999a; Garcia-Abreu et al. 1995). Compared to the ECM proteins in *glioblastoma multiforme* (GBM) cells, these cells seem to induce a stunted growth cone, folded laterally about itself, of the neurons onto these tumor cells completely different from the widely known illustrations of the growth cone of neurons cultured on normal astrocytes (our unpublished results). Further, if normal astrocytes have fundamental interactive properties with other brain cells including neurons, microglia, and endothelial cells, these properties are maintained, but with different characteristics if the astrocytes become tumorigenic. Astrocytes that become tumorigenic, such as GBMs, retain their ability to support neuritogenesis as young astrocytes (Faria et al. 2006). However, in vitro, GBM cells do not completely distinguish the neurite patterns of embryonic and postnatal neurons as effectively as do normal astrocytes, which are able to distinguish the different neurite growth potential characteristics of embryonic and neonatal neurons (Faria et al. 2006; Moura Neto et al. 1983). One explanation may be that the laminin organization in both young astrocytes and tumor glial cells was altered from a filamentous arrangement to a mixed punctuate/filamentous glial pattern when in coculture with neurons, suggesting that neurons modulate the organization of astrocytes and GBM laminin in the ECM (Faria et al. 2006). The effect of neurons on GBM protein expression was recently demonstrated, as GBM-derived connective tissue growth factor (CTGF) is negatively regulated, at the transcriptional level, by neurons, thus affecting the migration/invasion capacity of these GBM cells (Romao et al. 2013).

11.1.3 Implication of Cellular Forces: Responses to Cell Deformations During Migration

For cell–cell interaction and migration, cells use ECM proteins, integrin components of membranes, and intracellular proteins of the cytoskeleton, for instance. These responses to forces contribute to membrane deformation and allow the cells to adapt to different conditions imposed by the microenvironment.

A direct means of assessing these responses to forces is by measuring two elastic parameters of the cell membrane: its bending modulus κ (Gittes et al. 1993) and surface tension σ (Diz-Munoz et al. 2013). These parameters can be determined using a technique based on extracting the membrane tether from the cell by pulling on it with an attached microsphere trapped in optical tweezers (Bo and Waugh 1989). Analysis of the force-extension curve, together with the tether radius, yields these two elastic parameters and also information regarding the membrane–cytoskeleton interaction (Pontes et al. 2011). Tether pulling with optically trapped beads is a well-known method for such measurements (Gauthier et al. 2012).

Recent data suggest that cell specialization and differentiation can account for the existence of correlations between these elastic parameters and cell functions associated with mechanical deformation and/or force production (Gauthier et al. 2012). Pontes and collaborators (Pontes et al. 2013) tested this response-to-force hypothesis by measuring κ and σ for a variety of cell types in the brain. These authors observed that the elastic parameters for neurons are close to those for the isolated cell membrane (a membrane completely disconnected from the cytoskeleton), suggesting a weaker interaction between the membrane and the adjacent F-actin cortex. Astrocytes and GBM cells have the same ectodermal origin as neurons, but they are considerably more dynamic and have totally different functions. This is also observed in their greater elasticity compared to neurons. However, in spite of their differences from neurons, astrocytes and GBM closely resemble each other in their elastic parameters, which provides some evidence for the idea that they have the same origin and also similar functions, for example, for supporting neurons (Faria et al. 2006) and suggesting that GBMs keep a biochemical "memory" from the normal glia, to the membrane cellular properties, even after malignization.

11.2 Glioblastoma

11.2.1 Characteristics of Malignant Transformation

Glioblastoma (GBM), formerly termed "*glioblastoma multiforme*," the most malignant form of glioma, as described in different chapters of this book, is a highly vascularized brain tumor that displays rapid and invasive growth restricted to the central nervous system, and is also resistant to classical therapies, including surgery, irradiation, and chemotherapy (Kjellman et al. 2000). The excessive proliferation, diffuse infiltration of the surrounding brain tissue, and suppression of antitumor immune surveillance contribute to the high degree of morbidity and mortality of the patients (DeAngelis 2001). This complex phenomenon, known as invasion and metastasis, is triggered by signal transduction in the cell surface transmembrane adhesion molecules and the ECM (Maness and Schachner 2007; Schmid and Maness 2008; Siesser and Maness 2009). Other elements involved in the progression of the tumor are chemokines, small peptide mediators that play a role in many physiological and pathological processes, via their respective G protein-coupled receptor. For example, chemokine CXCL16 is highly expressed by glial tumor and stroma cells, whereas CXCR6 defines a subset of cells with stem cell character (Hattermann and Mentlein 2013; Hattermann et al. 2013). These data evoke the plasticity of the cell membrane to deform and adapt when subjected to forces of adhesion to the substrate, cell–cell interaction, and migration.

Interestingly, the transforming growth factor-β (TGF-β) signaling pathway plays a decisive role in tumor growth and cell motility, as it induces matrix metalloproteinase 2 (MMP2) expression and suppresses the tissue inhibitor of MMP2 (TIMP)-2 expression in glioma cell lines (Wick et al. 2001). However, further proteases such as dipeptidyl peptidase IV (DPP IV) and fibroblast activation protein (FAP, seprase) are also implicated in playing a decisive role for the spread and growth of gliomas (see Chap. 12) (Balaziova et al. 2011). In human malignant glioma, elevated levels of TGF-β are associated with high tumor grade, advanced tumor stages, and poor disease prognosis (Kjellman et al. 2000; Bruna et al. 2007). In addition to the effects on invasion and migration, TGF-β also has an important role in maintaining the population of cells within GBMs that exhibit stem cell-like properties, known as glioma-initiating cells (Anido et al. 2010; Penuelas et al. 2009). Cancer stem-like cells (CSC) from GBMs were first reported by Ignatova and coworkers in 2002; they isolated clonogenic, neurosphere-forming precursors from postsurgical human tissue samples (Ignatova et al. 2002). This cell population is responsible for driving tumor relapse, as it persists after standard treatments such as chemotherapy, radiotherapy, and surgery (Bao et al. 2006; Persano et al. 2013). Indeed, it is believed that malignant tumors are initiated and maintained by this population of tumor cells, which have similar biological properties to normal adult stem cells (Persano et al. 2013).

The progression of solid tumors critically depends on their ability to attract and form new blood vessels (Folkman et al. 1985), and accordingly, a increased degree of vascularization and vascular endothelial growth factor (VEGF) signaling are associated with the invasive characteristics and aggressiveness of GBM (Weis and Cheresh 2011). Anti-VEGF approaches, however, have shown modest effects on blocking GBM progression (Lima et al. 2012; Robles Irizarry et al. 2012), underscoring the importance of understanding endothelial cell signaling in the context of the GBM microenvironment to develop alternative anti-angiogenic therapies to block GBM progression and invasion. Tenascin C, another ECM element, is directly involved with neovascularization and impairs tubulogenesis of vessels. This ECM protein is present on the surface of GBM and could act on vessels (Alves et al. 2011).

Besides endothelial cells, microglial cells are also important in the promotion of tumor progression. In the GBM, microglia and tumor-associated macrophages may comprise up to 30 % of the tumor mass, which suggests that the presence of defense cells in the lesion is due to their anti-tumor activities. However, these cells are able to secrete pro-tumor factors, including MMPs, VEGF, and the stress-inducible protein 1 (STI1; Fonseca et al. 2012) which facilitate the proliferation and migration of tumor cells, increasing the growth and invasion of tumors in the brain parenchyma (Lima et al. 2012). A better understanding of the microglia–GBM interactions could contribute in the development of effective therapeutic strategies to combat GBMs.

11.2.2 Niche-Stem Cells and Adhesion of Tumor Cells

Cell–cell contact molecules have been shown to control the asymmetric division of stem cells (Rasin et al. 2007), and neural stem cells possess a homing mechanism similar to that which regulates blood-marrow stem cells. Blood vessels from the subventricular zone (SVZ) as well as ependymal cells express stromal cell-derived factor-1 (SDF-1, CXCL12), while neural stem cells express its receptor CXCR4 (see Chap. 7). SDF-1 acts mainly on activated progenitors (not on quiescent cells) and promote their adhesion to endothelial cells by upregulating integrin α6 and thus leading to laminin binding. Quiescent neural stem cells express low levels of α6β1 integrin and syndecan-1 (Kazanis et al. 2010). SDF-1 also enhances the expression of EGF receptor (EGFR, a receptor linked to stem cell proliferation) in neural progenitors.

Id proteins are inhibitors of the bHLH class of transcription factors, a family of proteins that function as primary inducers of cell fate determination and differentiation in mammals (Perk et al. 2005). These proteins have been shown to suppress neural commitment of embryonic stem cells (Ying et al. 2003) and also to control neural stem cell maintenance through the adhesion of these cells to the niche (Niola et al. 2012). In addition to the ECM components, the diffusible signals released by vascular and ependymal cells are also important in regulating neural progenitors (Mirzadeh et al. 2008; Shen et al. 2008; Androutsellis-Theotokis et al. 2009).

Recent studies have shown that brain tumors can arise both from differentiated cells (Friedmann-Morvinski et al. 2012; Radke et al. 2013) and from the cancer stem-like cells (Chen et al. 2012). In some GBM patients, a spatial relationship

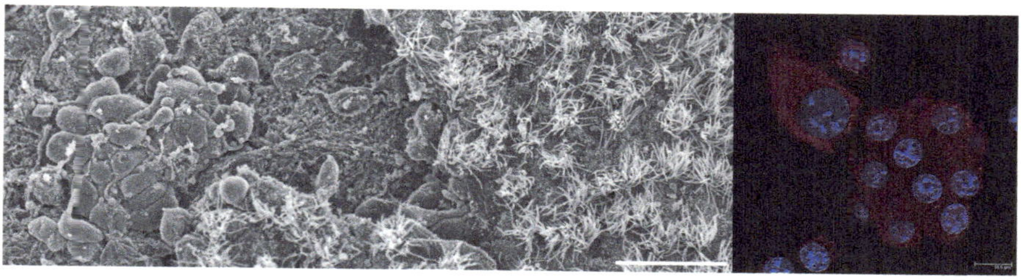

Fig. 11.1 The ventricular–subventricular zones as seen by scanning electron microscopy. In the VZ–SVZ region neural progenitors interact with several other tissue components that most likely participate in the control of proliferation and differentiation (**a**). Figure (**b**) shows an oncosphere—an aggregate of cancer stem cells that can be isolated and grown in culture, stained for Sox2 (red). Nuclei were stained with DAPI (blue). Bars: 25 μm (**a**); 10.6 μm (**b**)

between the tumor location and the neural stem cell niche is observed, while in other patients this association does not exist (Lim et al. 2007). A high incidence of multifocal GBM was observed in patients presenting lesions that contact both the SVZ and cortex regions. The authors suggested (Lim et al. 2007) that the SVZ microenvironment may be highly permissive to cell migration and proliferation and thus to tumor growth (Fig. 11.1). Focal adhesion kinase (FAK), a kinase involved in migration and proliferation, is overexpressed in GBM (Assoian and Klein 2008), as is osteopontin, which promotes motility and resistance to apoptosis in GBM cell lines (Yamaguchi et al. 2013). An RGD-integrin antagonist has been shown to inhibit adhesion and migration, and to induce anoikis in cultured GBM cells. These data explicitly illustrate the importance of studies of the interaction between GBM cells (whether or not they are stem cells) and ECM components for the understanding of tumorigenesis in the brain. The means by which adhesion proteins and other ECM components are regulated in concert with stem cells during tumorigenesis in the brain are still unknown.

The brain contains specific regions where neural stem cells can be found, and in these regions the generation of neurons and glial cells is sustained throughout life in animals (Alvarez-Buylla et al. 2002). These regions include the dentate gyrus of the hippocampus and the ventricular–subventricular zone of the lateral ventricles (Ihrie and Alvarez-Buylla 2011). The concept of a stem cell niche includes the idea that stem cells are kept quiescent, but can be reactivated when needed. When these cells reenter the cell cycle, they will divide asymmetrically and give rise to more-committed progenitors. These in turn will provide differentiated cells to compensate for lost, damaged, or aged cells. The neural stem cell lineage progression most likely involves the participation of niche-derived signals, which, together with cell autonomous signaling, orchestrate the events necessary for the maintenance of tissue homeostasis.

This concept also involves the idea that tumorigenesis may arise when niche-mediated control mechanisms fail to keep the cells quiescent or attached to the niche.

Neural stem cell behavior has been shown in many studies to be regulated by factors that are secreted by vascular and ependymal cells, and also by ECM proteins that are present in the basal lamina of vessels (see above-cited references). This supports the idea that, in addition to autonomous cell regulatory mechanisms niche signals are highly important for maintaining the balance of their quiescence, self-renewal, and differentiation.

11.2.3 Extracellular Vesicles and Their Content in Glioma Cell Biology

Glioblastoma cells secrete membrane vesicles, the exosomes, which may contain miRNA. Cells can transfer information to one another, even over long distances, by releasing extracellular vesicles (EVs) into the cell environment (Cocucci et al. 2009). The EVs are identified as microvesicles, microparticles, apoptotic blebs, exosomes, and others (Cocucci et al. 2009; Gyorgy et al. 2011). The two major types are microvesicles and exosomes (Simons and Raposo 2009; Raposo and Stoorvogel 2013). Microvesicles (or shedding vesicles) are 0.1–1 μm-diameter structures generated from outward protrusions of the plasma membrane, which detach from the cell surface by fission of their neck. Microvesicles participate in different aspects of tumor biology, as they are secreted in large numbers by tumor cells and function in events such as blood coagulation, angiogenesis, and metastasis (Diamant et al. 2004; Lima et al. 2013; Lima and Monteiro 2013).

11.2.3.1 Exosomes

Exosomes are small cup-shaped extracellular vesicles with a diameter from 30 to 100 nm. Exosomes originate from the endocytic pathway, generated upon fusion of multivesicular endosomes (MVBs) to the plasma membrane and the consequent release of intraluminal vesicles (ILVs) of MVBs into the extracellular environment. At this point, ILVs are referred to as exosomes (Simons and Raposo 2009; Raposo and Stoorvogel 2013).

Vesicles with the features of exosomes have been found in several cell types of both hematopoietic and non-hematopoietic origin. Under normal physiological conditions, many body fluids such as blood (Caby et al. 2005) and cerebrospinal fluid (Vella et al. 2007) also contain exosomes. Some pathogens induce their hosts to secrete exosomes in order to exploit these communication devices to maximize their spread between cells. This is the case for prions (Fevrier et al. 2004; Porto-Carreiro et al. 2005), β-amyloid peptides of Alzheimer's disease (Rajendran et al. 2006), and some retroviruses that exit from cells hijacked inside exosomes (Gould et al. 2003; Izquierdo-Useros et al. 2011). Tumors also secrete exosomes that contribute to the pathogenesis of cancer. Tumor-derived exosomes create an immunosuppressive microenvironment and facilitate the establishment of the pre-metastatic niche by enhancing angiogenesis, remodeling stromal cells, and promoting extracellular matrix degradation (Filipazzi et al. 2012; Yang and Robbins 2011). Brain tumors, especially glioblastoma, secrete large amounts of both microvesicles and exosomes, which serve as a means by which tumors manipulate their environment (Lai and Breakefield 2012).

11.2.3.2 Exosomes Contain RNA Species

A great advance in the field of exosome biology was the discovery of RNA species associated with exosomes. It has become evident that, in most cancers, abnormalities in noncoding genes can also contribute to cancer pathogenesis, in addition to alterations in protein-encoding genes. A class of noncoding small RNAs termed microRNAS (miRNAs) has been particularly implicated in tumorigenesis. Exosomes contain coding mRNA, microRNA (Ratajczak et al. 2006; Valadi et al. 2007), and a variety of other small, noncoding RNA sequences (Bellingham et al. 2012; Nolte-'t Hoen et al. 2012). Both the mRNA and microRNA present in exosomes can be delivered to another cell and can function in this new location (Valadi et al. 2007; Mittelbrunn et al. 2011; Montecalvo et al. 2012). Glioblastoma cells are exceptional in producing RNA-containing EVs. Extracellular vesicles released by this type of tumor contain large amounts of mRNA, microRNA, and angiogenic proteins (Skog et al. 2008). EVs from glioblastoma cells are taken up by normal brain endothelial cells of the host, and mRNA present in EVs can be translated in these receptor cells. The angiogenic proteins also found in EVs of glioblastoma can stimulate tubule formation by these endothelial cells. EVs purified from the serum of GBM patients stimulated the proliferation of a human glioma cell line in vitro. Skog and colleagues also found that serum from these patients contained microRNA sequences characteristic of glioblastoma, such as the mutant/variant of the epidermal growth factor receptor (EGFR)—EGFRνIII—(see Chaps. 3 and 6) indicating that EVs can be useful in diagnostic approaches.

11.2.3.3 miRNAs in Brain Cancer

The potential role of miRNAs in cancer initiation, progression, metastasis, and therapy opens new prospects for research in the field of GBM (Ciafre et al. 2005; Li et al. 2009), due to the deregulated expression of these molecules that seem to be very important in the genesis and development of many types of cancer (Hermeking 2007; Zhang et al. 2007). Also, some miRNAs may cause chemosensitivity and radioresistance in GBM (Shi et al. 2010; Chan et al. 2012; Ujifuku et al. 2010; Zhang et al. 2012). A cancer stem cell (CSC)-specific miRNA expression profile has been identified in cancer stem/progenitor cells (Chu et al. 2013; Di Stefano et al. 2011; Gonzalez-Gomez et al. 2011; Sharma and Wu 2013; Wilson et al. 2009). Based on the abundant evidence for the role of miRNAs in biological mechanisms and tumorigenesis, currently there is a consensus that miRNAs are promising molecular markers and therapeutic agents that could be used in parallel to current GBM treatment (for details see Chap. 4).

In conclusion, the exosomes of glioblastoma cells appear to be a useful way for a tumor to manipulate the environmental conditions in favor of its proliferation and dissemination. Further, exosomes may serve as vectors of information, since their content is a mirror of the tumor from which they originate (Kucharzewska et al. 2013). In addition, microvesicles and exosomes from tumor cells act as signaling platforms that diffuse to target cells in the microenvironment, modulating interactions of tumor cells with stromal, immune, or vascular cells, as well as priming the formation of metastasis. Because of their stability, they can be used as biomarkers for cancer detection and offer a range of molecular targets for the control of cell–cell interactions during tumor invasion (Martins et al. 2013).

11.3 The High Invasiveness of Gliomas: New Molecular Targets

During the tumor invasion, four independent and coordinated processes are present. First, the tumor mass detaches from the original location, involving downregulation of neural cell adhesion molecules (NCAMs) and E-cadherin, as well as upregulation of cell surface glycoprotein (CD44), which functions as an adhesion molecule (Thomas and Speight 2001). The second process is the interaction with the extracellular matrix (ECM), mediated by integrins and tenascin-C (Alves et al. 2011; Bellail et al. 2004), leading to the degradation and remodeling of the ECM, including the modulation of the metalloproteinase (MMP) activity. These proteases are responsible for the breakdown of the basal membrane, creating and maintaining a microenvironment that facilitates tumor cell survival (Platten et al. 2001). The most common molecules that allow glioma cells to adhere to the ECM are the integrins.

11.3.1 Integrins in Cell Adhesion and Invasion

The integrins are cell surface receptors composed by α and β subunits, which can be linked non-covalently, forming 24 different heterodimers, capable of mediating cell–cell interactions and cell-to-ECM attachment (Barczyk et al. 2010). They are involved in adhesion, migration, and survival processes of tumor cells, and establish the interaction between the microenvironment and tumor cells, including angiogenic endothelial cells. Integrins can bind directly to components of the ECM cell-membrane proteins, providing anchorage for cell motility and invasion (Guo and Giancotti 2004; Desgrosellier and Cheresh 2010). The integrin receptors have the ability to bind a wide variety of ligands (Niu and Chen 2011), and are grouped into subgroups based on ligand-binding properties or on their subunit composition. The β1 integrins, β2 integrins, and αv-containing integrins are the three largest groups (Fig. 11.2).

In addition to their role as cell adhesion molecules, binding of integrin to ECM ligands plays a crucial role as a transducer of bidirectional cell signaling, such as the NF-κB, PI3K/Akt, SRC, and RAS/MAPK pathways that may regulate survival, proliferation, invasion, and angiogenesis (Reardon and Cheresh 2011). Several studies have confirmed the existence of increased integrin interactions in glioma cells with the extracellular matrix, which may contribute to the migration and invasion of glioma cells (Reardon and Cheresh 2011; MacDonald et al. 2011; Vogetseder et al. 2013; Gladson and Cheresh 1991; Srikanth et al. 2013). It has been established that focal adhesion kinases (FAKs) and Src family kinases (SFKs) are key to the signal-transduction pathways triggered by integrins, which have been

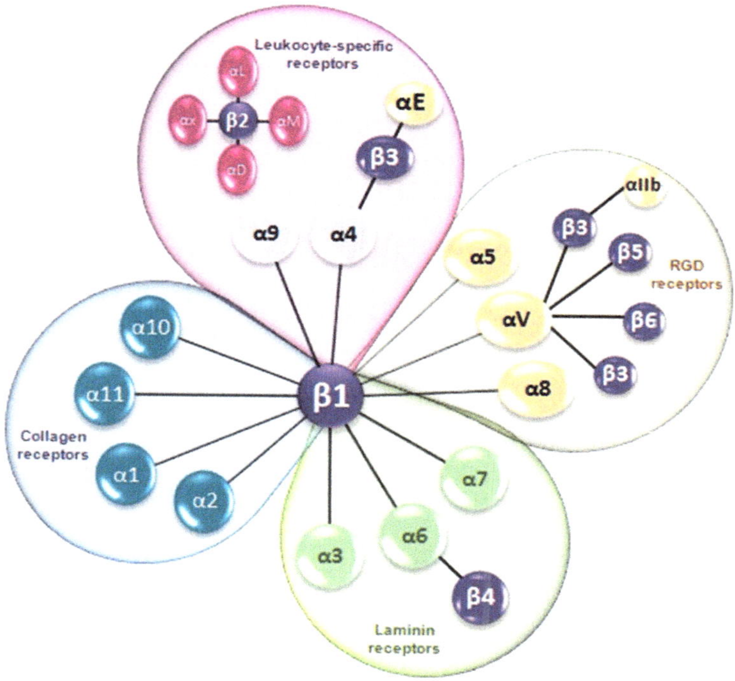

Fig. 11.2 The integrin subfamilies. There are 24 different integrin receptors in mammals. The integrins are composed by 18 α and 8 β subunits. This family has structural complexity, and also some subunits can combine with several different subunits. Integrins are the main receptors for ECM proteins, counter receptors (intercellular adhesion molecules, ICAMs, and vascular cell adhesion molecule, VCAM), and soluble molecules (fibrinogen). There are four different subfamilies: the RGD-binding integrins (*yellow*), laminin-binding integrins (*green*), leukocyte integrins (*pink*), and collagen integrins (*blue*)

involved in the regulation of cellular morphology, migration, proliferation, survival, and differentiation of tumor cells (Guo and Giancotti 2004).

The focal complexes are defined as smaller adhesion contacts found at the tip of extending protrusions, such as filopodia and lamellipodia (Petit and Thiery 2000). They are composed of clusters of integrins, a multiprotein complex that has anchor proteins for the actin cytoskeleton. These complexes can grow and stabilize to form the focal contacts, which provide biomechanical linkages between the actin network from the cytoskeleton and the extracellular binding sites (Hynes 2002; Demuth and Berens 2004). The focal contacts stabilize cellular structures; the focal complexes, on the other hand, show rapid turnover. These two modes of adhesion clearly demonstrate the dynamics of cell–ECM interaction as cells move along the substrate (Demuth and Berens 2004) (Fig. 11.3).

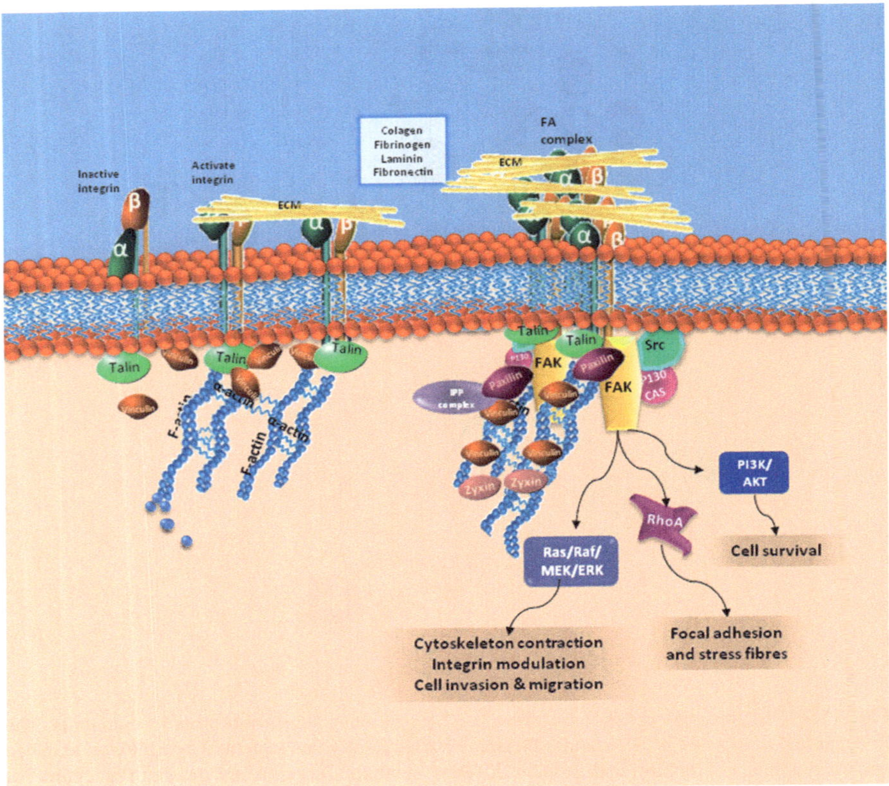

Fig. 11.3 Integrins are the major receptor for focal adhesions. Integrin binding promotes receptor clustering and the formation of focal adhesions. Many integrins that are not bound to the extracellular matrix (ECM) are present at the cell surface in an inactive conformation. Recruitment of intracellular proteins induces conformational changes in integrins, which allow the integrin to bind to specific ECM molecules. The maturation of focal adhesions involves clustering of active, ligand-bound integrins and the assembly of a multiprotein complex that is capable of linking integrins to the actin cytoskeleton and communicating with signaling pathways. Talin forms the initial contacts between integrin β-tails and the actin cytoskeleton. Vinculin cross-links with talin and actin thereby strengthen focal adhesions promoting focal adhesion growth. α-Actinin can interact with integrin β-tails, vinculin, talin, and actin to further reinforce the cross-linking of the cytoskeleton. Focal adhesion kinase (FAK) acts to integrate signals from extracellular cues, such as growth factor receptors and integrins, and from the upstream SRC family kinases, to control and coordinate adhesion dynamics and cell migration with survival signaling

11.3.2 Interactions Between the Extracellular Matrix Microenvironment and GBM Cells Mediated by Integrins

The most common molecules that allow glioma cells to adhere to ECM are the integrins, and integrins are also involved in many crucial pathways that contribute to the malignant GBM phenotype (Uhm et al. 1999). Therefore, integrin inhibitors might be ideal therapeutic tools for glioblastoma treatment. Many studies have suggested and confirmed the role of integrins as specialized cell adhesion molecules with the ability to integrate into different tissue microenvironments that modulate tumor growth and sensitivity to therapies (Weis et al. 2011).

Recently, it has been demonstrated that cell signaling generated by growth factors and oncogenes in transformed cells requires collaboration with specific integrins, especially during tumor initiation. In addition to their role as cell adhesion molecules, binding of integrins to ECM ligands plays a crucial role as a transducer of bidirectional cell signaling, such as increased expression of B-cell lymphoma 2 protein (BCl-2 or FlIP; also known as CFlAR), activation of the PI3K/Akt pathway, nuclear factor-κB (NF-κB) signaling, and p53 inhibition (Reardon and Cheresh 2011).

Many integrins including αvβ3, αvβ5, αvβ6, α2β1, α5β1, α6β1, and α6β4 have been implicated in tumor growth and invasion. Particularly for glioblastoma cells, some studies have confirmed that integrins, such as αvβ3, are specialized cell adhesion molecules with the ability to integrate environmental alterations, modulating the tumor growth (MacDonald et al. 2011).

The αV-integrins are overexpressed in brain metastases and tumor vasculature, especially αvβ3, αvβ8, and αvβ3 integrins, so inhibitors of αV-integrins are currently in clinical trials for glioblastoma treatment (Reardon and Cheresh 2011; Vogetseder et al. 2013; Gladson and Cheresh 1991). Integrin antagonists also block tumor angiogenesis and metastasis (Weis and Cheresh 2011; Weis et al. 2011). Thus, antagonists of these integrins may possibly suppress cell migration and invasion of primary and transformed cells, and induce apoptosis.

GBMs express integrin α3β1, both in vivo and in vitro (Fukushima et al. 1998). Treatment of cells with the α3β1 antibody decreases integrin expression on the cell surface, induces MMP-2 activity, and consequently increases tumor invasiveness (Chintala et al. 1996).

Interestingly, the αvβ8 integrin regulates the activation of Rho proteins, which are involved in GBM invasiveness, so that targeting the αvβ8 integrin–RhoGDI1 signaling axis regulates the activation of Rho GTPases. This could be an effective strategy for blocking GBM cell invasion (Reyes et al. 2013).

The heterodimer α6β1 plays a pivotal role in maintaining adhesion to the ventricular zone and contributes to tumor cell proliferation, survival, self-renewal, and in vivo growth (Lathia et al. 2010). Delamarre and collaborators (Delamarre et al. 2009) showed that U87MG cells express the α6β1 integrin, which increases cell spreading and attachment on laminin-111, leading to induction of proliferation and reduction of apoptosis due to serum starvation, and increased migration and invasion of GBM cells, in vitro and in vivo (Delamarre et al. 2009).

The presence of glioma stem(-like) cells (GSCs) in the tumor mass confers the ability to relapse and provides the chemoresistance phenotype. Some studies have reported that elimination of GSCs effectively slowed tumor progression (Lima et al. 2012; Lathia et al. 2010). In the brain, integrin α6, the receptor of the ECM protein laminin, which forms heterodimers with β1 and β4 integrins, has an important role in regulating the growth of neural stem cells and in maintaining adhesion to the ventricular zone, ensuring proper neural stem cell division. Therefore, targeting integrin α6 has severe consequences for the GSC phenotype and may provide a therapeutic target for GBMs (Lathia et al. 2010). Recently, studies conducted by Srikanth and collaborators showed that a self-assembling peptide amphiphile (PA), displaying the IKVAV (isoleucine-lysine-valine-alanine-valine) epitope, increases immobilization of β1-integrin at the GSC membrane, activating the integrin-linked kinase while inhibiting the focal adhesion kinase (FAK) (Srikanth et al. 2013).

Taken together, this information indicates that integrins are a promising therapeutic target in cancer, and a key to understanding the cellular processes of tumor growth. Antagonists of these integrins can suppress cell migration and invasion of primary and transformed cells.

11.4 How Do Current Standard Therapies Affect Glial Cells Within the Tumor Environment?

Surgery followed by a combination of radiotherapy and chemotherapy is the standard treatment for GBM (DeAngelis 2001; Lima et al. 2012). However, treatment outcomes did not change significantly over the last decade, and the 5-year survival rate of GBM patients is only 10 % (Stupp et al. 2009). The scenario becomes even grimmer if it is considered that 50–90 % of adult brain tumor patients who survive more than 6 months after radiotherapy develop some type of cognitive impairment, including severe declines in learning and memory (Nieder et al. 1999). Therefore, understanding how the heterogeneous cell populations respond to current therapies is necessary to improve treatment effectiveness and to offer a better quality of life to survivors.

Radiation-induced brain injury is usually divided into three phases: acute, early delayed, and late delayed. Early delayed (1–6 months after irradiation) may result in transient demyelination. Late-delayed injuries, which develop more than 6 months post-irradiation, include CNS edema, demyelination, vascular abnormalities, and white-matter loss. The cellular basis and the kinetics of these transformations are not completely understood. GBMs are significantly infiltrated by microglial (MG) cells, and microglial cells are considered the primary cellular mediators of radiation-induced neuroinflammation. Recently, Hua and collaborators showed that, in healthy adult rats, whole-brain irradiation (10 Gy) increased the proliferation of oligodendrocyte precursor cells, but only a few microglia underwent the transition to the classically "activated" microglial phenotype (e.g., enlarged cell body, reduced number of processes, and markers of phagocytosis) in response to irradiation (Hua et al. 2012). These findings are partially consistent with previous studies that observed increased expression of an immediate early gene (Egr1) in astrocytes and oligodendrocytes within a few hours after irradiation, but also showed an increase in the proportion of CD68+ cells, indicating strong activation of microglia (Vollmann et al. 2007).

Recently, our group has shown that equinatoxin II, a pore-forming toxin from the sea anemone *Actinia equina*, potentiates the cytotoxicity induced by temozolomide, a first-line GBM treatment, and by etoposide (VP-16), a second- or third-line GBM treatment. This effect is selective for GBM cells and occurs via inhibition of the PI3K/Akt pathway. In the same study, magnetic resonance imaging revealed that equinatoxin II potentiates the VP-16-induced inhibition of GBM growth in vivo. The conjugation of conventional chemotherapeutic drugs with equinatoxin II allows the use of lower chemotherapy doses in GBM patients and possibly reduces the adverse effects of chemotherapy (Kahn et al. 2012).

Conclusions

We commented here on the differences and similarities of glial cell functions in the healthy body and in an uncontrolled disease such as cancer. Several properties of normal glial cells, such as migration and interactions with cells present in the brain parenchyma, interaction with astrocytes, and exchange of molecules of normal and tumor astrocytes with neurons, microglia, and also endothelial cells from the vasculature, are preserved at least partially when astrocytes become an astrocytoma, a type of glioblastoma. Contemporary therapeutic opportunities for glioblastoma are discussed, in a search for new targets. A brain tumor such as glioblastoma is still likely to have a fatal outcome. Therapeutic alternatives must be developed in order to counterbalance invasive surgeries, which cannot definitively resolve the problem.

References

Abreu JG et al (2002) Connective-tissue growth factor (CTGF) modulates cell signalling by BMP and TGF-beta. Nat Cell Biol 4(8):599–604

Allen NJ, Barres BA (2009) Neuroscience: glia – more than just brain glue. Nature 457 (7230):675–677

Alvarez-Buylla A, Seri B, Doetsch F (2002) Identification of neural stem cells in the adult vertebrate brain. Brain Res Bull 57(6):751–758

Alves TR et al (2011) Tenascin-C in the extracellular matrix promotes the selection of highly proliferative and tubulogenesis-defective endothelial cells. Exp Cell Res 317(15):2073–2085

Androutsellis-Theotokis A et al (2009) Targeting neural precursors in the adult brain rescues injured dopamine neurons. Proc Natl Acad Sci U S A 106(32):13570–13575

Anido J et al (2010) TGF-beta receptor inhibitors target the CD44(high)/Id1(high) glioma-initiating cell population in human glioblastoma. Cancer Cell 18(6):655–668

Anton ES et al (1997) Role of GGF/neuregulin signaling in interactions between migrating neurons and radial glia in the developing cerebral cortex. Development 124(18):3501–3510

Assoian RK, Klein EA (2008) Growth control by intracellular tension and extracellular stiffness. Trends Cell Biol 18(7):347–352

Bae JJ et al (2011) Increased transforming growth factor-beta1 modulates glutamate receptor expression in the hippocampus. Int J Physiol Pathophysiol Pharmacol 3(1):9–20

Baghdassarian D et al (1993) Effects of transforming growth factor-beta 1 on the extracellular matrix and cytoskeleton of cultured astrocytes. Glia 7(3):193–202

Bahr M, Przyrembel C, Bastmeyer M (1995) Astrocytes from adult rat optic nerves are nonpermissive for regenerating retinal ganglion cell axons. Exp Neurol 131(2):211–220

Balaziova E et al (2011) Coupled expression of dipeptidyl peptidase-IV and fibroblast activation protein-alpha in transformed astrocytic cells. Mol Cell Biochem 354(1–2):283–289

Banker GA (1980) Trophic interactions between astroglial cells and hippocampal neurons in culture. Science 209(4458):809–810

Bao S et al (2006) Glioma stem cells promote radioresistance by preferential activation of the DNA damage response. Nature 444(7120):756–760

Barczyk M, Carracedo S, Gullberg D (2010) Integrins. Cell Tissue Res 339(1):269–280

Barres BA (2008) The mystery and magic of glia: a perspective on their roles in health and disease. Neuron 60(3):430–440

Bellail AC et al (2004) Microregional extracellular matrix heterogeneity in brain modulates glioma cell invasion. Int J Biochem Cell Biol 36(6):1046–1069

Bellingham SA, Coleman BM, Hill AF (2012) Small RNA deep sequencing reveals a distinct miRNA signature released in exosomes from prion-infected neuronal cells. Nucleic Acids Res 40(21):10937–10949

Bo L, Waugh RE (1989) Determination of bilayer membrane bending stiffness by tether formation from giant, thin-walled vesicles. Biophys J 55(3):509–517

Bottner M, Krieglstein K, Unsicker K (2000) The transforming growth factor-betas: structure, signaling, and roles in nervous system development and functions. J Neurochem 75(6) 2227–2240

Brionne TC et al (2003) Loss of TGF-beta 1 leads to increased neuronal cell death and microgliosis in mouse brain. Neuron 40(6):1133–1145

Bruna A et al (2007) High TGFbeta-Smad activity confers poor prognosis in glioma patients and promotes cell proliferation depending on the methylation of the PDGF-B gene. Cancer Cell 11 (2):147–160

Caby MP et al (2005) Exosomal-like vesicles are present in human blood plasma. Int Immunol 17 (7):879–887

Chan XH et al (2012) Targeting glioma stem cells by functional inhibition of a prosurvival oncomiR-138 in malignant gliomas. Cell Rep 2(3):591–602

Chen J et al (2012) A restricted cell population propagates glioblastoma growth after chemotherapy. Nature 488(7412):522–526

Chintala SK et al (1996) Modulation of matrix metalloprotease-2 and invasion in human glioma cells by alpha 3 beta 1 integrin. Cancer Lett 103(2):201–208

Chu PM et al (2013) Deregulated microRNAs identified in isolated glioblastoma stem cells: an overview. Cell Transplant 22(4):741–753

Ciafre SA et al (2005) Extensive modulation of a set of microRNAs in primary glioblastoma. Biochem Biophys Res Commun 334(4):1351–1358

Cocucci E, Racchetti G, Meldolesi J (2009) Shedding microvesicles: artefacts no more. Trends Cell Biol 19(2):43–51

de Sampaio e Spohr TC et al (2002) Neuro-glia interaction effects on GFAP gene: a novel role for transforming growth factor-beta1. Eur J Neurosci 16(11):2059–2069

DeAngelis LM (2001) Brain tumors. N Engl J Med 344(2):114–123

Delamarre E et al (2009) Expression of integrin alpha6beta1 enhances tumorigenesis in glioma cells. Am J Pathol 175(2):844–855

Demuth T, Berens ME (2004) Molecular mechanisms of glioma cell migration and invasion. J Neurooncol 70(2):217–228

Desgrosellier JS, Cheresh DA (2010) Integrins in cancer: biological implications and therapeutic opportunities. Nat Rev Cancer 10(1):9–22

Di Stefano B et al (2011) A microRNA-based system for selecting and maintaining the pluripotent state in human induced pluripotent stem cells. Stem Cells 29(11):1684–1695

Diamant M et al (2004) Cellular microparticles: new players in the field of vascular disease? Eur J Clin Invest 34(6):392–401

Diniz LP et al (2012) Astrocyte-induced synaptogenesis is mediated by transforming growth factor beta signaling through modulation of D-serine levels in cerebral cortex neurons. J Biol Chem 287(49):41432–41445
Diz-Munoz A, Fletcher DA, Weiner OD (2013) Use the force: membrane tension as an organizer of cell shape and motility. Trends Cell Biol 23(2):47–53
Faria J et al (2006) Interactive properties of human glioblastoma cells with brain neurons in culture and neuronal modulation of glial laminin organization. Differentiation 74(9–10):562–572
Fevrier B et al (2004) Cells release prions in association with exosomes. Proc Natl Acad Sci U S A 101(26):9683–9688
Filipazzi P et al (2012) Recent advances on the role of tumor exosomes in immunosuppression and disease progression. Semin Cancer Biol 22(4):342–349
Flanders KC et al (1993) Effects of TGF-betas and bFGF on astroglial cell growth and gene expression in vitro. Mol Cell Neurosci 4(5):406–417
Folkman J (1985) Tumor angiogenesis. In: George Klein SW (ed) Advances in cancer research. Academic, New York, NY, pp 175–203
Fonseca AC et al (2012) Microglial stress inducible protein 1 promotes proliferation and migration in human glioblastoma cells. Neuroscience 200:130–141
Friedmann-Morvinski D et al (2012) Dedifferentiation of neurons and astrocytes by oncogenes can induce gliomas in mice. Science 338(6110):1080–1084
Froes MM et al (1999) Gap-junctional coupling between neurons and astrocytes in primary central nervous system cultures. Proc Natl Acad Sci U S A 96(13):7541–7546
Fukushima Y et al (1998) Integrin alpha3beta1-mediated interaction with laminin-5 stimulates adhesion, migration and invasion of malignant glioma cells. Int J Cancer 76(1):63–72
Gagelin C et al (1995) Rapid TGF beta 1 effects on actin cytoskeleton of astrocytes: comparison with other factors and implications for cell motility. Glia 13(4):283–293
Garcia-Abreu J et al (1995) Regionally specific properties of midbrain glia: I. Interactions with midbrain neurons. J Neurosci Res 40(4):471–477
Garcia-Abreu J et al (2000) Contribution of heparan sulfate to the non-permissive role of the midline glia to the growth of midbrain neurites. Glia 29(3):260–272
Gauthier NC, Masters TA, Sheetz MP (2012) Mechanical feedback between membrane tension and dynamics. Trends Cell Biol 22(10):527–535
Giaume C (2010) Astroglial wiring is adding complexity to neuroglial networking. Front Neuroenergetics 2
Gittes F et al (1993) Flexural rigidity of microtubules and actin filaments measured from thermal fluctuations in shape. J Cell Biol 120(4):923–934
Gladson CL, Cheresh DA (1991) Glioblastoma expression of vitronectin and the alpha v beta 3 integrin. Adhesion mechanism for transformed glial cells. J Clin Invest 88(6):1924–1932
Gomes FC et al (1999a) Neurons induce GFAP gene promoter of cultured astrocytes from transgenic mice. Glia 26(2):97–108
Gomes FC, Paulin D, Moura Neto V (1999b) Glial fibrillary acidic protein (GFAP): modulation by growth factors and its implication in astrocyte differentiation. Braz J Med Biol Res 32(5):619–631
Gomes FC et al (1999c) Cerebellar astrocytes treated by thyroid hormone modulate neuronal proliferation. Glia 25(3):247–255
Gomes FC, Sousa Vde O, Romao L (2005) Emerging roles for TGF-beta1 in nervous system development. Int J Dev Neurosci 23(5):413–424
Gonzalez-Gomez P, Sanchez P, Mira H (2011) MicroRNAs as regulators of neural stem cell-related pathways in glioblastoma multiforme. Mol Neurobiol 44(3):235–249
Gould SJ, Booth AM, Hildreth JE (2003) The Trojan exosome hypothesis. Proc Natl Acad Sci U S A 100(19):10592–10597
Guo WJ, Giancotti FG (2004) Integrin signalling during tumour progression. Nat Rev Mol Cell Biol 5(10):816–826

Gyorgy B et al (2011) Membrane vesicles, current state-of-the-art: emerging role of extracellular vesicles. Cell Mol Life Sci 68(16):2667–2688

Hattermann K, Mentlein R (2013) An infernal trio: the chemokine CXCL12 and its receptors CXCR4 and CXCR7 in tumor biology. Ann Anat 195(2):103–110

Hattermann K et al (2013) The CXCL16-CXCR6 chemokine axis in glial tumors. J Neuroimmunol 260(1–2):47–54

Hermeking H (2007) p53 enters the microRNA world. Cancer Cell 12(5):414–418

Hua K et al (2012) Regionally distinct responses of microglia and glial progenitor cells to whole brain irradiation in adult and aging rats. PLoS One 7(12):e52728

Hynes RO (2002) Integrins: bidirectional, allosteric signaling machines. Cell 110(6):673–687

Ignatova TN et al (2002) Human cortical glial tumors contain neural stem-like cells expressing astroglial and neuronal markers in vitro. Glia 39(3):193–206

Ihrie RA, Alvarez-Buylla A (2011) Lake-front property: a unique germinal niche by the lateral ventricles of the adult brain. Neuron 70(4):674–686

Izquierdo-Useros N et al (2011) Exosomes and retroviruses: the chicken or the egg? Cell Microbiol 13(1):10–17

Javelaud D, Mauviel A (2005) Crosstalk mechanisms between the mitogen-activated protein kinase pathways and Smad signaling downstream of TGF-beta: implications for carcinogenesis. Oncogene 24(37):5742–5750

Kahn SA et al (2012) Equinatoxin II potentiates temozolomide- and etoposide-induced glioblastoma cell death. Curr Top Med Chem 12(19):2082–2093

Kazanis I et al (2010) Quiescence and activation of stem and precursor cell populations in the subependymal zone of the mammalian brain are associated with distinct cellular and extracellular matrix signals. J Neurosci 30(29):9771–9781

Kettenmann H, Verkhratsky A (2008) Neuroglia: the 150 years after. Trends Neurosci 31 (12):653–659

Kjellman C et al (2000) Expression of TGF-beta isoforms, TGF-beta receptors, and SMAD molecules at different stages of human glioma. Int J Cancer 89(3):251–258

Krieglstein K et al (1998) Glial cell line-derived neurotrophic factor requires transforming growth factor-beta for exerting its full neurotrophic potential on peripheral and CNS neurons. J Neurosci 18(23):9822–9834

Krieglstein K et al (2002) TGF-beta and the regulation of neuron survival and death. J Physiol Paris 96(1–2):25–30

Kriegstein A, Alvarez-Buylla A (2009) The glial nature of embryonic and adult neural stem cells. Annu Rev Neurosci 32:149–184

Kucharzewska P et al (2013) Exosomes reflect the hypoxic status of glioma cells and mediate hypoxia-dependent activation of vascular cells during tumor development. Proc Natl Acad Sci U S A 110(18):7312–7317

Lai CP, Breakefield XO (2012) Role of exosomes/microvesicles in the nervous system and use in emerging therapies. Front Physiol 3:228

Laping NJ et al (1994) Transforming growth factor-beta 1 induces neuronal and astrocyte genes: tubulin alpha 1, glial fibrillary acidic protein and clusterin. Neuroscience 58(3):563–572

Lathia JD et al (2010) Integrin alpha 6 regulates glioblastoma stem cells. Cell Stem Cell 6(5):421–432

Lesne S et al (2002) Transforming growth factor-beta1-modulated cerebral gene expression. J Cereb Blood Flow Metab 22(9):1114–1123

Li Y et al (2009) MicroRNA-21 targets LRRFIP1 and contributes to VM-26 resistance in glioblastoma multiforme. Brain Res 1286:13–18

Lim DA, Alvarez-Buylla A (1999) Interaction between astrocytes and adult subventricular zone precursors stimulates neurogenesis. Proc Natl Acad Sci U S A 96(13):7526–7531

Lim DA, Huang Y-C, Alvarez-Buylla A (2007) The adult neural stem cell niche: lessons for future neural cell replacement strategies. Neurosurg Clin N Am 18(1):81–92, ix

Lima LG, Monteiro RQ (2013) Activation of blood coagulation in cancer: implications for tumor progression. Biosci Rep 33
Lima FR et al (2007) Cellular prion protein expression in astrocytes modulates neuronal survival and differentiation. J Neurochem 103(6):2164–2176
Lima FR et al (2012) Glioblastoma: therapeutic challenges, what lies ahead. Biochim Biophys Acta 1826(2):338–349
Lima LG et al (2013) Intercellular transfer of tissue factor via the uptake of tumor-derived microvesicles. Thromb Res 132(4):450–456
MacDonald TJ, Aguilera D, Kramm CM (2011) Treatment of high-grade glioma in children and adolescents. Neuro Oncol 13(10):1049–1058
Mallat M et al (1986) Two simian virus 40 (SV40)-transformed cell lines from the mouse striatum and mesencephalon presenting astrocytic characters. II. Interactions with mesencephalic neurons. Brain Res 391(1):23–31
Maness PF, Schachner M (2007) Neural recognition molecules of the immunoglobulin superfamily: signaling transducers of axon guidance and neuronal migration. Nat Neurosci 10(1):19–26
Martinez R, Gomes FC (2002) Neuritogenesis induced by thyroid hormone-treated astrocytes is mediated by epidermal growth factor/mitogen-activated protein kinase-phosphatidylinositol 3-kinase pathways and involves modulation of extracellular matrix proteins. J Biol Chem 277 (51):49311–49318
Martinez R, Gomes FC (2005) Proliferation of cerebellar neurons induced by astrocytes treated with thyroid hormone is mediated by a cooperation between cell contact and soluble factors and involves the epidermal growth factor-protein kinase a pathway. J Neurosci Res 80(3):341–349
Martins VR, Dias MS, Hainaut P (2013) Tumor-cell-derived microvesicles as carriers of molecular information in cancer. Curr Opin Oncol 25(1):66–75
Massague J, Gomis RR (2006) The logic of TGFbeta signaling. FEBS Lett 580(12):2811–2820
Mendes FA et al (2003) Concentration-dependent actions of glial chondroitin sulfate on the neuritic growth of midbrain neurons. Brain Res Dev Brain Res 142(2):111–119
Mirzadeh Z et al (2008) Neural stem cells confer unique pinwheel architecture to the ventricular surface in neurogenic regions of the adult brain. Cell Stem Cell 3(3):265–278
Mittelbrunn M et al (2011) Unidirectional transfer of microRNA-loaded exosomes from T cells to antigen-presenting cells. Nat Commun 2:282
Montecalvo A et al (2012) Mechanism of transfer of functional microRNAs between mouse dendritic cells via exosomes. Blood 119(3):756–766
Moura Neto V et al (1983) Microheterogeneity of tubulin proteins in neuronal and glial cells from the mouse brain in culture. EMBO J 2(8):1243–1248
Moura Neto V et al (1985) Astrocytic cerebellar cell clones synthesize the beta' isoforms of the beta-tubulin protein family. Neuroscience 16(2):333–341
Moura Neto V et al (1986) Two simian virus 40 (SV40)-transformed cell lines from the mouse striatum and mesencephalon presenting astrocytic characters. I. Immunological and pharmacological properties. Brain Res 391(1):11–22
Moustakas A, Heldin CH (2005) Non-Smad TGF-beta signals. J Cell Sci 118(Pt 16):3573–3584
Nieder C et al (1999) Late radiation toxicity after whole brain radiotherapy: the influence of antiepileptic drugs. Am J Clin Oncol 22(6):573–579
Niola F et al (2012) Id proteins synchronize stemness and anchorage to the niche of neural stem cells. Nature 14(5):477–487
Niu G, Chen X (2011) Why integrin as a primary target for imaging and therapy. Theranostics 1:30–47
Nolte-'t Hoen EN et al (2012) Deep sequencing of RNA from immune cell-derived vesicles uncovers the selective incorporation of small non-coding RNA biotypes with potential regulatory functions. Nucleic Acids Res 40(18):9272–9285
Oberheim NA, Goldman SA, Nedergaard M (2012) Heterogeneity of astrocytic form and function. Methods Mol Biol 814:23–45

Penuelas S et al (2009) TGF-beta increases glioma-initiating cell self-renewal through the induction of LIF in human glioblastoma. Cancer Cell 15(4):315–327

Perk J, Iavarone A, Benezra R (2005) Id family of helix-loop-helix proteins in cancer. Nat Rev Cancer 5(8):603–614

Persano L et al (2013) Glioblastoma cancer stem cells: role of the microenvironment and therapeutic targeting. Biochem Pharmacol 85(5):612–622

Petit V, Thiery JP (2000) Focal adhesions: structure and dynamics. Biol Cell 92(7):477–494

Platten M, Wick W, Weller M (2001) Malignant glioma biology: role for TGF-beta in growth, motility, angiogenesis, and immune escape. Microsc Res Tech 52(4):401–410

Pontes B et al (2011) Cell cytoskeleton and tether extraction. Biophys J 101(1):43–52

Pontes B et al (2013) Membrane elastic properties and cell function. PLoS One 8(7):e67708

Porto-Carreiro I et al (2005) Prions and exosomes: from PrPc trafficking to PrPsc propagation. Blood Cells Mol Dis 35(2):143–148

Radke J, Bortolussi G, Pagenstecher A (2013) Akt and c-Myc induce stem-cell markers in mature primary p53(-)/(-) astrocytes and render these cells gliomagenic in the brain of immunocompetent mice. PLoS One 8(2):e56691

Rajendran L et al (2006) Alzheimer's disease beta-amyloid peptides are released in association with exosomes. Proc Natl Acad Sci U S A 103(30):11172–11177

Raposo G, Stoorvogel W (2013) Extracellular vesicles: exosomes, microvesicles, and friends. J Cell Biol 200(4):373–383

Rasin MR et al (2007) Numb and Numbl are required for maintenance of cadherin-based adhesion and polarity of neural progenitors. Nat Neurosci 10(7):819–827

Ratajczak J et al (2006) Membrane-derived microvesicles: important and underappreciated mediators of cell-to-cell communication. Leukemia 20(9):1487–1495

Reardon DA, Cheresh D (2011) Cilengitide: a prototypic integrin inhibitor for the treatment of glioblastoma and other malignancies. Genes Cancer 2(12):1159–1165

Reyes SB et al (2013) alphavbeta8 integrin interacts with RhoGDI1 to regulate Rac1 and Cdc42 activation and drive glioblastoma cell invasion. Mol Biol Cell 24(4):474–482

Robles Irizarry L et al (2012) Therapeutic targeting of VEGF in the treatment of glioblastoma. Expert Opin Ther Targets 16(10):973–984

Romao LF et al (2008) Glutamate activates GFAP gene promoter from cultured astrocytes through TGF-beta1 pathways. J Neurochem 106(2):746–756

Romao LF et al (2013) Connective tissue growth factor (CTGF/CCN2) is negatively regulated during neuron-glioblastoma interaction. PLoS One 8(1):e55605

Schmid RS, Maness PF (2008) L1 and NCAM adhesion molecules as signaling coreceptors in neuronal migration and process outgrowth. Curr Opin Neurobiol 18(3):245–250

Sharma A, Wu JC (2013) MicroRNA expression profiling of human-induced pluripotent and embryonic stem cells. Methods Mol Biol 936:247–256

Shen Q et al (2008) Adult SVZ stem cells lie in a vascular niche: a quantitative analysis of niche cell-cell interactions. Cell Stem Cell 3(3):289–300

Shi L et al (2010) MiR-21 protected human glioblastoma U87MG cells from chemotherapeutic drug temozolomide induced apoptosis by decreasing Bax/Bcl-2 ratio and caspase-3 activity. Brain Res 1352:255–264

Siesser PF, Maness PF (2009) L1 cell adhesion molecules as regulators of tumor cell invasiveness. Cell Adh Migr 3(3):275–277

Simons M, Raposo G (2009) Exosomes–vesicular carriers for intercellular communication. Curr Opin Cell Biol 21(4):575–581

Skog J et al (2008) Glioblastoma microvesicles transport RNA and proteins that promote tumour growth and provide diagnostic biomarkers. Nat Cell Biol 10(12):1470–1476

Sousa Vde O et al (2004) Glial fibrillary acidic protein gene promoter is differently modulated by transforming growth factor-beta 1 in astrocytes from distinct brain regions. Eur J Neurosci 19 (7):1721–1730

Spohr TC et al (2008) Lysophosphatidic acid receptor-dependent secondary effects via astrocytes promote neuronal differentiation. J Biol Chem 283(12):7470–7479

Spohr TC et al (2011) Astrocytes treated by lysophosphatidic acid induce axonal outgrowth of cortical progenitors through extracellular matrix protein and epidermal growth factor signaling pathway. J Neurochem 119(1):113–123

Spohr TC et al (2012) Sphingosine 1-phosphate-primed astrocytes enhance differentiation of neuronal progenitor cells. J Neurosci Res 90(10):1892–1902

Srikanth M et al (2013) Nanofiber-mediated inhibition of focal adhesion kinase sensitizes glioma stemlike cells to epidermal growth factor receptor inhibition. Neuro Oncol 15(3):319–329

Stupp R et al (2009) Effects of radiotherapy with concomitant and adjuvant temozolomide versus radiotherapy alone on survival in glioblastoma in a randomised phase III study: 5-year analysis of the EORTC-NCIC trial. Lancet Oncol 10(5):459–466

Thomas GJ, Speight PM (2001) Cell adhesion molecules and oral cancer. Crit Rev Oral Biol Med 12(6):479–498

Uhm JH et al (1999) Vitronectin, a glioma-derived extracellular matrix protein, protects tumor cells from apoptotic death. Clin Cancer Res 5(6):1587–1594

Ujifuku K et al (2010) miR-195, miR-455-3p and miR-10a(*) are implicated in acquired temozolomide resistance in glioblastoma multiforme cells. Cancer Lett 296(2):241–248

Unsicker K, Strelau J (2000) Functions of transforming growth factor-beta isoforms in the nervous system. Cues based on localization and experimental in vitro and in vivo evidence. Eur J Biochem 267(24):6972–6975

Valadi H et al (2007) Exosome-mediated transfer of mRNAs and microRNAs is a novel mechanism of genetic exchange between cells. Nat Cell Biol 9(6):654–659

Vella LJ et al (2007) Packaging of prions into exosomes is associated with a novel pathway of PrP processing. J Pathol 211(5):582–590

Vogetseder A et al (2013) alphav-Integrin isoform expression in primary human tumors and brain metastases. Int J Cancer 133(10):2362–2371

Vollmann H et al (2007) Differential expression of egr1 and activation of microglia following irradiation in the rat brain. Strahlenther Onkol 183(5):248–255

Weis SM, Cheresh D (2011) αV integrins in angiogenesis and cancer. Cold Spring Harb Perspect Med 1(1):a006478

Wick W, Platten M, Weller M (2001) Glioma cell invasion: regulation of metalloproteinase activity by TGF-beta. J Neurooncol 53(2):177–185

Wilson KD et al (2009) MicroRNA profiling of human-induced pluripotent stem cells. Stem Cells Dev 18(5):749–758

Wyss-Coray T et al (1995) Increased central nervous system production of extracellular matrix components and development of hydrocephalus in transgenic mice overexpressing transforming growth factor-beta 1. Am J Pathol 147(1):53–67

Yamaguchi Y et al (2013) Thrombin-cleaved fragments of osteopontin are overexpressed in malignant glial tumors and provide a molecular niche with survival advantage. J Biol Chem 288(5):3097–3111

Yang C, Robbins PD (2011) The roles of tumor-derived exosomes in cancer pathogenesis. Clin Dev Immunol 2011:842849

Ying Q-L et al (2003) BMP induction of Id proteins suppresses differentiation and sustains embryonic stem cell self-renewal in collaboration with STAT3. Cell 115(3):281–292

Zhang B et al (2007) microRNAs as oncogenes and tumor suppressors. Dev Biol 302(1):1–12

Zhang C et al (2009) Co-suppression of miR-221/222 cluster suppresses human glioma cell growth by targeting p27kip1 in vitro and in vivo. Int J Oncol 34(6):1653–1660

Zhang QQ et al (2012) MicroRNA-195 plays a tumor-suppressor role in human glioblastoma cells by targeting signaling pathways involved in cellular proliferation and invasion. Neuro Oncol 14(3):278–287

Glioma-Associated Proteases

12

Petr Busek, Martin Prevorovsky, Evzen Krepela, and Aleksi Sedo

Abstract

Proteolytic enzymes constitute 2–3 % of all known human genes and represent an important tool for the control of the biological functions of proteins. During gliomagenesis, the complex regulation of proteases in transformed and stromal cells is impaired as a result of several factors, and proteases critically contribute to the hallmarks of gliomas. Proteins regulated by proteases include components of the extracellular matrix, local mediators, cell surface receptors, ion channels and adhesion molecules, cytoskeletal proteins, components of the intracellular signaling cascades, and regulators of the cell cycle. Via the proteolytic modifications of these substrates and/or by non-proteolytic mechanisms, the extracellular as well as intracellular proteases contribute to the increased invasiveness of glioma cells, promote the self-renewal and proliferation of glioma stem-like cells, and facilitate tumor neovascularization. The role of proteases in glioma progression is therefore multifaceted and complex. Glioma-associated proteases represent attractive therapeutic targets and several approaches were proposed including the inhibition of the extracellular proteases involved in glioma invasiveness as well as the inhibition of the proteasome and γ-secretase. However, a more precise understanding of the pathogenetic role of proteases in individual glioma patients at different stages of the disease is necessary as indicated by the relatively low therapeutic efficacy of the protease inhibitors in the initial clinical trials. Identification of the key protease-dependent processes in individual glioma patients, the most effective modes of protease targeting, and optimal delivery schedules and routes seem to be crucial to improve the therapeutic outcomes.

P. Busek (✉) • E. Krepela • A. Sedo (✉)
First Faculty of Medicine, Institute of Biochemistry and Experimental Oncology, Charles University in Prague, Prague, Czech Republic
e-mail: busekpetr@seznam.cz; aleksi@cesnet.cz

M. Prevorovsky
Faculty of Sciences, Department of Cell Biology, Charles University in Prague, Prague, Czech Republic

A. Sedo and R. Mentlein (eds.), *Glioma Cell Biology*,
DOI 10.1007/978-3-7091-1431-5_12, © Springer-Verlag Wien 2014

Keywords
Glioma • Tumorigenesis • Angiogenesis • Invasion • Apoptosis • Microenvironment • Therapeutic targeting • Tumor stroma

Abbreviations

ADAM	A disintegrin and metalloproteinase
ADAMTS	A disintegrin and metalloproteinase with thrombospondin motifs
bFGF	Basic fibroblast growth factor
BDNF	Brain-derived neurotrophic factor
BMP	Bone morphogenic protein
CAM	Cell adhesion molecule
CDKN2A	Cyclin-dependent kinase inhibitor 2A
CXCL12	Chemokine (C-X-C Motif) Ligand 12 (SDF stromal cell-derived factor)
CXCL16	Chemokine (C-X-C Motif) Ligand 16
DISC	Death-inducing signaling complex
ECE	Endothelin converting enzyme
ECM	Extracellular matrix
EGFR	Epidermal growth factor receptor
ERK	Extracellular signal-regulated kinase
FAK	Focal adhesion kinase
GCP-II	Glutamate carboxypeptidase-II (N-acetyl-L-aspartyl-L-glutamate peptidase I, NAALADase I, PSMA, prostate-specific membrane antigen)
GrB	Granzyme-B
HB-EGF	Heparin-binding EGF-like growth factor
HGF	Hepatocyte growth factor
HIF	Hypoxia inducible factor
IDH	Isocitrate dehydrogenase
IFN	Interferon
IGF	Insulin-like growth factor
MAPK	Mitogen-activated protein kinase
miRNA	Micro RNA
MMP	Matrix metalloproteinase
MT-MMP	Membrane-type MMP
PAI	Plasminogen activator inhibitor
PAR	Protease-activated receptor
PI3K	Phosphatidylinositol-4,5-bisphosphate 3-kinase
PTEN	Phosphatase and tensin homolog
RNAi	RNA interference
RTK	Receptor tyrosine kinase
Shh	Sonic hedgehog

SUMO	Small ubiquitin-like modifier
TF	Tissue factor (fIII)
TFPI	Tissue factor pathway inhibitor
TGF	Transforming growth factor
TIMP	Tissue inhibitor of matrix metalloproteinase
TNF	Tumor necrosis factor
tPA	Tissue-type plasminogen activator
TRAIL	TNF-related apoptosis-inducing ligand
uPA	Urokinase-type plasminogen activator
uPAR	Urokinase-type plasminogen activator receptor
USP	Ubiquitin-specific protease
VEGF	Vascular endothelial growth factor
VEGFR	Vascular endothelial growth factor receptor

12.1 Proteolytic Enzymes: General Overview

With a total of 585 proteases listed in the degradome database (http://degradome.uniovi.es/dindex.html), proteases constitute approximately 2–3 % of all known human genes (Puente et al. 2003). Biochemically, proteases belong to hydrolases that cleave the covalent bonds linking amino acids in the polypeptide backbone releasing smaller protein fragments and/or individual amino acids (Hooper 2002), or removing posttranslationally attached proteins such as ubiquitin or SUMO (Clague et al. 2012; Hickey et al. 2012). Based on the nature of their active site and the mechanism of action, five main classes of proteases are distinguished, i.e., aspartic, cysteine, metallo, serine, and threonine, with additional classes found in lower organisms (http://merops.sanger.ac.uk/, Rawlings et al. 2012). The specificity of the peptide bond cleavage varies widely: some proteases are highly specific (e.g., blood coagulation proteases, caspases) whereas others (e.g., proteasome, cathepsins, proteinase K) cleave a wide variety of substrates at several positions.

Besides its role in protein digestion and catabolism, the proteolytic cleavage represents a very important and mostly irreversible mechanism of protein function regulation on the posttranslational level (Clague et al. 2012; Hickey et al. 2012; Turk et al. 2012a). Proteolysis can lead to

1. Protein activation [e.g., zymogen activation, protein maturation by proprotein convertases such as furin (Seidah and Chretien 1999; Maret et al. 2012)],
2. Protein inactivation [e.g., cleavage of a specific inhibitor of caspase-activated DNAse by caspase-3 during apoptosis (Enari et al. 1998)],
3. Adjustment of the biological activity of the protein [e.g., chemokine processing by matrix metalloproteinases or aminopeptidases (Wolf et al. 2008)],

4. Changes affecting the turnover of the protein (e.g., changed susceptibility to degradation after the removal of ubiquitin by deubiquitinating enzymes (Wilkinson 2009), or protein cleavage that enables further proteolytic processing).

Proteolysis thus determines the spatiotemporal bioavailability of proteins either quantitatively (e.g., degradation of structural or nutritive proteins) or qualitatively by highly specific ("limited") proteolysis of particular peptide bonds, fine-tuning the biological activities of regulatory peptides (Konkoy and Davis 1996).

The activity of proteases can be controlled at several levels. In addition to the regulation of the expression level via transcriptional and posttranscriptional mechanisms and protein degradation, several proteases are synthesized in an inactive form (zymogen) that requires proteolytic cleavage in order to be converted to the active form. This activation is frequently accomplished by proteolysis in a multiprotein complex and represents a well-established regulatory step for, e.g., matrix metalloproteinases (MMPs) (Kessenbrock et al. 2010), plasminogen, urokinase-type plasminogen activator (uPA), and caspases (Donepudi and Grutter 2002). The activity of proteases is further regulated by their sequestration to specialized cellular compartments such as nucleus (Clague et al. 2012; Geng et al. 2012), endoplasmic reticulum, Golgi apparatus, endosomes, lysosomes, or secretory granules (Shen and Prywes 2004; Colbert et al. 2009; Lemberg 2011; Krzewski and Coligan 2012; Turk et al. 2012b; Bergbold and Lemberg 2013; Hattori and Tsujimoto 2013; Repnik et al. 2013; Seidah et al. 2013), mitochondria (Bulteau and Bayot 2011; Anand et al. 2013), and invadosomes (Brisson et al. 2012), where they meet with a stabilizing and optimal reaction microenvironment as well as with the target substrates. Besides this, proteases are controlled by a number of endogenous extracellular (secretory) and intracellular inhibitors. While α2-macroglobulin and α1-antitrypsin have the ability to inhibit a broad range of proteases, other endogenous inhibitors are more selective for individual proteases or protease classes, although there is some overlap in their specificities (Turk 2006; Mason and Joyce 2011). The serpins such as plasminogen activator inhibitors (PAI) 1 and 2 or α2-antiplasmin (Law et al. 2006) or the Kunitz-type inhibitors such as tissue factor pathway inhibitor (TFPI) predominantly inhibit serine proteases, whereas cystatins (Turk et al. 2008) and calpastatin inhibit the cysteine proteases cathepsins and calpains, respectively (Turk 2006; Turk et al. 2008; Cox 2009; Mason and Joyce 2011; Campbell and Davies 2012). Some intracellular serpins, such as serpinB3, serpinB4, and serpinB9, can function as cross-class inhibitors inhibiting serine as well as cysteine proteases (Law et al. 2006; Izuhara et al. 2008). The activity of MMPs, ADAMs ("A Disintegrin And Metalloproteinase"), and ADAMTS ("A Disintegrin And Metalloproteinase with Thrombospondin Motifs") is regulated by the four members of the tissue inhibitor of matrix metalloproteinase (TIMP) family TIMP1-4, which form tight 1:1 complexes with the target proteases (Murphy 2011). TIMP1–4 differ somewhat in their ability to inhibit individual metalloproteases and in their expression pattern. TIMP1 has a restricted inhibitory potential as it does not inhibit the membrane-type MMPs (MT-MMPs); it is widely expressed extracranially, but its expression in the brain is confined to the regions with persistent neuronal plasticity such as the hippocampus, the olfactory bulb, and the cerebellum (Rivera et al. 2010; Murphy 2011). TIMP2 is constitutively expressed in many tissues and is the most abundantly expressed TIMP in the brain at least in the rat (Fager and Jaworski 2000). TIMP2 plays a dual role in regulating metalloproteases; besides being a protease inhibitor, TIMP2 is critically involved in the complex process of MMP2 activation. TIMP2 forms a trimolecular complex with MMP2 and MT1-MMP (MMP14) thereby enabling efficient activation of MMP2 by MMP14 on the cell surface (Strongin et al. 1995). These results of in vitro studies are further supported by the observation that TIMP2 knockout mice have impaired MMP2 activation, which can be rescued by exogenous TIMP2 (Caterina et al. 2000; Wang et al. 2000). TIMP3 is expressed in several tissues and its ablation in mice leads to emphysema-like damage of the lungs and increased apoptosis in the mammary gland suggesting its important role in regulating metalloproteases (Murphy 2011). Expression of the recently described TIMP4 is restricted to heart, kidney, pancreas, colon, testes, adipose tissue, and brain (Melendez-Zajgla et al. 2008). The data on TIMP4 function are limited, but the protein is suspected to have pro-tumorigenic activities (Melendez-Zajgla et al. 2008). It is important to stress that the majority of the endogenous protease inhibitors exhibits several physiological and pathological roles that are independent of their ability to inhibit the activity of proteases (Rivera et al. 2010), which must be considered when interpreting their effects on protease-mediated processes in the experimental models.

In addition to the protease genes, several proteins highly homologous to proteases but devoid of the proteolytic activity due to amino-acid substitution (s) in the active site are encoded in the human genome (Puente et al. 2003). For example, human ADAMs from the metzincin subgroup of the zinc protease superfamily comprise 19 members, of which approximately half are enzymatically inactive (Seals and Courtneidge 2003; Klein and Bischoff 2011). The inactive protease homologues may have important regulatory functions in sequestering inhibitors or protease substrates, or participate on protein–protein interactions that are unrelated to their "evolutionary original" proteolytic role. This latter aspect is also typical for several "bona fide" proteases, as they frequently contain non-protease domains that enable non-hydrolytic protein–protein interactions (Del Rosso et al. 2002; Rozanov et al. 2004; Mina-Osorio 2008; Sakamoto and Seiki 2009; Dufour et al. 2010; Redondo-Munoz et al. 2010). Thus, a number of proteases can serve as receptors (Mina-Osorio 2008; Klein and Bischoff 2011; Raj et al. 2013), signaling molecules (Rogove et al. 1999; Sower et al. 1999; Beffert et al. 2006; LaRusch et al. 2010; Strongin 2010), or modulator molecules directly interacting with and influencing other components of the signal transduction pathways (Sumitomo et al. 2000).

Proteases act intracellularly, in the extracellular space as well as in body fluids and secretions. The intracellular proteases are indispensable for proper protein maturation [e.g., proprotein convertases (Seidah and Chretien 1999)], contribute to the cytoskeletal remodeling [e.g., calpains (Franco and Huttenlocher 2005)] and regulation of transcription (Best et al. 2002; Chapman 2004), remove misfolded, damaged, or unneeded proteins (lysosomal proteases, proteasomes) (Das et al. 2012; Kaminskyy and Zhivotovsky 2012; Viry et al. 2014), and initiate and execute apoptosis (Pop and Salvesen 2009; Goldschneider and Mehlen 2010). Proteolysis also takes place within biological membranes (Lemberg 2011; Bergbold and Lemberg 2013). This unique mechanism mediated by, e.g., presenilins or rhomboids frequently leads to the release of effector peptides, often with a signaling or transcription factor activity, from transmembrane proteins such as Notch (Lemberg 2011). Proteases acting in the extracellular space are either secreted (e.g., uPA, MMPs 2 and 9, cathepsins, ADAMTS) or plasma membrane bound (e.g., MT1-MMP, ADAMs). Nevertheless, the secreted proteases can also associate with the cell surface by binding to specific membrane receptors as demonstrated for uPA, MMP2, MMP9, and procathepsin-B and their respective receptors uPAR, α-v-β-3 integrin, CD44, and annexin II (Brooks et al. 1996; Yu and Stamenkovic 1999; Mohamed and Sloane 2006; Eden et al. 2011). This binding enhances the proteolytic activity not only by the spatial concentration of the protease but frequently also decreases the effects of endogenous inhibitors.

The widespread distribution of proteases reflects their participation in almost every physiological process on the level of individual cells as well as on the level of the whole organism. Proteases exert control of cell behavior including its metabolism, signal sensing and transduction, proliferation and death, as well as participate among others in angiogenesis, blood clotting, and immune defense (see Lopez-Otin and Bond (2008) for review).

Proteases from various classes are expressed in the brain and fulfil numerous biological functions. The serine proteases neurotrypsin and neuropsin were originally identified due to their abundant expression in the neural tissue, but other more "traditional" serine proteases such as thrombin, plasmin, tPA, and trypsin 4 are also locally produced in the brain (see Wang et al. (2008b), Almonte and Sweatt (2011) and references therein). Similarly, metalloproteases such as MMP2, MMP9, several ADAMs, and ADAMTS4 and 5 are expressed in the central nervous system (for review see Yang et al. 2006; Rivera et al. 2010). Extracellular as well as intracellular brain proteases play an important role in neuronal signaling. Besides the processing, conversion, and inactivation of neuropeptides and growth factors such as pro-BDNF (brain-derived neurotrophic factor) (Hallberg et al. 2005; Almonte and Sweatt 2011), proteases trigger intracellular signaling cascades through "protease-activated receptors" (PAR) 1–4. These receptors are expressed on the surface of neurons, microglia, and astrocytes and their proteolytic cleavage exposes an amino-terminal part of the molecule that acts as a tethered intramolecular ligand for the same receptor and activates signal transduction through G proteins (see Noorbakhsh et al. (2003) for review). Ligand-gated ion channels may also be cleaved by proteases with subsequent changes in their turnover or signaling. For example, the intracellular calcium-activated cysteine proteases calpains cleave the NR2A subunit of the NMDA receptors (Guttmann et al. 2002) as well as degrade GRIP (glutamate receptor-interacting protein), a molecule important for the morphological and functional organization of the synapses (Lu et al. 2001). Likewise, the extracellular serine protease tPA is thought to be directly involved in glutaminergic transmission as it is induced by neuronal activity (Qian et al. 1993) and facilitates NMDA receptor-mediated signaling by cleaving the NR1 subunit (Nicole et al. 2001). Neuronal neurotransmitter receptors are also regulated by a spatiotemporally limited activation of caspases. In hippocampal neurons, NMDA receptor signaling causes caspase-3 activation by the mitochondrial pathway. The activated caspase-3 subsequently cleaves and inactivates the serine-threonine protein kinase Akt1, which blocks AMPA receptor internalization and inhibits long term depression (LTD) (Li et al. 2010b; Li and Sheng 2012). Similarly, caspase-1 was demonstrated to function as a negative regulator of the AMPA receptor signaling, resulting in the inhibition of long term potentiation (LTP) (Lu et al. 2006a). Other proteases with signaling roles in the brain include secretases that are indispensable for the Notch-1 signaling by liberating its C-terminal intracellular domain (Prox et al. 2012) and the ubiquitin-proteasome system, which impacts, e.g., on the Notch, Wnt, BMP, or Shh signal transduction pathways by regulating their intracellular effectors (Lehman 2009). The neuropsin-mediated cleavage of fibronectin (Almonte and Sweatt 2011) and of the L1 cell adhesion molecule (L1CAM, Shimizu et al. 1998) as well as the processing of beta-dystroglycan and N-cadherin by MMP9 (Dziembowska and Wlodarczyk 2012) are other examples of the remodeling of the extracellular matrix and the cleavage of cell adhesion molecules by which several proteases play an important role in LTP, synapse remodeling, and neuronal plasticity [for further details see, e.g., Tomimatsu et al. (2002), Reif et al. (2007), Stephan et al. (2008), Wright and Harding (2009), Shiosaka and Ishikawa (2011), Dziembowska and Wlodarczyk (2012), Li and Sheng (2012)].

In addition, proteases are essential factors during brain development. MMPs such as MMP3 and MMP9 contribute to neurite outgrowth, axon guidance, and neuronal migration (Vaillant et al. 1999, 2003; Van Hove et al. 2012). A similar role was proposed for the plasminogen activator-plasmin system (Seeds et al. 1997) and caspases (Williams et al. 2006; Westphal et al. 2010). ADAMs, most notably ADAM 10, 22, and 23, seem to have a complex role during central nervous system (CNS) development affecting the proliferation, neuronal migration, axon growth, and differentiation of neuronal progenitors, and their absence in mice results in serious CNS defects and early mortality (reviewed in Yang et al. 2006). Likewise, cathepsins were demonstrated to be indispensable for the proper development of the CNS. Combined deletion of the cysteine cathepsins B and L leads to neuronal loss with profound brain atrophy as well as early mortality in mice (Felbor et al. 2002). The deficiency of the aspartate protease cathepsin-D in mice was reported to cause disturbed myelin structure (Mutka et al. 2010), accumulation of ceroid, astrogliosis, and early mortality (Shevtsova et al. 2010) paralleling the phenotype of the congenital neuronal ceroid lipofuscinosis caused by cathepsin-D deficiency in humans (Siintola et al. 2006; Fritchie et al. 2009). The phenotype in cathepsin-K knockout mice is less severe and involves changes in hippocampal cytoarchitecture as well as learning and memory deficits (Dauth et al. 2011).

Imbalances in the proteolytic activities lead to several pathologies (Lopez-Otin and Bond 2008) including inflammatory, neurodegenerative, and malignant diseases affecting the CNS. Proteases were thus shown to participate on the pathogenesis of Alzheimer's (Bernstein 2005) and Parkinson's disease (Crocker et al. 2005) as well as the pathogenesis of ischemic (Morancho et al. 2010), traumatic (Knoblach and Faden 2005), inflammatory, and infectious (Kieseier and Bernal 2005) states.

The concept of the role of proteolytic enzymes in malignant diseases dates back to the half of the last century (Fischer 1946). At that time, proteases were viewed as effectors facilitating tumor cell dissemination by cleaving the protein components of the extracellular matrix. This is certainly the case for some of the proteases localized extracellularly or concentrated in the specialized regions of plasma membrane where their activity, together with the adhesion molecules, mediates interactions of the cancer cells with their surroundings (Stylli et al. 2008). However, experimental work in the ensuing years led to a more complex picture emphasizing the importance of proteases for several aspects of malignancy (reviewed in Nomura and Katunuma 2005; van Hinsbergh et al. 2006; Lopez-Otin and Matrisian 2007; Kessenbrock et al. 2010). For example, the membrane-bound MT1-MMP (MMP14) can mediate intracellular proteolysis of a centrosomal protein pericentrin-2, which leads to chromosomal instability and aneuploidy (Golubkov and Strongin 2007). Similar role at the early stages of tumor development was suggested for other proteases as well. Caspases, the initiators and executors of apoptotic cell death, are generally viewed as "guardians" protecting the organism from cancer. However, the activation of the DNA fragmentation factor by caspase-3 may lead to increased mutation frequency, genome instability, and chromosomal translocations when the apoptotic program is not completed (Aplan 2006; Lovric and Hawkins 2010). In addition, the caspase-3-mediated activation of prostaglandin E2 production and secretion by the tumor cells exposed to cytotoxic therapies can in fact enhance tumor repopulation by the surviving tumor cells (Huang et al. 2011) and thus eventually contribute to the cancer recurrence and progression. This together with other reports supports the notion that proteases in addition to tissue invasion and metastasis participate in cancer initiation, generation of genomic instability in tumor cells (Radisky et al. 2005), tumor angiogenesis, immune escape, dysregulation of apoptosis (Abraham et al. 2005), and cell proliferation, as well as on the infiltration of the tumors by immune cells (reviewed in McCawley and Matrisian 2001; Kessenbrock et al. 2010).

On the other hand, several proteases have rather anti-tumorigenic effects, e.g., by inactivating bioactive peptides, generating anti-angiogenic fragments from the extracellular matrix, inhibiting cell growth or modulating the inflammatory response (Lopez-Otin and Matrisian 2007). In other cases the net pro- or anti-oncogenic activity of a protease may critically depend on its cellular source as well as on other microenvironmental factors. For example, MMP9 may promote tumor growth and invasiveness as well as enhance angiogenesis by liberating VEGF (vascular endothelial growth factor) (Bergers et al. 2000), whereas the MMP9 mediated degradation of other extracellular matrix proteins such as type IV collagen or plasminogen produces potent suppressors of angiogenesis tumstatin and angiostatin, respectively (Hamano et al. 2003; Rege et al. 2005).

The role of proteases in cancer progression is therefore multifaceted and only partially elucidated. In this chapter we summarize the current knowledge of the expression of proteases in gliomas and discuss their regulation and potential mechanisms whereby they contribute to several aspects of glioma pathogenesis. Finally, the possible exploitation of proteases as therapeutic targets is also reviewed.

12.2 Dysregulation of Proteolytic Enzymes in Gliomas

Several studies (summarized in Table 12.1) addressed the expression and possible function of proteolytic enzymes in gliomas. Of these, the extracellularly localized proteases participating on glioma invasiveness and angiogenesis were most extensively studied and were the subject of a number of excellent reviews (Chintala et al. 1999; Fillmore et al. 2001; VanMeter et al. 2001; Binder and Berger 2002; Levicar et al. 2003b; Rao 2003; Lakka et al. 2005; Mentlein et al. 2012).

A very comprehensive view of the differential protease expression in gliomas can be obtained from the data in The Cancer Genome Atlas Research Network (2008). These data clearly show a consistent signature of up- or downregulated proteases and their homologues (Fig. 12.1a) corroborating the over- (e.g., MMP14, MMP2, MMP7, MMP9, tPA, uPA) or under-expression (e.g., kallikreins, ADAMTS8) observed in previous smaller studies. Importantly, it reveals the dysregulation of expression of numerous proteases that have been little or not at all explored in relation to glioma pathogenesis so far, making them candidates for future studies (e.g., CFI, CPVL, DPEP1, ENPEP, LTF, MEST, MAP1D, ubiquitin-specific proteases).

Several of the consistently upregulated proteases are implicated in the immune response (ADAMDEC1, C1RL, C1S, CFI, CPVL, LTF, PSMB8, PSMB9, SPPL2A) probably reflecting the proinflammatory state induced by gliomas and high infiltration by microglia/macrophages (Gabrusiewicz et al. 2011) and other immune cells. Similarly, proteases involved in the remodeling of the extracellular matrix and angiogenesis (ADAMTS5, MMP14, MMP2, MMP7, MMP9, tPA, uPA) are highly upregulated. The consistently downregulated proteases on the other hand comprise proteases or their homologues typical for the normally functioning neuronal tissue, i.e., involved in neurotransmission, ion channel regulation, and axonal growth (e.g., AMZ1, ASPA, DPP10, PCSK2, TRHDE, AGTPBP1, CPE, KLK8, PCSK1, ACY3, ASRGL1, DPP6, DPYSL2, DPYSL4, FOLH1, NELF, THOP1). In addition, several proteases with a proposed tumor suppressor function (e.g., ADAM11, PRSS3, ADAM23, ADAMTS8, CRMP1, KLK10) are downregulated.

Interestingly, a subset of proteases seems to be preferentially dysregulated in association with the individual glioblastoma molecular subtypes that were recently proposed by Verhaak et al. (2010) (Fig. 12.1b). Indeed, several proteases and protease inhibitors (e.g., ADAM12, caspases, cathepsins, uPA, cystatin A, serpins) are part of the Verhaak mesenchymal transcriptomic signature (Verhaak et al. 2010). In addition, a number of proteases are in fact upregulated solely or most markedly in this subtype (Fig. 12.1b), which might correspond to the higher

Table 12.1 Proteases with possible pathogenetic role in gliomas

Protease	Expression compared to nontumorous brain	Note, reference(s)
Aspartic proteases		
Cathepsin-D	Increased	(Robson et al. 1990; Sivaparvathi et al. 1996b); expressed in reactive macrophages (Robson et al. 1990); higher expression correlates with shorter patient survival (Fukuda et al. 2005); in contrast, Warich et al. observed decreased expression with increased anaplasia (Warich et al. 1995); may act as a mitogen and contribute to glioma invasiveness (Levicar et al. 2002).
Presenilin-1	Expression detected by IHC	(Miake et al. 1999), presenilin-1 is the enzymatic component of the γ-secretase complex, cleaves CD44 in glioma cells (Murakami et al. 2003), inhibitors interfere with the Notch and neurotrophin signaling and inhibit glioma growth (Fan et al. 2010; Gilbert et al. 2010; Chen et al. 2010; Hu et al. 2011; Chu et al. 2013; Kristoffersen et al. 2013; Zou et al. 2013; Forsyth et al. 2014; Saito et al. 2014).
Presenilin-2	Increased	Enzymatic component of the γ-secretase complex, depletion by RNAi in glioma cells leads to decreased growth and invasiveness (Liu et al. 2012).
Renin	Increased	(Ariza et al. 1988); inhibition of renin leads to decreased cell growth in vitro (Juillerat-Jeanneret et al. 2004).
Cysteine proteases		
Calpain-1	Increased	(Ray et al. 2002)
Calpain-2	Increased	(Ray et al. 2002); calpain-2 is required for the invasion of glioma cells (Jang et al. 2010).

Cathepsin-B	Increased	(Rempel et al. 1994; Mikkelsen et al. 1995; Sivaparvathi et al. 1995; Demchik et al. 1999; Strojnik et al. 2000; Konduri et al. 2001; Wang et al. 2005; Colin et al. 2009; Gole et al. 2009, 2012); glioblastoma cells secrete cathepsin-B (McCormick 1993; Formolo et al. 2011); endothelial cells of blood vessels also express cathepsin-B (Mikkelsen et al. 1995; Sivaparvathi et al. 1995; Wang et al. 2005); cathepsin-B activity is higher in the invasive vs. noninvasive glioma cells and is expressed at the invasive front of the cells (Gole et al. 2009); expression of its inhibitor cystatin E/M is silenced by promoter methylation in gliomas (Qiu et al. 2008). Downregulation of cathepsin-B expression impairs the tumorigenic and invasive potential of glioblastoma cells (Mohanam et al. 2001) and suppresses the glioblastoma-induced neoangiogenesis (Yanamandra et al. 2004). Higher expression of cathepsin-B correlated with shorter patient survival (Strojnik et al. 2000, 2005; Colin et al. 2009).
Cathepsin-H	Increased	Anti-cathepsin-H antibodies inhibit invasion of glioblastoma cells in vitro (Sivaparvathi et al. 1996b).
Cathepsin-L	Increased	(Sivaparvathi et al. 1996b; Strojnik et al. 2005; Gole et al. 2012); glioblastoma cells secrete cathepsin-L (Formolo et al. 2011); VEGF transcriptionally upregulates the expression of cathepsin-L in glioblastoma cells (Keerthivasan et al. 2007). Anti-cathepsin-L antibodies inhibit in vitro invasion of glioblastoma cells (Sivaparvathi et al. 1996b); downregulation of cathepsin-L expression impairs invasion and promotes apoptosis of glioma cells (Levicar et al. 2003b).
Cathepsin-S	Increased	(Flannery et al. 2003, 2006); higher expression of cathepsin-S correlated with shorter patient survival (Flannery et al. 2006).

(continued)

Table 12.1 (continued)

Protease	Expression compared to nontumorous brain	Note, reference(s)
Procaspase-3	Increased	(Ray et al. 2002; Bodey et al. 2004; Tirapelli et al. 2010); higher expression of cleaved caspase-3 correlated with longer patient survival (Kobayashi et al. 2007); basal caspase-3 activity promoted migration and invasiveness of glioblastoma cells (Gdynia et al. 2007).
Procaspase-8	Equivocal results	(Xiao et al. 2002; Bodey et al. 2004; Ashley et al. 2005; Bellail et al. 2010); basal caspase-8 activity promoted migration of and invasiveness glioblastoma cells (Gdynia et al. 2007); possible association between a polymorphism of propaspase-8 and increased risk of glioma (Bethke et al. 2008).
Ubiquitin-specific protease (USP) 2a	Increased	(Tao et al. 2013)
USP7	Decreased	(Ruano et al. 2008)
USP8	Expression detected by IHC	USP8 deubiquitinates AIP4, an ubiquitin E3 ligase, which ubiquitinates and destabilizes FLIPs (CASP8 and FADD-like apoptosis regulator) to confer TRAIL sensitivity of glioma cells (Panner et al. 2010).
USP9x	Expression detected by WB and qRT-PCR	USP9x, as an interactant of SOX-2, is required for the growth and survival of glioma cells (Cox et al. 2013).
USP10	Increased	Higher expression of USP10 is correlated with shorter patient survival (Grunda et al. 2006).
USP15	Increased	*USP15* gene amplification observed in 2 % of glioblastomas confers poor prognosis (Eichhorn et al. 2012); USP15 binds to the type I TGF-β receptor complex, deubiquitinates, and stabilizes the type I TGF-β receptor, leading to an enhanced TGF-β signaling and oncogenic capacity of glioma-initiating cells (Eichhorn et al. 2012).
USP22	Expression detected by WB and RT-PCR	USP22 is required for the proliferation and survival of glioma cells (Li et al. 2013).

Metalloproteases		
ADAM3A		Homozygous deletion of *ADAM3A* gene in 16 % podiatric high grade gliomas (Barrow et al. 2011).
ADAM8	Increased	Expression associated with worse clinical outcome (He et al. 2012a); ADAM8 enzymatic activity promotes glioma invasion through matrigel (Wildeboer et al. 2006).
ADAM10	Increased	Acts as a sheddase for membrane-bound molecules such as N-CAM (Kohutek et al. 2009), CD44 (Murai et al. 2004), L1CAM (Mohanan et al. 2013), CXCL16 (Ludwig et al. 2005), and Coxsackievirus and Adenovirus Receptor (CAR) (Houri et al. 2013), thereby promoting glioma motility and growth; important for the renewal of glioma stem-like cells (Bulstrode et al. 2012); involved in the impaired immune recognition of glioma stem cells (Wolpert et al. 2014).
ADAM12	Increased	Expressed by oligodendrocytes and oligodendrogliomas (Kanakis et al. 2013), may be involved in proHB-EGF (heparin-binding EGF-like growth factor) release in glioblastoma (Kodama et al. 2004).
ADAM17	Ambiguous data	Decreased expression reported by Comincini et al. (2009), increased according to TCGA data; mediates hypoxia-induced glioma invasiveness (Zheng et al. 2007); may contribute to stemness maintenance through the Notch signaling (Chen et al. 2013b); promotes the malignant phenotype of glioma cells (Zheng et al. 2012); introduction into astrocytes induced a neoplastic phenotype by deregulating EGFR signaling (Katakowski et al. 2009); involved in the impaired immune recognition of glioma stem cells (Wolpert et al. 2014).
ADAM19	Increased	ADAM19 enzymatic activity promotes glioma invasion through matrigel (Wildeboer et al. 2006).
ADAM22	Decreased	Enzymatically inactive, inhibits glioma proliferation via its disintegrin domain (D'Abaco et al. 2006), receptor for the Leucine-Rich Glioma-Inactivated (LGI) proteins (Kegel et al. 2013).

(continued)

Table 12.1 (continued)

Protease	Expression compared to nontumorous brain	Note, reference(s)
ADAM33		An association between ADAM33 polymorphisms and the risk of glioma was reported (Schwartzbaum et al. 2005; Backes et al. 2013).
ADAMTS4	Increased	(Held-Feindt et al. 2006); cleaves brevican in vitro (Matthews et al. 2000).
ADAMTS5/11	Increased	Cleaves brevican in vitro (Held-Feindt et al. 2006).
ADAMTS8	Decreased	(Dunn et al. 2006), proposed tumor suppressor (Choi et al. 2014) and angiogenesis inhibitor in other systems (Vazquez et al. 1999).
ATP23 peptidase (XRCC6BP1)		Lack of amplification associated with better glioma patient survival (Fischer et al. 2010).
Carboxypeptidase-E	Decreased	Lower immunohistochemical staining associated with shorter patient survival; downregulated by hypoxia; overexpression promotes glioma cell proliferation but inhibits migration in vitro (Horing et al. 2012).
CSN5/JAB1	Increased	Probable protease subunit of the COP9 signalosome complex, possible prognostic factor (He et al. 2012b).
Endothelin-converting enzyme-1 (ECE-1)	Expression detected	Glioma and endothelial cells express ECE-1 together with other components of the endothelin system in situ suggesting local endothelin synthesis in gliomas (Egidy et al. 2000; Naidoo et al. 2005).
Glutamate carboxypeptidase II (GCP-II, PSMA)		Expressed by endothelial cells in gliomas (Chang et al. 1999; Wernicke et al. 2011).
Matrix metallopeptidase (MMP)1 (collagenase-1)	Increased	(Stojic et al. 2008; Hagemann et al. 2012), expression associated with worse survival (Zhang et al. 2011; Xu et al. 2013), polymorphism in the promotor region linked to glioblastoma (McCready et al. 2005), possible role in glioma invasion (Gessler et al. 2011); mediates the NO-induced increase of glioma cell motility (Pullen and Fillmore 2010), promotes tumor-induced angiogenesis (Pullen et al. 2012).

MMP2 (gelatinase-A)	Increased	(Sawaya et al. 1996; Forsyth et al. 1999; Munaut et al. 2003), expressed by glioma and less by endothelial cells (Raithatha et al. 2000), increased expression at the invasive edge compared to tumor core in 40 % of glioblastoma cases in a study by Kim et al. (2011). See main text for further details.
MMP3 (stromelysin-1)	High in childhood astrocytoma (Bodey et al. 2000)	Variable expression reported in small cohorts of glioblastoma patients (Nakagawa et al. 1994; Vince et al. 1999); may be involved in glioma invasion (Laurent et al. 2003; Mercapide et al. 2003; Jin et al. 2013; Zheng et al. 2013); may activate MMP2 (Miyazaki et al. 1992).
MMP7 (matrilysin)	Variable expression	(Vince et al. 1999; Thorns et al. 2003; Rome et al. 2007; Xie et al. 2011); weak expression in oligodendrocytic regions (Thorns et al. 2003), expressed in some glioma cell lines in vitro (Nakano et al. 1993, 1995); polymorphism in the promoter region possibly associated with astrocytoma development (Lu et al. 2006b).
MMP9 (gelatinase-B)	Increased	(Rao et al. 1993b; Komatsu et al. 2004); expressed by glioma and various stromal cells, plays role in angiogenesis, glioma proliferation, and invasiveness (Rao et al. 1993b, 1996; Forsyth et al. 1999; Vince et al. 1999; Pagenstecher et al. 2001; Munaut et al. 2003; Bhoopathi et al. 2010; Sun et al. 2013); induces vascular remodeling via VEGF liberation (Du et al. 2008); possible cerebrospinal fluid marker for primary and secondary CNS cancers (Friedberg et al. 1998); higher expression associated with shorter survival in glioblastoma patients.
MMP10 (stromelysin-2)		Higher expression in gemistocytic astrocytes compared to oligodendrocytic regions (Thorns et al. 2003).
MMP11 (stromelysin-3)	Increased	Expressed by tumor infiltrating macrophages, endothelial cells, and some glioma cells (Thorns et al. 2003; Stojic et al. 2008).
MMP12 (macrophage elastase)		Mediates tenascin C-induced increase in glioma invasiveness (Sarkar et al. 2006); upregulated during bevacizumab treatment (Lucio-Eterovic et al. 2009).

(continued)

Table 12.1 (continued)

Protease	Expression compared to nontumorous brain	Note, reference(s)
MMP13 (collagenase-3)	Increased	Higher expression associated with worse prognosis in glioma (Wang et al. 2012), involved in glioma invasiveness (Inoue et al. 2010), mediates promigratory effects of GDNF (glial cell line-derived neurotrophic factor) and leptin (Yeh et al. 2009; Lu et al. 2010).
MMP14 (MT1-MMP)	Increased	(Yamamoto et al. 1996; Lampert et al. 1998; Fillmore et al. 2001; Guo et al. 2005; Wang et al. 2013b); cleaves and activates proMMP2 (Fillmore et al. 2001), promotes neovascularization (Hiraoka et al. 1998; Deryugina et al. 2002), important for the infiltration of glioma cells along white matter (Belien et al. 1999); increased expression at the invasive front in gliomas (Guo et al. 2005); induces COX-2 expression in glioma cells (Annabi et al. 2009); associated with glioma patient survival (Wang et al. 2013b), expressed in microglia (Markovic et al. 2009).
MMP15 (MT2-MMP)	Increased	(Lampert et al. 1998; Nakada et al. 1999a; Fillmore et al. 2001); mediates increased invasiveness of glioma cells exposed to CXCL12 (Zhang et al. 2005).
MMP16 (MT3-MMP)	Increased	(Fillmore et al. 2001); involved in glioma invasiveness (Li et al. 2013a).
MMP19	Increased	More expressed at the invasion margin of human glioblastomas (Stojic et al. 2008), involved in glioma invasiveness (Lettau et al. 2010).
Neprilysin	Increased	(Monod et al. 1992), expressed by glioma cells in vitro (Monod et al. 1989), cleaves brain natriuretic peptide (Medeiros Mdos et al. 1991), substance P (Endo et al. 1989) and basic fibroblast growth factor (bFGF) (Goodman et al. 2006).
YME1-like 1 (YME1L1)	Decreased	Associated with patient survival, identified as a landscape gene consistently altered in the TCGA glioma dataset (Bredel et al. 2009).

Serine proteases		
Dipeptidyl peptidase IV	Increased	Expressed by vascular and scattered mononuclear-like cells in gliomas, by glioma cells in glioblastomas (Mares et al. 2012); variable expression in glioma cell lines, may inhibit glioma cell growth through an enzymatic activity independent mechanism (Stremenova et al. 2007; Busek et al. 2012).
Furin	Increased	Necessary for the Bcl2-mediated increase of glioma invasiveness (Wick et al. 2004); decreased growth and invasiveness after inhibition of furin in glioma cells, role in the activation of MT1-MMP (MMP14), TGF-β and IGF-I receptor (Mercapide et al. 2002); in contrast, processing of N-cadherin by furin leads to decreased migration (Maret et al. 2012).
Hepatocyte growth factor activator		Promotes glioma migration, enhances growth, and increases vascular density of experimental tumors (Uchinokura et al. 2006); expression of its endogenous inhibitor (hepatocyte growth factor activator inhibitor type-2/placental bikunin, HAI-2/PB) is decreased in glioblastomas (Hamasuna et al. 2001).
Kallikrein related peptidases (KLK)	Decreased	Dysregulated KLK1, 13, and 15 may predict worse overall-survival in glioblastoma (Girgis et al. 2012); KLK7 detected by RT-PCR in 50 % glioblastomas, may contribute to glioma cell invasion (Prezas et al. 2006); ambiguous data for KLK6 expression-decreased in glioblastoma compared to low grade tumors (IHC) (Strojnik et al. 2009), increased KLK6 (qRT-PCR) (Talieri et al. 2012); higher KLK6 IHC staining associated with worse patient survival, KLK6 promotes the resistance of glioma cells to apoptosis in vitro (Drucker et al. 2013).
Matriptase	Increased	(Clark et al. 2010); overexpression of an endogenous matriptase inhibitor decreased glioma invasion (Miyata et al. 2007).
Neutrophil elastase	Increased	Elastase-positive polymorphonuclear neutrophils are recruited to the infiltrative margin of high grade gliomas (Iwatsuki et al. 2000).

(continued)

Table 12.1 (continued)

Protease	Expression compared to nontumorous brain	Note, reference(s)
PCSK5A		Increases migration of glioma cells in vitro by cleaving N-cadherin (Maret et al. 2012).
PCSK6		Increased in infiltrating glioma cells, may contribute to the invasive phenotype of glioma cells (Delic et al. 2012).
Plasminogen		Produced by glioma cells, contributes to MMP activation (Le et al. 2003; Tsatas and Kaye 2003).
Prolyl oligopeptidase		Possible role in protein secretion in glioma cells (Schulz et al. 2005).
Rhomboid domain containing 1		Silencing by RNAi inhibits glioma cell growth in vitro (Wei et al. 2014).
Seprase (FAP)	Increased	Expressed in gliomas in vivo (Stremenova et al. 2007; Mentlein et al. 2011) and glioma cell lines in vitro (Rettig et al. 1986; Mentlein et al. 2011; Busek et al. 2012); possible role in the migration of glioma cells within the brain extracellular matrix (Mentlein et al. 2011).
Thrombin		Thrombin-positive glioma cells detected in glioblastomas by IHC; thrombin induces proliferation of glioma cells (Ogiichi et al. 2000) and induces VEGF expression in vitro (Yamahata et al. 2002); a direct thrombin inhibitor argatroban reduces glioma growth in animal models (Hua et al. 2005b; Yamaguchi et al. 2013).
Tissue-type plasminogen activator (tPA)	Equivocal results	(Sawaya et al. 1991; Kinder et al. 1993; Caccamo et al. 1994; Landau et al. 1994; Arai et al. 1998; Goh et al. 2005; Salmaggi et al. 2006)
Urokinase-type plasminogen activator (uPA)	Increased	(Kinder et al. 1993; Caccamo et al. 1994; Landau et al. 1994; Arai et al. 1998); possible prognostic factor in glioma (Hsu et al. 1995); uPA is expressed in tumor and vasculature cells in high grade gliomas (Yamamoto et al. 1994a; Zhang et al. 2000), activates other proteases directly (Zhao et al. 2008a) or via plasminogen activation; activates hepatocyte growth factor (HGF) (Naldini et al. 1992); binding of uPA to its cellular receptor uPAR promotes its proteolytic activation as

		well as triggers signaling through uPAR (reviewed in Tsatas and Kaye 2003; Eden et al. 2011); mediates the proinvasive effect of basic fibroblast growth factor and transforming growth factor (TGF)-α (Mori et al. 2000).
Threonine proteases		
Gamma-glutamyltransferase	Increased	Expressed in the vasculature and glioma cells (Mares et al. 2012), possible role in glioma growth and GSH metabolism (Schafer et al. 2001).
Proteasome		Decreased chymotrypsin-like and peptidylglutamyl peptide hydrolyzing activity of purified S20 proteasome in 67 % tumors compared to peritumoral tissue, changed subunit composition (Piccinini et al. 2005); decreased LMP2 subunit (Mehling et al. 2007); excessive proteasome-mediated degradation of NF1 (Neurofibromin 1) contributes to its functional inactivation in gliomas (McGillicuddy et al. 2009); decreased proteasome activity in glioma cancer stem cells may promote Notch signaling by stabilizing the Notch intracellular domain (Lagadec et al. 2014) and may be useful for tracking and possibly targeting of glioma stem-like cells (Vlashi et al. 2009); proteasome inhibitors are tested in clinical trials in glioma patients (see Sect. 12.5; Vlachostergios et al. 2013).
Taspase	Increased	(Scrideli et al. 2008); functional role not yet determined, cleaves MLL (mixed lineage leukemia; Lysine (K)-Specific Methyltransferase 2A), a protein required for the maintenance of HOX gene expression in glioma stem cells (Gallo et al. 2013).

Note: Increased expression of procaspase-9 (Ray et al. 2002; Waltereit and Weller 2002; Bodey et al. 2004; Johnson et al. 2007; Fukushima et al. 2008) and expression of other procaspases [procaspase-10 (Xiao et al. 2002; Ashley et al. 2005), procaspase-6 (Bodey et al. 2004)] were detected in gliomas and glioma cell lines

IHC immunohistochemistry, *WB* western blot, *qRT-PCR* real-time reverse transcription PCR

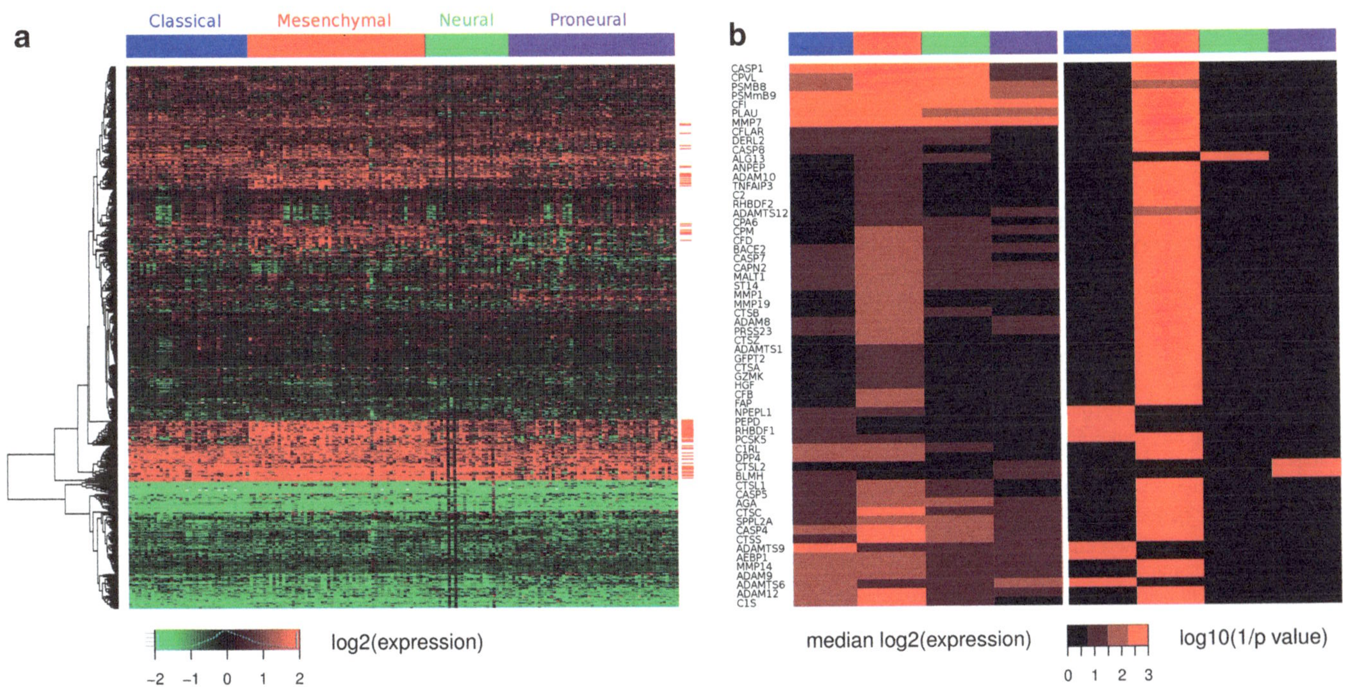

Fig. 12.1 Dysregulation of protease expression in glioblastomas. (**a**) Hierarchical clustering of 528 protease gene expression profiles (*rows*) in 173 glioblastoma patients comprised in the Verhaak dataset (Verhaak et al. 2010, TCGA, Agilent platform). Data were normalized to the expression levels of healthy controls (median values of 10 control samples). The glioblastoma molecular subtype, as defined by Verhaak et al. (2010), is indicated at the *top* of the heatmap. Proteases displaying significant subtype-specific upregulation are indicated on the *right side* of the heatmap. (**b**) A subset of proteases from panel (**a**) that exert ≥2-fold upregulation of their median mRNA levels in a particular glioblastoma subtype compared to healthy controls and expression significantly higher ($p \leq 0.01$) compared to the rest of the patients. Glioblastoma subtypes are indicated as in panel (**a**)

overall necrosis with ensuing inflammatory infiltrate and expression of angiogenic genes in the mesenchymal subtype (Verhaak et al. 2010).

Albeit the transcriptomic TCGA data give very rich and complex information on the dysregulation of proteolytic systems based on over 500 glioblastoma cases, it is important to realize their limitations. As detailed above, the transcriptional regulation of protease expression represents only one possible level of control mechanisms, as proteolytic systems are often regulated on the posttranscriptional and posttranslational levels. In addition, although the TCGA data corroborate the heterogeneity of glioblastomas also with respect to the expression of proteases, they do not allow distinguishing the contribution of transformed and stromal cells and may therefore point to a changed cellular composition in different glioma subtypes rather than differential expression in the transformed cells. Irrespective of these limitations, they are a valuable source for selecting proteases for studies of glioma pathogenesis and identification of future therapeutic targets.

12.3 Signals and Mechanisms Affecting Expression and Activity of Proteases in Gliomas

The mechanisms leading to protease dysregulation in gliomas are complex and still only partially understood. However, several aberrant signaling pathways and mechanisms characteristic for gliomagenesis drive the tumor progression in part by affecting the proteolytic machinery (Fig. 12.2).

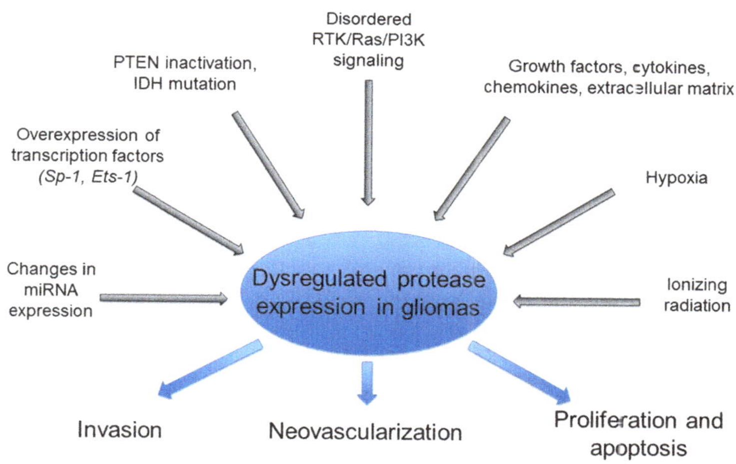

Fig. 12.2 Examples of gliomagenesis relevant factors and events affecting the local proteolytic balance, and the biological consequences of dysregulated protease expression. *PTEN* phosphatase and tensin homolog, *IDH* isocitrate dehydrogenase, *RTK* receptor tyrosine kinase

12.3.1 Mechanisms Leading to Dysregulation of Protease Expression

The receptor tyrosine kinase (RTK)/Ras/PI3K signaling pathway is dysregulated in over 80 % of gliomas (Van Meir et al. 2010) and promotes cell proliferation, survival, and invasion. The most frequent alteration is *EGFR* gene amplification, overexpression, or the presence of a constitutively active mutant EGFR*vIII* variant; these are present in approximately half of the patients and are a characteristic phenotypic feature of the primary glioblastomas (Ohgaki and Kleihues 2007 Van Meir et al. 2010). It is worth mentioning that the active MMP9 was detected in 69 % of primary and just 14 % of secondary glioblastomas and its expression was strongly associated with the expression of EGFR*vIII* (Choe et al. 2002). Indeed, the EGFR signaling was demonstrated to increase the expression of several proteases in glioma cells in vitro including MMP9 (Kang et al. 2005; Zhao et al. 2010), MMP1 (Anand et al. 2011; Li et al. 2011), MMP2 (Lal et al. 2002; Park et al. 2006), MMP14 (Van Meter et al. 2004), and uPA (Amos et al. 2010). Besides that, EGFR signaling influences the subunit composition of the proteasome which was speculated to be associated with the radioresistance of glioma cells (Kim et al. 2008b). Protease expression can also be influenced by other alterations in the abovementioned RTK/Ras/PI3K signaling pathway such as PTEN (phosphatase and tensin homolog) deletion or Ras mutations, which occur in 37 % and 2 % of gliomas, respectively (Van Meir et al. 2010). PTEN inactivation in gliomas correlates with higher MMP9 expression (Comincini et al. 2009) and in a similar way as the activation of the downstream PI3K promotes the production of MMP2 and MMP9 in glioma cells in vitro (Koul et al. 2001; Kubiatowski et al. 2001; Park et al. 2002; Furukawa et al. 2006; Kwiatkowska et al. 2011). Similarly, constitutively active form of Ras was demonstrated to increase uPA expression and promote cell invasiveness in normal human astrocytes transformed by the introduction of human telomerase in combination with inactivation of p53/pRb by E6/E7 (Zhao et al. 2008b). The aberrant EGFR/Ras/PI3K pathway is also associated with the increased expression of tissue factor, a crucial cofactor for the initiation of the proteolytic blood clotting cascade (Rong et al. 2009; Magnus et al. 2010).

Glioblastoma microenvironment typically contains increased concentration of a number of cytokines including growth factors and chemokines, as well as local mediators such as adenosine or NO, which were demonstrated to influence the expression of proteases in glioma cells (Table 12.2).

Protease expression with the subsequent promotion of invasiveness is also affected by the interaction of glioma cells with the components of the extracellular matrix (ECM). Park et al. (2002) showed that binding of the hyaluronic acid to glioma cells activates the focal adhesion kinase (FAK)-ERK 1/2 signaling pathway and leads to increased MMP9 secretion. Similarly, tenascin C increases the invasiveness of glioma cells by stimulating the expression of MMP12 (Sarkar et al. 2006).

Table 12.2 Examples of local mediators affecting the expression of proteases in glioma cells

Local mediator	Upregulated protease(s)
VEGF	Cathepsin-L (Keerthivasan et al. 2007)
HGF	uPA (Moriyama et al. 1999), MMP2, MMP14 (Moriyama et al. 1996; Hamasuna et al. 1999)
CXCL12	MMP15 (Zhang et al. 2005), MMP9 (Kenig et al. 2010)
CXCL1	MMP2 (Zhou et al. 2005)
TGF-β	MMP19 (Lettau et al. 2010), MMP9 (Dziembowska et al. 2007; Ye et al. 2012), MMP7 (Nakano et al. 1993), ADAMTS4, ADAMTS5 (Held-Feindt et al. 2006)
bFGF	uPA (Mori et al. 2000)
TNF-α[a]	MMP9 (Esteve et al. 1998, 2002), MMP19 (Lettau et al. 2010)
IL-1	MMP9 (Esteve et al. 1998, 2002), uPA (Kasza and Koj 2002), ADAMTS4 (Held-Feindt et al. 2006)
IL-6	MMP2 (Li et al. 2010b)
Oncostatin M	uPA (Krona et al. 2007)
IFN-γ	MMP19[a] (Lettau et al. 2010)
TRAIL	MMP9 (Kim et al. 2008a)
Adenosine	MMP9 (Gessi et al. 2010)
NO	MMP1 (Pullen and Fillmore 2010)

[a]Suppression of constitutive MMP2 expression was reported by Qin et al. (1998)

The Sp1 transcription factor promotes the expression of MMP2 (Qin et al. 1999; Guan et al. 2012), ADAM17 (Szalad et al. 2009), and cathepsin-B (Yan et al. 2000) in glioma cells and is highly expressed in the majority of gliomas. Similarly, the Ets-1 transcription factor (Oikawa and Yamada 2003) is overexpressed in glioblastomas (Kitange et al. 1999a) and mediates the transcription of MMP9 (Sahin et al. 2005), cathepsin-B (Yan et al. 2000) and uPA (Kitange et al. 1999b; Nakada et al. 1999b). Several MMP genes also contain the AP-1 element and their transcription may therefore be activated by Jun and Fos transcription factors (Westermarck and Kahari 1999) in response to a variety of signals from the extracellular milieu (Chakraborti et al. 2003).

Intratumoral hypoxia, a typical feature of glioblastoma microenvironment affecting multiple signaling pathways, as well as mutations of IDH (isocitrate dehydrogenase) lead to the stabilization and thereby activation of the HIF1-α subunit of the transcription factor HIF (hypoxia inducible factor), which accelerates glioma progression in part through the upregulation of MMP2, MMP9, and ADAM17 (Brat et al. 2004; Fujiwara et al. 2007; Zheng et al. 2007; Fu et al. 2012).

Recently (see also Chap. 4) alterations in the expression of microRNAs (miRNAs) were described in glioblastomas (Moller et al. 2013) and the mechanisms by which miRNAs may influence the phenotype of glioma cells may also involve the regulation of proteases. miRNAs may directly target the protease mRNA, or exert an indirect effect by modulating the pathways regulating the expression of proteases or their inhibitors. In glioblastoma specimens, miR-211 expression is silenced by promotor hypermethylation and its expression level inversely correlates with MMP9 expression (Asuthkar et al. 2012). Similar negative correlation was also observed between MMP9 and miR-491-5p (Yan et al. 2011). Importantly, both miRNAs were demonstrated to downregulate MMP9 in glioma cells in vitro resulting in decreased invasiveness (Yan et al. 2011; Asuthkar et al. 2012). Other examples of the effects of miRNAs on the proteolytic balance include the negative regulation of MMP16 by miR-146b-5p (Xia et al. 2009; Li et al. 2013a), MMP3 by miR-152 (Zheng et al. 2013), and ADAM17 by miR-145 (Lu et al. 2013). The expression of MMP2 and MMP9 in glioma cells in vitro is further influenced by miR-7, which is downregulated in gliomas, and affects the proteases indirectly by downregulating FAK (Wu et al. 2011). In addition to these tumor suppressor miRNAs, protease targets were also demonstrated for the oncogenic miRNAs. miR-10b is overexpressed in gliomas (Moller et al. 2013) and induces glioma cell invasiveness by targeting the homeobox DNA-binding domain containing transcription factor HOXD10, a negative regulator of MMP14 and uPAR (Sun et al. 2011).

Interestingly, the expression of proteases involved in the remodeling of ECM was recently demonstrated to be induced by the therapeutic interventions used to treat gliomas, thereby somewhat paradoxically contributing to the tumor recurrence. Glioma cells exposed to ionizing radiation exhibit enhanced invasiveness caused by several mechanisms including the upregulation of proteases (Wild-Bode et al. 2001; Park et al. 2006; Badiga et al. 2011; Shankar et al. 2014). A rapid increase in the expression and activation of MMP2, MMP9, and MMP14 after sublethal irradiation was demonstrated in glioma cells in vitro and an MMP inhibitor o-phenantroline was able to reduce the increased invasiveness of the irradiated glioma cells (Wild-Bode et al. 2001). Similarly, treatment with a plasmid silencing the expression of MMP2 resulted in the reversal of the proinvasive effects of irradiation and increased the radiosensitivity of glioma cells (Badiga et al. 2011). The radiation-mediated enhancement of protease expression is probably p53 independent (Wild-Bode et al. 2001) and is suppressed in the presence of the active PTEN (Park et al. 2006). In the case of MMP2, radiation increases its transcription as a result of the activation of the Src and EGFR signaling with the ensuing activation of p38, PI3K, and Akt (Park et al. 2006). Bevacizumab, a monoclonal antibody targeting VEGF, was similarly shown to promote the invasiveness of glioma cells (Lucio-Eterovic et al. 2009; de Groot et al. 2010). Glioma cells exposed to bevacizumab upregulated MMP12, MMP9, MMP2, as well as plasminogen in vitro and a similar upregulation was observed in experimental tumors in mice treated with bevacizumab contributing to their more infiltrative growth (Lucio-Eterovic et al. 2009). This increase of MMP2 and MMP14 expression was also detected in glioma patients treated with bevacizumab (de Groot et al. 2010; Furuta et al. 2014) and the increase of MMP9 in the urine was in fact suggested as a marker for bevacizumab failure in glioma patients (Takano et al. 2010).

12.3.2 Mechanisms Leading to Dysregulation of Protease Activation

Besides the dysregulation of the protease gene expression, the increased activity of extracellular proteases is frequently a result of an inappropriate zymogen activation caused by the protease cofactor overexpression. The cell surface receptors urokinase-type plasminogen activator receptor (uPAR, CD87) (Yamamoto et al. 1994b) and tissue factor (TF, fIII) (Hamada et al. 1996), a crucial cofactor for the initiation of blood coagulation cascade, are both upregulated in gliomas. uPAR is a multifunctional glycosylphosphatidylinositol (GPI)-linked membrane protein that binds several extracellular ligands including uPA (Eden et al. 2011). The binding of uPA to uPAR not only promotes the activation of uPA leading to enhanced pericellular proteolysis but also directly regulates glioma cell adhesion and migration through uPAR (see Mohanam et al. (1999), Eden et al. (2011) and Sect. 12.4.1.3). TF is a transmembrane receptor and cofactor of the coagulation factor VIIa and its overexpression contributes to the procoagulative state in gliomas (Rong et al. 2006). Similarly to uPAR, TF promotes the malignant phenotype of glioma cells by activating intracellular signaling independent of its role in protease activation (Dutzmann et al. 2010; Gessler et al. 2010).

Zymogens may further be inappropriately activated as a result of the dysregulation of the local "proteolytic context" (reviewed in Mason and Joyce 2011). Interestingly, some aspects of the complex proteolytic systems resemble the basic features of the signaling cascades such as signal amplification, cross talk, and a purposeful adaptation. Within the broader, functionally interrelated proteolytic networks, the interactions of individual proteases are frequently reciprocal (Fig. 12.3). For example, uPA can both activate and be activated by plasmin and cathepsin-B (Mason and Joyce 2011) and some proteases facilitate the activity of other proteases by inactivating their inhibitors [e.g., MMPs can inactivate a range of serpins (Kessenbrock et al. 2010) and cathepsin-B cleaves some TIMPs and serpins (Mason and Joyce 2011)]. The glioma proteases cathepsin-B, uPA, and a number of the MMPs seem to occupy critical nodes within the complex proteolytic systems due to the broad range of their respective activators, the ability to activate numerous other proteases and to inactivate several protease inhibitors (Mason and Joyce 2011).

Altered availability of the endogenous protease inhibitors may be another mechanism contributing to the dysregulation of proteolytic activity in gliomas, the best example being the imbalance between the cysteine cathepsins and cystatins that is thought to contribute to tumor invasion. In gliomas, cystatin-E/M downregulation was reported in 78 % of cases and the reduced expression correlated with its promoter hypermethylation (Qiu et al. 2008). Cystatin-C is also downregulated in glioblastomas compared to the low grade tumors and may be a predictive factor for patient survival (Nakabayashi et al. 2005). Further, Gole et al. (2012) showed that at the tumor margin the expression of cystatin-B is decreased, while cathepsin-B is, compared to the central part of the tumor, redistributed into the extracellular space. Such pathologic regulation of the availability of the cathepsin activity may lead to increased invasiveness (Gole

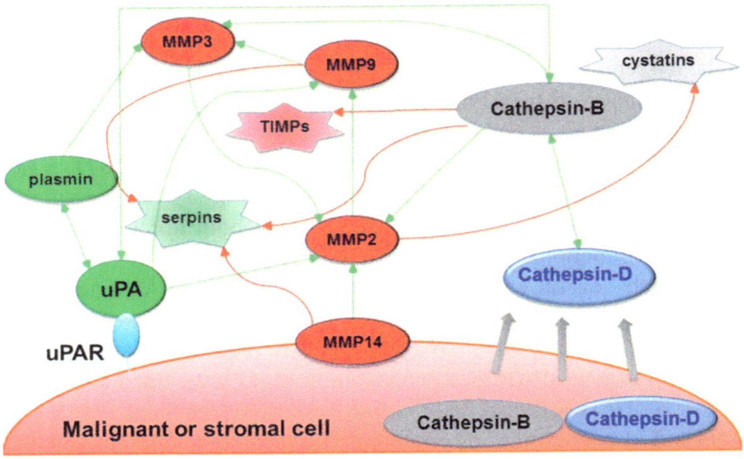

Fig. 12.3 The proteolytic network of extracellular proteases implicated in glioma progression. Proteases from various classes interact in a multidirectional network through the proteolytic activation of zymogens (*green lines*) and inactivation (*red lines*) of endogenous protease inhibitors. Cysteine proteases in *black*, metalloproteases in *red*, serine proteases in *green*, aspartate proteases in *blue*; the protease inhibitors of individual protease classes are shown in the corresponding shades. *MMP* matrix metalloproteinase, *uPA* urokinase-type plasminogen activator, *uPAR* urokinase-type plasminogen activator receptor

et al. 2012). These results are supported by the findings that ectopic expression of the cathepsin-B inhibitor cystatin-C reduces in vitro glioma invasion as well as in vivo tumor formation (Konduri et al. 2002). Tissue factor pathway inhibitor-2 (TFPI-2) is a Kunitz-type serine protease inhibitor that inhibits a variety of serine proteases and is a potential tumor suppressor in several malignancies (Sierko et al. 2007). In gliomas, TFPI-2 is probably silenced by promoter hypermethylation (Konduri et al. 2003) and its expression is inversely correlated with tumor grade. Moreover, the protein was demonstrated to inhibit glioma cell invasion in vitro suggesting that its loss may be directly linked to glioma aggressiveness (Rao et al. 2001).

An opposite role is evident for the serpin plasminogen activator inhibitor 1 (PAI-1), the major inhibitor of the fibrinolytic system. Paradoxically, PAI-1 expression is associated with worse prognosis in several tumors (Van De Craen et al. 2012). In gliomas, a grade-dependent upregulation of PAI-1 is consistently reported with the most intense expression in the areas of vascular proliferation and perinecrotic areas (Rao et al. 1993a; Kono et al. 1994; Landau et al. 1994; Arai et al. 1998; Muracciole et al. 2002; Colin et al. 2009); in addition, high PAI-1 expression (Muracciole et al. 2002) and serum levels (Iwadate et al. 2008) are negative prognostic factors in glioma patients. This seeming paradox likely reflects the non-protease-mediated functions of PAI-1 (Van De Craen et al. 2012) that include its effects on cell adhesion and migration (Bryan et al. 2008; Paugh et al. 2008) and promotion of angiogenesis (Hjortland et al. 2004). The data for other endogenous protease inhibitors are equivocal. Changes in the expression of TIMPs were reported in gliomas by several groups. The most consistent data exist for the extracellular matrix-associated TIMP3, which is regarded as a tumor suppressor. *TIMP3* promoter is hypermethylated in gliomas (Liu et al. 2010) and the hypermethylation correlates with the loss of TIMP3 expression in secondary glioblastomas; in addition, the region on chromosome 22 containing the *TIMP3* gene is frequently deleted in a large proportion of secondary and some primary glioblastomas (Nakamura et al. 2005). Other mechanisms that may lead to TIMP3 downregulation involve its targeting by miR21 (Gabriely et al. 2008) and the frequent loss of the cell cycle regulator P14ARF encoded by the CDKN2A locus. P14ARF acts as a tumor suppressor via p53 stabilization but was recently shown to inhibit human glioblastoma-induced angiogenesis by upregulating the expression of TIMP3 (Zerrouqi et al. 2012). TIMP3 inhibits invasion and promotes apoptosis in several tumor cells (Baker et al. 1999). However in a glioma model, the presence of TIMP3 did not significantly affect the antitumor efficacy of an oncolytic adenovirus (Lamfers et al. 2005). For TIMP1, 2, and 4 some studies indicate decreased expression in glioblastomas compared to normal brain (Nakagawa et al. 1994; Mohanam et al. 1995; Aaberg-Jessen et al. 2009) and inhibitory effects on glioma cells (Merzak et al. 1994; Nakano et al. 1995; Matsuzawa et al. 1996; Hoshi et al. 2000; Groft et al. 2001; Nakada et al. 2001; Takahashi et al. 2002; LuW et al. 2004), whereas others suggest that these TIMPs are produced in excess and may participate on the tumor progression (Nakagawa et al. 1995; Lampert et al. 1998; Kachra et al. 1999; Saxena et al. 1999; Pagenstecher et al. 2001; Lu et al. 2004; Blazquez et al. 2008a; Rorive et al. 2010; Crocker et al. 2011). These conflicting results may reflect not only the different methodologies and the heterogeneity of the studied tumors but also the ambiguous role of TIMPs in tumor pathogenesis. This is best illustrated by TIMP2 that promotes the activation of MMP2 [see Sect. 12.1 and Strongin et al. (1995)], may activate the promigratory signaling pathways by binding to MMP14 on the cell membrane (Sounni et al. 2010), and has anti-angiogenic effects independent of MMPs (see Stetler-Stevenson and Seo (2005) for review).

The production of the membrane-bound and secreted proteases in stromal cells, such as reactive astrocytes, microglia, endothelial, and infiltrating immune cells, is an additional factor, which increases the complexity of their biological impact in gliomas (Fig. 12.4) (Rivera et al. 2010). Because of their localization in the extracellular space, these stromal proteases may augment the proteolytic systems deployed by the glioma cells. Examples of this cooperation include the activation of glioma MMP2 by microglial MMP14, which promotes glioma growth (Markovic et al. 2009) or activation of MMP2 in reactive astrocytes by glioma cell-derived plasmin (Le et al. 2003). Interestingly, forced expression of MMP14 directly in glioma cells induced their death (Markovic et al. 2009) further strengthening the ordered and highly context-dependent role of proteases.

With the exception of the activation of the caspase cascade, the reports on the mechanisms regulating the expression and activity of intracellular proteases implicated in glioma pathogenesis are scarce. Hypermethylation of the caspase-

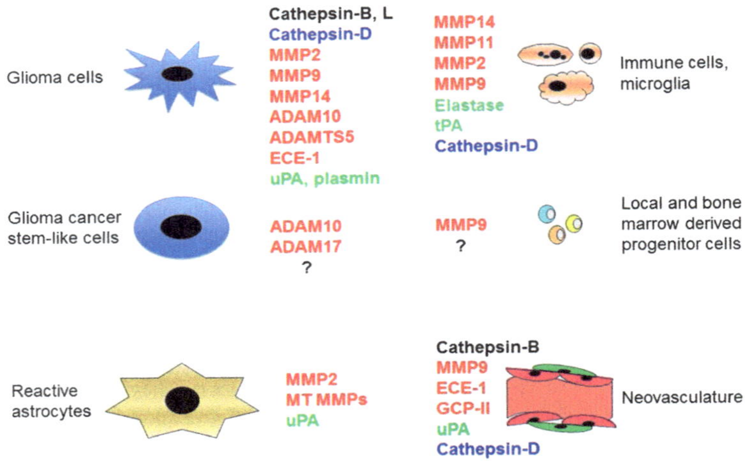

Fig. 12.4 Examples of the extracellular proteases expressed by individual constituents of the glioma microenvironment. It is important to note that the list is non-exhaustive as the expression of several proteases was not yet analyzed in all cell types and the expression in individual cell types may dynamically change in response to microenvironmental stimuli. Cysteine proteases in *black*, metalloproteases in *red*, serine proteases in *green*, aspartate proteases in *blue*. *MMP* matrix metalloproteinase, *uPA* urokinase-type plasminogen activator, *tPA* tissue-type plasminogen activator, *ECE* endothelin converting enzyme, *MT MMPs* membrane type MMPs, *GCP-II* glutamate carboxypeptidase II

8 promoter, which leads to the block of the extrinsic apoptotic pathway, was observed in over 50 % of glioblastomas by Skiriute et al. (2012) and was associated with shorter patient survival. Several mechanisms were proposed to regulate the activity of the cysteine proteases calpains, although their in vivo relevance mostly remains to be determined (see Franco and Huttenlocher (2005) for review). Elevation of the intracellular concentration of Ca^{2+}, autocatalytic cleavage of the enzymes, and the interaction with phospholipids and the endogenous inhibitor calpastatin were demonstrated to influence the activity of calpains (Franco and Huttenlocher 2005). Interestingly, calpain-2 is phosphorylated by ERK after the activation of the EGFR signaling, which is essential for EGF-induced motility in fibroblasts (Glading et al. 2000, 2004), and may thus play an important role in the motility of glioma cells as well (see Sect. 12.4.1.5).

In conclusion, multiple proteases show differential expression in glioma tissue as a result of a number of different pathogenetic mechanisms eventually leading to the dysregulation of the proteolytic homeostasis in gliomas. In addition, the dysregulation of the proteolytic activity in the extracellular space can be triggered by various mechanisms originating from changes in the stromal as well as the malignant cells. The relative importance of the various mechanisms may change during glioma progression and in some cases the proteolytic homeostasis can be impaired even without an overt overexpression of the protease(s). Conversely, the overexpression of a protease by itself need not be sufficient to change the proteolytic balance in the presence of counteracting compensatory mechanisms in the tumor microenvironment or in tumor cells.

12.4 Protease-Mediated Protein Processing in Gliomas and Participation of Proteases on the Biological Hallmarks of Gliomas

Extracellularly localized and plasma membrane-bound proteases are important tools used both by glioma cells as well as by the stromal cells to shape the glioma microenvironment. Their substrates include structural protein components of the extracellular matrix (ECM), zymogens of other proteases, and regulatory molecules such as chemokines, cytokines, and cell surface receptors or plasma membrane-bound proteins that may be released into the extracellular space (Fig. 12.5). These proteases therefore frequently serve as direct (e.g., via protease-activated receptors) or indirect (e.g., by modifying or releasing signaling molecules) mediators of intercellular communication. The biological effects of their proteolytic activities are manifold; proteases contribute to the profound remodeling of the unique ECM of human brain observed in glioblastoma (Bellail et al. 2004) and promote glioma cell dissemination. Cleavage of the ECM not only removes the physical barriers to glioma dissemination but also releases growth factors sequestered in the ECM (e.g., VEFG, TGF-β) and produces fragments of ECM proteins with new biological activities (e.g., anti-angiogenic peptides) that influence glioma cell proliferation,

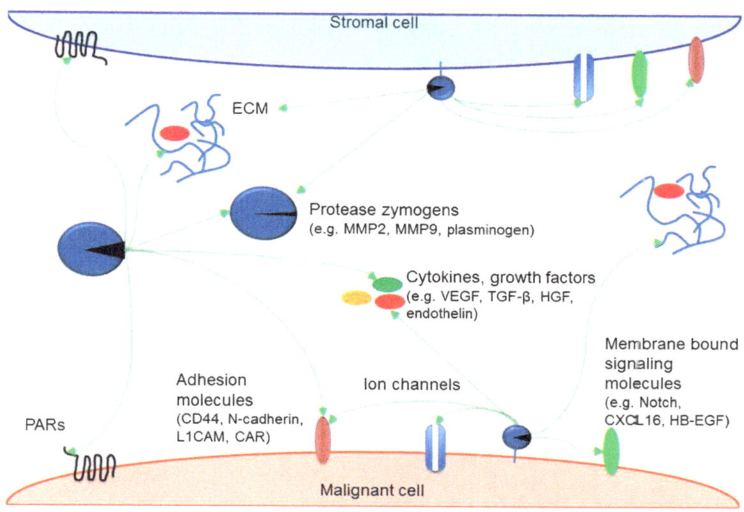

Fig. 12.5 Proteins regulated by plasma membrane-bound and extracellular proteases in the glioma microenvironment. *CAR* Coxsackievirus and Adenovirus Receptor, *ECM* extracellular matrix, *HB-EGF* heparin-binding EGF-like growth factor, *HGF* hepatocyte growth factor, *PAR* protease-activated receptors

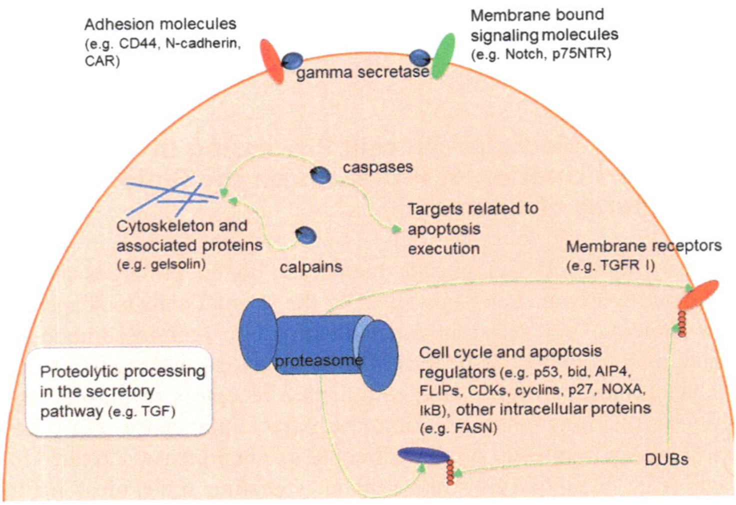

Fig. 12.6 Proteins regulated by intracellular proteases contributing to glioma pathogenesis. *CAR* Coxsackievirus and Adenovirus Receptor, *CDK* cyclin dependent kinases, *FASN* fatty acid synthase, *IkB* inhibitors of NFκB, *p75NTR* p75 neurotrophin receptor, *DUBs* deubiquitinating enzymes

resistance to apoptosis, and neovascularization. Finally, by shedding cell surface molecules responsible for the activation of the immune system such as NKG2D ligands (Eisele et al. 2006; Wolpert et al. 2014) and activating latent TGF-β (Huber et al. 1992; Leitlein et al. 2001), proteases aid glioma cells to escape from the immune surveillance.

In contrast with the extracellularly localized proteases, the role of the intracellular proteases is largely restricted to the expressing cells, possibly with the exception of some proteases present in the secretory pathway and exosomes or released into the extracellular space due to apoptotic or necrotic disintegration of the cells. The substrates of these proteases relevant for the progression of glioblastomas are less explored, but include the cytoskeletal and organellar proteins, components of the signal transduction pathways, proteins regulating and executing vesicular transport, gene expression and replication, autophagy, and apoptosis. Importantly, the proteasome together with the deubiquitinating enzymes (DUBs) are critical regulators of many cellular proteins implicated in glioma pathogenesis (Fig. 12.6).

12.4.1 Proteases and Glioma Invasion

The widespread infiltration of the surrounding brain tissue by glioma cells is a long recognized hallmark of gliomas (Bramwell 1888) and in gliomas with IDH mutation, individual invading glioma cells were demonstrated to be dispersed throughout the brain (Sahm et al. 2012). The locoregional treatments (i.e., surgical removal of the tumor with subsequent radiation therapy) therefore almost invariably fail and the tumor recurs usually within 2–3 cm of the original location.

Several reports establish a critical role of the extracellular and membrane-bound metalloproteases (especially MMP2, MMP9, MMP14, ADAMTS), the serine protease u-PA, and secreted lysosomal cysteine proteases including cathepsin-B in glioma invasiveness (reviewed in Binder and Berger 2002; Levicar et al. 2003b; Rao 2003; Mentlein et al. 2012). Indeed, MMP2 and MMP9 (Kim et al. 2011), ADAM17 (Chen et al. 2013a), as well as uPA and its receptor uPAR (Yamamoto et al. 1994a; Zhang et al. 2000; Colin et al. 2009) are abundantly expressed at the invasive edge. In addition, the urokinase-type plasminogen activator receptor (uPAR) and cathepsin-B are expressed by glioma cells infiltrating the surrounding brain tissue (Mikkelsen et al. 1995).

The activation of proteolytic systems seems to be sufficient as well as necessary for glioma cell invasiveness and involves proteases expressed by the malignant as well as stromal cells (Le et al. 2003). MMP14 is a key enzyme endowing glioma cells with the ability to spread and migrate on myelin (Paganetti et al. 1988; Amberger et al. 1994). Interestingly, even non-glioma cells such as fibroblasts could gain the capacity to migrate on an otherwise unpermissive myelin substrate as well as to invade the myelinated optic nerve fibers after the introduction of MMP14 (Belien et al. 1999). Similarly, introduction of ADAM17 into an astrocytic cell line caused upregulation of several invasion and angiogenesis-promoting genes and resulted in increased invasiveness and formation of high grade brain tumors upon intracranial implantation (Katakowski et al. 2009). MMP2 (gelatinase-A) is overexpressed in the majority of glioblastomas (Yamamoto et al. 1996; Kunishio et al. 2003; Hagemann et al. 2012) and a number of studies demonstrated that MMP2 may be one of the shared downstream components critical for the execution of the cell invasion program. MMP2 is produced as an inactive zymogen that is activated in the extracellular space by a complex mechanism that involves MMP14 and TIMP2 (see Kessenbrock et al. (2010) and references therein). Blocking MMP2 by RNAi (Kargiotis et al. 2008; Badiga et al. 2011), inhibitors (Noha et al. 2000; Nuti et al. 2011), or interference with the pathways that drive its expression (e.g., by modulating miRNA (Nan et al. 2010; Pan et al. 2012), the PI3K-Akt (Koul et al. 2001; Kubiatowski et al. 2001; Kwiatkowska et al. 2011; Jung et al. 2013) and other signaling pathways (Blazquez et al. 2008b; Kamino et al. 2011; Guan et al. 2012) decrease the invasiveness of glioma cells. Similar indispensable proinvasive role was demonstrated by targeting uPA and its receptors uPAR, MMP9, cathepsin-B (Kondraganti et al. 2000; Gondi et al. 2004a, b; Lakka et al. 2004) and cathepsin-L (Levicar et al. 2003b) with an additive effect when multiple proteases were targeted.

Recently, a number of other proteases including the ones localized predominantly intracellularly were also shown to contribute to glioma invasiveness. These include the metallo- [MMP1 (Gessler et al. 2011), MMP3 (Mercapide et al. 2003), MMP13 (Inoue et al. 2010), MMP15 (Zhang et al. 2005), MMP16 (Li et al. 2013a), MMP19 (Lettau et al. 2010), MMP26 (Deng et al. 2010), ADAMs 8 and

19 (Wildeboer et al. 2006), ADAMTS 4 and 5 (Held-Feindt et al. 2006)] as well as serine [(Miyata et al. 2007), FAP (Mentlein et al. 2011), PCSK6 (Delic et al. 2012), PCSK5A (Maret et al. 2012), furin (Mercapide et al. 2002; Wick et al. 2004), HGFA (Uchinokura et al. 2006)], and cysteine proteases [cathepsin-H (Sivaparvathi et al. 1996a), cathepsin-L (Sivaparvathi et al. 1996c; Levicar et al. 2003a), calpain-2 (Jang et al. 2010)].

The pathways and mechanisms by which proteases contribute to glioma invasion are diverse, strongly context-dependent, and in several cases highly redundant and overlapping.

12.4.1.1 Extracellular Matrix Cleavage

By cleaving the ECM, proteases remove the physical barriers hindering glioma invasion. Hyaluronic acid, tenascins, and the proteoglycans lecticans are major components of the brain ECM (Bellail et al. 2004). The fibrillar proteins typical for the ECM in other organs were previously thought to be largely restricted to the vicinity of blood vessels (collagen IV, fibronectin, laminin) or completely absent (collagen I), but there is increasing evidence that they are produced by glioma cells in situ (Paulus et al. 1994; Senner et al. 2008; Huijbers et al. 2010; Payne and Huang 2013; Serres et al. 2013). The brain-specific components and mesenchymal fibrillar proteins are subject to proteolytic degradation by several proteases. For example, the lecticans are cleaved by numerous ADAMTS, in particular the aggrecanases ADAMTS4 and ADAMTS5 (Held-Feindt et al. 2006) that are typically elevated in brain tumors, but contribution of several other MMPs including MMP19 is likely (Lettau et al. 2010). The broad range of extracellular components cleaved by MMPs (Yong et al. 2001; Hagemann et al. 2012) and aspartate and cysteine cathepsins, together with the profound dysregulation of their activities in brain tumors, makes these proteases most likely candidates involved in the degradation of ECM.

12.4.1.2 Modification of the Function of Cell Adhesion Molecules

CD44 is a major hyaluronic acid cell receptor on glioma cells and promotes their migration and invasion upon proteolytic cleavage. MMP9 binds to and cleaves CD44 (Chetty et al. 2012); in addition, CD44 can be cleaved by ADAM10 in response to CD44 ligation (Murai et al. 2004). The released extracellular fragment of CD44 was shown to directly promote glioma migration and invasion (Chetty et al. 2012), whereas the cell membrane-bound CD44 fragment is cleaved by γ-secretase. The liberated intracellular domain translocates to the cell nucleus where it acts as a transcription factor (Murakami et al. 2003) and may play a role in glioma cell adhesion (Chetty et al. 2012). Other examples of proteolytically processed cell adhesion molecules implicated in glioma migration are the ADAM 10 substrates N-cadherin (Kohutek et al. 2009) and L1CAM (CD171) (Yang et al. 2011).

12.4.1.3 Direct or Indirect Activation of the Migration-Promoting Signaling Pathways

Protease expression is associated with the activation of motility-promoting signaling cascades. Motility is decreased in glioma cells after the downregulation of uPA by RNAi in parallel with the disorganization of the cytoskeleton, downregulation of the small GTPase of the Rho-subfamily Cdc42, and decreased PI3K/Akt phosphorylation (Chandrasekar et al. 2003). The underlying mechanisms may include binding of uPA to uPAR, which converts the latter into a membrane-bound chemokine and triggers intracellular signaling (Eden et al. 2011). In addition, uPA may proteolytically activate HGF (Naldini et al. 1992), a cell motility enhancing growth factor (Moriyama et al. 1996). The tissue factor (TF) is robustly upregulated in glioma cells in response to hypoxia (Rong et al. 2009) and may trigger the local activation of the coagulation cascade proteases such as fVIIa or thrombin that subsequently initiate the promigratory signaling through the protease-activated receptors (PARs) (Gessler et al. 2010). Similarly to uPA, thrombin facilitates the maturation of HGF by activating the protease HGF activator (Uchinokura et al. 2006). Metalloproteases such as MMP14 were also demonstrated to promote promigratory MAPK signaling possibly through EGFR transactivation as well as by EGFR-independent mechanisms (Gingras et al. 2001; Langlois et al. 2007).

Zheng et al. (2007) further showed that ADAM17, a sheddase involved in the release of membrane-bound growth factors and cytokines, is implicated in the hypoxia-mediated increase in glioma invasiveness by activating the EGFR signaling pathway. Proteases such as MMP9 are also able to promote glioma cell migration by cleaving the IGFII-IGFBP complexes thereby releasing the cell motility enhancing IGF-II (Insulin-like growth factor-II, Rorive et al. (2008) and references therein). The cytokine TGF-β released from the extracellular matrix or from infiltrating microglia is another mediator that strongly promotes glioma invasiveness (Wesolowska et al. 2008; Ye et al. 2012). TGF-β is produced in a latent form and several proteases including plasmin, MMP2, thrombin, MMP14, and furin-like proteases are involved in its maturation (Leitlein et al. 2001; Jenkins 2008).

12.4.1.4 Removal of Migration Inhibitors and Proteolytic Modification of Extracellular Matrix-Promoting Glioma Migration

The brain ECM and myelin in particular are inhibitory to cellular migration (Caroni and Schwab 1988; Schwab and Caroni 1988). bNI-220, one of the most potent CNS myelin inhibitory proteins, was shown to be cleaved and inactivated by MMP14 (Belien et al. 1999), which promoted glioma cell invasiveness. Another inhibitory molecule is brevican (Yamada et al. 1997), a CNS-specific proteoglycan of the lectican family, that is upregulated in gliomas (Gary et al. 1998; Viapiano et al. 2005; Viapiano and Matthews 2006) and requires proteolysis for the promotion of glioma invasion. The studies by Viapiano et al. (2008) and Hu et al. (2008) demonstrate that the transfection of full length brevican that is resistant to the ADAMTS-mediated cleavage does not promote the malignant phenotype of glioma cells, whereas proteolytic cleavage or an N-terminal fragment of the brevican molecule promotes glioma cell adhesion, migration, and invasiveness as well as in vivo tumor growth (Hu et al. 2008; Viapiano et al. 2008). Although the downstream mechanisms remain to be established, the resulting N-terminal fragment of brevican seems to promote EGFR activation and expression of cell adhesion molecules and fibronectin (Hu et al. 2008). The EGFR activation may further promote glioma invasiveness by increasing protease expression (see Sect. 12.3). Yet another example of a secreted ECM molecule modified by proteolysis is the TGF-β-induced protein (betaig-h3). This protein has pleiotropic effects on cell–cell and cell–extracellular matrix interactions (Thapa et al. 2007) and influences adhesion and migration. In a study by Kim et al. (2012), betaig-h3 inhibited glioma cell invasion in vitro and its MMP9-mediated cleavage was suggested to be in part responsible for the proinvasive effects of MMP9 (Kim et al. 2012).

12.4.1.5 Cytoskeleton and Membrane Protrusion Remodeling

The dynamics of the actin fibers and the remodeling of the focal adhesions are critical determinants of cellular motility. Intracellular proteases such as the cysteine proteases of the calpain family importantly contribute to this process by the cleavage of the adhesion complex proteins (e.g., tallin, FAK, and paxillin) and of the actin regulators (e.g., cortactin and the small GTPase RhoA) (Carragher and Frame 2002; Franco and Huttenlocher 2005). In neuronal cells, EGF and brain-derived neurotrophic factor (BDNF) lead to calpain-2 phosphorylation via MAPK signaling (Zadran et al. 2010) and this posttranslational modification is necessary for the promigratory effects of EGF signaling (Glading et al. 2000, 2004; Cuevas et al. 2003). In gliomas, calpain-2 promotes cell migration after the binding of betaig-h3 (TGF-β-induced protein) to the α-5-β-1 integrin on the surface of glioma cells (Ma et al. 2012) and was shown to be required for glioma invasiveness by regulating the invadopodia dynamics and MMP2 secretion (Jang et al. 2010; Lal et al. 2012).

Somewhat surprisingly, low levels of active caspase-3 likewise promote glioma migration and invasion probably by the processing of gelsolin (Gdynia et al. 2007), a protein involved in actin remodeling (Silacci et al. 2004).

12.4.2 Proteases and Glioma Cell Proliferation and Apoptosis

Results of recent studies provide evidence that multiple different proteolytic mechanisms contribute to the cell cycle deregulation in glioma cells. They include the proliferative signaling mediated by the presenilin-generated carboxyterminal fragment of the overexpressed Notch receptor (Stockhausen et al. 2010; Capaccione and Pine 2013), by increased activation or activity of the protease-activated receptors (Hayakawa et al. 2007; Gessler et al. 2010), by the upregulation of some ADAM family members possibly activating the EGFR-PI3K-AKT signaling pathway (Bulstrode et al. 2012; Zheng et al. 2012; Chen et al. 2013b), and by the reduced degradation of the overexpressed or mutated tyrosine kinase receptors such as EGFR and MET by the ubiquitin-proteasome system (Hede et al. 2013). The protein components of the cell cycle machinery as well as their regulators are targeted for degradation by the ubiquitin-proteasome system (Vlachostergios et al. 2012; Hede et al. 2013). However, several specific cysteine isopeptidases called deubiquitinating enzymes (DUBs), which remove the ubiquitin chains from the ubiquitinated proteins, reverse the function of ubiquitination as an integral component of the core cell cycle machinery and of the cell cycle check points (Clague et al. 2012; Fraile et al. 2012; Cox et al. 2013; Li et al. 2013a). Besides their involvement in the control of cell cycle progression, DUBs participate also in the regulation of the signaling pathways, DNA damage repair, and apoptosis (Panner et al. 2010; Ramakrishna et al. 2011; Clague et al. 2012; Eichhorn et al. 2012; Fraile et al. 2012). It was recently demonstrated that many DUBs are targets of reversible redox regulation (Kulathu et al. 2013; Lee et al. 2013) and thus their functions may be influenced by the inherent redox disturbances in cancer.

Defects in the initiation and execution of apoptosis represent a hallmark of malignant cells including transformed astrocytes and proteolytic enzymes are indispensable players in the apoptotic pathways. Although some glioma cell lines are sensitive to apoptosis induction by TRAIL (tumor necrosis factor-α-related apoptosis-inducing ligand) and anti-Fas agonistic antibodies, many others including those with the stem cell features show resistance against the extrinsic death receptor (DR)-mediated apoptosis (Xiao et al. 2002; Yang et al. 2007; Capper et al. 2009; Bellail et al. 2010; Tao et al. 2012). Despite the high frequency of *CASP8* gene promoter, hypermethylation together with the absence of procaspase-8 protein expression in gliomas (Ashley et al. 2005; Martinez et al. 2007; Skiriute et al. 2012), reexpression of procaspase-8 in response to the DNA methyltransferase inhibitor 5-aza-2′-deoxycytidine treatment was not sufficient to restore TRAIL sensitivity in glioma cells. These results suggest the participation of additional factors responsible for the resistance to TRAIL (Capper et al. 2009). Indeed, other studies showed that the resistance of glioma cells against apoptosis induction by activated DRs is often caused by the combination of several factors including the downregulation of procaspase-8 and -10 expression and the inhibition of their activation at the death-inducing signaling complex (DISC) by c-FLIP$_L$ and c-FLIP$_S$, RIP, and PED/PEA-15 proteins (Xiao et al. 2002; Yang et al. 2007; Bellail et al. 2010; Panner et al. 2010). There is evidence that glioma cells can be killed by the granzyme-B (GrB)/perforin apoptosis pathway triggered by tumor-infiltrating lymphocytes which establish immunological synapses with tumorigenic cells (Hishii et al. 1999; Barcia et al. 2009). The synapsing CD8$^+$GrB$^+$ T cells showed GrB polarization towards tumor cells, which displayed a pycnotic or fragmented nucleus and a high positivity of the cleaved caspase-3, together indicating induced entry of the tumor cells into the execution phase of apoptosis (Barcia et al. 2009). Despite the very low percentage of synapsing cytotoxic T lymphocytes in all examined glioma cases, suggesting a deficient immune response (Barcia et al. 2009), a recent immunohistochemical study revealed that the higher expression of cleaved caspase-3 in gliomas was associated with longer survival of surgically treated glioma patients (Kobayashi et al. 2007). Tumor microenvironment of gliomas frequently contains hypoxic and acidic regions. Since hypoxia and acidic stress induce overexpression of serpinB9 in glioma stem cells (Li et al. 2009; Hjelmeland et al. 2011), this serpin, being a physiological irreversible inhibitor of GrB (Rousalova and Krepela 2010), may protect them from the GrB/perforin-mediated immune killing. Compared to the normal brain cortex tissue and primary neurons, glioma cells and tumors are hypersensitive to the (cytochrome-c + dATP)-mediated induction of the apoptosome-driven activation of procaspase-9 and -3 (Johnson et al. 2007). The authors attributed the differential sensitivity of the apoptosome apparatus activation to high Apaf-1 mRNA and protein levels in the tumor tissue compared with low Apaf-1 levels in the normal adjacent brain tissue. These differences in Apaf-1 levels correlated with differences in the levels of the transcription factor E2F1, an activator of Apaf-1 transcription, which was overexpressed in gliomas and bound to Apaf-1 promoter specifically in the tumor tissue (Johnson et al. 2007). A systems medicine model was recently suggested to predict the susceptibility of gliomas to apoptosis execution via the apoptosome-dependent procaspase activation. Based on the Apaf-1, procaspase-9, procaspase-3, Smac, and XIAP protein expression levels, the mathematical model predicted the sensitivity of glioma cell lines to temozolomide in vitro. Even more importantly, higher sensitivity to apoptosis induction was predicted by the model for the patients with longer progression free survival (Murphy et al. 2013).

Taken together, the proteolytic balance represents a prominent homeostatic regulator affecting both the proliferation and apoptosis of glioma cells.

12.4.3 Proteases and Glioma Neovascularization, Formation of Necrotic Areas and Pseudopalisades

Glioblastomas rank among the most vascularized neoplasms and typically contain areas of microvascular proliferation and a large number of dysplastic vessels. In fact, endothelial cells in glioblastomas are responsible for a substantial proportion of the extracellular matrix degrading proteolytic activity within the glioma micro-environment. Using CNS-1 cells in a rat glioma model, Regina et al. demonstrated that endothelial cells were the dominant source of the MMP9 activity in the tumors (Regina et al. 2003). MMP2, MMP9 (Rao et al. 1996; Vince et al. 1999; Raithatha et al. 2000), cathepsin-B (Mikkelsen et al. 1995; Sivaparvathi et al. 1995; Wang et al. 2005), and uPA (Yamamoto et al. 1994a) are expressed by vascular structures in gliomas. Interestingly, whereas endothelial cells in the intracranial rat glioma tumors strongly upregulated MMP9, this upregulation was absent in the subcutaneous tumors, suggesting that complex microenvironmental signals are important for the induction of proteases in endothelial cells (Regina et al. 2003).

Mechanistically, membrane-bound and extracellular proteases are well-known regulators as well as executors of the process of glioma neovascularization (Lakka et al. 2005). The vascular structures are generated with the participation of local as well as bone marrow-derived cells by several mutually nonexclusive mechanisms including angiogenesis, vasculogenesis, vascular cooption, vascular mimicry, and transdifferentiation of glioma stem cells into vascular cell types (Weis and Cheresh

2011). During angiogenesis, the endothelial cell-produced matrix metalloproteinases, cysteine proteases such as cathepsin-B, and serine proteases such as uPA and plasmin mediate the degradation of the basal membrane, thus permitting the invasion of endothelial cells and sprouting of new vessels (Lakka et al. 2005). The proteolytic remodeling of extracellular matrix also leads to the release of proangiogenic growth factors. By releasing sequestered VEGF, MMP9 was shown to participate on the angiogenic switch in pancreatic cancer (Bergers et al. 2000) and a similar mechanism seems to operate in glioblastomas (Du et al. 2008). In an animal model, ablation of tumor cell derived and stromal MMP9 impeded vascular remodeling and recruitment of endothelial and pericyte progenitors (i.e., vasculogenesis) (Du et al. 2008). Interestingly, MMP9 on bone marrow-derived $CD45^{+}$ tumor infiltrating cells was sufficient to induce angiogenesis by increasing the bioavailability of VEGF in this model (Du et al. 2008). The expression of MMP9 was also able to reverse the anti-angiogenic and growth inhibitory effect of SPARC overexpression in meduloblastoma cells (Bhoopathi et al. 2010). Thus the presence of a supercritical amount of MMP9 in the glioma microenvironment, most likely irrespective of its cellular source, seems to be necessary and sufficient to induce vascular remodeling and angiogenesis.

Another mode of glioma neovascularization involves the accumulation of glioma cells along preexisting vessels (vascular cooption) with an ensuing intravascular thrombosis, which leads to necrosis and hypoxia-induced angiogenesis (Rong et al. 2009). Proteases play various roles during this process. The enzymatic activity of matrix metalloproteinases contributes to the disruption of the endothelial cell barrier function (Ishihara et al. 2008; Rivera et al. 2010) and vessel destabilization. The serine proteases of the blood coagulation cascade thus gain access to the increased levels of tissue factor (TF) expressed by glioma cells, which initiates blood clotting and occlusion of the vessel (reviewed in Rong et al. 2009). Angiogenesis is activated by the resulting tissue hypoxia. In addition thrombin generated by the coagulation cascade stimulates endothelial cells by upregulating the α-v-β-3 integrin (Tsopanoglou et al. 2002), promotes the production of VEGF in glioma cells through PAR-2 signaling (Yamahata et al. 2002; Dutra-Oliveira et al. 2012), and leads to the activation of the proangiogenic HGF (Abounader and Laterra 2005; Uchinokura et al. 2006). The pathologic activation of the proteolytic coagulation cascade was also suggested to foster glioma stem cells in the perivascular niche (Garnier et al. 2010; Magnus et al. 2010).

Other proteases may be involved in glioma neovascularization by mechanisms not directly involving the degradation of the ECM. Endothelin converting enzyme (ECE) is expressed by glioma cells and glioma vasculature together with other components of the endothelin system (Egidy et al. 2000; Naidoo et al. 2005). ECE proteolytically converts a precursor protein into endothelin-1, which promotes vascularization due to its proproliferative and promigratory effects on endothelial cells (reviewed in Kaur et al. 2005). Clinical studies targeting ET-1 in glioblastoma are underway (Phuphanich et al. 2008) and an alternative approach could involve ECE inhibition (Berger et al. 2005). Glutamate carboxypeptidase II (prostate-specific membrane antigen, PSMA) is expressed in the neovasculature of several malignancies (Chang et al. 1999) including glioblastoma (Wernicke et al. 2011). In addition to being a maker of tumor neovascularization (Chang et al. 1999), PSMA seems to be functionally important as its absence or inhibition was demonstrated to prevent angiogenesis through the modification of integrin signaling in endothelial cells (Conway et al. 2006; Grant et al. 2012).

Proteases are driving the formation of the pseudopalisades, a typical morphological feature of glioblastomas. These highly cellular regions typically surround necrotic areas, are hypoxic, exhibit significant gelatinolytic activity (Brat et al. 2004), and abundant expression of cathepsin-B and PAI-1 (Colin et al. 2009). The increased cellularity of these structures is not due to the increased cell proliferation but rather reflects an increased migratory capacity of the glioma cells (Brat et al. 2004), which is promoted by proteases. Hypoxia, EGFR activation, and PTEN loss increase the expression of tissue factor in glioma cells (Rong et al. 2009) and the interaction of this protease cofactor with the coagulation fVIIa together with the activation of the protease-activated receptors (PARs) subsequently promotes their migration (reviewed in Dutzmann et al. 2010).

Noteworthy, proteases also exert anti-angiogenic effects (Rege et al. 2005; Lopez-Otin and Matrisian 2007; Ribatti 2009). ADAMTS1 and 8 were identified based on their sequence homology with the angioinhibitory thrombospondin-1 and similarly to thrombospondin-1 inhibit endothelial cell proliferation and angiogenesis (Vazquez et al. 1999), possibly by the sequestration of VEGF (Luque et al. 2003) or via the release of anti-angiogenic peptides from thrombospondin-1 and 2 (Lee et al. 2006). Interestingly, ADAMTS8 is downregulated in the majority of gliomas by a so far unknown mechanism (Dunn et al. 2006) and may therefore contribute to the excessive neovascularization of these tumors. A cell surface protease neprilysin (neutral endopeptidase 24.11, CD10) likewise inhibits angiogenesis by cleaving and inactivating the basic fibroblast growth factor (Goodman et al. 2006). Somewhat unexpectedly, anti-angiogenic effects are also observed in the case of proteases traditionally viewed as promoters of tumor progression. Several endogenous angiogenesis inhibitors such as endostatin, tumstatin, and angiostatin are proteolytic fragments produced from the ECM or plasminogen (Rege et al. 2005; Ribatti 2009). Indeed, MMP9 is crucial for the liberation of the C-terminal domain from collagen-IV producing tumstatin (Hamano et al. 2003) and the furin-mediated activation of MMP14 is crucial for the generation of vasculostatin from the membrane-bound brain-specific angiogenesis inhibitor 1 (BAI1) (Cork et al. 2012). Supporting the functional importance of these processes, the absence of MMP9 in mice leads to decreased levels of circulating tumstatin and increased growth of experimental tumors (Hamano et al. 2003). Similarly, the absence of MMP12 and MMP19 is associated with enhanced tumor growth owing to the promotion of angiogenesis (Houghton et al. 2006; Jost et al. 2006). Even more surprisingly, a proteolytically produced fragment of MMP2 corresponding to its hemopexin domain blocks glioma growth by inhibiting angiogenesis as well as proliferation and migration of glioma cells (Bello et al. 2001). This might explain the bewildering results of a recent study that showed a reduced in vivo growth of glioma cells overexpressing MMP2 and a more pronounced destabilization of the tumor vasculature together with higher tumor proliferation in its absence (Tremblay et al. 2011).

Proteases probably participate in the anti-angiogenic treatment-induced changes of the tumor vessels towards a more mature morphology and function. This vascular "normalization" (Goel et al. 2011) improves the perfusion and oxygenation of the tumors and is an important mechanism supporting the effectiveness of radiotherapy and cytotoxic therapies (Jain 2005; Sato 2011). In a mouse glioma model, VEGFR2 blockade promoted pericyte recruitment and a temporary restoration of the vascular morphology accompanied by the remodeling of the abnormally thick basal membrane. Collagenase IV enzymatic activity in the perivascular areas was significantly increased by the anti-VEGFR2 treatment, and the thinning of the basal membrane was prevented by a coadministration of a matrix metalloprotease inhibitor, further confirming the important role of the proteases in the process (Winkler et al. 2004).

In conclusion, the effects of proteases on angiogenesis are diverse as proteases may promote or inhibit the formation of new vessels depending on the stage of tumor development and the presence of their substrates in the tumor microenvironment.

12.5 Exploitation of Glioma-Associated Proteases as Possible Markers and Therapeutic Targets: Failures and Promises

Numerous studies examined the association between the expression of proteases and clinicopathologic data and expression of several proteases was even suggested to predict patient survival.

Patients with anaplastic astrocytomas with positive immunoreactivity for the cleaved caspase-3 in more than 10 % tumor cells had better survival rate in a study by Kobayashi et al. (2007). Similarly, higher immunohistochemical staining for carboxypeptidase-E was associated with longer survival (Horing et al. 2012). On the contrary, higher expression of ADAM8 (He et al. 2012a), CSN5 (Jab1) (He et al. 2012b), MMP13 (Wang et al. 2012), kallikrein 6 (Drucker et al. 2013), uPA (Bindal et al. 1994; Hsu et al. 1995), MMP1 (Zhang et al. 2011; Xu et al. 2013), and cathepsin-B (Colin et al. 2009) are negative prognostic factors in gliomas. However, these potentially clinically applicable results are mostly based on small scale studies or explorative microarray analyses and thus will require validation in larger patient cohorts before these candidate proteases can be accepted as reliable molecular markers in neurooncology (Jansen et al. 2010; Weller et al. 2013). Analysis of the TCGA data shows a possible prognostic significance of the proteases USP15 (Eichhorn et al. 2012) and YME1-like-1 (Bredel et al. 2009). These promising microarray-based results will need to be converted into standardized clinically applicable assays enabling the risk assessment for the individual glioma patient.

According to some authors, proteases may also be exploited for the visualization of the tumor tissue. Complete surgical resection of the tumor is an important factor influencing the prognosis of glioma patients and several methods such as microscopic surgery, intraoperative MRI, or 5-ALA fluorescence-guided resection are currently used to improve the identification and surgical removal of the tumor tissue (Black et al. 1997; Stummer et al. 2006). A recent work by Cutter et al. (2012) provides evidence that the differential expression of proteases such as cysteine cathepsins in the tumor tissue compared to normal brain may be utilized to differentiate tumor borders. The authors used fluorescently quenched activity-based probes that were constructed from suicide protease inhibitors. Proteolytic cleavage of such probes leads to the liberation of the quencher and the covalent attachment of the fluorochrome to the protease active site. Topical application of such probes rapidly visualized tumors in a preclinical glioma model as well as in the resected human glioma tissue (Cutter et al. 2012) suggesting its possible utility in guiding glioma resection.

Dysregulated proteolytic activity and often a direct role in glioma progression make proteases promising therapeutic targets. Several concepts may be envisioned for the therapeutic exploitation of glioma-associated proteases.

In addition to directly targeting the "prooncogenic" functions of glioma-associated proteases, the enzymatic activity of certain proteases may be used for the selective activation of a prodrug in the tumor tissue (tumor-activated prodrugs) (Atkinson et al. 2008). This approach relies on the highly elevated expression of the target protease in the tumor and its ability to efficiently convert the prodrug into a highly cytotoxic compound. The serine protease prostate-specific antigen (PSA) and the MMPs were studied in this context in several tumor types (Atkinson et al. 2008), but to the best of our knowledge, this approach has not been tested in gliomas. Similarly, the expression of proteases restricted to or highly upregulated in gliomas may be utilized as a means of targeted delivery of conventional chemo- or radiotherapy. GCP-II (PSMA, prostate-specific membrane antigen) is being tested as a therapeutic target in prostate cancer (Mullard 2013), but its expression in the vasculature of several tumors including gliomas make it a possible target in gliomas as well (Wernicke et al. 2011). Recently, the pronounced expression of uPA in gliomas was suggested for the selective targeting of the tumors by a recombinant uPA-directed oncolytic virus that markedly improved survival in a rodent glioma model (Hasegawa et al. 2010).

Downregulation of the target protease by RNA interference or the inhibition of the upstream signaling cascades driving protease expression are potential therapeutic approaches, the latter possibly being a contributing factor to the effects of therapies targeting receptor tyrosine kinases and the downstream signaling pathways (Lakka et al. 2000; Mao et al. 2012). Indeed, data from preclinical glioma models suggest that inhibition or genetic ablation of proteases or their respective critical activators in glioma cells leads to the induction of apoptosis (Gondi et al. 2009; Chetty et al. 2010), suppression of cell proliferation, invasiveness and inhibition of tumor-induced angiogenesis, as well as tumorigenicity in experimental animals (Rao 2003; Lakka et al. 2005; Badiga et al. 2011; Veeravalli et al. 2012; Kesanakurti et al. 2013).

Low-molecular-weight protease inhibitors have become an effective strategy in the treatment of cardiovascular and infectious diseases (reviewed in Turk 2006) and the hope is that they could also improve the outcomes in cancer patients.

Given the important role of MMPs in glioma invasiveness, MMP inhibitors were among the first protease inhibitors tested in glioblastomas. Four clinical trials testing the broad spectrum MMP inhibitor marimastat in combination with temozolomide or radiotherapy in patients with grade III and IV tumors were published. Overall, the therapeutic effects were small and the treatment was accompanied by musculoskeletal toxicity in a large proportion of patients (Groves et al. 2002, 2006; Larson et al. 2002; Levin et al. 2006). Possibly, more specific MMP inhibitors and/or optimized dosing and timing of administration may prove more efficacious in future studies, in particular considering the potential of MMP inhibitors to block the proinvasive effects of radiotherapy and anti-angiogenic treatments (Lucio-Eterovic et al. (2009) and Sect. 12.3.1).

More recently, proteasome inhibition was suggested as a potential therapeutic modality in gliomas. Proteasome is responsible for the proteolytic removal of regulatory proteins with growth inhibitory activity such as the CDK inhibitor p27 (Piva et al. 1999), the inhibitors of NFκB (IkB), and proteins p53, BID (Unterkircher et al. 2011), NOXA (Ohshima-Hosoyama et al. 2011), or activated caspases-8 and -3 (Kim et al. 2004) (see also Fig. 12.6). Proteasomal inhibition therefore induces the accumulation of these proteins; in addition, it enhances the expression of the apoptosis-inducing receptors such as Fas (Tani et al. 2001) or the TRAIL receptor DR5 (Hetschko et al. 2008) and triggers the endoplasmic reticulum stress (Kardosh et al. 2008). By these complex mechanisms proteasome inhibitors cause growth arrest and sensitization to proapoptotic stimuli as demonstrated in glioma cell lines (Kitagawa et al. 1999; Wagenknecht et al. 1999, 2000; Tani et al. 2001; Yin et al. 2005; Pedeboscq et al. 2008; Zanotto-Filho et al. 2012) and in patient-derived primary cells (Koschny et al. 2007). Proteasome inhibitors also enhanced the effectiveness of other chemotherapeutics (Ceruti et al. 2006), radiation (Ng et al. 2009), and TRAIL (La Ferla-Bruhl et al. 2007; Hetschko et al. 2008; Balyasnikova et al. 2011; Jane et al. 2011; Kahana et al. 2011; Seol 2011; Unterkircher et al. 2011) and had synergistic inhibitory effects on glioma stem-like cells in combination with the histonedeacetylase inhibitors (Asklund et al. 2012). In preclinical animal models, the antitumor activity was observed for lactacystin (Legnani et al. 2006; Wang et al. 2013a). However, studies using the new proteasome inhibitor salinosporamide A (Vlashi et al. 2010) or the most frequently utilized proteasome inhibitor approved for clinical application in other cancers bortezomib (Labussiere et al. 2008) failed at reducing the tumor growth in a xenotransplantation glioma model. Phase I clinical trials demonstrated safety of bortezomib in glioma patients (Kubicek et al. 2009; Phuphanich et al. 2010), but a phase II trial in patients with recurrent glioblastoma failed to show its efficacy in combination with vorinostat (Friday et al. 2012). Several other phase I and II studies of bortezomib in combination with other therapeutic modalities (Table 12.3) as well as testing of new proteasome inhibitors with improved in vivo efficacy (Roth et al. 2009) are underway (Vlachostergios et al. 2013). Another possibility to target the ubiquitin-proteosome system lies in the modulation of the deubiquinating enzymes (DUBs, reviewed in Hussain et al. (2009), Fraile et al. (2012)]. By cleaving the isopeptide bonds joining ubiquitin to other proteins, DUBs stabilize

Table 12.3 Clinical trials targeting proteases in gliomas

Target	Agent	Phase and trial identifier	Remarks, reference
MMPs	Marimastat	II, III	Limited efficacy, musculoskeletal side effects (Groves et al. 2002, 2006; Larson et al. 2002; Levin et al. 2006)
Coagulation cascade	Low molecular weight heparin (LMWH)	II, NCT00028678, III, NCT00135876	Similar 12 month mortality rate in patients receiving a preventive dose of dalteparin compared to placebo (Perry et al. 2010), similar survival in patients using dalteparin compared to historical controls (Robins et al. 2008), trend for improved progression free survival and 1 year overall survival with concomitant chemoradiotherapy with LMWH (Zincircioglu et al. 2012)
Proteasome (presumably)	Nelfinavir	I, NCT01020292, NCT00915694, I/II NCT00694837	Mechanism probably involves proteasome inhibition (Pyrko et al. 2007).
Proteasome	Ritonavir/ lopinavir	II, NCT01095094	Minimal activity at the dose tested, did not prolong progression free survival at 6 months as a single agent in heavily pretreated patients (Ahluwalia et al. 2011)
Proteasome	Bortezomib	I, NCT00006773, NCT00994500, NCT00544284, NCT01435395	Ongoing
Proteasome	Bortezomib	II, NCT00108069, NCT00611325, NCT00990652, NCT00641706, NCT00998010	Ongoing
γ-Secretase	RO4929097	I//II, NCT01088763	Ongoing

http://clinicaltrials.gov/

a number of proteins essential for the cell cycle regulation, chromatin remodeling, DNA damage repair, and signaling pathways including the membrane-bound receptors such as EGFR (Fraile et al. 2012; Liu et al. 2014). In the case of DUBs associated with the proteasome (i.e., UCHL5, USP14, and POH1), however, the inhibition leads to the block of proteasomal degradation of polyubiquitinated proteins. b-AP15, an inhibitor of the proteasome-associated DUBs, was recently demonstrated to have antitumor activity against a range of cancer cells with glioma cell lines of the NCI60 panel being particularly sensitive (D'Arcy et al. 2011). With few exceptions (see Table 12.1), the role of DUBs in glioma pathogenesis is nevertheless largely unknown and more research is needed before using DUB inhibitors or approaches aiming at restoring the activities of tumor-suppressor DUBs (Fraile et al. 2012) in glioma therapy.

Some protease inhibitors were found to have novel, unexpected antitumor activities. The aspartate protease inhibitors widely used in the treatment of HIV infection such as nelfinavir and ritonavir were demonstrated to inhibit glioma growth. The mechanisms are currently not fully elucidated but presumably involve the induction of endoplasmic reticulum stress, interference with the activity of Hsp90, and reduction of MMP expression (Laurent et al. 2004; Liuzzi et al. 2004; Pyrko et al. 2007) (see Kast et al. (2013) for review). Clinical studies in glioma patients were rapidly initiated (Table 12.3) as these drugs have a favorable safety profile and have already been in clinical use.

Inhibitors of the γ-secretase (GSIs) seem to be another promising class of drugs (see Golde et al. (2013) for review). Originally designed for the treatment of Alzheimer's disease, GSIs are currently tested in several cancers including gliomas because of their inhibitory effect on the Notch signaling (Golde et al. 2013; Takebe et al. 2014). This pathway typically promotes the self-renewal and inhibits the differentiation of stem cells including glioma stem cells (Chen et al. 2010; Hu et al. 2011; Kristoffersen et al. 2013). Gliomas of the proneural subtype may be in particular sensitive to the inhibition of γ-secretase, as the Notch pathway seems to be typically activated in this subtype (Saito et al. 2014, #2290). In addition, the GSIs may also block the neurotrophin-induced glioma invasion (Wang et al. 2008a). In line with these results, GSIs inhibited glioma growth in preclinical models and enhanced the effects of temozolomide (Gilbert et al. 2010). Despite these promising results, there are several important areas that remain to be explored, such as possible off-target effects of GSIs on signal peptidase and the effects of GSI on angiogenesis and immune functions (Golde et al. 2013). For example, a recent report concluded that GSIs reduced the growth of xenotransplanted gliomas, but somewhat surprisingly enhanced the proliferation of endothelial cells in response to VEGF and increased the vascular density in the tumors (Zou et al. 2013). The vasculature was dysfunctional as the perfusion and oxygenation of the treated tumors was decreased similarly to observations in other cancers (see Liu et al. (2013) for review). In addition, γ-secretase was suggested to function as a tumor suppressor in squamous cell carcinoma (Li et al. 2007). Indeed, a prematurely halted clinical trial in patients with Alzheimer's disease (Doody et al. 2013) reported increased incidence of skin cancer. Inhibition of γ-secretase may have negative effect on the proliferation and self-renewal of normal stem cells in several organs (van Es et al. 2005; Hu et al. 2011). Despite a phase I trial with a γ-secretase inhibitor MK-0752 in pediatric patients with recurrent CNS malignancies demonstrated good tolerability (Fouladi et al. 2011), the obvious complexity of the effects of GSIs on the tumor microenvironment still precludes the straightforward claim that γ-secretase represents a suitable target for antiglioma treatment and requires further studies with careful evaluation of its therapeutic efficacy and safety.

Other proteases were also suggested as potential targets in glioma treatment mostly based on preclinical data. Inhibition of the sheddase ADAM10 was demonstrated to block glioma growth by interfering with the abovementioned Notch signaling pathway (Floyd et al. 2012). In addition, the inhibition of ADAM10 may improve the ability of the immune system to eradicate glioma stem cell (Wolpert et al. 2014) and also block glioma cell migration and adhesion (Table 12.1, see Moss et al. (2008) for general review). Targeting of the extracellularly localized cathepsin-B by cell impermeable inhibitors (Withana et al. 2012) or its elimination by RNAi were also proposed, although at least in gliomas, the effects of downregulating cathepsin-B were most pronounced when combined with the targeting of other extracellular proteolytic systems (Gondi and Rao 2013). The proteolytic activation of the immunosuppressive TGF-β in gliomas may represent another example of a potential functional target of protease inhibitors (Huber et al. 1992; Leitlein et al. 2001).

Given the known high risk of thromboembolic events in patients with high grade gliomas and preclinical studies suggesting possible anticancer effects of low-molecular-weight heparin and direct thrombin inhibitors, inhibition of the blood coagulation proteases may seem beneficial in glioma patients (Ornstein et al. 2002; Hua et al. 2005a, b). The results from the so far performed clinical trials with low-molecular-weight heparins are however ambiguous (Perry et al. 2010), although some hints of potential benefit can be found (Robins et al. 2008; Zincircioglu et al. 2012). Given the observed increased risk of bleeding associated with their use (Perry et al. 2010) and the limited data on their clinical efficacy, the role of anticoagulation as an adjuvant antitumor therapy remains to be established (Perry 2010).

Despite the wealth of encouraging data in the preclinical studies, protease inhibitors were so far largely ineffective in the majority of the clinical trials in cancer patients (Coussens et al. 2002), including patients with gliomas (Groves et al. 2002, 2006; Larson et al. 2002; Levin et al. 2006; Ahluwalia et al. 2011). There are a number of causes that may have contributed to these failures:

1. Malignancies frequently "hijack" physiologically occurring mechanisms including the deployment of proteases; many "on-target" side effects should therefore be anticipated due to the interference with their physiological functions (see Sect. 12.1). This risk is illustrated by the musculoskeletal side effects observed in the clinical studies using MMP inhibitors (Coussens et al. 2002). In addition, several proteases act as tumor suppressors or produce anti-angiogenic mediators (see Sect. 12.4.3, Bello et al. (2001), Rege et al. (2005), Lopez-Otin and Matrisian (2007), Cork et al. (2012)] and their inhibition may thus facilitate tumorigenesis.
2. The complexity, functional overlap, and redundancy of the proteolytic cascades in gliomas suggest that targeting of several different types of proteases may be necessary to achieve the desired biological effects. This not only complicates the design of low molecular weight inhibitors capable of targeting several biochemically distinct protease classes but also brings about an increased risk of "on-target" side effects.
3. Incomplete understanding of the role of proteases in tumor progression and spatiotemporal and interindividual differences in the dependence of the tumors

on the protease-mediated events. The variability of protease expression (Fig. 12.1, Jaworski et al. 2010) and their differing functional role at various stages of tumor progression requires careful characterization and selection of appropriate patients for clinical trials as well as for subsequent use in clinical practice.

4. Limitations of current preclinical glioma models in determining the clinical efficacy of anti-protease therapies. The in vitro systems and long term propagated glioma cell lines grown in serum-containing media have numerous limitations (Schulte et al. 2011) and may produce misleading results. For example, an adenoviral-mediated overexpression of PAI-1 in a glioma cell line reduced their invasiveness through Matrigel, but not in brain tissue (Hjortland et al. 2003). Similarly, several studies with the proteasome inhibitor bortezomib in glioma cell lines indicated its growth inhibitory effects, which was not recapitulated in more complex in vivo models or so far completed clinical studies (Friday et al. 2012; Vlachostergios et al. 2013). Reliable preclinical models recapitulating the heterogeneity of tumor cells (Snuderl et al. 2011), enabling the assessment of the effects of therapeutic interventions on the stromal cells (Jones and Holland 2012), as well as the complex host-tumor pro-tumorigenic and anti-tumorigenic interactions are needed.
5. Absence of appropriate biomarkers of achieving target protease inhibition in the clinical setting. Identification of the biologically active dose is necessary before performing large phase III trials. Furthermore, the design of clinical trials represents another challenge as the effect of agents that do not directly kill cancer cells cannot be measured by reduction of tumor size (Coussens et al. 2002).
6. Last but not least: some of the compounds targeting proteases were rushed into clinical trials at an unprecedented pace and in several cases in suboptimal settings. Enormous need to improve the outcome of otherwise untreatable diseases, but also patent issues and competition may have contributed to such decisions in the past, but these factors are unfortunately likely to play a role in the future trials as well (Coussens et al. 2002).

Conclusion

Proteases are important regulators of the biological functions of proteins under both physiological and pathological conditions. Their dysregulation in gliomas is a consequence of complex changes in the glioma microenvironment and involves transformed glioma cells as well as various cell types present in the tumor stroma. Despite some common patterns, glioblastomas are, similarly to other parameters, heterogeneous also in terms of protease expression and activity. Proteases participate on almost every aspect of gliomagenesis acting both as regulators and executors and may therefore represent promising therapeutic targets. However, the role of individual proteases at various stages of the disease progression is usually complex, highly context dependent, and redundant. This might have contributed to the failures of the initial attempts on therapeutic targeting of proteases in gliomas as well as in other cancers. Designing effective and safe anti-glioma strategies exploiting proteases will therefore require a more in-depth understanding of their physiological and pathophysiological roles. Equally important will be the identification of critical protease-mediated processes in individual glioma subtypes and careful selection of subgroups of glioma patients most likely to benefit from therapies interfering with the functions of proteases.

Acknowledgment This work was supported by grants IGA 12237-5/2011, PRVOUK-P27/LF1/1, UNCE 204013 and by the project "BIOCEV—Biotechnology and Biomedicine Centre of the Academy of Sciences and Charles University" (CZ.1.05/1.1.00/02.0109), from the European Regional Development Fund.

References

Aaberg-Jessen C, Christensen K, Offenberg H, Bartels A, Dreehsen T, Hansen S, Schroder HD, Brunner N, Kristensen BW (2009) Low expression of tissue inhibitor of metalloproteinases-1 (TIMP-1) in glioblastoma predicts longer patient survival. J Neurooncol 95(1):117–128

Abounader R, Laterra J (2005) Scatter factor/hepatocyte growth factor in brain tumor growth and angiogenesis. Neuro Oncol 7(4):436–451

Abraham R, Schafer J, Rothe M, Bange J, Knyazev P, Ullrich A (2005) Identification of MMP-15 as an anti-apoptotic factor in cancer cells. J Biol Chem 280(40):34123–34132

Ahluwalia MS, Patton C, Stevens G, Tekautz T, Angelov L, Vogelbaum MA, Weil RJ, Chao S, Elson P, Suh JH, Barnett GH, Peereboom DM (2011) Phase II trial of ritonavir/lopinavir in patients with progressive or recurrent high-grade gliomas. J Neurooncol 102(2):317–321

Almonte AG, Sweatt JD (2011) Serine proteases, serine protease inhibitors, and protease-activated receptors: roles in synaptic function and behavior. Brain Res 1407:107–122

Amberger VR, Paganetti PA, Seulberger H, Eldering JA, Schwab ME (1994) Characterization of a membrane-bound metalloendoprotease of rat C6 glioblastoma cells. Cancer Res 54(15): 4017–4025

Amos S, Redpath GT, Dipierro CG, Carpenter JE, Hussaini IM (2010) Epidermal growth factor receptor-mediated regulation of urokinase plasminogen activator expression and glioblastoma invasion via C-SRC/MAPK/AP-1 signaling pathways. J Neuropathol Exp Neurol 69(6): 582–592

Anand M, Van Meter TE, Fillmore HL (2011) Epidermal growth factor induces matrix metalloproteinase-1 (MMP-1) expression and invasion in glioma cell lines via the MAPK pathway. J Neurooncol 104(3):679–687

Anand R, Langer T, Baker MJ (2013) Proteolytic control of mitochondrial function and morphogenesis. Biochim Biophys Acta 1833(1):195–204

Annabi B, Laflamme C, Sina A, Lachambre MP, Beliveau R (2009) A MT1-MMP/NF-kappaB signaling axis as a checkpoint controller of COX-2 expression in CD133+ U87 glioblastoma cells. J Neuroinflammation 6:8

Aplan PD (2006) Causes of oncogenic chromosomal translocation. Trends Genet 22(1):46–55

Arai Y, Kubota T, Nakagawa T, Kabuto M, Sato K, Kobayashi H (1998) Production of urokinase-type plasminogen activator (u-PA) and plasminogen activator inhibitor-1 (PAI-1) in human brain tumours. Acta Neurochir 140(4):377–386

Ariza A, Fernandez LA, Inagami T, Kim JH, Manuelidis EE (1988) Renin in glioblastoma multiforme and its role in neovascularization. Am J Clin Pathol 90(4):437–441

Ashley DM, Riffkin CD, Muscat AM, Knight MJ, Kaye AH, Novak U, Hawkins CJ (2005) Caspase 8 is absent or low in many ex vivo gliomas. Cancer 104(7):1487–1496

Asklund T, Kvarnbrink S, Holmlund C, Wibom C, Bergenheim T, Henriksson R, Hedman H (2012) Synergistic killing of glioblastoma stem-like cells by bortezomib and HDAC inhibitors. Anticancer Res 32(7):2407–2413

Asuthkar S, Velpula KK, Chetty C, Gorantla B, Rao JS (2012) Epigenetic regulation of miRNA-211 by MMP-9 governs glioma cell apoptosis, chemosensitivity and radiosensitivity. Oncotarget 3(11):1439–1454

Atkinson JM, Siller CS, Gill JH (2008) Tumour endoproteases: the cutting edge of cancer drug delivery? Br J Pharmacol 153(7):1344–1352

Backes DM, Siddiq A, Cox DG, Calboli FC, Gaziano JM, Ma J, Stampfer M, Hunter DJ, Camargo CA, Michaud DS (2013) Single-nucleotide polymorphisms of allergy-related genes and risk of adult glioma. J Neurooncol 113(2):229–238

Badiga AV, Chetty C, Kesanakurti D, Are D, Gujrati M, Klopfenstein JD, Dinh DH, Rao JS (2011) MMP-2 siRNA inhibits radiation-enhanced invasiveness in glioma cells. PLoS One 6(6): e20614

Baker AH, George SJ, Zaltsman AB, Murphy G, Newby AC (1999) Inhibition of invasion and induction of apoptotic cell death of cancer cell lines by overexpression of TIMP-3. Br J Cancer 79(9–10):1347–1355

Balyasnikova IV, Ferguson SD, Han Y, Liu F, Lesniak MS (2011) Therapeutic effect of neural stem cells expressing TRAIL and bortezomib in mice with glioma xenografts. Cancer Lett 310(2):148–159

Barcia C Jr, Gomez A, Gallego-Sanchez JM, Perez-Valles A, Castro MG, Lowenstein PR, Barcia C Sr, Herrero MT (2009) Infiltrating CTLs in human glioblastoma establish immunological synapses with tumorigenic cells. Am J Pathol 175(2):786–798

Barrow J, Adamowicz-Brice M, Cartmill M, MacArthur D, Lowe J, Robson K, Brundler MA, Walker DA, Coyle B, Grundy R (2011) Homozygous loss of ADAM3A revealed by genome-wide analysis of pediatric high-grade glioma and diffuse intrinsic pontine gliomas. Neuro Oncol 13(2):212–222

Beffert U, Durudas A, Weeber EJ, Stolt PC, Giehl KM, Sweatt JD, Hammer RE, Herz J (2006) Functional dissection of Reelin signaling by site-directed disruption of Disabled-1 adaptor binding to apolipoprotein E receptor 2: distinct roles in development and synaptic plasticity. J Neurosci 26(7):2041–2052

Belien AT, Paganetti PA, Schwab ME (1999) Membrane-type 1 matrix metalloprotease (MT1-MMP) enables invasive migration of glioma cells in central nervous system white matter. J Cell Biol 144(2):373–384

Bellail AC, Hunter SB, Brat DJ, Tan C, Van Meir EG (2004) Microregional extracellular matrix heterogeneity in brain modulates glioma cell invasion. Int J Biochem Cell Biol 36(6): 1046–1069

Bellail AC, Tse MC, Song JH, Phuphanich S, Olson JJ, Sun SY, Hao C (2010) DR5-mediated DISC controls caspase-8 cleavage and initiation of apoptosis in human glioblastomas. J Cell Mol Med 14(6A):1303–1317

Bello L, Lucini V, Carrabba G, Giussani C, Machluf M, Pluderi M, Nikas D, Zhang J, Tomei G, Villani RM, Carroll RS, Bikfalvi A, Black PM (2001) Simultaneous inhibition of glioma angiogenesis, cell proliferation, and invasion by a naturally occurring fragment of human metalloproteinase-2. Cancer Res 61(24):8730–8736

Bergbold N, Lemberg MK (2013) Emerging role of rhomboid family proteins in mammalian biology and disease. Biochim Biophys Acta 1828(12):2840–2848

Berger Y, Dehmlow H, Blum-Kaelin D, Kitas EA, Loffler BM, Aebi JD, Juillerat-Jeanneret L (2005) Endothelin-converting enzyme-1 inhibition and growth of human glioblastoma cells. J Med Chem 48(2):483–498

Bergers G, Brekken R, McMahon G, Vu TH, Itoh T, Tamaki K, Tanzawa K, Thorpe P, Itohara S, Werb Z, Hanahan D (2000) Matrix metalloproteinase-9 triggers the angiogenic switch during carcinogenesis. Nat Cell Biol 2(10):737–744

Bernstein H-G (2005) Proteases and Alzheimer's disease: present knowledge and emerging concepts of therapy. In: Lendeckel U, Hooper N (eds) Proteases in the brain, vol 3. Springer, New York, NY, pp 1–23

Best JL, Ganiatsas S, Agarwal S, Changou A, Salomoni P, Shirihai O, Meluh PB, Pandolfi PP, Zon LI (2002) SUMO-1 protease-1 regulates gene transcription through PML. Mol Cell 10(4): 843–855

Bethke L, Sullivan K, Webb E, Murray A, Schoemaker M, Auvinen A, Kiuru A, Salminen T, Johansen C, Christensen HC, Muir K, McKinney P, Hepworth S, Dimitropoulou P, Lophatananon A, Feychting M, Lonn S, Ahlbom A, Malmer B, Henriksson R, Swerdlow A, Houlston R (2008) The common D302H variant of CASP8 is associated with risk of glioma. Cancer Epidemiol Biomarkers Prev 17(4):987–989

Bhoopathi P, Chetty C, Gujrati M, Dinh DH, Rao JS, Lakka SS (2010) The role of MMP-9 in the anti-angiogenic effect of secreted protein acidic and rich in cysteine. Br J Cancer 102(3): 530–540

Bindal AK, Hammoud M, Shi WM, Wu SZ, Sawaya R, Rao JS (1994) Prognostic significance of proteolytic enzymes in human brain tumors. J Neurooncol 22(2):101–110

Binder DK, Berger MS (2002) Proteases and the biology of glioma invasion. J Neurooncol 56(2): 149–158

Black PM, Moriarty T, Alexander E III, Stieg P, Woodard EJ, Gleason PL, Martin CH, Kikinis R, Schwartz RB, Jolesz FA (1997) Development and implementation of intraoperative magnetic resonance imaging and its neurosurgical applications. Neurosurgery 41(4):831–842, discussion 842–835

Blazquez C, Carracedo A, Salazar M, Lorente M, Egia A, Gonzalez-Feria L, Haro A, Velasco G, Guzman M (2008a) Down-regulation of tissue inhibitor of metalloproteinases-1 in gliomas: a new marker of cannabinoid antitumoral activity? Neuropharmacology 54(1):235–243

Blazquez C, Salazar M, Carracedo A, Lorente M, Egia A, Gonzalez-Feria L, Haro A, Velasco G, Guzman M (2008b) Cannabinoids inhibit glioma cell invasion by down-regulating matrix metalloproteinase-2 expression. Cancer Res 68(6):1945–1952

Bodey B, Bodey B Jr, Siegel SE, Kaiser HE (2000) Matrix metalloproteinase expression in childhood astrocytomas. Anticancer Res 20(5A):3287–3292

Bodey B, Bodey V, Siegel SE, Nasir A, Coppola D, Hakam A, Kaiser HE (2004) Immunocytochemical detection of members of the caspase cascade of apoptosis in high-grade astrocytomas. In Vivo 18(5):593–602

Bramwell B (1888) Intracranial tumours. Pentland, Edinburgh

Brat DJ, Castellano-Sanchez AA, Hunter SB, Pecot M, Cohen C, Hammond EH, Devi SN, Kaur B, Van Meir EG (2004) Pseudopalisades in glioblastoma are hypoxic, express extracellular matrix proteases, and are formed by an actively migrating cell population. Cancer Res 64(3):920–927

Bredel M, Scholtens DM, Harsh GR, Bredel C, Chandler JP, Renfrow JJ, Yadav AK, Vogel H, Scheck AC, Tibshirani R, Sikic BI (2009) A network model of a cooperative genetic landscape in brain tumors. JAMA 302(3):261–275

Brisson L, Reshkin SJ, Gore J, Roger S (2012) pH regulators in invadosomal functioning: proton delivery for matrix tasting. Eur J Cell Biol 91(11–12):847–860

Brooks PC, Stromblad S, Sanders LC, von Schalscha TL, Aimes RT, Stetler-Stevenson WG, Quigley JP, Cheresh DA (1996) Localization of matrix metalloproteinase MMP-2 to the surface of invasive cells by interaction with integrin alpha v beta 3. Cell 85(5):683–693

Bryan L, Paugh BS, Kapitonov D, Wilczynska KM, Alvarez SM, Singh SK, Milstien S, Spiegel S, Kordula T (2008) Sphingosine-1-phosphate and interleukin-1 independently regulate plasminogen activator inhibitor-1 and urokinase-type plasminogen activator receptor expression in glioblastoma cells: implications for invasiveness. Mol Cancer Res 6(9):1469–1477

Bulstrode H, Jones LM, Siney EJ, Sampson JM, Ludwig A, Gray WP, Willaime-Morawek S (2012) A-Disintegrin and Metalloprotease (ADAM) 10 and 17 promote self-renewal of brain tumor sphere forming cells. Cancer Lett 326(1):79–87

Bulteau AL, Bayot A (2011) Mitochondrial proteases and cancer. Biochim Biophys Acta 1807(6): 595–601

Busek P, Stremenova J, Sromova L, Hilser M, Balaziova E, Kosek D, Trylcova J, Strnad H, Krepela E, Sedo A (2012) Dipeptidyl peptidase-IV inhibits glioma cell growth independent of its enzymatic activity. Int J Biochem Cell Biol 44(5):738–747

Caccamo DV, Keohane ME, McKeever PE (1994) Plasminogen activators and inhibitors in gliomas: an immunohistochemical study. Mod Pathol 7(1):99–104

Campbell RL, Davies PL (2012) Structure-function relationships in calpains. Biochem J 447(3): 335–351

Cancer Genome Atlas Research Network (2008) Comprehensive genomic characterization defines human glioblastoma genes and core pathways. Nature 455(7216):1061–1068

Capaccione KM, Pine SR (2013) The Notch signaling pathway as a mediator of tumor survival. Carcinogenesis 34(7):1420–1430

Capper D, Gaiser T, Hartmann C, Habel A, Mueller W, Herold-Mende C, von Deimling A, Siegelin MD (2009) Stem-cell-like glioma cells are resistant to TRAIL/Apo2L and exhibit down-regulation of caspase-8 by promoter methylation. Acta Neuropathol 117(4):445–456

Caroni P, Schwab ME (1988) Two membrane protein fractions from rat central myelin with inhibitory properties for neurite growth and fibroblast spreading. J Cell Biol 106(4):1281–1288

Carragher NO, Frame MC (2002) Calpain: a role in cell transformation and migration. Int J Biochem Cell Biol 34(12):1539–1543

Caterina JJ, Yamada S, Caterina NC, Longenecker G, Holmback K, Shi J, Yermovsky AE, Engler JA, Birkedal-Hansen H (2000) Inactivating mutation of the mouse tissue inhibitor of metalloproteinases-2(Timp-2) gene alters proMMP-2 activation. J Biol Chem 275(34): 26416–26422

Ceruti S, Mazzola A, Abbracchio MP (2006) Proteasome inhibitors potentiate etoposide-induced cell death in human astrocytoma cells bearing a mutated p53 isoform. J Pharmacol Exp Ther 319(3):1424–1434

Chakraborti S, Mandal M, Das S, Mandal A, Chakraborti T (2003) Regulation of matrix metalloproteinases: an overview. Mol Cell Biochem 253(1–2):269–285

Chandrasekar N, Mohanam S, Gujrati M, Olivero WC, Dinh DH, Rao JS (2003) Downregulation of uPA inhibits migration and PI3k/Akt signaling in glioblastoma cells. Oncogene 22(3): 392–400

Chang SS, Reuter VE, Heston WD, Bander NH, Grauer LS, Gaudin PB (1999) Five different anti-prostate-specific membrane antigen (PSMA) antibodies confirm PSMA expression in tumor-associated neovasculature. Cancer Res 59(13):3192–3198

Chapman HA (2004) Cathepsins as transcriptional activators? Dev Cell 6(5):610–611

Chen J, Kesari S, Rooney C, Strack PR, Chen J, Shen H, Wu L, Griffin JD (2010) Inhibition of notch signaling blocks growth of glioblastoma cell lines and tumor neurospheres. Genes Cancer 1(8):822–835

Chen X, Chen L, Chen J, Hu W, Gao H, Xie B, Wang X, Yin Z, Li S, Wang X (2013a) ADAM17 promotes U87 glioblastoma stem cell migration and invasion. Brain Res 1538:151–158

Chen X, Chen L, Zhang R, Yi Y, Ma Y, Yan K, Jiang X, Wang X (2013b) ADAM17 regulates self-renewal and differentiation of U87 glioblastoma stem cells. Neurosci Lett 537:44–49

Chetty C, Lakka SS, Bhoopathi P, Gondi CS, Veeravalli KK, Fassett D, Klopfenstein JD, Dinh DH, Gujrati M, Rao JS (2010) Urokinase plasminogen activator receptor and/or matrix metalloproteinase-9 inhibition induces apoptosis signaling through lipid rafts in glioblastoma xenograft cells. Mol Cancer Ther 9(9):2605–2617

Chetty C, Vanamala SK, Gondi CS, Dinh DH, Gujrati M, Rao JS (2012) MMP-9 induces CD44 cleavage and CD44 mediated cell migration in glioblastoma xenograft cells. Cell Signal 24(2): 549–559

Chintala SK, Tonn JC, Rao JS (1999) Matrix metalloproteinases and their biological function in human gliomas. Int J Dev Neurosci 17(5–6):495–502

Choe G, Park JK, Jouben-Steele L, Kremen TJ, Liau LM, Vinters HV, Cloughesy TF, Mischel PS (2002) Active matrix metalloproteinase 9 expression is associated with primary glioblastoma subtype. Clin Cancer Res 8(9):2894–2901

Choi GC, Li J, Wang Y, Li L, Zhong L, Ma B, Su X, Ying J, Xiang T, Rha SY, Yu J, Sung JJ, Tsao SW, Chan AT, Tao Q (2014) The metalloprotease ADAMTS8 displays antitumor properties through antagonizing EGFR-MEK-ERK signaling and is silenced in carcinomas by CpG methylation. Mol Cancer Res 12(2):228–238

Chu Q, Orr BA, Semenkow S, Bar EE, Eberhart CG (2013) Prolonged inhibition of glioblastoma xenograft initiation and clonogenic growth following in vivo Notch blockade. Clin Cancer Res 19(12):3224–3233

Clague MJ, Coulson JM, Urbe S (2012) Cellular functions of the DUBs. J Cell Sci 125 Pt 2): 277–286

Clark EB, Jovov B, Rooj AK, Fuller CM, Benos DJ (2010) Proteolytic cleavage of human acid-sensing ion channel 1 by the serine protease matriptase. J Biol Chem 285(35):27130–27143

Colbert JD, Matthews SP, Miller G, Watts C (2009) Diverse regulatory roles for lysosomal proteases in the immune response. Eur J Immunol 39(11):2955–2965

Colin C, Voutsinos-Porche B, Nanni I, Fina F, Metellus P, Intagliata D, Baeza N, Bouvier C, Delfino C, Loundou A, Chinot O, Lah T, Kos J, Martin PM, Ouafik L, Figarella-Branger D (2009) High expression of cathepsin B and plasminogen activator inhibitor type-1 are strong predictors of survival in glioblastomas. Acta Neuropathol 118(6):745–754

Comincini S, Paolillo M, Barbieri G, Palumbo S, Sbalchiero E, Azzalin A, Russo MA, Schinelli S (2009) Gene expression analysis of an EGFR indirectly related pathway identified PTEN and MMP9 as reliable diagnostic markers for human glial tumor specimens. J Biomed Biotechnol 2009:924565

Conway RE, Petrovic N, Li Z, Heston W, Wu D, Shapiro LH (2006) Prostate-specific membrane antigen regulates angiogenesis by modulating integrin signal transduction. Mol Cell Biol 26(14):5310–5324

Cork SM, Kaur B, Devi NS, Cooper L, Saltz JH, Sandberg EM, Kaluz S, Van Meir EG (2012) A proprotein convertase/MMP-14 proteolytic cascade releases a novel 40 kDa vasculostatin from tumor suppressor BAI1. Oncogene 31(50):5144–5152

Coussens LM, Fingleton B, Matrisian LM (2002) Matrix metalloproteinase inhibitors and cancer: trials and tribulations. Science 295(5564):2387–2392

Cox JL (2009) Cystatins and cancer. Front Biosci (Landmark Ed) 14:463–474

Cox JL, Wilder PJ, Gilmore JM, Wuebben EL, Washburn MP, Rizzino A (2013) The SOX2-interactome in brain cancer cells identifies the requirement of MSI2 and USP9X for the growth of brain tumor cells. PLoS One 8(5):e62857

Crocker S, Smith P, Park D (2005) Calpain proteolysis and the etiology of Parkinson's disease: an emerging hypothesis. In: Lendeckel U, Hooper N (eds) Proteases in the brain, vol 3. Springer, New York, NY, pp 25–61

Crocker M, Ashley S, Giddings I, Petrik V, Hardcastle A, Aherne W, Pearson A, Bell BA, Zacharoulis S, Papadopoulos MC (2011) Serum angiogenic profile of patients with glioblastoma identifies distinct tumor subtypes and shows that TIMP-1 is a prognostic factor. Neuro Oncol 13(1):99–108

Cuevas BD, Abell AN, Witowsky JA, Yujiri T, Johnson NL, Kesavan K, Ware M, Jones PL, Weed SA, DeBiasi RL, Oka Y, Tyler KL, Johnson GL (2003) MEKK1 regulates calpain-dependent proteolysis of focal adhesion proteins for rear-end detachment of migrating fibroblasts. EMBO J 22(13):3346–3355

Cutter JL, Cohen NT, Wang J, Sloan AE, Cohen AR, Panneerselvam A, Schluchter M, Blum G, Bogyo M, Basilion JP (2012) Topical application of activity-based probes for visualization of brain tumor tissue. PLoS One 7(3):e33060

D'Abaco GM, Ng K, Paradiso L, Godde NJ, Kaye A, Novak U (2006) ADAM22, expressed in normal brain but not in high-grade gliomas, inhibits cellular proliferation via the disintegrin domain. Neurosurgery 58(1):179–186, discussion 179-186

D'Arcy P, Brnjic S, Olofsson MH, Fryknas M, Lindsten K, De Cesare M, Perego P, Sadeghi B, Hassan M, Larsson R, Linder S (2011) Inhibition of proteasome deubiquitinating activity as a new cancer therapy. Nat Med 17(12):1636–1640

Das G, Shravage BV, Baehrecke EH (2012) Regulation and function of autophagy during cell survival and cell death. Cold Spring Harb Perspect Biol 4(6). doi:10.1101/cshperspect.a008813

Dauth S, Sirbulescu RF, Jordans S, Rehders M, Avena L, Oswald J, Lerchl A, Saftig P, Brix K (2011) Cathepsin K deficiency in mice induces structural and metabolic changes in the central nervous system that are associated with learning and memory deficits. BMC Neurosci 12:74

de Groot JF, Fuller G, Kumar AJ, Piao Y, Eterovic K, Ji Y, Conrad CA (2010) Tumor invasion after treatment of glioblastoma with bevacizumab: radiographic and pathologic correlation in humans and mice. Neuro Oncol 12(3):233–242

Del Rosso M, Fibbi G, Schmitt M (2002) Non-enzymatic activities of proteases: from scepticism to reality. Biol Chem 383(1):1–4

Delic S, Lottmann N, Jetschke K, Reifenberger G, Riemenschneider MJ (2012) Identification and functional validation of CDH11, PCSK6 and SH3GL3 as novel glioma invasion-associated candidate genes. Neuropathol Appl Neurobiol 38(2):201–212

Demchik LL, Sameni M, Nelson K, Mikkelsen T, Sloane BF (1999) Cathepsin B and glioma invasion. Int J Dev Neurosci 17(5–6):483–494

Deng Y, Li W, Li Y, Yang H, Xu H, Liang S, Zhang L, Li Y (2010) Expression of Matrix Metalloproteinase-26 promotes human glioma U251 cell invasion in vitro and in vivo. Oncol Rep 23(1):69–78

Deryugina EI, Soroceanu L, Strongin AY (2002) Up-regulation of vascular endothelial growth factor by membrane-type 1 matrix metalloproteinase stimulates human glioma xenograft growth and angiogenesis. Cancer Res 62(2):580–588

Donepudi M, Grutter MG (2002) Structure and zymogen activation of caspases. Biophys Chem 101–102:145–153

Doody RS, Raman R, Farlow M, Iwatsubo T, Vellas B, Joffe S, Kieburtz K, He F, Sun X, Thomas RG, Aisen PS, C. Alzheimer's Disease Cooperative Study Steering, Siemers E, Sethuraman G, Mohs R, G. Semagacestat Study (2013) A phase 3 trial of semagacestat for treatment of Alzheimer's disease. N Engl J Med 369(4):341–350

Drucker KL, Paulsen AR, Giannini C, Decker PA, Blaber SI, Blaber M, Uhm JH, O'Neill BP, Jenkins RB, Scarisbrick IA (2013) Clinical significance and novel mechanism of action of kallikrein 6 in glioblastoma. Neuro Oncol 15(3):305–318

Du R, Lu KV, Petritsch C, Liu P, Ganss R, Passegue E, Song H, Vandenberg S, Johnson RS, Werb Z, Bergers G (2008) HIF1alpha induces the recruitment of bone marrow-derived vascular modulatory cells to regulate tumor angiogenesis and invasion. Cancer Cell 13(3): 206–220

Dufour A, Zucker S, Sampson NS, Kuscu C, Cao J (2010) Role of matrix metalloproteinase-9 dimers in cell migration: design of inhibitory peptides. J Biol Chem 285(45):35944–35956

Dunn JR, Reed JE, du Plessis DG, Shaw EJ, Reeves P, Gee AL, Warnke P, Walker C (2006) Expression of ADAMTS-8, a secreted protease with antiangiogenic properties, is downregulated in brain tumours. Br J Cancer 94(8):1186–1193

Dutra-Oliveira A, Monteiro RQ, Mariano-Oliveira A (2012) Protease-activated receptor-2 (PAR2) mediates VEGF production through the ERK1/2 pathway in human glioblastoma cell lines. Biochem Biophys Res Commun 421(2):221–227

Dutzmann S, Gessler F, Harter PN, Gerlach R, Mittelbronn M, Seifert V, Kogel D (2010) The pro-migratory and pro-invasive role of the procoagulant tissue factor in malignant gliomas. Cell Adh Migr 4(4):515–522

Dziembowska M, Wlodarczyk J (2012) MMP9: a novel function in synaptic plasticity. Int J Biochem Cell Biol 44(5):709–713

Dziembowska M, Danilkiewicz M, Wesolowska A, Zupanska A, Chouaib S, Kaminska B (2007) Cross-talk between Smad and p38 MAPK signalling in transforming growth factor beta signal transduction in human glioblastoma cells. Biochem Biophys Res Commun 354(4):1101–1106

Eden G, Archinti M, Furlan F, Murphy R, Degryse B (2011) The urokinase receptor interactome. Curr Pharm Des 17(19):1874–1889

Egidy G, Eberl LP, Valdenaire O, Irmler M, Majdi R, Diserens AC, Fontana A, Janzer RC, Pinet F, Juillerat-Jeanneret L (2000) The endothelin system in human glioblastoma. Lab Invest 80(11): 1681–1689

Eichhorn PJ, Rodon L, Gonzalez-Junca A, Dirac A, Gili M, Martinez-Saez E, Aura C, Barba I, Peg V, Prat A, Cuartas I, Jimenez J, Garcia-Dorado D, Sahuquillo J, Bernards R, Baselga J, Seoane J (2012) USP15 stabilizes TGF-beta receptor I and promotes oncogenesis through the activation of TGF-beta signaling in glioblastoma. Nat Med 18(3):429–435

Eisele G, Wischhusen J, Mittelbronn M, Meyermann R, Waldhauer I, Steinle A, Weller M, Friese MA (2006) TGF-beta and metalloproteinases differentially suppress NKG2D ligand surface expression on malignant glioma cells. Brain 129(Pt 9):2416–2425

Enari M, Sakahira H, Yokoyama H, Okawa K, Iwamatsu A, Nagata S (1998) A caspase-activated DNase that degrades DNA during apoptosis, and its inhibitor ICAD. Nature 391(6662):43–50

Endo S, Yokosawa H, Ishii S (1989) Involvement of endopeptidase-24.11 in degradation of substance P by glioma cells. Neuropeptides 14(3):177–184

Esteve PO, Tremblay P, Houde M, St-Pierre Y, Mandeville R (1998) In vitro expression of MMP-2 and MMP-9 in glioma cells following exposure to inflammatory mediators. Biochim Biophys Acta 1403(1):85–96

Esteve PO, Chicoine E, Robledo O, Aoudjit F, Descoteaux A, Potworowski EF, St-Pierre Y (2002) Protein kinase C-zeta regulates transcription of the matrix metalloproteinase-9 gene induced by IL-1 and TNF-alpha in glioma cells via NF-kappa B. J Biol Chem 277(38):35150–35155

Fager N, Jaworski DM (2000) Differential spatial distribution and temporal regulation of tissue inhibitor of metalloproteinase mRNA expression during rat central nervous system development. Mech Dev 98(1–2):105–109

Fan X, Khaki L, Zhu TS, Soules ME, Talsma CE, Gul N, Koh C, Zhang J, Li YM, Maciaczyk J, Nikkhah G, Dimeco F, Piccirillo S, Vescovi AL, Eberhart CG (2010) NOTCH pathway blockade depletes CD133-positive glioblastoma cells and inhibits growth of tumor neurospheres and xenografts. Stem Cells 28(1):5–16

Felbor U, Kessler B, Mothes W, Goebel HH, Ploegh HL, Bronson RT, Olsen BR (2002) Neuronal loss and brain atrophy in mice lacking cathepsins B and L. Proc Natl Acad Sci U S A 99(12): 7883–7888

Fillmore HL, VanMeter TE, Broaddus WC (2001) Membrane-type matrix metalloproteinases (MT-MMPs): expression and function during glioma invasion. J Neurooncol 53(2):187–202

Fischer A (1946) Mechanism of the proteolytic activity of malignant tissue cells. Nature 157:442

Fischer U, Leidinger P, Keller A, Folarin A, Ketter R, Graf N, Lenhof HP, Meese E (2010) Amplicons on chromosome 12q13-21 in glioblastoma recurrences. Int J Cancer 126(11): 2594–2602

Flannery T, Gibson D, Mirakhur M, McQuaid S, Greenan C, Trimble A, Walker B, McCormick D, Johnston PG (2003) The clinical significance of cathepsin S expression in human astrocytomas. Am J Pathol 163(1):175–182

Flannery T, McQuaid S, McGoohan C, McConnell RS, McGregor G, Mirakhur M, Hamilton P, Diamond J, Cran G, Walker B, Scott C, Martin L, Ellison D, Patel C, Nicholson C, Mendelow D, McCormick D, Johnston PG (2006) Cathepsin S expression: an independent prognostic factor in glioblastoma tumours–a pilot study. Int J Cancer 119(4):854–860

Floyd DH, Kefas B, Seleverstov O, Mykhaylyk O, Dominguez C, Comeau L, Plank C, Purow B (2012) Alpha-secretase inhibition reduces human glioblastoma stem cell growth in vitro and in vivo by inhibiting Notch. Neuro Oncol 14(10):1215–1226

Formolo CA, Williams R, Gordish-Dressman H, MacDonald TJ, Lee NH, Hathout Y (2011) Secretome signature of invasive glioblastoma multiforme. J Proteome Res 10(7):3149–3159

Forsyth PA, Wong H, Laing TD, Rewcastle NB, Morris DG, Muzik H, Leco KJ, Johnston RN, Brasher PM, Sutherland G, Edwards DR (1999) Gelatinase-A (MMP-2), gelatinase-B

(MMP-9) and membrane type matrix metalloproteinase-1 (MT1-MMP) are involved in different aspects of the pathophysiology of malignant gliomas. Br J Cancer 79(11–12):1828–1835

Forsyth PA, Krishna N, Lawn S, Valadez JG, Qu X, Fenstermacher DA, Fournier M, Potthast L, Chinnaiyan P, Gibney GT, Zeinieh M, Barker PA, Carter BD, Cooper MK, Kenchappa RS (2014) p75 neurotrophin receptor cleavage by alpha- and gamma-secretases is required for neurotrophin mediated proliferation of brain tumor initiating cells. J Biol Chem 289(12): 8067–8085

Fouladi M, Stewart CF, Olson J, Wagner LM, Onar-Thomas A, Kocak M, Packer RJ, Goldman S, Gururangan S, Gajjar A, Demuth T, Kun LE, Boyett JM, Gilbertson RJ (2011) Phase I trial of MK-0752 in children with refractory CNS malignancies: a pediatric brain tumor consortium study. J Clin Oncol 29(26):3529–3534

Fraile JM, Quesada V, Rodriguez D, Freije JM, Lopez-Otin C (2012) Deubiquitinases in cancer: new functions and therapeutic options. Oncogene 31(19):2373–2388

Franco SJ, Huttenlocher A (2005) Regulating cell migration: calpains make the cut. J Cell Sci 118 (Pt 17):3829–3838

Friday BB, Anderson SK, Buckner J, Yu C, Giannini C, Geoffroy F, Schwerkoske J, Mazurczak M, Gross H, Pajon E, Jaeckle K, Galanis E (2012) Phase II trial of vorinostat in combination with bortezomib in recurrent glioblastoma: a north central cancer treatment group study. Neuro Oncol 14(2):215–221

Friedberg MH, Glantz MJ, Klempner MS, Cole BF, Perides G (1998) Specific matrix metalloproteinase profiles in the cerebrospinal fluid correlated with the presence of malignant astrocytomas, brain metastases, and carcinomatous meningitis. Cancer 82(5):923–930

Fritchie K, Siintola E, Armao D, Lehesjoki AE, Marino T, Powell C, Tennison M, Booker JM, Koch S, Partanen S, Suzuki K, Tyynela J, Thorne LB (2009) Novel mutation and the first prenatal screening of cathepsin D deficiency (CLN10). Acta Neuropathol 117(2):201–208

Fu Y, Zheng Y, Li K, Huang R, Zheng S, An N, Liang A (2012) Mutations in isocitrate dehydrogenase 2 accelerate glioma cell migration via matrix metalloproteinase-2 and 9. Biotechnol Lett 34(3):441–446

Fujiwara S, Nakagawa K, Harada H, Nagato S, Furukawa K, Teraoka M, Seno T, Oka K, Iwata S, Ohnishi T (2007) Silencing hypoxia-inducible factor-1alpha inhibits cell migration and invasion under hypoxic environment in malignant gliomas. Int J Oncol 30(4):793–802

Fukuda ME, Iwadate Y, Machida T, Hiwasa T, Nimura Y, Nagai Y, Takiguchi M, Tanzawa H, Yamaura A, Seki N (2005) Cathepsin D is a potential serum marker for poor prognosis in glioma patients. Cancer Res 65(12):5190–5194

Fukushima S, Kato S, Maeda M, Shigemori M (2008) Caspase-9 pathway activation by inhibiting endogenous fibroblast growth factor signaling in human glioma cells. Int J Oncol 32(2): 467–473

Furukawa K, Kumon Y, Harada H, Kohno S, Nagato S, Teraoka M, Fujiwara S, Nakagawa K, Hamada K, Ohnishi T (2006) PTEN gene transfer suppresses the invasive potential of human malignant gliomas by regulating cell invasion-related molecules. Int J Oncol 29(1):73–81

Furuta T, Nakada M, Misaki K, Sato Y, Hayashi Y, Nakanuma Y, Hamada JI (2014) Molecular analysis of a recurrent glioblastoma treated with bevacizumab. Brain Tumor Pathol 31(1): 32–39

Gabriely G, Wurdinger T, Kesari S, Esau CC, Burchard J, Linsley PS, Krichevsky AM (2008) MicroRNA 21 promotes glioma invasion by targeting matrix metalloproteinase regulators. Mol Cell Biol 28(17):5369–5380

Gabrusiewicz K, Ellert-Miklaszewska A, Lipko M, Sielska M, Frankowska M, Kaminska B (2011) Characteristics of the alternative phenotype of microglia/macrophages and its modulation in experimental gliomas. PLoS One 6(8):e23902

Gallo M, Ho J, Coutinho FJ, Vanner R, Lee L, Head R, Ling EK, Clarke ID, Dirks PB (2013) A tumorigenic MLL-homeobox network in human glioblastoma stem cells. Cancer Res 73(1): 417–427

Garnier D, Milsom C, Magnus N, Meehan B, Weitz J, Yu J, Rak J (2010) Role of the tissue factor pathway in the biology of tumor initiating cells. Thromb Res 125(Suppl 2):S44–S50

Gary SC, Kelly GM, Hockfield S (1998) BEHAB/brevican: a brain-specific lectican implicated in gliomas and glial cell motility. Curr Opin Neurobiol 8(5):576–581

Gdynia G, Grund K, Eckert A, Bock BC, Funke B, Macher-Goeppinger S, Sieber S, Herold-Mende C, Wiestler B, Wiestler OD, Roth W (2007) Basal caspase activity promotes migration and invasiveness in glioblastoma cells. Mol Cancer Res 5(12):1232–1240

Geng F, Wenzel S, Tansey WP (2012) Ubiquitin and proteasomes in transcription. Annu Rev Biochem 81:177–201

Gessi S, Sacchetto V, Fogli E, Merighi S, Varani K, Baraldi PG, Tabrizi MA, Leung E, Maclennan S, Borea PA (2010) Modulation of metalloproteinase-9 in U87MG glioblastoma cells by A3 adenosine receptors. Biochem Pharmacol 79(10):1483–1495

Gessler F, Voss V, Dutzmann S, Seifert V, Gerlach R, Kogel D (2010) Inhibition of tissue factor/protease-activated receptor-2 signaling limits proliferation, migration and invasion of malignant glioma cells. Neuroscience 165(4):1312–1322

Gessler F, Voss V, Seifert V, Gerlach R, Kogel D (2011) Knockdown of TFPI-2 promotes migration and invasion of glioma cells. Neurosci Lett 497(1):49–54

Gilbert CA, Daou MC, Moser RP, Ross AH (2010) Gamma-secretase inhibitors enhance temozolomide treatment of human gliomas by inhibiting neurosphere repopulation and xenograft recurrence. Cancer Res 70(17):6870–6879

Gingras D, Bousquet-Gagnon N, Langlois S, Lachambre MP, Annabi B, Beliveau R (2001) Activation of the extracellular signal-regulated protein kinase (ERK) cascade by membrane-type-1 matrix metalloproteinase (MT1-MMP). FEBS Lett 507(2):231–236

Girgis AH, Bui A, White NM, Yousef GM (2012) Integrated genomic characterization of the kallikrein gene locus in cancer. Anticancer Res 32(3):957–963

Glading A, Chang P, Lauffenburger DA, Wells A (2000) Epidermal growth factor receptor activation of calpain is required for fibroblast motility and occurs via an ERK/MAP kinase signaling pathway. J Biol Chem 275(4):2390–2398

Glading A, Bodnar RJ, Reynolds IJ, Shiraha H, Satish L, Potter DA, Blair HC, Wells A (2004) Epidermal growth factor activates m-calpain (calpain II), at least in part, by extracellular signal-regulated kinase-mediated phosphorylation. Mol Cell Biol 24(6):2499–2512

Goel S, Duda DG, Xu L, Munn LL, Boucher Y, Fukumura D, Jain RK (2011) Normalization of the vasculature for treatment of cancer and other diseases. Physiol Rev 91(3):1071–1121

Goh KY, Poon WS, Chan DT, Ip CP (2005) Tissue plasminogen activator expression in meningiomas and glioblastomas. Clin Neurol Neurosurg 107(4):296–300

Golde TE, Koo EH, Felsenstein KM, Osborne BA, Miele L (2013) gamma-Secretase inhibitors and modulators. Biochim Biophys Acta 1828(12):2898–2907

Goldschneider D, Mehlen P (2010) Dependence receptors: a new paradigm in cell signaling and cancer therapy. Oncogene 29(13):1865–1882

Gole B, Duran Alonso MB, Dolenc V, Lah T (2009) Post-translational regulation of cathepsin B, but not of other cysteine cathepsins, contributes to increased glioblastoma cell invasiveness in vitro. Pathol Oncol Res 15(4):711–723

Gole B, Huszthy PC, Popovic M, Jeruc J, Ardebili YS, Bjerkvig R, Lah TT (2012) The regulation of cysteine cathepsins and cystatins in human gliomas. Int J Cancer 131(8):1779–1789

Golubkov VS, Strongin AY (2007) Proteolysis-driven oncogenesis. Cell Cycle 6(2):147–150

Gondi CS, Rao JS (2013) Cathepsin B as a cancer target. Expert Opin Ther Targets 17(3):281–291

Gondi CS, Lakka SS, Dinh DH, Olivero WC, Gujrati M, Rao JS (2004a) Downregulation of uPA, uPAR and MMP-9 using small, interfering, hairpin RNA (siRNA) inhibits glioma cell invasion, angiogenesis and tumor growth. Neuron Glia Biol 1(2):165–176

Gondi CS, Lakka SS, Yanamandra N, Olivero WC, Dinh DH, Gujrati M, Tung CH, Weissleder R, Rao JS (2004b) Adenovirus-mediated expression of antisense urokinase plasminogen activator receptor and antisense cathepsin B inhibits tumor growth, invasion, and angiogenesis in gliomas. Cancer Res 64(12):4069–4077

Gondi CS, Talluri L, Dinh DH, Gujrati M, Rao JS (2009) RNAi-mediated downregulation of MMP-2 activates the extrinsic apoptotic pathway in human glioma xenograft cells. Int J Oncol 35(4):851–859

Goodman OB Jr, Febbraio M, Simantov R, Zheng R, Shen R, Silverstein RL, Nanus DM (2006) Neprilysin inhibits angiogenesis via proteolysis of fibroblast growth factor-2. J Biol Chem 281 (44):33597–33605

Grant CL, Caromile LA, Ho V, Durrani K, Rahman MM, Claffey KP, Fong GH, Shapiro LH (2012) Prostate specific membrane antigen (PSMA) regulates angiogenesis independently of VEGF during ocular neovascularization. PLoS One 7(7):e41285

Groft LL, Muzik H, Rewcastle NB, Johnston RN, Knauper V, Lafleur MA, Forsyth PA, Edwards DR (2001) Differential expression and localization of TIMP-1 and TIMP-4 in human gliomas. Br J Cancer 85(1):55–63

Groves MD, Puduvalli VK, Hess KR, Jaeckle KA, Peterson P, Yung WK, Levin VA (2002) Phase II trial of temozolomide plus the matrix metalloproteinase inhibitor, marimastat, in recurrent and progressive glioblastoma multiforme. J Clin Oncol 20(5):1383–1388

Groves MD, Puduvalli VK, Conrad CA, Gilbert MR, Yung WK, Jaeckle K, Liu V, Hess KR, Aldape KD, Levin VA (2006) Phase II trial of temozolomide plus marimastat for recurrent anaplastic gliomas: a relationship among efficacy, joint toxicity and anticonvulsant status. J Neurooncol 80(1):83–90

Grunda JM, Nabors LB, Palmer CA, Chhieng DC, Steg A, Mikkelsen T, Diasio RB, Zhang K, Allison D, Grizzle WE, Wang W, Gillespie GY, Johnson MR (2006) Increased expression of thymidylate synthetase (TS), ubiquitin specific protease 10 (USP10) and survivin is associated with poor survival in glioblastoma multiforme (GBM). J Neurooncol 80(3):261–274

Guan H, Cai J, Zhang N, Wu J, Yuan J, Li J, Li M (2012) Sp1 is upregulated in human glioma, promotes MMP-2-mediated cell invasion and predicts poor clinical outcome. Int J Cancer 130(3):593–601

Guo P, Imanishi Y, Cackowski FC, Jarzynka MJ, Tao HQ, Nishikawa R, Hirose T, Hu B, Cheng SY (2005) Up-regulation of angiopoietin-2, matrix metalloprotease-2, membrane type 1 metalloprotease, and laminin 5 gamma 2 correlates with the invasiveness of human glioma. Am J Pathol 166(3):877–890

Guttmann RP, Sokol S, Baker DL, Simpkins KL, Dong Y, Lynch DR (2002) Proteolysis of the N-methyl-d-aspartate receptor by calpain in situ. J Pharmacol Exp Ther 302(3):1023–1030

Hagemann C, Anacker J, Ernestus RI, Vince GH (2012) A complete compilation of matrix metalloproteinase expression in human malignant gliomas. World J Clin Oncol 3(5):67–79

Hallberg M, Grevès P, Nyberg F (2005) Neuropeptide processing. In: Lendeckel U, Hooper N (eds) Proteases in the brain, vol 3. Springer, New York, NY, pp 203–234

Hamada K, Kuratsu J, Saitoh Y, Takeshima H, Nishi T, Ushio Y (1996) Expression of tissue factor correlates with grade of malignancy in human glioma. Cancer 77(9):1877–1883

Hamano Y, Zeisberg M, Sugimoto H, Lively JC, Maeshima Y, Yang C, Hynes RO, Werb Z, Sudhakar A, Kalluri R (2003) Physiological levels of tumstatin, a fragment of collagen IV alpha3 chain, are generated by MMP-9 proteolysis and suppress angiogenesis via alphaV beta3 integrin. Cancer Cell 3(6):589–601

Hamasuna R, Kataoka H, Moriyama T, Itoh H, Seiki M, Koono M (1999) Regulation of matrix metalloproteinase-2 (MMP-2) by hepatocyte growth factor/scatter factor (HGF/SF) in human glioma cells: HGF/SF enhances MMP-2 expression and activation accompanying up-regulation of membrane type-1 MMP. Int J Cancer 82(2):274–281

Hamasuna R, Kataoka H, Meng JY, Itoh H, Moriyama T, Wakisaka S, Koono M (2001) Reduced expression of hepatocyte growth factor activator inhibitor type-2/placental bikunin (HAI-2/PB) in human glioblastomas: implication for anti-invasive role of HAI-2/PB in glioblastoma cells. Int J Cancer 93(3):339–345

Hasegawa Y, Kinoh H, Iwadate Y, Onimaru M, Ueda Y, Harada Y, Saito S, Furuya A, Saegusa T, Morodomi Y, Hasegawa M, Saito S, Aoki I, Saeki N, Yonemitsu Y (2010) Urokinase-targeted fusion by oncolytic Sendai virus eradicates orthotopic glioblastomas by pronounced synergy with interferon-beta gene. Mol Ther 18(10):1778–1786

Hattori A, Tsujimoto M (2013) Endoplasmic reticulum aminopeptidases: biochemistry, physiology and pathology. J Biochem 154(3):219–228

Hayakawa Y, Kurimoto M, Nagai S, Kurosaki K, Tsuboi Y, Hamada H, Hayashi N, Endo S (2007) Thrombin-induced cell proliferation and platelet-derived growth factor-AB release from A172 human glioblastoma cells. J Thromb Haemost 5(11):2219–2226

He S, Ding L, Cao Y, Li G, Deng J, Tu Y, Wang B (2012a) Overexpression of a disintegrin and metalloprotease 8 in human gliomas is implicated in tumor progression and prognosis. Med Oncol 29(3):2032–2037

He SM, Zhao ZW, Wang Y, Zhao JP, Wang L, Hou F, Gao GD (2012b) Potential role of Jun activation domain-binding protein 1 and phosphorylated p27 expression in prognosis of glioma. Brain Tumor Pathol 29(1):3–9

Hede SM, Savov V, Weishaupt H, Sangfelt O, Swartling FJ (2013) Oncoprotein stabilization in brain tumors. Oncogene. doi:10.1038/onc.2013.445

Held-Feindt J, Paredes EB, Blomer U, Seidenbecher C, Stark AM, Mehdorn HM, Mentlein R (2006) Matrix-degrading proteases ADAMTS4 and ADAMTS5 (disintegrins and metalloproteinases with thrombospondin motifs 4 and 5) are expressed in human glioblastomas. Int J Cancer 118(1):55–61

Hetschko H, Voss V, Seifert V, Prehn JH, Kogel D (2008) Upregulation of DR5 by proteasome inhibitors potently sensitizes glioma cells to TRAIL-induced apoptosis. FEBS J 275(8): 1925–1936

Hickey CM, Wilson NR, Hochstrasser M (2012) Function and regulation of SUMO proteases. Nat Rev Mol Cell Biol 13(12):755–766

Hiraoka N, Allen E, Apel IJ, Gyetko MR, Weiss SJ (1998) Matrix metalloproteinases regulate neovascularization by acting as pericellular fibrinolysins. Cell 95(3):365–377

Hishii M, Kurnick JT, Ramirez-Montagut T, Pandolfi F (1999) Studies of the mechanism of cytolysis by tumour-infiltrating lymphocytes. Clin Exp Immunol 116(3):388–394

Hjelmeland AB, Wu Q, Heddleston JM, Choudhary GS, MacSwords J, Lathia JD, McLendon R, Lindner D, Sloan A, Rich JN (2011) Acidic stress promotes a glioma stem cell phenotype. Cell Death Differ 18(5):829–840

Hjortland GO, Bjornland K, Pettersen S, Garman-Vik SS, Emilsen E, Nesland JM, Fodstad O, Engebraaten O (2003) Modulation of glioma cell invasion and motility by adenoviral gene transfer of PAI-1. Clin Exp Metastasis 20(4):301–309

Hjortland GO, Lillehammer T, Somme S, Wang JB, Halvorsen T, Juell S, Hirschberg H, Fodstad O, Engebraaten O (2004) Plasminogen activator inhibitor-1 increases the expression of VEGF in human glioma cells. Exp Cell Res 294(1):130–139

Hooper NM (2002) Proteases: a primer. Essays Biochem 38:1–8

Horing E, Harter PN, Seznec J, Schittenhelm J, Buhring HJ, Bhattacharyya S, von Hattingen E, Zachskorn C, Mittelbronn M, Naumann U (2012) The "go or grow" potential of gliomas is linked to the neuropeptide processing enzyme carboxypeptidase E and mediated by metabolic stress. Acta Neuropathol 124(1):83–97

Hoshi M, Harada A, Kawase T, Uyemura K, Yazaki T (2000) Antitumoral effects of defective herpes simplex virus-mediated transfer of tissue inhibitor of metalloproteinases-2 gene in malignant glioma U87 in vitro: consequences for anti-cancer gene therapy. Cancer Gene Ther 7(5):799–805

Houghton AM, Grisolano JL, Baumann ML, Kobayashi DK, Hautamaki RD, Nehring LC, Cornelius LA, Shapiro SD (2006) Macrophage elastase (matrix metalloproteinase-12) suppresses growth of lung metastases. Cancer Res 66(12):6149–6155

Houri N, Huang KC, Nalbantoglu J (2013) The Coxsackievirus and Adenovirus Receptor (CAR) undergoes ectodomain shedding and regulated intramembrane proteolysis (RIP). PLoS One 8(8):e73296

Hsu DW, Efird JT, Hedley-Whyte ET (1995) Prognostic role of urokinase-type plasminogen activator in human gliomas. Am J Pathol 147(1):114–123

Hu B, Kong LL, Matthews RT, Viapiano MS (2008) The proteoglycan brevican binds to fibronectin after proteolytic cleavage and promotes glioma cell motility. J Biol Chem 283(36):24848–24859

Hu YY, Zheng MH, Cheng G, Li L, Liang L, Gao F, Wei YN, Fu LA, Han H (2011) Notch signaling contributes to the maintenance of both normal neural stem cells and patient-derived glioma stem cells. BMC Cancer 11:82

Hua Y, Tang L, Keep RF, Schallert T, Fewel ME, Muraszko KM, Hoff JT, Xi G (2005a) The role of thrombin in gliomas. J Thromb Haemost 3(9):1917–1923

Hua Y, Tang LL, Fewel ME, Keep RF, Schallert T, Muraszko KM, Hoff JT, Xi GH (2005b) Systemic use of argatroban reduces tumor mass, attenuates neurological deficits and prolongs survival time in rat glioma models. Acta Neurochir Suppl 95:403–406

Huang Q, Li F, Liu X, Li W, Shi W, Liu FF, O'Sullivan B, He Z, Peng Y, Tan AC, Zhou L, Shen J, Han G, Wang XJ, Thorburn J, Thorburn A, Jimeno A, Raben D, Bedford JS, Li CY (2011) Caspase 3-mediated stimulation of tumor cell repopulation during cancer radiotherapy. Nat Med 17(7):860–866

Huber D, Philipp J, Fontana A (1992) Protease inhibitors interfere with the transforming growth factor-beta-dependent but not the transforming growth factor-beta-independent pathway of tumor cell-mediated immunosuppression. J Immunol 148(1):277–284

Huijbers IJ, Iravani M, Popov S, Robertson D, Al-Sarraj S, Jones C, Isacke CM (2010) A role for fibrillar collagen deposition and the collagen internalization receptor endo180 in glioma invasion. PLoS One 5(3):e9808

Hussain S, Zhang Y, Galardy PJ (2009) DUBs and cancer: the role of deubiquitinating enzymes as oncogenes, non-oncogenes and tumor suppressors. Cell Cycle 8(11):1688–1697

Inoue A, Takahashi H, Harada H, Kohno S, Ohue S, Kobayashi K, Yano H, Tanaka J, Ohnishi T (2010) Cancer stem-like cells of glioblastoma characteristically express MMP-13 and display highly invasive activity. Int J Oncol 37(5):1121–1131

Ishihara H, Kubota H, Lindberg RL, Leppert D, Gloor SM, Errede M, Virgintino D, Fontana A, Yonekawa Y, Frei K (2008) Endothelial cell barrier impairment induced by glioblastomas and transforming growth factor beta2 involves matrix metalloproteinases and tight junction proteins. J Neuropathol Exp Neurol 67(5):435–448

Iwadate Y, Hayama M, Adachi A, Matsutani T, Nagai Y, Hiwasa T, Saeki N (2008) High serum level of plasminogen activator inhibitor-1 predicts histological grade of intracerebral gliomas. Anticancer Res 28(1B):415–418

Iwatsuki K, Kumara E, Yoshimine T, Nakagawa H, Sato M, Hayakawa T (2000) Elastase expression by infiltrating neutrophils in gliomas. Neurol Res 22(5):465–468

Izuhara K, Ohta S, Kanaji S, Shiraishi H, Arima K (2008) Recent progress in understanding the diversity of the human ov-serpin/clade B serpin family. Cell Mol Life Sci 65(16):2541–2553

Jain RK (2005) Normalization of tumor vasculature: an emerging concept in antiangiogenic therapy. Science 307(5706):58–62

Jane EP, Premkumar DR, Pollack IF (2011) Bortezomib sensitizes malignant human glioma cells to TRAIL, mediated by inhibition of the NF-{kappa}B signaling pathway. Mol Cancer Ther 10(1):198–208

Jang HS, Lal S, Greenwood JA (2010) Calpain 2 is required for glioblastoma cell invasion: regulation of matrix metalloproteinase 2. Neurochem Res 35(11):1796–1804

Jansen M, Yip S, Louis DN (2010) Molecular pathology in adult gliomas: diagnostic, prognostic, and predictive markers. Lancet Neurol 9(7):717–726

Jaworski DM, Stradecki HM, Penar PL, Pendlebury WW, Pennington CJ, Edwards DR, Broaddus WC, Fillmore FL (2010) Degradome profiling fails to identify a unique protease signature in primary malignant brain tumors. Neuro Oncol 12(Suppl 4):iv1–iv7

Jenkins G (2008) The role of proteases in transforming growth factor-beta activation. Int J Biochem Cell Biol 40(6–7):1068–1078

Jin X, Jin X, Sohn YW, Yin J, Kim SH, Joshi K, Nam DH, Nakano I, Kim H (2013) Blockade of EGFR signaling promotes glioma stem-like cell invasiveness by abolishing ID3-mediated inhibition of p27(KIP1) and MMP3 expression. Cancer Lett 328(2):235–242

Johnson CE, Huang YY, Parrish AB, Smith MI, Vaughn AE, Zhang Q, Wright KM, Van Dyke T, Wechsler-Reya RJ, Kornbluth S, Deshmukh M (2007) Differential Apaf-1 levels allow cytochrome c to induce apoptosis in brain tumors but not in normal neural tissues. Proc Natl Acad Sci U S A 104(52):20820–20825

Jones TS, Holland EC (2012) Standard of care therapy for malignant glioma and its effect on tumor and stromal cells. Oncogene 31(16):1995–2006

Jost M, Folgueras AR, Frerart F, Pendas AM, Blacher S, Houard X, Berndt S, Munaut C, Cataldo D, Alvarez J, Melen-Lamalle L, Foidart JM, Lopez-Otin C, Noel A (2006) Earlier onset of tumoral angiogenesis in matrix metalloproteinase-19-deficient mice. Cancer Res 66(10):5234–5241

Juillerat-Jeanneret L, Celerier J, Chapuis Bernasconi C, Nguyen G, Wostl W, Maerki HP, Janzer RC, Corvol P, Gasc JM (2004) Renin and angiotensinogen expression and functions in growth and apoptosis of human glioblastoma. Br J Cancer 90(5):1059–1068

Jung CH, Kim EM, Park JK, Hwang SG, Moon SK, Kim WJ, Um HD (2013) Bmal1 suppresses cancer cell invasion by blocking the phosphoinositide 3-kinase-Akt-MMP-2 signaling pathway. Oncol Rep 29(6):2109–2113

Kachra Z, Beaulieu E, Delbecchi L, Mousseau N, Berthelet F, Moumdjian R, Del Maestro R, Beliveau R (1999) Expression of matrix metalloproteinases and their inhibitors in human brain tumors. Clin Exp Metastasis 17(7):555–566

Kahana S, Finniss S, Cazacu S, Xiang C, Lee HK, Brodie S, Goldstein RS, Roitman V, Slavin S, Mikkelsen T, Brodie C (2011) Proteasome inhibitors sensitize glioma cells and glioma stem cells to TRAIL-induced apoptosis by PKCepsilon-dependent downregulation of AKT and XIAP expressions. Cell Signal 23(8):1348–1357

Kamino M, Kishida M, Kibe T, Ikoma K, Iijima M, Hirano H, Tokudome M, Chen L, Koriyama C, Yamada K, Arita K, Kishida S (2011) Wnt-5a signaling is correlated with infiltrative activity in human glioma by inducing cellular migration and MMP-2. Cancer Sci 102(3):540–548

Kaminskyy V, Zhivotovsky B (2012) Proteases in autophagy. Biochim Biophys Acta 1824(1): 44–50

Kanakis D, Lendeckel U, Theodosiou P, Dobrowolny H, Mawrin C, Keilhoff G, Bukowska A, Dietzmann K, Bogerts B, Bernstein HG (2013) ADAM 12: a putative marker of oligodendrogliomas? Dis Markers 34(2):81–91

Kang CS, Pu PY, Li YH, Zhang ZY, Qiu MZ, Huang Q, Wang GX (2005) An in vitro study on the suppressive effect of glioma cell growth induced by plasmid-based small interference RNA (siRNA) targeting human epidermal growth factor receptor. J Neurooncol 74(3):267–273

Kardosh A, Golden EB, Pyrko P, Uddin J, Hofman FM, Chen TC, Louie SG, Petasis NA, Schonthal AH (2008) Aggravated endoplasmic reticulum stress as a basis for enhanced glioblastoma cell killing by bortezomib in combination with celecoxib or its non-coxib analogue, 2,5-dimethyl-celecoxib. Cancer Res 68(3):843–851

Kargiotis O, Chetty C, Gondi CS, Tsung AJ, Dinh DH, Gujrati M, Lakka SS, Kyritsis AP, Rao JS (2008) Adenovirus-mediated transfer of siRNA against MMP-2 mRNA results in impaired invasion and tumor-induced angiogenesis, induces apoptosis in vitro and inhibits tumor growth in vivo in glioblastoma. Oncogene 27(35):4830–4840

Kast RE, Boockvar JA, Bruning A, Cappello F, Chang WW, Cvek B, Dou QP, Duenas-Gonzalez-A, Efferth T, Focosi D, Ghaffari SH, Karpel-Massler G, Ketola K, Khoshnevisan A, Keizman D, Magne N, Marosi C, McDonald K, Munoz M, Paranjpe A, Pourgholami MH, Sardi I, Sella A, Srivenugopal KS, Tuccori M, Wang W, Wirtz CR, Halatsch ME (2013) A conceptually new treatment approach for relapsed glioblastoma: coordinated undermining of survival paths with nine repurposed drugs (CUSP9) by the International Initiative for Accelerated Improvement of Glioblastoma Care. Oncotarget 4(4):502–530

Kasza A, Koj A (2002) Cytokines regulate plasminogen activation system in astrocytoma cells. J Physiol Pharmacol 53(1):95–104
Katakowski M, Jiang F, Zheng X, Gutierrez JA, Szalad A, Chopp M (2009) Tumorigenicity of cortical astrocyte cell line induced by the protease ADAM17. Cancer Sci 100(9):1597–1604
Kaur B, Khwaja FW, Severson EA, Matheny SL, Brat DJ, Van Meir EG (2005) Hypoxia and the hypoxia-inducible-factor pathway in glioma growth and angiogenesis. Neuro Oncol 7(2): 134–153
Keerthivasan S, Keerthivasan G, Mittal S, Chauhan SS (2007) Transcriptional upregulation of human cathepsin L by VEGF in glioblastoma cells. Gene 399(2):129–136
Kegel L, Aunin E, Meijer D, Bermingham JR (2013) LGI proteins in the nervous system. ASN Neuro 5(3):167–181
Kenig S, Alonso MB, Mueller MM, Lah TT (2010) Glioblastoma and endothelial cells cross-talk, mediated by SDF-1, enhances tumour invasion and endothelial proliferation by increasing expression of cathepsins B, S, and MMP-9. Cancer Lett 289(1):53–61
Kesanakurti D, Chetty C, Dinh DH, Gujrati M, Rao JS (2013) Role of MMP-2 in the regulation of IL-6/Stat3 survival signaling via interaction with alpha5beta1 integrin in glioma. Oncogene 32(3):327–340
Kessenbrock K, Plaks V, Werb Z (2010) Matrix metalloproteinases: regulators of the tumor microenvironment. Cell 141(1):52–67
Kieseier B, Bernal F (2005) Proteases in CNS infection and inflammation. In: Lendeckel U, Hooper N (eds) Proteases in the brain, vol 3. Springer, New York, NY, pp 109–137
Kim S, Choi K, Kwon D, Benveniste EN, Choi C (2004) Ubiquitin-proteasome pathway as a primary defender against TRAIL-mediated cell death. Cell Mol Life Sci 61(9):1075–1081
Kim JH, Choi C, Benveniste EN, Kwon D (2008a) TRAIL induces MMP-9 expression via ERK activation in human astrocytoma cells. Biochem Biophys Res Commun 377(1):195–199
Kim K, Brush JM, Watson PA, Cacalano NA, Iwamoto KS, McBride WH (2008b) Epidermal growth factor receptor vIII expression in U87 glioblastoma cells alters their proteasome composition, function, and response to irradiation. Mol Cancer Res 6(3):426–434
Kim CS, Jung S, Jung TY, Jang WY, Sun HS, Ryu HH (2011) Characterization of invading glioma cells using molecular analysis of leading-edge tissue. J Korean Neurosurg Soc 50(3):157–165
Kim YH, Kwon HJ, Kim DS (2012) Matrix metalloproteinase 9 (MMP-9)-dependent processing of betaig-h3 protein regulates cell migration, invasion, and adhesion. J Biol Chem 287(46): 38957–38969
Kinder DH, Berger MS, Mueller BA, Silber JR (1993) Urokinase plasminogen-activator is elevated in human astrocytic gliomas relative to normal adjacent brain. Oncol Res 5(10–11): 409–414
Kitagawa H, Tani E, Ikemoto H, Ozaki I, Nakano A, Omura S (1999) Proteasome inhibitors induce mitochondria-independent apoptosis in human glioma cells. FEBS Lett 443(2):181–186
Kitange G, Kishikawa M, Nakayama T, Naito S, Iseki M, Shibata S (1999a) Expression of the Ets-1 proto-oncogene correlates with malignant potential in human astrocytic tumors. Mod Pathol 12(6):618–626
Kitange G, Shibata S, Tokunaga Y, Yagi N, Yasunaga A, Kishikawa M, Naito S (1999b) Ets-1 transcription factor-mediated urokinase-type plasminogen activator expression and invasion in glioma cells stimulated by serum and basic fibroblast growth factors. Lab Invest 79(4):407–416
Klein T, Bischoff R (2011) Active metalloproteases of the A Disintegrin and Metalloprotease (ADAM) family: biological function and structure. J Proteome Res 10(1):17–33
Knoblach S, Faden A (2005) Proteases in traumatic brain injury. In: Lendeckel U, Hooper N (eds) Proteases in the brain, vol 3. Springer, New York, NY, pp 79–108
Kobayashi T, Masumoto J, Tada T, Nomiyama T, Hongo K, Nakayama J (2007) Prognostic significance of the immunohistochemical staining of cleaved caspase-3, an activated form of caspase-3, in gliomas. Clin Cancer Res 13(13):3868–3874
Kodama T, Ikeda E, Okada A, Ohtsuka T, Shimoda M, Shiomi T, Yoshida K, Nakada M, Ohuchi E, Okada Y (2004) ADAM12 is selectively overexpressed in human glioblastomas

and is associated with glioblastoma cell proliferation and shedding of heparin-binding epidermal growth factor. Am J Pathol 165(5):1743–1753

Kohutek ZA, diPierro CG, Redpath GT, Hussaini IM (2009) ADAM-10-mediated N-cadherin cleavage is protein kinase C-alpha dependent and promotes glioblastoma cell migration. J Neurosci 29(14):4605–4615

Komatsu K, Nakanishi Y, Nemoto N, Hori T, Sawada T, Kobayashi M (2004) Expression and quantitative analysis of matrix metalloproteinase-2 and -9 in human gliomas. Brain Tumor Pathol 21(3):105–112

Kondraganti S, Mohanam S, Chintala SK, Kin Y, Jasti SL, Nirmala C, Lakka SS, Adachi Y, Kyritsis AP, Ali-Osman F, Sawaya R, Fuller GN, Rao JS (2000) Selective suppression of matrix metalloproteinase-9 in human glioblastoma cells by antisense gene transfer impairs glioblastoma cell invasion. Cancer Res 60(24):6851–6855

Konduri S, Lakka SS, Tasiou A, Yanamandra N, Gondi CS, Dinh DH, Olivero WC, Gujrati M, Rao JS (2001) Elevated levels of cathepsin B in human glioblastoma cell lines. Int J Oncol 19(3):519–524

Konduri SD, Yanamandra N, Siddique K, Joseph A, Dinh DH, Olivero WC, Gujrati M, Kouraklis G, Swaroop A, Kyritsis AP, Rao JS (2002) Modulation of cystatin C expression impairs the invasive and tumorigenic potential of human glioblastoma cells. Oncogene 21(57): 8705–8712

Konduri SD, Srivenugopal KS, Yanamandra N, Dinh DH, Olivero WC, Gujrati M, Foster DC, Kisiel W, Ali-Osman F, Kondraganti S, Lakka SS, Rao JS (2003) Promoter methylation and silencing of the tissue factor pathway inhibitor-2 (TFPI-2), a gene encoding an inhibitor of matrix metalloproteinases in human glioma cells. Oncogene 22(29):4509–4516

Konkoy CS, Davis TP (1996) Ectoenzymes as sites of peptide regulation. Trends Pharmacol Sci 17(8):288–294

Kono S, Rao JS, Bruner JM, Sawaya R (1994) Immunohistochemical localization of plasminogen-activator inhibitor type-1 in human brain-tumors. J Neuropathol Exp Neurol 53(3):256–262

Koschny R, Holland H, Sykora J, Haas TL, Sprick MR, Ganten TM, Krupp W, Bauer M, Ahnert P, Meixensberger J, Walczak H (2007) Bortezomib sensitizes primary human astrocytoma cells of WHO grades I to IV for tumor necrosis factor-related apoptosis-inducing ligand-induced apoptosis. Clin Cancer Res 13(11):3403–3412

Koul D, Parthasarathy R, Shen R, Davies MA, Jasser SA, Chintala SK, Rao JS, Sun Y, Benvenisite EN, Liu TJ, Yung WK (2001) Suppression of matrix metalloproteinase-2 gene expression and invasion in human glioma cells by MMAC/PTEN. Oncogene 20(46):6669–6678

Kristoffersen K, Villingshoj M, Poulsen HS, Stockhausen MT (2013) Level of Notch activation determines the effect on growth and stem cell-like features in glioblastoma multiforme neurosphere cultures. Cancer Biol Ther 14(7):625–637

Krona A, Aman P, Orndal C, Josefsson A (2007) Oncostatin M-induced genes in human astrocytomas. Int J Oncol 31(6):1457–1463

Krzewski K, Coligan JE (2012) Human NK cell lytic granules and regulation of their exocytosis. Front Immunol 3:335

Kubiatowski T, Jang T, Lachyankar MB, Salmonsen R, Nabi RR, Quesenberry PJ, Litofsky NS, Ross AH, Recht LD (2001) Association of increased phosphatidylinositol 3-kinase signaling with increased invasiveness and gelatinase activity in malignant gliomas. J Neurosurg 95(3): 480–488

Kubicek GJ, Werner-Wasik M, Machtay M, Mallon G, Myers T, Ramirez M, Andrews D, Curran WJ Jr, Dicker AP (2009) Phase I trial using proteasome inhibitor bortezomib and concurrent temozolomide and radiotherapy for central nervous system malignancies. Int J Radiat Oncol Biol Phys 74(2):433–439

Kulathu Y, Garcia FJ, Mevissen TE, Busch M, Arnaudo N, Carroll KS, Barford D, Komander D (2013) Regulation of A20 and other OTU deubiquitinases by reversible oxidation. Nat Commun 4:1569

Kunishio K, Okada M, Matsumoto Y, Nagao S (2003) Matrix metalloproteinase-2 and -9 expression in astrocytic tumors. Brain Tumor Pathol 20(2):39–45

Kwiatkowska A, Kijewska M, Lipko M, Hibner U, Kaminska B (2011) Downregulation of Akt and FAK phosphorylation reduces invasion of glioblastoma cells by impairment of MT1-MMP shuttling to lamellipodia and downregulates MMPs expression. Biochim Biophys Acta 1813(5):655–667

La Ferla-Bruhl K, Westhoff MA, Karl S, Kasperczyk H, Zwacka RM, Debatin KM, Fulda S (2007) NF-kappaB-independent sensitization of glioblastoma cells for TRAIL-induced apoptosis by proteasome inhibition. Oncogene 26(4):571–582

Labussiere M, Pinel S, Delfortrie S, Plenat F, Chastagner P (2008) Proteasome inhibition by bortezomib does not translate into efficacy on two malignant glioma xenografts. Oncol Rep 20(5):1283–1287

Lagadec C, Vlashi E, Frohnen P, Alhiyari Y, Chan M, Pajonk F (2014) The RNA-binding protein Musashi-1 regulates proteasome subunit expression in breast cancer- and glioma-initiating cells. Stem Cells 32(1):135–144

Lakka SS, Jasti SL, Kyritsis AP, Yung WK, Ali-Osman F, Nicolson GL, Rao JS (2000) Regulation of MMP-9 (type IV collagenase) production and invasiveness in gliomas by the extracellular signal-regulated kinase and jun amino-terminal kinase signaling cascades. Clin Exp Metastasis 18(3):245–252

Lakka SS, Gondi CS, Yanamandra N, Olivero WC, Dinh DH, Gujrati M, Rao JS (2004) Inhibition of cathepsin B and MMP-9 gene expression in glioblastoma cell line via RNA interference reduces tumor cell invasion, tumor growth and angiogenesis. Oncogene 23(27):4681–4689

Lakka SS, Gondi CS, Rao JS (2005) Proteases and glioma angiogenesis. Brain Pathol 15(4): 327–341

Lal A, Glazer CA, Martinson HM, Friedman HS, Archer GE, Sampson JH, Riggins GJ (2002) Mutant epidermal growth factor receptor up-regulates molecular effectors of tumor invasion. Cancer Res 62(12):3335–3339

Lal S, La Du J, Tanguay RL, Greenwood JA (2012) Calpain 2 is required for the invasion of glioblastoma cells in the zebrafish brain microenvironment. J Neurosci Res 90(4):769–781

Lamfers ML, Gianni D, Tung CH, Idema S, Schagen FH, Carette JE, Quax PH, Van Beusechem VW, Vandertop WP, Dirven CM, Chiocca EA, Gerritsen WR (2005) Tissue inhibitor of metalloproteinase-3 expression from an oncolytic adenovirus inhibits matrix metalloproteinase activity in vivo without affecting antitumor efficacy in malignant glioma. Cancer Res 65(20): 9398–9405

Lampert K, Machein U, Machein MR, Conca W, Peter HH, Volk B (1998) Expression of matrix metalloproteinases and their tissue inhibitors in human brain tumors. Am J Pathol 153(2): 429–437

Landau BJ, Kwaan HC, Verrusio EN, Brem SS (1994) Elevated levels of urokinase-type plasminogen activator and plasminogen activator inhibitor type-1 in malignant human brain tumors. Cancer Res 54(4):1105–1108

Langlois S, Nyalendo C, Di Tomasso G, Labrecque L, Roghi C, Murphy G, Gingras D, Beliveau R (2007) Membrane-type 1 matrix metalloproteinase stimulates cell migration through epidermal growth factor receptor transactivation. Mol Cancer Res 5(6):569–583

Larson DA, Prados M, Lamborn KR, Smith V, Sneed PK, Chang S, Nicholas KM, Wara WM, Devriendt D, Kunwar S, Berger M, McDermott MW (2002) Phase II study of high central dose gamma knife radiosurgery and marimastat in patients with recurrent malignant glioma. Int J Radiat Oncol Biol Phys 54(5):1397–1404

LaRusch GA, Mahdi F, Shariat-Madar Z, Adams G, Sitrin RG, Zhang WM, McCrae KR, Schmaier AH (2010) Factor XII stimulates ERK1/2 and Akt through uPAR, integrins, and the EGFR to initiate angiogenesis. Blood 115(24):5111–5120

Laurent M, Martinerie C, Thibout H, Hoffman MP, Verrecchia F, Le Bouc Y, Mauviel A, Kleinman HK (2003) NOVH increases MMP3 expression and cell migration in glioblastoma cells via a PDGFR-alpha-dependent mechanism. FASEB J 17(13):1919–1921

Laurent N, de Bouard S, Guillamo JS, Christov C, Zini R, Jouault H, Andre P, Lotteau V, Peschanski M (2004) Effects of the proteasome inhibitor ritonavir on glioma growth in vitro and in vivo. Mol Cancer Ther 3(2):129–136

Law RH, Zhang Q, McGowan S, Buckle AM, Silverman GA, Wong W, Rosado CJ, Langendorf CG, Pike RN, Bird PI, Whisstock JC (2006) An overview of the serpin superfamily. Genome Biol 7(5):216

Le DM, Besson A, Fogg DK, Choi KS, Waisman DM, Goodyer CG, Rewcastle B, Yong VW (2003) Exploitation of astrocytes by glioma cells to facilitate invasiveness: a mechanism involving matrix metalloproteinase-2 and the urokinase-type plasminogen activator-plasmin cascade. J Neurosci 23(10):4034–4043

Lee NV, Sato M, Annis DS, Loo JA, Wu L, Mosher DF, Iruela-Arispe ML (2006) ADAMTS1 mediates the release of antiangiogenic polypeptides from TSP1 and 2. EMBO J 25(22): 5270–5283

Lee JG, Baek K, Soetandyo N, Ye Y (2013) Reversible inactivation of deubiquitinases by reactive oxygen species in vitro and in cells. Nat Commun 4:1568

Legnani FG, Pradilla G, Thai QA, Fiorindi A, Recinos PF, Tyler BM, Gaini SM, DiMeco F, Brem H, Olivi A (2006) Lactacystin exhibits potent anti-tumor activity in an animal model of malignant glioma when administered via controlled-release polymers. J Neurooncol 77(3): 225–232

Lehman NL (2009) The ubiquitin proteasome system in neuropathology. Acta Neuropathol 118(3):329–347

Leitlein J, Aulwurm S, Waltereit R, Naumann U, Wagenknecht B, Garten W, Weller M, Platten M (2001) Processing of immunosuppressive pro-TGF-beta 1,2 by human glioblastoma cells involves cytoplasmic and secreted furin-like proteases. J Immunol 166(12):7238–7243

Lemberg MK (2011) Intramembrane proteolysis in regulated protein trafficking. Traffic 12(9): 1109–1118

Lettau I, Hattermann K, Held-Feindt J, Brauer R, Sedlacek R, Mentlein R (2010) Matrix metalloproteinase-19 is highly expressed in astroglial tumors and promotes invasion of glioma cells. J Neuropathol Exp Neurol 69(3):215–223

Levicar N, Strojnik T, Kos J, Dewey RA, Pilkington GJ, Lah TT (2002) Lysosomal enzymes, cathepsins in brain tumour invasion. J Neurooncol 58(1):21–32

Levicar N, Dewey RA, Daley E, Bates TE, Davies D, Kos J, Pilkington GJ, Lah TT (2003a) Selective suppression of cathepsin L by antisense cDNA impairs human brain tumor cell invasion in vitro and promotes apoptosis. Cancer Gene Ther 10(2):141–151

Levicar N, Nuttall RK, Lah TT (2003b) Proteases in brain tumour progression. Acta Neurochir (Wien) 145(9):825–838

Levin VA, Phuphanich S, Yung WK, Forsyth PA, Maestro RD, Perry JR, Fuller GN, Baillet M (2006) Randomized, double-blind, placebo-controlled trial of marimastat in glioblastoma multiforme patients following surgery and irradiation. J Neurooncol 78(3):295–302

Li Z, Sheng M (2012) Caspases in synaptic plasticity. Mol Brain 5:15

Li T, Wen H, Brayton C, Das P, Smithson LA, Fauq A, Fan X, Crain BJ, Price DL, Golde TE, Eberhart CG, Wong PC (2007) Epidermal growth factor receptor and notch pathways participate in the tumor suppressor function of gamma-secretase. J Biol Chem 282(44):32264–32273

Li Z, Bao S, Wu Q, Wang H, Eyler C, Sathornsumetee S, Shi Q, Cao Y, Lathia J, McLendon RE, Hjelmeland AB, Rich JN (2009) Hypoxia-inducible factors regulate tumorigenic capacity of glioma stem cells. Cancer Cell 15(6):501–513

Li R, Li G, Deng L, Liu Q, Dai J, Shen J, Zhang J (2010a) IL-6 augments the invasiveness of U87MG human glioblastoma multiforme cells via up-regulation of MMP-2 and fascin-1. Oncol Rep 23(6):1553–1559

Li Z, Jo J, Jia JM, Lo SC, Whitcomb DJ, Jiao S, Cho K, Sheng M (2010b) Caspase-3 activation via mitochondria is required for long-term depression and AMPA receptor internalization. Cell 141(5):859–871

Li M, Mukasa A, Inda MM, Zhang J, Chin L, Cavenee W, Furnari F (2011) Guanylate binding protein 1 is a novel effector of EGFR-driven invasion in glioblastoma. J Exp Med 208(13): 2657–2673

Li Y, Wang Y, Yu L, Sun C, Cheng D, Yu S, Wang Q, Yan Y, Kang C, Jin S, An T, Shi C, Xu J, Wei C, Liu J, Sun J, Wen Y, Zhao S, Kong Y (2013a) miR-146b-5p inhibits glioma migration and invasion by targeting MMP16. Cancer Lett 339(2):260–269

Li ZH, Yu Y, Du C, Fu H, Wang J, Tian Y (2013b) RNA interference-mediated USP22 gene silencing promotes human brain glioma apoptosis and induces cell cycle arrest. Oncol Lett 5 (4):1290–1294

Liu BL, Cheng JX, Zhang W, Zhang X, Wang R, Lin H, Huo JL, Cheng H (2010) Quantitative detection of multiple gene promoter hypermethylation in tumor tissue, serum, and cerebrospinal fluid predicts prognosis of malignant gliomas. Neuro Oncol 12(6):540–548

Liu B, Wang L, Shen LL, Shen MZ, Guo XD, Wang T, Liang QC, Wang C, Zheng J, Li Y, Jia LT, Zhang H, Gao GD (2012) RNAi-mediated inhibition of presenilin 2 inhibits glioma cell growth and invasion and is involved in the regulation of Nrg1/ErbB signaling. Neuro Oncol 14(8): 994–1006

Liu Z, Zanata SM, Kim J, Peterson MA, Di Vizio D, Chirieac LR, Pyne S, Agostini M, Freeman MR, Loda M (2013) The ubiquitin-specific protease USP2a prevents endocytosis-mediated EGFR degradation. Oncogene 32(13):1660–1669

Liu Z, Fan F, Wang A, Zheng S, Lu Y (2014) Dll4-Notch signaling in regulation of tumor angiogenesis. J Cancer Res Clin Oncol 140(4):525–536

Liuzzi GM, Mastroianni CM, Latronico T, Mengoni F, Fasano A, Lichtner M, Vullo V, Riccio P (2004) Anti-HIV drugs decrease the expression of matrix metalloproteinases in astrocytes and microglia. Brain 127(Pt 2):398–407

Lopez-Otin C, Bond JS (2008) Proteases: multifunctional enzymes in life and disease. J Biol Chem 283(45):30433–30437

Lopez-Otin C, Matrisian LM (2007) Emerging roles of proteases in tumour suppression. Nat Rev Cancer 7(10):800–808

Lovric MM, Hawkins CJ (2010) TRAIL treatment provokes mutations in surviving cells. Oncogene 29(36):5048–5060

Lu X, Wyszynski M, Sheng M, Baudry M (2001) Proteolysis of glutamate receptor-interacting protein by calpain in rat brain: implications for synaptic plasticity. J Neurochem 77(6): 1553–1560

Lu KV, Jong KA, Rajasekaran AK, Cloughesy TF, Mischel PS (2004) Upregulation of tissue inhibitor of metalloproteinases (TIMP)-2 promotes matrix metalloproteinase (MMP)-2-activation and cell invasion in a human glioblastoma cell line. Lab Invest 84(1):8–20

LuW, ZX, Hong B, Liu J, Yue Z (2004) Suppression of invasion in human U87 glioma cells by adenovirus-mediated co-transfer of TIMP-2 and PTEN gene. Cancer Lett 214(2):205–213

Lu C, Wang Y, Furukawa K, Fu W, Ouyang X, Mattson MP (2006a) Evidence that caspase-1 is a negative regulator of AMPA receptor-mediated long-term potentiation at hippocampal synapses. J Neurochem 97(4):1104–1110

Lu Z, Wang Y, Zhang Q, Zhang X, Wang S, Xie H, Li Y, Jiao B, Zhang J (2006b) Association between the functional polymorphism in the matrix metalloproteinase-7 promoter and susceptibility to adult astrocytoma. Brain Res 1118(1):6–12

Lu DY, Leung YM, Cheung CW, Chen YR, Wong KL (2010) Glial cell line-derived neurotrophic factor induces cell migration and matrix metalloproteinase-13 expression in glioma cells. Biochem Pharmacol 80(8):1201–1209

Lu Y, Chopp M, Zheng X, Katakowski M, Buller B, Jiang F (2013) MiR-145 reduces ADAM17 expression and inhibits in vitro migration and invasion of glioma cells. Oncol Rep 29(1):67–72

Lucio-Eterovic AK, Piao Y, de Groot JF (2009) Mediators of glioblastoma resistance and invasion during antivascular endothelial growth factor therapy. Clin Cancer Res 15(14):4589–4599

Ludwig A, Schulte A, Schnack C, Hundhausen C, Reiss K, Brodway N, Held-Feindt J, Mentlein R (2005) Enhanced expression and shedding of the transmembrane chemokine CXCL16 by reactive astrocytes and glioma cells. J Neurochem 93(5):1293–1303

Luque A, Carpizo DR, Iruela-Arispe ML (2003) ADAMTS1/METH1 inhibits endothelial cell proliferation by direct binding and sequestration of VEGF165. J Biol Chem 278(26): 23656–23665

Ma J, Cui W, He SM, Duan YH, Heng LJ, Wang L, Gao GD (2012) Human U87 astrocytoma cell invasion induced by interaction of betaig-h3 with integrin alpha5beta1 involves calpain-2. PLoS One 7(5):e37297

Magnus N, Garnier D, Rak J (2010) Oncogenic epidermal growth factor receptor up-regulates multiple elements of the tissue factor signaling pathway in human glioma cells. Blood 116(5): 815–818

Mao H, Lebrun DG, Yang J, Zhu VF, Li M (2012) Deregulated signaling pathways in glioblastoma multiforme: molecular mechanisms and therapeutic targets. Cancer Invest 30(1):48–56

Mares V, Stremenova J, Lisa V, Kozakova H, Marek J, Syrucek M, Soula O, Sedo A (2012) Compartment- and malignance-dependent up-regulation of gamma-glutamyltranspeptidase and dipetidylpeptidase-IV activity in human brain gliomas. Histol Histopathol 27(7):931–940

Maret D, Sadr MS, Sadr ES, Colman DR, Del Maestro RF, Seidah NG (2012) Opposite roles of furin and PC5A in N-cadherin processing. Neoplasia 14(10):880–892

Markovic DS, Vinnakota K, Chirasani S, Synowitz M, Raguet H, Stock K, Sliwa M, Lehmann S, Kalin R, van Rooijen N, Holmbeck K, Heppner FL, Kiwit J, Matyash V, Lehnardt S, Kaminska B, Glass R, Kettenmann H (2009) Gliomas induce and exploit microglial MT1-MMP expression for tumor expansion. Proc Natl Acad Sci U S A 106(30):12530–12535

Martinez R, Setien F, Voelter C, Casado S, Quesada MP, Schackert G, Esteller M (2007) CpG island promoter hypermethylation of the pro-apoptotic gene caspase-8 is a common hallmark of relapsed glioblastoma multiforme. Carcinogenesis 28(6):1264–1268

Mason SD, Joyce JA (2011) Proteolytic networks in cancer. Trends Cell Biol 21(4):228–237

Matsuzawa K, Fukuyama K, Hubbard SL, Dirks PB, Rutka JT (1996) Transfection of an invasive human astrocytoma cell line with a TIMP-1 cDNA: modulation of astrocytoma invasive potential. J Neuropathol Exp Neurol 55(1):88–96

Matthews RT, Gary SC, Zerillo C, Pratta M, Solomon K, Arner EC, Hockfield S (2000) Brain-enriched hyaluronan binding (BEHAB)/brevican cleavage in a glioma cell line is mediated by a disintegrin and metalloproteinase with thrombospondin motifs (ADAMTS) family member. J Biol Chem 275(30):22695–22703

McCawley LJ, Matrisian LM (2001) Matrix metalloproteinases: they're not just for matrix anymore! Curr Opin Cell Biol 13(5):534–540

McCormick D (1993) Secretion of cathepsin B by human gliomas in vitro. Neuropathol Appl Neurobiol 19(2):146–151

McCready J, Broaddus WC, Sykes V, Fillmore HL (2005) Association of a single nucleotide polymorphism in the matrix metalloproteinase-1 promoter with glioblastoma. Int J Cancer 117(5):781–785

McGillicuddy LT, Fromm JA, Hollstein PE, Kubek S, Beroukhim R, De Raedt T, Johnson BW, Williams SM, Nghiemphu P, Liau LM, Cloughesy TF, Mischel PS, Parret A, Seiler J, Moldenhauer G, Scheffzek K, Stemmer-Rachamimov AO, Sawyers CL, Brennan C, Messiaen L, Mellinghoff IK, Cichowski K (2009) Proteasomal and genetic inactivation of the NF1 tumor suppressor in gliomagenesis. Cancer Cell 16(1):44–54

Medeiros Mdos S, Balmforth AJ, Vaughan PF, Turner AJ (1991) Hydrolysis of atrial and brain natriuretic peptides by the human astrocytoma clone D384 and the neuroblastoma line SH-SY5Y. Neuroendocrinology 54(3):295–302

Mehling M, Simon P, Mittelbronn M, Meyermann R, Ferrone S, Weller M, Wiendl H (2007) WHO grade associated downregulation of MHC class I antigen-processing machinery components in human astrocytomas: does it reflect a potential immune escape mechanism? Acta Neuropathol 114(2):111–119

Melendez-Zajgla J, Del Pozo L, Ceballos G, Maldonado V (2008) Tissue inhibitor of metalloproteinases-4. The road less traveled. Mol Cancer 7:85

Mentlein R, Hattermann K, Hemion C, Jungbluth A, Held-Feindt J (2011) Expression and role of the cell surface protease seprase/fibroblast activation protein-alpha (FAP-alpha) in astroglial tumors. Biol Chem 392(3):199–207

Mentlein R, Hattermann K, Held-Feindt J (2012) Lost in disruption: role of proteases in glioma invasion and progression. Biochim Biophys Acta 1825(2):178–185

Mercapide J, Lopez De Cicco R, Bassi DE, Castresana JS, Thomas G, Klein-Szanto AJ (2002) Inhibition of furin-mediated processing results in suppression of astrocytoma cell growth and invasiveness. Clin Cancer Res 8(6):1740–1746

Mercapide J, Lopez De Cicco R, Castresana JS, Klein-Szanto AJ (2003) Stromelysin-1/matrix metalloproteinase-3 (MMP-3) expression accounts for invasive properties of human astrocytoma cell lines. Int J Cancer 106(5):676–682

Merzak A, Parker C, Koochekpour S, Sherbet GV, Pilkington GJ (1994) Overexpression of the 18A2/mts1 gene and down-regulation of the TIMP-2 gene in invasive human glioma cell lines in vitro. Neuropathol Appl Neurobiol 20(6):614–619

Miake H, Tsuchiya K, Nakamura A, Ikeda K, Levesque L, Fraser PE, St-George Hyslop PH, Mizusawa H, Uchihara T (1999) Glial expression of presenilin epitopes in human brain with cerebral infarction and in astrocytoma. Acta Neuropathol 98(4):337–340

Mikkelsen T, Yan PS, Ho KL, Sameni M, Sloane BF, Rosenblum ML (1995) Immunolocalization of cathepsin B in human glioma: implications for tumor invasion and angiogenesis. J Neurosurg 83(2):285–290

Mina-Osorio P (2008) The moonlighting enzyme CD13: old and new functions to target. Trends Mol Med 14(8):361–371

Miyata S, Fukushima T, Kohama K, Tanaka H, Takeshima H, Kataoka H (2007) Roles of Kunitz domains in the anti-invasive effect of hepatocyte growth factor activator inhibitor type 1 in human glioblastoma cells. Hum Cell 20(4):100–106

Miyazaki K, Umenishi F, Funahashi K, Koshikawa N, Yasumitsu H, Umeda M (1992) Activation of TIMP-2/progelatinase A complex by stromelysin. Biochem Biophys Res Commun 185(3): 852–859

Mohamed MM, Sloane BF (2006) Cysteine cathepsins: multifunctional enzymes in cancer. Nat Rev Cancer 6(10):764–775

Mohanam S, Wang SW, Rayford A, Yamamoto M, Sawaya R, Nakajima M, Liotta LA, Nicolson GL, Stetler-Stevenson WG, Rao JS (1995) Expression of tissue inhibitors of metalloproteinases: negative regulators of human glioblastoma invasion in vivo. Clin Exp Metastasis 13(1):57–62

Mohanam S, Gladson CL, Rao CN, Rao JS (1999) Biological significance of the expression of urokinase-type plasminogen activator receptors (uPARs) in brain tumors. Front Biosci 4: D178–D187

Mohanam S, Jasti SL, Kondraganti SR, Chandrasekar N, Lakka SS, Kin Y, Fuller GN, Yung AW, Kyritsis AP, Dinh DH, Olivero WC, Gujrati M, Ali-Osman F, Rao JS (2001) Down-regulation of cathepsin B expression impairs the invasive and tumorigenic potential of human glioblastoma cells. Oncogene 20(28):3665–3673

Mohanan V, Temburni MK, Kappes JC, Galileo DS (2013) L1CAM stimulates glioma cell motility and proliferation through the fibroblast growth factor receptor. Clin Exp Metastasis 30(4):507–520

Moller HG, Rasmussen AP, Andersen HH, Johnsen KB, Henriksen M, Duroux M (2013) A systematic review of microRNA in glioblastoma multiforme: micro-modulators in the mesenchymal mode of migration and invasion. Mol Neurobiol 47(1):131–144

Monod L, Diserens AC, Jongeneel CV, Carrel S, Ronco P, Verroust P, de Tribolet N (1989) Human glioma cell lines expressing the common acute lymphoblastic leukemia antigen (cALLa) have neutral endopeptidase activity. Int J Cancer 44(5):948–951

Monod L, Hamou MF, Ronco P, Verroust P, de Tribolet N (1992) Expression of cALLa/NEP on gliomas: a possible marker of malignancy. Acta Neurochir (Wien) 114(1–2):3–7

Morancho A, Rosell A, Garcia-Bonilla L, Montaner J (2010) Metalloproteinase and stroke infarct size: role for anti-inflammatory treatment? Ann NY Acad Sci 1207:123–133

Mori T, Abe T, Wakabayashi Y, Hikawa T, Matsuo K, Yamada Y, Kuwano M, Hori S (2000) Up-regulation of urokinase-type plasminogen activator and its receptor correlates with enhanced invasion activity of human glioma cells mediated by transforming growth factor-alpha or basic fibroblast growth factor. J Neurooncol 46(2):115–123

Moriyama T, Kataoka H, Seguchi K, Tsubouchi H, Koono M (1996) Effects of hepatocyte growth factor (HGF) on human glioma cells in vitro: HGF acts as a motility factor in glioma cells. Int J Cancer 66(5):678–685

Moriyama T, Kataoka H, Hamasuna R, Yoshida E, Sameshima T, Iseda T, Yokogami K, Nakano S, Koono M, Wakisaka S (1999) Simultaneous up-regulation of urokinase-type plasminogen activator (uPA) and uPA receptor by hepatocyte growth factor/scatter factor in human glioma cells. Clin Exp Metastasis 17(10):873–879

Moss ML, Stoeck A, Yan W, Dempsey PJ (2008) ADAM10 as a target for anti-cancer therapy. Curr Pharm Biotechnol 9(1):2–8

Mullard A (2013) Maturing antibody-drug conjugate pipeline hits 30. Nat Rev Drug Discov 12(5): 329–332

Munaut C, Noel A, Hougrand O, Foidart JM, Boniver J, Deprez M (2003) Vascular endothelial growth factor expression correlates with matrix metalloproteinases MT1-MMP, MMP-2 and MMP-9 in human glioblastomas. Int J Cancer 106(6):848–855

Muracciole X, Romain S, Dufour H, Palmari J, Chinot O, Ouafik L, Grisoli F, Branger DF, Martin PM (2002) PAI-1 and EGFR expression in adult glioma tumors: toward a molecular prognostic classification. Int J Radiat Oncol Biol Phys 52(3):592–598

Murai T, Miyazaki Y, Nishinakamura H, Sugahara KN, Miyauchi T, Sako Y, Yanagida T, Miyasaka M (2004) Engagement of CD44 promotes Rac activation and CD44 cleavage during tumor cell migration. J Biol Chem 279(6):4541–4550

Murakami D, Okamoto I, Nagano O, Kawano Y, Tomita T, Iwatsubo T, De Strooper B, Yumoto E, Saya H (2003) Presenilin-dependent gamma-secretase activity mediates the intramembranous cleavage of CD44. Oncogene 22(10):1511–1516

Murphy G (2011) Tissue inhibitors of metalloproteinases. Genome Biol 12(11):233

Murphy AC, Weyhenmeyer B, Schmid J, Kilbride SM, Rehm M, Huber HJ, Senft C, Weissenberger J, Seifert V, Dunst M, Mittelbronn M, Kogel D, Prehn JH, Murphy BM (2013) Activation of executioner caspases is a predictor of progression-free survival in glioblastoma patients: a systems medicine approach. Cell Death Dis 4:e629

Mutka AL, Haapanen A, Kakela R, Lindfors M, Wright AK, Inkinen T, Hermansson M, Rokka A, Corthals G, Jauhiainen M, Gillingwater TH, Ikonen E, Tyynela J (2010) Murine cathepsin D deficiency is associated with dysmyelination/myelin disruption and accumulation of cholesteryl esters in the brain. J Neurochem 112(1):193–203

Naidoo V, Naidoo S, Mahabeer R, Raidoo DM (2005) Localization of the endothelin system in human diffuse astrocytomas. Cancer 104(5):1049–1057

Nakabayashi H, Hara M, Shimuzu K (2005) Clinicopathologic significance of cystatin C expression in gliomas. Hum Pathol 36(9):1008–1015

Nakada M, Nakamura H, Ikeda E, Fujimoto N, Yamashita J, Sato H, Seiki M, Okada Y (1999a) Expression and tissue localization of membrane-type 1, 2, and 3 matrix metalloproteinases in human astrocytic tumors. Am J Pathol 154(2):417–428

Nakada M, Yamashita J, Okada Y, Sato H (1999b) Ets-1 positively regulates expression of urokinase-type plasminogen activator (uPA) and invasiveness of astrocytic tumors. J Neuropathol Exp Neurol 58(4):329–334

Nakada M, Kita D, Futami K, Yamashita J, Fujimoto N, Sato H, Okada Y (2001) Roles of membrane type 1 matrix metalloproteinase and tissue inhibitor of metalloproteinases 2 in invasion and dissemination of human malignant glioma. J Neurosurg 94(3):464–473

Nakagawa T, Kubota T, Kabuto M, Sato K, Kawano H, Hayakawa T, Okada Y (1994) Production of matrix metalloproteinases and tissue inhibitor of metalloproteinases-1 by human brain tumors. J Neurosurg 81(1):69–77

Nakagawa T, Kubota T, Kabuto M, Sato K, Arai Y, Kodera T (1995) Production of tissue inhibitor of metalloproteinases-1 (TIMP-1) by human astrocytic tumors. Neurol Med Chir (Tokyo) 35(10):728–731

Nakamura M, Ishida E, Shimada K, Kishi M, Nakase H, Sakaki T, Konishi N (2005) Frequent LOH on 22q12.3 and TIMP-3 inactivation occur in the progression to secondary glioblastomas. Lab Invest 85(2):165–175

Nakano A, Tani E, Miyazaki K, Furuyama J, Matsumoto T (1993) Expressions of matrilysin and stromelysin in human glioma cells. Biochem Biophys Res Commun 192(3):999–1003

Nakano A, Tani E, Miyazaki K, Yamamoto Y, Furuyama J (1995) Matrix metalloproteinases and tissue inhibitors of metalloproteinases in human gliomas. J Neurosurg 83(2):298–307

Naldini L, Tamagnone L, Vigna E, Sachs M, Hartmann G, Birchmeier W, Daikuhara Y, Tsubouchi H, Blasi F, Comoglio PM (1992) Extracellular proteolytic cleavage by urokinase is required for activation of hepatocyte growth factor/scatter factor. EMBO J 11(13): 4825–4833

Nan Y, Han L, Zhang A, Wang G, Jia Z, Yang Y, Yue X, Pu P, Zhong Y, Kang C (2010) MiRNA-451 plays a role as tumor suppressor in human glioma cells. Brain Res 1359:14–21

Ng K, Nitta M, Hu L, Kesari S, Kung A, D'Andrea A, Chen CC (2009) A small interference RNA screen revealed proteasome inhibition as strategy for glioblastoma therapy. Clin Neurosurg 56: 107–118

Nicole O, Docagne F, Ali C, Margaill I, Carmeliet P, MacKenzie ET, Vivien D, Buisson A (2001) The proteolytic activity of tissue-plasminogen activator enhances NMDA receptor-mediated signaling. Nat Med 7(1):59–64

Noha M, Yoshida D, Watanabe K, Teramoto A (2000) Suppression of cell invasion on human malignant glioma cell lines by a novel matrix-metalloproteinase inhibitor SI-27: in vitro study. J Neurooncol 48(3):217–223

Nomura T, Katunuma N (2005) Involvement of cathepsins in the invasion, metastasis and proliferation of cancer cells. J Med Invest 52(1–2):1–9

Noorbakhsh F, Vergnolle N, Hollenberg MD, Power C (2003) Proteinase-activated receptors in the nervous system. Nat Rev Neurosci 4(12):981–990

Nuti E, Casalini F, Santamaria S, Gabelloni P, Bendinelli S, Da Pozzo E, Costa B, Marinelli L, La Pietra V, Novellino E, Margarida Bernardo M, Fridman R, Da Settimo F, Martini C, Rossello A (2011) Synthesis and biological evaluation in U87MG glioma cells of (ethynylthiophene)sulfonamido-based hydroxamates as matrix metalloproteinase inhibitors. Eur J Med Chem 46(7):2617–2629

Oguchi T, Hirashima Y, Nakamura S, Endo S, Kurimoto M, Takaku A (2000) Tissue factor and cancer procoagulant expressed by glioma cells participate in their thrombin-mediated proliferation. J Neurooncol 46(1):1–9

Ohgaki H, Kleihues P (2007) Genetic pathways to primary and secondary glioblastoma. Am J Pathol 170(5):1445–1453

Ohshima-Hosoyama S, Davare MA, Hosoyama T, Nelon LD, Keller C (2011) Bortezomib stabilizes NOXA and triggers ROS-associated apoptosis in medulloblastoma. J Neurooncol 105(3):475–483

Oikawa T, Yamada T (2003) Molecular biology of the Ets family of transcription factors. Gene 303:11–34

Ornstein DL, Meehan KR, Zacharski LR (2002) The coagulation system as a target for the treatment of human gliomas. Semin Thromb Hemost 28(1):19–28

Paganetti PA, Caroni P, Schwab ME (1988) Glioblastoma infiltration into central nervous system tissue in vitro: involvement of a metalloprotease. J Cell Biol 107(6 Pt 1):2281–2291

Pagenstecher A, Wussler EM, Opdenakker G, Volk B, Campbell IL (2001) Distinct expression patterns and levels of enzymatic activity of matrix metalloproteinases and their inhibitors in primary brain tumors. J Neuropathol Exp Neurol 60(6):598–612

Pan SJ, Zhan SK, Pei BG, Sun QF, Bian LG, Sun BM (2012) MicroRNA-149 inhibits proliferation and invasion of glioma cells via blockade of AKT1 signaling. Int J Immunopathol Pharmacol 25(4):871–881

Panner A, Crane CA, Weng C, Feletti A, Fang S, Parsa AT, Pieper RO (2010) Ubiquitin-specific protease 8 links the PTEN-Akt-AIP4 pathway to the control of FLIPS stability and TRAIL sensitivity in glioblastoma multiforme. Cancer Res 70(12):5046–5053

Park MJ, Kim MS, Park IC, Kang HS, Yoo H, Park SH, Rhee CH, Hong SI, Lee SH (2002) PTEN suppresses hyaluronic acid-induced matrix metalloproteinase-9 expression in U87MG glioblastoma cells through focal adhesion kinase dephosphorylation. Cancer Res 62(21):6318–6322

Park CM, Park MJ, Kwak HJ, Lee HC, Kim MS, Lee SH, Park IC, Rhee CH, Hong SI (2006) Ionizing radiation enhances matrix metalloproteinase-2 secretion and invasion of glioma cells through Src/epidermal growth factor receptor-mediated p38/Akt and phosphatidylinositol 3-kinase/Akt signaling pathways. Cancer Res 66(17):8511–8519

Paugh BS, Paugh SW, Bryan L, Kapitonov D, Wilczynska KM, Gopalan SM, Rokita H, Milstien S, Spiegel S, Kordula T (2008) EGF regulates plasminogen activator inhibitor-1 (PAI-1) by a pathway involving c-Src, PKCdelta, and sphingosine kinase 1 in glioblastoma cells. FASEB J 22(2):455–465

Paulus W, Huettner C, Tonn JC (1994) Collagens, integrins and the mesenchymal drift in glioblastomas: a comparison of biopsy specimens, spheroid and early monolayer cultures. Int J Cancer 58(6):841–846

Payne LS, Huang PH (2013) The pathobiology of collagens in glioma. Mol Cancer Res 11(10):1129–1140

Pedeboscq S, L'Azou B, Passagne I, De Giorgi F, Ichas F, Pometan JP, Cambar J (2008) Cytotoxic and apoptotic effects of bortezomib and gefitinib compared to alkylating agents on human glioblastoma cells. J Exp Ther Oncol 7(2):99–111

Perry JR (2010) Anticoagulation of malignant glioma patients in the era of novel antiangiogenic agents. Curr Opin Neurol 23(6):592–596

Perry JR, Julian JA, Laperriere NJ, Geerts W, Agnelli G, Rogers LR, Malkin MG, Sawaya R, Baker R, Falanga A, Parpia S, Finch T, Levine MN (2010) PRODIGE: a randomized placebo-controlled trial of dalteparin low-molecular-weight heparin thromboprophylaxis in patients with newly diagnosed malignant glioma. J Thromb Haemost 8(9):1959–1965

Phuphanich S, Carson KA, Grossman SA, Lesser G, Olson J, Mikkelsen T, Desideri S, Fisher JD, C. N. S. C. New Approaches to Brain Tumor Therapy (2008) Phase I safety study of escalating doses of atrasentan in adults with recurrent malignant glioma. Neuro Oncol 10(4):617–623

Phuphanich S, Supko JG, Carson KA, Grossman SA, Burt Nabors L, Mikkelsen T, Lesser G, Rosenfeld S, Desideri S, Olson JJ (2010) Phase 1 clinical trial of bortezomib in adults with recurrent malignant glioma. J Neurooncol 100(1):95–103

Piccinini M, Rinaudo MT, Anselmino A, Ramondetti C, Buccinna B, Fiano V, Ghimenti C, Schiffer D (2005) Characterization of the 20S proteasome in human glioblastomas. Anticancer Res 25(5):3203–3210

Piva R, Cancelli I, Cavalla P, Bortolotto S, Dominguez J, Draetta GF, Schiffer D (1999) Proteasome-dependent degradation of p27/kip1 in gliomas. J Neuropathol Exp Neurol 58(7):691–696

Pop C, Salvesen GS (2009) Human caspases: activation, specificity, and regulation. J Biol Chem 284(33):21777–21781

Prezas P, Scorilas A, Yfanti C, Viktorov P, Agnanti N, Diamandis E, Talieri M (2006) The role of human tissue kallikreins 7 and 8 in intracranial malignancies. Biol Chem 387(12):1607–1612

Prox J, Rittger A, Saftig P (2012) Physiological functions of the amyloid precursor protein secretases ADAM10, BACE1, and presenilin. Exp Brain Res 217(3–4):331–341

Puente XS, Sanchez LM, Overall CM, Lopez-Otin C (2003) Human and mouse proteases: a comparative genomic approach. Nat Rev Genet 4(7):544–558

Pullen NA, Fillmore HL (2010) Induction of matrix metalloproteinase-1 and glioma cell motility by nitric oxide. J Neurooncol 96(2):201–209

Pullen NA, Anand M, Cooper PS, Fillmore HL (2012) Matrix metalloproteinase-1 expression enhances tumorigenicity as well as tumor-related angiogenesis and is inversely associated with TIMP-4 expression in a model of glioblastoma. J Neurooncol 106(3):461–471

Pyrko P, Kardosh A, Wang W, Xiong W, Schonthal AH, Chen TC (2007) HIV-1 protease inhibitors nelfinavir and atazanavir induce malignant glioma death by triggering endoplasmic reticulum stress. Cancer Res 67(22):10920–10928

Qian Z, Gilbert ME, Colicos MA, Kandel ER, Kuhl D (1993) Tissue-plasminogen activator is induced as an immediate-early gene during seizure, kindling and long-term potentiation. Nature 361(6411):453–457

Qin H, Moellinger JD, Wells A, Windsor LJ, Sun Y, Benveniste EN (1998) Transcriptional suppression of matrix metalloproteinase-2 gene expression in human astroglioma cells by TNF-alpha and IFN-gamma. J Immunol 161(12):6664–6673

Qin H, Sun Y, Benveniste EN (1999) The transcription factors Sp1, Sp3, and AP-2 are required for constitutive matrix metalloproteinase-2 gene expression in astroglioma cells. J Biol Chem 274(41):29130–29137

Qiu J, Ai L, Ramachandran C, Yao B, Gopalakrishnan S, Fields CR, Delmas AL, Dyer LM, Melnick SJ, Yachnis AT, Schwartz PH, Fine HA, Brown KD, Robertson KD (2008) Invasion suppressor cystatin E/M (CST6): high-level cell type-specific expression in normal brain and epigenetic silencing in gliomas. Lab Invest 88(9):910–925

Radisky DC, Levy DD, Littlepage LE, Liu H, Nelson CM, Fata JE, Leake D, Godden EL, Albertson DG, Nieto MA, Werb Z, Bissell MJ (2005) Rac1b and reactive oxygen species mediate MMP-3-induced EMT and genomic instability. Nature 436(7047):123–127

Raithatha SA, Muzik H, Muzik H, Rewcastle NB, Johnston RN, Edwards DR, Forsyth PA (2000) Localization of gelatinase-A and gelatinase-B mRNA and protein in human gliomas. Neuro Oncol 2(3):145–150

Raj VS, Mou H, Smits SL, Dekkers DH, Muller MA, Dijkman R, Muth D, Demmers JA, Zaki A, Fouchier RA, Thiel V, Drosten C, Rottier PJ, Osterhaus AD, Bosch BJ, Haagmans BL (2013) Dipeptidyl peptidase 4 is a functional receptor for the emerging human coronavirus-EMC. Nature 495(7440):251–254

Ramakrishna S, Suresh B, Baek KH (2011) The role of deubiquitinating enzymes in apoptosis. Cell Mol Life Sci 68(1):15–26

Rao JS (2003) Molecular mechanisms of glioma invasiveness: the role of proteases. Nat Rev Cancer 3(7):489–501

Rao JS, Rayford A, Morantz RA, Festoff BW, Sawaya R (1993a) Increased levels of plasminogen-activator inhibitor-1 (Pai-1) in human brain-tumors. J Neurooncol 17(3):215–221

Rao JS, Steck PA, Mohanam S, Stetler-Stevenson WG, Liotta LA, Sawaya R (1993b) Elevated levels of M(r) 92,000 type IV collagenase in human brain tumors. Cancer Res 53(10 Suppl): 2208–2211

Rao JS, Yamamoto M, Mohaman S, Gokaslan ZL, Fuller GN, Stetler-Stevenson WG, Rao VH, Liotta LA, Nicolson GL, Sawaya RE (1996) Expression and localization of 92 kDa type IV collagenase/gelatinase B (MMP-9) in human gliomas. Clin Exp Metastasis 14(1):12–18

Rao CN, Lakka SS, Kin Y, Konduri SD, Fuller GN, Mohanam S, Rao JS (2001) Expression of tissue factor pathway inhibitor 2 inversely correlates during the progression of human gliomas. Clin Cancer Res 7(3):570–576

Rawlings ND, Barrett AJ, Bateman A (2012) MEROPS: the database of proteolytic enzymes, their substrates and inhibitors. Nucleic Acids Res 40(Database issue):D343–D350

Ray SK, Patel SJ, Welsh CT, Wilford GG, Hogan EL, Banik NL (2002) Molecular evidence of apoptotic death in malignant brain tumors including glioblastoma multiforme: upregulation of calpain and caspase-3. J Neurosci Res 69(2):197–206

Redondo-Munoz J, Ugarte-Berzal E, Terol MJ, Van den Steen PE, Hernandez del Cerro M, Roderfeld M, Roeb E, Opdenakker G, Garcia-Marco JA, Garcia-Pardo A (2010) Matrix metalloproteinase-9 promotes chronic lymphocytic leukemia b cell survival through its hemopexin domain. Cancer Cell 17(2):160–172

Rege TA, Fears CY, Gladson CL (2005) Endogenous inhibitors of angiogenesis in malignant gliomas: nature's antiangiogenic therapy. Neuro Oncol 7(2):106–121

Regina A, Demeule M, Berube A, Moumdjian R, Berthelet F, Beliveau R (2003) Differences in multidrug resistance phenotype and matrix metalloproteinases activity between endothelial cells from normal brain and glioma. J Neurochem 84(2):316–324

Reif R, Sales S, Hettwer S, Dreier B, Gisler C, Wolfel J, Luscher D, Zurlinden A, Stephan A, Ahmed S, Baici A, Ledermann B, Kunz B, Sonderegger P (2007) Specific cleavage of agrin by neurotrypsin, a synaptic protease linked to mental retardation. FASEB J 21(13):3468–3478

Rempel SA, Rosenblum ML, Mikkelsen T, Yan PS, Ellis KD, Golembieski WA, Sameni M, Rozhin J, Ziegler G, Sloane BF (1994) Cathepsin B expression and localization in glioma progression and invasion. Cancer Res 54(23):6027–6031

Repnik U, Cesen MH, Turk B (2013) The endolysosomal system in cell death and survival. Cold Spring Harb Perspect Biol 5(1):a008755

Rettig WJ, Chesa PG, Beresford HR, Feickert HJ, Jennings MT, Cohen J, Oettgen HF, Old LJ (1986) Differential expression of cell surface antigens and glial fibrillary acidic protein in human astrocytoma subsets. Cancer Res 46(12 Pt 1):6406–6412

Ribatti D (2009) Endogenous inhibitors of angiogenesis: a historical review. Leuk Res 33(5): 638–644

Rivera S, Khrestchatisky M, Kaczmarek L, Rosenberg GA, Jaworski DM (2010) Metzincin proteases and their inhibitors: foes or friends in nervous system physiology? J Neurosci 30(46):15337–15357

Robins HI, O'Neill A, Gilbert M, Olsen M, Sapiente R, Berkey B, Mehta M (2008) Effect of dalteparin and radiation on survival and thromboembolic events in glioblastoma multiforme: a phase II ECOG trial. Cancer Chemother Pharmacol 62(2):227–233

Robson DK, Ironside JW, Reid WA, Bogue PR (1990) Immunolocalization of cathepsin D in the human central nervous system and central nervous system neoplasms. Neuropathol Appl Neurobiol 16(1):39–44

Rogove AD, Siao C, Keyt B, Strickland S, Tsirka SE (1999) Activation of microglia reveals a non-proteolytic cytokine function for tissue plasminogen activator in the central nervous system. J Cell Sci 112(Pt 22):4007–4016

Rome C, Arsaut J, Taris C, Couillaud F, Loiseau H (2007) MMP-7 (matrilysin) expression in human brain tumors. Mol Carcinog 46(6):446–452

Rong Y, Durden DL, Van Meir EG, Brat DJ (2006) 'Pseudopalisading' necrosis in glioblastoma: a familiar morphologic feature that links vascular pathology, hypoxia, and angiogenesis. J Neuropathol Exp Neurol 65(6):529–539

Rong Y, Brat DJ (2009) Vaso-occlusive mechanisms that initiate hypoxia and necrosis in glioblastoma: the role of thrombosis and tissue factor. In: Teicher BA (ed) CNS cancer. Humana, Totowa, NJ, pp 507–528 (This is a book chapter- please see http://link.springer.com/chapter/10.1007%2F978-1-60327-553-8_22)

Rorive S, Berton A, D'Haene N, Takacs CN, Debeir O, Decaestecker C, Salmon I (2008) Matrix metalloproteinase-9 interplays with the IGFBP2-IGFII complex to promote cell growth and motility in astrocytomas. Glia 56(15):1679–1690

Rorive S, Lopez XM, Maris C, Trepant AL, Sauvage S, Sadeghi N, Roland I, Decaestecker C, Salmon I (2010) TIMP-4 and CD63: new prognostic biomarkers in human astrocytomas. Mod Pathol 23(10):1418–1428

Roth P, Kissel M, Herrmann C, Eisele G, Leban J, Weller M, Schmidt F (2009) SC68896, a novel small molecule proteasome inhibitor, exerts antiglioma activity in vitro and in vivo. Clin Cancer Res 15(21):6609–6618

Rousalova I, Krepela E (2010) Granzyme B-induced apoptosis in cancer cells and its regulation (review). Int J Oncol 37(6):1361–1378

Rozanov DV, Sikora S, Godzik A, Postnova TI, Golubkov V, Savinov A, Tomlinson S, Strongin AY (2004) Non-proteolytic, receptor/ligand interactions associate cellular membrane type-1 matrix metalloproteinase with the complement component C1q. J Biol Chem 279(48): 50321–50328

Ruano Y, Mollejo M, Camacho FI, Rodriguez de Lope A, Fiano C, Ribalta T, Martinez P, Hernandez-Moneo JL, Melendez B (2008) Identification of survival-related genes of the phosphatidylinositol 3'-kinase signaling pathway in glioblastoma multiforme. Cancer 112(7):1575–1584

Sahin A, Velten M, Pietsch T, Knuefermann P, Okuducu AF, Hahne JC, Wernert N (2005) Inactivation of Ets 1 transcription factor by a specific decoy strategy reduces rat C6 glioma cell proliferation and mmp-9 expression. Int J Mol Med 15(5):771–776

Sahm F, Capper D, Jeibmann A, Habel A, Paulus W, Troost D, von Deimling A (2012) Addressing diffuse glioma as a systemic brain disease with single-cell analysis. Arch Neurol 69(4): 523–526

Saito N, Fu J, Zheng S, Yao J, Wang S, Liu DD, Yuan Y, Sulman EP, Lang FF, Colman H, Verhaak RG, Yung WK, Koul D (2014) A high Notch pathway activation predicts response to gamma secretase inhibitors in proneural subtype of glioma tumor-initiating cells. Stem Cells 32(1):301–312. doi:10.1002/stem.1528

Sakamoto T, Seiki M (2009) Cytoplasmic tail of MT1-MMP regulates macrophage motility independently from its protease activity. Genes Cells 14(5):617–626

Salmaggi A, Croci D, Prina P, Cajola L, Pollo B, Marras CE, Ciusani E, Silvani A, Boiardi A, Sciacca FL (2006) Production and post-surgical modification of VEGF, tPA and PAI-1 in patients with glioma. Cancer Biol Ther 5(2):204–209

Sarkar S, Nuttall RK, Liu S, Edwards DR, Yong VW (2006) Tenascin-C stimulates glioma cell invasion through matrix metalloproteinase-12. Cancer Res 66(24):11771–11780

Sato Y (2011) Persistent vascular normalization as an alternative goal of anti-angiogenic cancer therapy. Cancer Sci 102(7):1253–1256

Sawaya R, Ramo OJ, Shi ML, Mandybur G (1991) Biological significance of tissue plasminogen activator content in brain tumors. J Neurosurg 74(3):480–486

Sawaya RE, Yamamoto M, Gokaslan ZL, Wang SW, Mohanam S, Fuller GN, McCutcheon IE, Stetler-Stevenson WG, Nicolson GL, Rao JS (1996) Expression and localization of 72 kDa type IV collagenase (MMP-2) in human malignant gliomas in vivo. Clin Exp Metastasis 14(1): 35–42

Saxena A, Shriml LM, Dean M, Ali IU (1999) Comparative molecular genetic profiles of anaplastic astrocytomas/glioblastomas multiforme and their subsequent recurrences. Oncogene 18(6):1385–1390

Schafer C, Fels C, Brucke M, Holzhausen HJ, Bahn H, Wellman M, Visvikis A, Fischer P, Rainov NG (2001) Gamma-glutamyl transferase expression in higher-grade astrocytic glioma. Acta Oncol 40(4):529–535

Schulte A, Gunther HS, Phillips HS, Kemming D, Martens T, Kharbanda S, Soriano RH, Modrusan Z, Zapf S, Westphal M, Lamszus K (2011) A distinct subset of glioma cell lines with stem cell-like properties reflects the transcriptional phenotype of glioblastomas and overexpresses CXCR4 as therapeutic target. Glia 59(4):590–602

Schulz I, Zeitschel U, Rudolph T, Ruiz-Carrillo D, Rahfeld JU, Gerhartz B, Bigl V, Demuth HU, Rossner S (2005) Subcellular localization suggests novel functions for prolyl endopeptidase in protein secretion. J Neurochem 94(4):970–979

Schwab ME, Caroni P (1988) Oligodendrocytes and CNS myelin are nonpermissive substrates for neurite growth and fibroblast spreading in vitro. J Neurosci 8(7):2381–2393

Schwartzbaum J, Ahlbom A, Malmer B, Lonn S, Brookes AJ, Doss H, Debinski W, Henriksson R, Feychting M (2005) Polymorphisms associated with asthma are inversely related to glioblastoma multiforme. Cancer Res 65(14):6459–6465

Scrideli CA, Carlotti CG Jr, Okamoto OK, Andrade VS, Cortez MA, Motta FJ, Lucio-Eterovic AK, Neder L, Rosemberg S, Oba-Shinjo SM, Marie SK, Tone LG (2008) Gene expression profile analysis of primary glioblastomas and non-neoplastic brain tissue: identification of potential target genes by oligonucleotide microarray and real-time quantitative PCR. J Neurooncol 88(3):281–291

Seals DF, Courtneidge SA (2003) The ADAMs family of metalloproteases: multidomain proteins with multiple functions. Genes Dev 17(1):7–30

Seeds NW, Siconolfi LB, Haffke SP (1997) Neuronal extracellular proteases facilitate cell migration, axonal growth, and pathfinding. Cell Tissue Res 290(2):367–370

Seidah NG, Chretien M (1999) Proprotein and prohormone convertases: a family of subtilases generating diverse bioactive polypeptides. Brain Res 848(1–2):45–62

Seidah NG, Sadr MS, Chretien M, Mbikay M (2013) The multifaceted proprotein convertases: their unique, redundant, complementary, and opposite functions. J Biol Chem 288(30): 21473–21481

Senner V, Ratzinger S, Mertsch S, Grassel S, Paulus W (2008) Collagen XVI expression is upregulated in glioblastomas and promotes tumor cell adhesion. FEBS Lett 582(23–24): 3293–3300

Seol DW (2011) p53-Independent up-regulation of a TRAIL receptor DR5 by proteasome inhibitors: a mechanism for proteasome inhibitor-enhanced TRAIL-induced apoptosis. Biochem Biophys Res Commun 416(1–2):222–225

Serres E, Debarbieux F, Stanchi F, Maggiorella L, Grall D, Turchi L, Burel-Vandenbos F, Figarella-Branger D, Virolle T, Rougon G, Van Obberghen-Schilling E (2013) Fibronectin expression in glioblastomas promotes cell cohesion, collective invasion of basement membrane in vitro and orthotopic tumor growth in mice. Oncogene. doi:10.1038/onc.2013.305

Shankar A, Kumar S, Iskander A, Varma NR, Janic B, Decarvalho A, Mikkelsen T, Frank JA, Ali MM, Knight RA, Brown S, Arbab AS (2014) Subcurative radiation significantly increases proliferation, invasion, and migration of primary GBM in vivo. Chin J Cancer 33(3):148–158

Shen J, Prywes R (2004) Dependence of site-2 protease cleavage of ATF6 on prior site-1 protease digestion is determined by the size of the luminal domain of ATF6. J Biol Chem 279(41): 43046–43051

Shevtsova Z, Garrido M, Weishaupt J, Saftig P, Bahr M, Luhder F, Kugler S (2010) CNS-expressed cathepsin D prevents lymphopenia in a murine model of congenital neuronal ceroid lipofuscinosis. Am J Pathol 177(1):271–279

Shimizu C, Yoshida S, Shibata M, Kato K, Momota Y, Matsumoto K, Shiosaka T, Midorikawa R, Kamachi T, Kawabe A, Shiosaka S (1998) Characterization of recombinant and brain neuropsin, a plasticity-related serine protease. J Biol Chem 273(18):11189–11196

Shiosaka S, Ishikawa Y (2011) Neuropsin–a possible modulator of synaptic plasticity. J Chem Neuroanat 42(1):24–29

Sierko E, Wojtukiewicz MZ, Kisiel W (2007) The role of tissue factor pathway inhibitor-2 in cancer biology. Semin Thromb Hemost 33(7):653–659

Siintola E, Partanen S, Stromme P, Haapanen A, Haltia M, Maehlen J, Lehesjoki AE, Tyynela J (2006) Cathepsin D deficiency underlies congenital human neuronal ceroid-lipofuscinosis. Brain 129(Pt 6):1438–1445

Silacci P, Mazzolai L, Gauci C, Stergiopulos N, Yin HL, Hayoz D (2004) Gelsolin superfamily proteins: key regulators of cellular functions. Cell Mol Life Sci 61(19–20):2614–2623

Sivaparvathi M, Sawaya R, Wang SW, Rayford A, Yamamoto M, Liotta LA, Nicolson GL, Rao JS (1995) Overexpression and localization of cathepsin B during the progression of human gliomas. Clin Exp Metastasis 13(1):49–56

Sivaparvathi M, Sawaya R, Gokaslan ZL, Chintala SK, Rao JS (1996a) Expression and the role of cathepsin H in human glioma progression and invasion. Cancer Lett 104(1):121–126

Sivaparvathi M, Sawaya R, Chintala SK, Go Y, Gokaslan ZL, Rao JS (1996b) Expression of cathepsin D during the progression of human gliomas. Neurosci Lett 208(3):171–174

Sivaparvathi M, Yamamoto M, Nicolson GL, Gokaslan ZL, Fuller GN, Liotta LA, Sawaya R, Rao JS (1996c) Expression and immunohistochemical localization of cathepsin L during the progression of human gliomas. Clin Exp Metastasis 14(1):27–34

Skiriute D, Vaitkiene P, Saferis V, Asmoniene V, Skauminas K, Deltuva VP, Tamasauskas A (2012) MGMT, GATA6, CD81, DR4, and CASP8 gene promoter methylation in glioblastoma. BMC Cancer 12:218

Snuderl M, Fazlollahi L, Le LP, Nitta M, Zhelyazkova BH, Davidson CJ, Akhavanfard S, Cahill DP, Aldape KD, Betensky RA, Louis DN, Iafrate AJ (2011) Mosaic amplification of multiple receptor tyrosine kinase genes in glioblastoma. Cancer Cell 20(6):810–817

Sounni NE, Rozanov DV, Remacle AG, Golubkov VS, Noel A, Strongin AY (2010) Timp-2 binding with cellular MT1-MMP stimulates invasion-promoting MEK/ERK signaling in cancer cells. Int J Cancer 126(5):1067–1078

Sower LE, Payne DA, Meyers R, Carney DH (1999) Thrombin peptide, TP508, induces differential gene expression in fibroblasts through a nonproteolytic activation pathway. Exp Cell Res 247(2):422–431

Stephan A, Mateos JM, Kozlov SV, Cinelli P, Kistler AD, Hettwer S, Rulicke T, Streit P, Kunz B, Sonderegger P (2008) Neurotrypsin cleaves agrin locally at the synapse. FASEB J 22(6): 1861–1873

Stetler-Stevenson WG, Seo DW (2005) TIMP-2: an endogenous inhibitor of angiogenesis. Trends Mol Med 11(3):97–103

Stockhausen MT, Kristoffersen K, Poulsen HS (2010) The functional role of Notch signaling in human gliomas. Neuro Oncol 12(2):199–211

Stojic J, Hagemann C, Haas S, Herbold C, Kuhnel S, Gerngras S, Roggendorf W, Roosen K, Vince GH (2008) Expression of matrix metalloproteinases MMP-1, MMP-11 and MMP-19 is correlated with the WHO-grading of human malignant gliomas. Neurosci Res 60(1):40–49

Stremenova J, Krepela E, Mares V, Trim J, Dbaly V, Marek J, Vanickova Z, Lisa V, Yea C, Sedo A (2007) Expression and enzymatic activity of dipeptidyl peptidase-IV in human astrocytic tumours are associated with tumour grade. Int J Oncol 31(4):785–792

Strojnik T, Zajc I, Bervar A, Zidanik B, Golouh R, Kos J, Dolenc V, Lah T (2000) Cathepsin B and its inhibitor stefin A in brain tumors. Pflugers Arch 439(3 Suppl):R122–R123

Strojnik T, Kavalar R, Trinkaus M, Lah TT (2005) Cathepsin L in glioma progression: comparison with cathepsin B. Cancer Detect Prev 29(5):448–455

Strojnik T, Kavalar R, Zajc I, Diamandis EP, Oikonomopoulou K, Lah TT (2009) Prognostic impact of CD68 and kallikrein 6 in human glioma. Anticancer Res 29(8):3269–3279

Strongin AY (2010) Proteolytic and non-proteolytic roles of membrane type-1 matrix metalloproteinase in malignancy. Biochim Biophys Acta 1803(1):133–141

Strongin AY, Collier I, Bannikov G, Marmer BL, Grant GA, Goldberg GI (1995) Mechanism of cell surface activation of 72-kDa type IV collagenase. Isolation of the activated form of the membrane metalloprotease. J Biol Chem 270(10):5331–5338

Stummer W, Pichlmeier U, Meinel T, Wiestler OD, Zanella F, Reulen HJ, A. L.-G. S. Group (2006) Fluorescence-guided surgery with 5-aminolevulinic acid for resection of malignant glioma: a randomised controlled multicentre phase III trial. Lancet Oncol 7(5):392–401

Stylli SS, Kaye AH, Lock P (2008) Invadopodia: at the cutting edge of tumour invasion. J Clin Neurosci 15(7):725–737

Sumitomo M, Shen R, Walburg M, Dai J, Geng Y, Navarro D, Boileau G, Papandreou CN, Giancotti FG, Knudsen B, Nanus DM (2000) Neutral endopeptidase inhibits prostate cancer cell migration by blocking focal adhesion kinase signaling. J Clin Invest 106(11):1399–1407

Sun L, Yan W, Wang Y, Sun G, Luo H, Zhang J, Wang X, You Y, Yang Z, Liu N (2011) MicroRNA-10b induces glioma cell invasion by modulating MMP-14 and uPAR expression via HOXD10. Brain Res 1389:9–18

Sun C, Wang Q, Zhou H, Yu S, Simard AR, Kang C, Li Y, Kong Y, An T, Wen Y, Shi F, Hao J (2013) Antisense MMP-9 RNA inhibits malignant glioma cell growth in vitro and in vivo. Neurosci Bull 29(1):83–93

Szalad A, Katakowski M, Zheng X, Jiang F, Chopp M (2009) Transcription factor Sp1 induces ADAM17 and contributes to tumor cell invasiveness under hypoxia. J Exp Clin Cancer Res 28:129

Takahashi M, Fukami S, Iwata N, Inoue K, Itohara S, Itoh H, Haraoka J, Saido T (2002) In vivo glioma growth requires host-derived matrix metalloproteinase 2 for maintenance of angioarchitecture. Pharmacol Res 46(2):155–163

Takano S, Mashiko R, Osuka S, Ishikawa E, Ohneda O, Matsumura A (2010) Detection of failure of bevacizumab treatment for malignant glioma based on urinary matrix metalloproteinase activity. Brain Tumor Pathol 27(2):89–94

Takebe N, Nguyen D, Yang SX (2014) Targeting Notch signaling pathway in cancer: clinical development advances and challenges. Pharmacol Ther 141(2):140–149

Talieri M, Zoma M, Devetzi M, Scorilas A, Ardavanis A (2012) Kallikrein-related peptidase 6 (KLK6)gene expression in intracranial tumors. Tumour Biol 33(5):1375–1383

Tani E, Kitagawa H, Ikemoto H, Matsumoto T (2001) Proteasome inhibitors induce Fas-mediated apoptosis by c-Myc accumulation and subsequent induction of FasL message in human glioma cells. FEBS Lett 504(1–2):53–58

Tao J, Qiu B, Zhang D, Wang Y (2012) Expression levels of Fas/Fas-L mRNA in human brain glioma stem cells. Mol Med Rep 5(5):1202–1206

Tao BB, He H, Shi XH, Wang CL, Li WQ, Li B, Dong Y, Hu GH, Hou LJ, Luo C, Chen JX, Chen HR, Yu YH, Sun QF, Lu YC (2013) Up-regulation of USP2a and FASN in gliomas correlates strongly with glioma grade. J Clin Neurosci 20(5):717–720

Thapa N, Lee BH, Kim IS (2007) TGFBIp/betaig-h3 protein: a versatile matrix molecule induced by TGF-beta. Int J Biochem Cell Biol 39(12):2183–2194

Thorns V, Walter GF, Thorns C (2003) Expression of MMP-2, MMP-7, MMP-9, MMP-10 and MMP-11 in human astrocytic and oligodendroglial gliomas. Anticancer Res 23(5A): 3937–3944

Tirapelli LF, Bolini PH, Tirapelli DP, Peria FM, Becker AN, Saggioro FP, Carlotti CG Jr (2010) Caspase-3 and Bcl-2 expression in glioblastoma: an immunohistochemical study. Arq Neuropsiquiatr 68(4):603–607

Tomimatsu Y, Idemoto S, Moriguchi S, Watanabe S, Nakanishi H (2002) Proteases involved in long-term potentiation. Life Sci 72(4–5):355–361

Tremblay P, Beaudet MJ, Tremblay E, Rueda N, Thomas T, Vallieres L (2011) Matrix metalloproteinase 2 attenuates brain tumour growth, while promoting macrophage recruitment and vascular repair. J Pathol 224(2):222–233

Tsatas D, Kaye AH (2003) The role of the plasminogen activation cascade in glioma cell invasion: a review. J Clin Neurosci 10(2):139–145

Tsopanoglou NE, Andriopoulou P, Maragoudakis ME (2002) On the mechanism of thrombin-induced angiogenesis: involvement of alphavbeta3-integrin. Am J Physiol Cell Physiol 283(5): C1501–C1510

Turk B (2006) Targeting proteases: successes, failures and future prospects. Nat Rev Drug Discov 5(9):785–799

Turk V, Stoka V, Turk D (2008) Cystatins: biochemical and structural properties, and medical relevance. Front Biosci 13:5406–5420

Turk B, Turk D, Turk V (2012a) Protease signalling: the cutting edge. EMBO J 31(7):1630–1643

Turk V, Stoka V, Vasiljeva O, Renko M, Sun T, Turk B, Turk D (2012b) Cysteine cathepsins: from structure, function and regulation to new frontiers. Biochim Biophys Acta 1824(1):68–88

Uchinokura S, Miyata S, Fukushima T, Itoh H, Nakano S, Wakisaka S, Kataoka H (2006) Role of hepatocyte growth factor activator (HGF activator) in invasive growth of human glioblastoma cells in vivo. Int J Cancer 118(3):583–592

Unterkircher T, Cristofanon S, Vellanki SH, Nonnenmacher L, Karpel-Massler G, Wirtz CR, Debatin KM, Fulda S (2011) Bortezomib primes glioblastoma, including glioblastoma stem cells, for TRAIL by increasing tBid stability and mitochondrial apoptosis. Clin Cancer Res 17(12):4019–4030

Vaillant C, Didier-Bazes M, Hutter A, Belin MF, Thomasset N (1999) Spatiotemporal expression patterns of metalloproteinases and their inhibitors in the postnatal developing rat cerebellum. J Neurosci 19(12):4994–5004

Vaillant C, Meissirel C, Mutin M, Belin MF, Lund LR, Thomasset N (2003) MMP-9 deficiency affects axonal outgrowth, migration, and apoptosis in the developing cerebellum. Mol Cell Neurosci 24(2):395–408

Van De Craen B, Declerck PJ, Gils A (2012) The biochemistry, physiology and pathological roles of PAI-1 and the requirements for PAI-1 inhibition in vivo. Thromb Res 130(4):576–585

van Es JH, van Gijn ME, Riccio O, van den Born M, Vooijs M, Begthel H, Cozijnsen M, Robine S, Winton DJ, Radtke F, Clevers H (2005) Notch/gamma-secretase inhibition turns proliferative cells in intestinal crypts and adenomas into goblet cells. Nature 435(7044):959–963

van Hinsbergh VW, Engelse MA, Quax PH (2006) Pericellular proteases in angiogenesis and vasculogenesis. Arterioscler Thromb Vasc Biol 26(4):716–728

Van Hove I, Lemmens K, Van de Velde S, Verslegers M, Moons L (2012) Matrix metalloproteinase-3 in the central nervous system: a look on the bright side. J Neurochem 123(2):203–216

Van Meir EG, Hadjipanayis CG, Norden AD, Shu HK, Wen PY, Olson JJ (2010) Exciting new advances in neuro-oncology: the avenue to a cure for malignant glioma. CA Cancer J Clin 60(3):166–193

Van Meter TE, Broaddus WC, Rooprai HK, Pilkington GJ, Fillmore HL (2004) Induction of membrane-type-1 matrix metalloproteinase by epidermal growth factor-mediated signaling in gliomas. Neuro Oncol 6(3):188–199

VanMeter TE, Rooprai HK, Kibble MM, Fillmore HL, Broaddus WC, Pilkington GJ (2001) The role of matrix metalloproteinase genes in glioma invasion: co-dependent and interactive proteolysis. J Neurooncol 53(2):213–235

Vazquez F, Hastings G, Ortega MA, Lane TF, Oikemus S, Lombardo M, Iruela-Arispe ML (1999) METH-1, a human ortholog of ADAMTS-1, and METH-2 are members of a new family of proteins with angio-inhibitory activity. J Biol Chem 274(33):23349–23357

Veeravalli KK, Ponnala S, Chetty C, Tsung AJ, Gujrati M, Rao JS (2012) Integrin alpha9beta1-mediated cell migration in glioblastoma via SSAT and Kir4.2 potassium channel pathway. Cell Signal 24(1):272–281

Verhaak RG, Hoadley KA, Purdom E, Wang V, Qi Y, Wilkerson MD, Miller CR, Ding L, Golub T, Mesirov JP, Alexe G, Lawrence M, O'Kelly M, Tamayo P, Weir BA, Gabriel S, Winckler W, Gupta S, Jakkula L, Feiler HS, Hodgson JG, James CD, Sarkaria JN, Brennan C, Kahn A, Spellman PT, Wilson RK, Speed TP, Gray JW, Meyerson M, Getz G, Perou CM, Hayes DN (2010) Integrated genomic analysis identifies clinically relevant subtypes of glioblastoma characterized by abnormalities in PDGFRA, IDH1, EGFR, and NF1. Cancer Cell 17(1):98–110

Viapiano MS, Matthews RT (2006) From barriers to bridges: chondroitin sulfate proteoglycans in neuropathology. Trends Mol Med 12(10):488–496

Viapiano MS, Bi WL, Piepmeier J, Hockfield S, Matthews RT (2005) Novel tumor-specific isoforms of BEHAB/brevican identified in human malignant gliomas. Cancer Res 65(15):6726–6733

Viapiano MS, Hockfield S, Matthews RT (2008) BEHAB/brevican requires ADAMTS-mediated proteolytic cleavage to promote glioma invasion. J Neurooncol 88(3):261–272

Vince GH, Wagner S, Pietsch T, Klein R, Goldbrunner RH, Roosen K, Tonn JC (1999) Heterogeneous regional expression patterns of matrix metalloproteinases in human malignant gliomas. Int J Dev Neurosci 17(5–6):437–445

Viry E, Baginska J, Berchem G, Noman MZ, Medves S, Chouaib S, Janji B (2014) Autophagic degradation of GZMB/granzyme B: a new mechanism of hypoxic tumor cell escape from natural killer cell-mediated lysis. Autophagy 10(1):173–175

Vlachostergios PJ, Voutsadakis IA, Papandreou CN (2012) The ubiquitin-proteasome system in glioma cell cycle control. Cell Div 7(1):18

Vlachostergios PJ, Voutsadakis IA, Papandreou CN (2013) Mechanisms of proteasome inhibitor-induced cytotoxicity in malignant glioma. Cell Biol Toxicol 29(4):199–211

Vlashi E, Kim K, Lagadec C, Donna LD, McDonald JT, Eghbali M, Sayre JW, Stefani E, McBride W, Pajonk F (2009) In vivo imaging, tracking, and targeting of cancer stem cells. J Natl Cancer Inst 101(5):350–359

Vlashi E, Mattes M, Lagadec C, Donna LD, Phillips TM, Nikolay P, McBride WH, Pajonk F (2010) Differential effects of the proteasome inhibitor NPI-0052 against glioma cells. Transl Oncol 3(1):50–55

Wagenknecht B, Hermisson M, Eitel K, Weller M (1999) Proteasome inhibitors induce p53/p21-independent apoptosis in human glioma cells. Cell Physiol Biochem 9(3):117–125

Wagenknecht B, Hermisson M, Groscurth P, Liston P, Krammer PH, Weller M (2000) Proteasome inhibitor-induced apoptosis of glioma cells involves the processing of multiple caspases and cytochrome c release. J Neurochem 75(6):2288–2297

Waltereit R, Weller M (2002) The role of caspases 9 and 9-short (9S) in death ligand- and drug-induced apoptosis in human astrocytoma cells. Brain Res Mol Brain Res 106(1–2):42–49

Wang Z, Juttermann R, Soloway PD (2000) TIMP-2 is required for efficient activation of proMMP-2 in vivo. J Biol Chem 275(34):26411–26415

Wang M, Tang J, Liu S, Yoshida D, Teramoto A (2005) Expression of cathepsin B and micro-vascular density increases with higher grade of astrocytomas. J Neurooncol 71(1):3–7

Wang L, Rahn JJ, Lun X, Sun B, Kelly JJ, Weiss S, Robbins SM, Forsyth PA, Senger DL (2008a) Gamma-secretase represents a therapeutic target for the treatment of invasive glioma mediated by the p75 neurotrophin receptor. PLoS Biol 6(11):e289

Wang Y, Luo W, Reiser G (2008b) Trypsin and trypsin-like proteases in the brain: proteolysis and cellular functions. Cell Mol Life Sci 65(2):237–252

Wang J, Li Y, Wang J, Li C, Yu K, Wang Q (2012) Increased expression of matrix metalloproteinase-13 in glioma is associated with poor overall survival of patients. Med Oncol 29(4):2432–2437

Wang H, Zhang S, Zhong J, Zhang J, Luo Y, Pengfei G (2013a) The proteasome inhibitor lactacystin exerts its therapeutic effects on glioma via apoptosis: an in vitro and in vivo study. J Int Med Res 41(1):72–81

Wang L, Yuan J, Tu Y, Mao X, He S, Fu G, Zong J, Zhang Y (2013b) Co-expression of MMP-14 and MMP-19 predicts poor survival in human glioma. Clin Transl Oncol 15(2):139–145

Warich M, von Bossanyi P, Dietzmann K (1995) Expression of cathepsin D in human astrocytic neoplasias. Gen Diagn Pathol 141(2):93–96

Wei X, Lv T, Chen D, Guan J (2014) Lentiviral vector mediated delivery of RHBDD1 shRNA down regulated the proliferation of human glioblastoma cells. Technol Cancer Res Treat 13(1): 87–93

Weis SM, Cheresh DA (2011) Tumor angiogenesis: molecular pathways and therapeutic targets. Nat Med 17(11):1359–1370

Weller M, Pfister SM, Wick W, Hegi ME, Reifenberger G, Stupp R (2013) Molecular neuro-oncology in clinical practice: a new horizon. Lancet Oncol 14(9):e370–e379

Wernicke AG, Edgar MA, Lavi E, Liu H, Salerno P, Bander NH, Gutin PH (2011) Prostate-specific membrane antigen as a potential novel vascular target for treatment of glioblastoma multiforme. Arch Pathol Lab Med 135(11):1486–1489

Wesolowska A, Kwiatkowska A, Slomnicki L, Dembinski M, Master A, Sliwa M, Franciszkiewicz K, Chouaib S, Kaminska B (2008) Microglia-derived TGF-beta as an important regulator of glioblastoma invasion–an inhibition of TGF-beta-dependent effects by shRNA against human TGF-beta type II receptor. Oncogene 27(7):918–930

Westermarck J, Kahari VM (1999) Regulation of matrix metalloproteinase expression in tumor invasion. FASEB J 13(8):781–792

Westphal D, Sytnyk V, Schachner M, Leshchyns'ka I (2010) Clustering of the neural cell adhesion molecule (NCAM) at the neuronal cell surface induces caspase-8- and -3-dependent changes of the spectrin meshwork required for NCAM-mediated neurite outgrowth. J Biol Chem 285(53): 42046–42057

Wick W, Wild-Bode C, Frank B, Weller M (2004) BCL-2-induced glioma cell invasiveness depends on furin-like proteases. J Neurochem 91(6):1275–1283

Wild-Bode C, Weller M, Rimner A, Dichgans J, Wick W (2001) Sublethal irradiation promotes migration and invasiveness of glioma cells: implications for radiotherapy of human glioblastoma. Cancer Res 61(6):2744–2750

Wildeboer D, Naus S, Amy Sang QX, Bartsch JW, Pagenstecher A (2006) Metalloproteinase disintegrins ADAM8 and ADAM19 are highly regulated in human primary brain tumors and their expression levels and activities are associated with invasiveness. J Neuropathol Exp Neurol 65(5):516–527

Wilkinson KD (2009) DUBs at a glance. J Cell Sci 122(Pt 14):2325–2329

Williams DW, Kondo S, Krzyzanowska A, Hiromi Y, Truman JW (2006) Local caspase activity directs engulfment of dendrites during pruning. Nat Neurosci 9(10):1234–1236

Winkler F, Kozin SV, Tong RT, Chae SS, Booth MF, Garkavtsev I, Xu L, Hicklin DJ, Fukumura D, di Tomaso E, Munn LL, Jain RK (2004) Kinetics of vascular normalization by VEGFR2 blockade governs brain tumor response to radiation: role of oxygenation, angiopoietin-1, and matrix metalloproteinases. Cancer Cell 6(6):553–563

Withana NP, Blum G, Sameni M, Slaney C, Anbalagan A, Olive MB, Bidwell BN, Edgington L, Wang L, Moin K, Sloane BF, Anderson RL, Bogyo MS, Parker BS (2012) Cathepsin B inhibition limits bone metastasis in breast cancer. Cancer Res 72(5):1199–1209

Wolf M, Albrecht S, Marki C (2008) Proteolytic processing of chemokines: implications in physiological and pathological conditions. Int J Biochem Cell Biol 40(6–7):1185–1198

Wolpert F, Tritschler I, Steinle A, Weller M, Eisele G (2014) A disintegrin and metalloproteinases 10 and 17 modulate the immunogenicity of glioblastoma-initiating cells. Neuro Oncol 16(3): 382–391

Wright JW, Harding JW (2009) Contributions of matrix metalloproteinases to neural plasticity, habituation, associative learning and drug addiction. Neural Plast 2009:579382

Wu DG, Wang YY, Fan LG, Luo H, Han B, Sun LH, Wang XF, Zhang JX, Cao L, Wang XR, You YP, Liu N (2011) MicroRNA-7 regulates glioblastoma cell invasion via targeting focal adhesion kinase expression. Chin Med J (Engl) 124(17):2616–2621

Xia H, Qi Y, Ng SS, Chen X, Li D, Chen S, Ge R, Jiang S, Li G, Chen Y, He ML, Kung HF, Lai L, Lin MC (2009) microRNA-146b inhibits glioma cell migration and invasion by targeting MMPs. Brain Res 1269:158–165

Xiao C, Yang BF, Asadi N, Beguinot F, Hao C (2002) Tumor necrosis factor-related apoptosis-inducing ligand-induced death-inducing signaling complex and its modulation by c-FLIP and PED/PEA-15 in glioma cells. J Biol Chem 277(28):25020–25025

Xie H, Xue YX, Liu LB, Wang P, Liu YH, Ying HQ (2011) Expressions of matrix metalloproteinase-7 and matrix metalloproteinase-14 associated with the activation of extracellular signal-regulated kinase1/2 in human brain gliomas of different pathological grades. Med Oncol 28(Suppl 1):S433–S438

Xu Y, Zhong Z, Yuan J, Zhang Z, Wei Q, Song W, Chen H (2013) Collaborative overexpression of matrix metalloproteinase-1 and vascular endothelial growth factor-C predicts adverse prognosis in patients with gliomas. Cancer Epidemiol 37(5):697–702

Yamada H, Fredette B, Shitara K, Hagihara K, Miura R, Ranscht B, Stallcup WB, Yamaguchi Y (1997) The brain chondroitin sulfate proteoglycan brevican associates with astrocytes ensheathing cerebellar glomeruli and inhibits neurite outgrowth from granule neurons. J Neurosci 17(20):7784–7795

Yamaguchi Y, Shao Z, Sharif S, Du XY, Myles T, Merchant M, Harsh G, Glantz M, Recht L, Morser J, Leung LL (2013) Thrombin-cleaved fragments of osteopontin are overexpressed in malignant glial tumors and provide a molecular niche with survival advantage. J Biol Chem 288(5):3097–3111

Yamahata H, Takeshima H, Kuratsu J, Sarker KP, Tanioka K, Wakimaru N, Nakata M, Kitajima I, Maruyama I (2002) The role of thrombin in the neo-vascularization of malignant gliomas: an intrinsic modulator for the up-regulation of vascular endothelial growth factor. Int J Oncol 20(5):921–928

Yamamoto M, Sawaya R, Mohanam S, Bindal AK, Bruner JM, Oka K, Rao VH, Tomonaga M, Nicolson GL, Rao JS (1994a) Expression and localization of urokinase-type plasminogen activator in human astrocytomas in vivo. Cancer Res 54(14):3656–3661

Yamamoto M, Sawaya R, Mohanam S, Rao VH, Bruner JM, Nicolson GL, Rao JS (1994b) Expression and localization of urokinase-type plasminogen activator receptor in human gliomas. Cancer Res 54(18):5016–5020

Yamamoto M, Mohanam S, Sawaya R, Fuller GN, Seiki M, Sato H, Gokaslan ZL, Liotta LA, Nicolson GL, Rao JS (1996) Differential expression of membrane-type matrix metalloproteinase and its correlation with gelatinase A activation in human malignant brain tumors in vivo and in vitro. Cancer Res 56(2):384–392

Yan S, Berquin IM, Troen BR, Sloane BF (2000) Transcription of human cathepsin B is mediated by Sp1 and Ets family factors in glioma. DNA Cell Biol 19(2):79–91

Yan W, Zhang W, Sun L, Liu Y, You G, Wang Y, Kang C, You Y, Jiang T (2011) Identification of MMP-9 specific microRNA expression profile as potential targets of anti-invasion therapy in glioblastoma multiforme. Brain Res 1411:108–115

Yanamandra N, Gumidyala KV, Waldron KG, Gujrati M, Olivero WC, Dinh DH, Rao JS, Mohanam S (2004) Blockade of cathepsin B expression in human glioblastoma cells is associated with suppression of angiogenesis. Oncogene 23(12):2224–2230

Yang P, Baker KA, Hagg T (2006) The ADAMs family: coordinators of nervous system development, plasticity and repair. Prog Neurobiol 79(2):73–94

Yang BF, Xiao C, Li H, Yang SJ (2007) Resistance to Fas-mediated apoptosis in malignant tumours is rescued by KN-93 and cisplatin via downregulation of c-FLIP expression and phosphorylation. Clin Exp Pharmacol Physiol 34(12):1245–1251

Yang M, Li Y, Chilukuri K, Brady OA, Boulos MI, Kappes JC, Galileo DS (2011) L1 stimulation of human glioma cell motility correlates with FAK activation. J Neurooncol 105(1):27–44

Ye XZ, Xu SL, Xin YH, Yu SC, Ping YF, Chen L, Xiao HL, Wang B, Yi L, Wang QL, Jiang XF, Yang L, Zhang P, Qian C, Cui YH, Zhang X, Bian XW (2012) Tumor-associated microglia/macrophages enhance the invasion of glioma stem-like cells via TGF-beta1 signaling pathway. J Immunol 189(1):444–453

Yeh WL, Lu DY, Lee MJ, Fu WM (2009) Leptin induces migration and invasion of glioma cells through MMP-13 production. Glia 57(4):454–464

Yin D, Zhou H, Kumagai T, Liu G, Ong JM, Black KL, Koeffler HP (2005) Proteasome inhibitor PS-341 causes cell growth arrest and apoptosis in human glioblastoma multiforme (GBM). Oncogene 24(3):344–354

Yong VW, Power C, Forsyth P, Edwards DR (2001) Metalloproteinases in biology and pathology of the nervous system. Nat Rev Neurosci 2(7):502–511

Yu Q, Stamenkovic I (1999) Localization of matrix metalloproteinase 9 to the cell surface provides a mechanism for CD44-mediated tumor invasion. Genes Dev 13(1):35–48

Zadran S, Jourdi H, Rostamiani K, Qin Q, Bi X, Baudry M (2010) Brain-derived neurotrophic factor and epidermal growth factor activate neuronal m-calpain via mitogen-activated protein kinase-dependent phosphorylation. J Neurosci 30(3):1086–1095

Zanotto-Filho A, Braganhol E, Battastini AM, Moreira JC (2012) Proteasome inhibitor MG132 induces selective apoptosis in glioblastoma cells through inhibition of PI3K/Akt and NFkappaB pathways, mitochondrial dysfunction, and activation of p38-JNK1/2 signaling. Invest New Drugs 30(6):2252–2262

Zerrouqi A, Pyrzynska B, Febbraio M, Brat DJ, Van Meir EG (2012) P14ARF inhibits human glioblastoma-induced angiogenesis by upregulating the expression of TIMP3. J Clin Invest 122(4):1283–1295

Zhang X, Fei Z, Bu X, Zhen H, Zhang Z, Gu J, Chen Y (2000) Expression and significance of urokinase type plasminogen activator gene in human brain gliomas. J Surg Oncol 74(2):90–94

Zhang J, Sarkar S, Yong VW (2005) The chemokine stromal cell derived factor-1 (CXCL12) promotes glioma invasiveness through MT2-matrix metalloproteinase. Carcinogenesis 26(12): 2069–2077

Zhang Y, Zhan H, Xu W, Yuan Z, Lu P, Zhan L, Li Q (2011) Upregulation of matrix metalloproteinase-1 and proteinase-activated receptor-1 promotes the progression of human gliomas. Pathol Res Pract 207(1):24–29

Zhao Y, Lyons CE Jr, Xiao A, Templeton DJ, Sang QA, Brew K, Hussaini IM (2008a) Urokinase directly activates matrix metalloproteinases-9: a potential role in glioblastoma invasion. Biochem Biophys Res Commun 369(4):1215–1220

Zhao Y, Xiao A, Dipierro CG, Abdel-Fattah R, Amos S, Redpath GT, Carpenter JE, Pieper RO, Hussaini IM (2008b) H-Ras increases urokinase expression and cell invasion in genetically modified human astrocytes through Ras/Raf/MEK signaling pathway. Glia 56(8):917–924

Zhao Y, Xiao A, diPierro CG, Carpenter JE, Abdel-Fattah R, Redpath GT, Lopes MB, Hussaini IM (2010) An extensive invasive intracranial human glioblastoma xenograft model: role of high level matrix metalloproteinase 9. Am J Pathol 176(6):3032–3049

Zheng X, Jiang F, Katakowski M, Kalkanis SN, Hong X, Zhang X, Zhang ZG, Yang H, Chopp M (2007) Inhibition of ADAM17 reduces hypoxia-induced brain tumor cell invasiveness. Cancer Sci 98(5):674–684

Zheng X, Jiang F, Katakowski M, Lu Y, Chopp M (2012) ADAM17 promotes glioma cell malignant phenotype. Mol Carcinog 51(2):150–164

Zheng X, Chopp M, Lu Y, Buller B, Jiang F (2013) MiR-15b and miR-152 reduce glioma cell invasion and angiogenesis via NRP-2 and MMP-3. Cancer Lett 329(2):146–154

Zhou Y, Zhang J, Liu Q, Bell R, Muruve DA, Forsyth P, Arcellana-Panlilio M, Robbins S, Yong VW (2005) The chemokine GRO-alpha (CXCL1) confers increased tumorigenicity to glioma cells. Carcinogenesis 26(12):2058–2068

Zincircioglu SB, Kaplan MA, Isikdogan A, Cil T, Karadayi B, Dirier A, Kucukoner M, Inal A, Yildiz I, Aggil F, Donmez O, Urakci Z, Pekkolay Z, Firat U (2012) Contribution of low-molecular weight heparin addition to concomitant chemoradiotherapy in the treatment of glioblastoma multiforme. J BUON 17(1):124–127

Zou Y, Cao Y, Yue Z, Liu J (2013) Gamma-secretase inhibitor DAPT suppresses glioblastoma growth via uncoupling of tumor vessel density from vessel function. Clin Exp Med 13(4): 271–278

Part III

Interactions: Concepts and Tools in Glioma Research

Experimental Models of Glioma

13

Paolo Malatesta, Filippo Calzolari, and Irene Appolloni

Abstract

The development and refinement of animal models of gliomagenesis has been fundamental to test hypotheses concerning the etiology of gliomas and their molecular and cellular pathogenesis. During the last few decades, modeling has gained in complexity and is nowadays mostly relying on the cell type-specific modulation of the expression of candidate oncogenes and oncosuppressors. Despite such technological advances, the recent appreciation of the molecular heterogeneity underlying human high-grade glioma variability revealed the need for a deeper characterization of the available models. It is now clear that most of the existing animal systems mimic one of the human molecular classes, known as "proneural," leaving the other groups underrepresented. While there is thus the need for an expansion of the range of available models, existing ones have already proven useful as translational research platforms, allowing preliminary assessment of the efficacy of classical and innovative therapeutic approaches. In this contribution, we provide a general view of the field and synthesize our understanding of the biology of the most thoroughly studied model family, that of platelet-derived growth factor (PDGF)-induced gliomas.

P. Malatesta (✉)
Department of Experimental Medicine (DIMES), University of Genoa, Genoa, Italy

IRCCS AOU San Martino – IST, Genoa, Italy
e-mail: paolo.malatesta@unige.it

F. Calzolari (✉)
Institute for Stem Cell Research, National Research Center for Environment and Health, Neuherberg, Germany
e-mail: filippo.calzolari@yahoo.it

I. Appolloni
IRCCS AOU San Martino – IST, Genoa, Italy

A. Sedo and R. Mentlein (eds.), *Glioma Cell Biology*,
DOI 10.1007/978-3-7091-1431-5_13, © Springer-Verlag Wien 2014

Keywords
Glioma • RTK signaling • Oncogene • Tumor suppressor • Orthotopic transplant • Retrovirus • Neural stem cells

Abbreviations

bFGF	basic fibroblast growth factor
CSF	Colony-stimulating factor
EGF	Epidermal growth factor
EGFR	EGF receptor
GFAP	Glial fibrillary acidic protein
GIC	Glioma-initiating cells
NF1	Neurofibromin-1
OPC	Oligodendrocyte progenitor cell
PDGF	Platelet-derived growth factor
PDGFR	PDGF receptor
PI3K	Phosphatidylinositide-3-kinase
PTEN	Phosphatase and tensin homolog deleted from chromosome 10
Rb	Retinoblastoma protein
RG	Radial glia
RGC	Radial glia cell
RTK	Receptor tyrosine kinase

13.1 Animal Models

Finding the "right" animal model often represents the most critical aspect for research in life science. When very basic biological phenomena are investigated, the "right" model can be in principle any organism, provided not to be unlucky enough to hit on an exception to the general rule. When it is a specific human disease to be modeled, the figure is quite different. First, the disease might be due to a peculiarity of our species, shared by few or no other animal. More subtly, the model might resemble some aspects of the human pathology but differ on others, depending on the intrinsic differences between species. Consequently, there does not exist a "perfect" animal model of a disease, and it is rather better to rely on different models, each one mimicking a specific aspect and suitable for a specific aim.

In the case of gliomas, models have been used to investigate, among other things, the molecular insults that may be responsible for tumor initiation or progression, the identity of the cells of origin, and how tumor cells could be specifically targeted while leaving healthy cells unarmed. Different subjects are better inquired by using different models: in some cases, employing a glioma model generated by introducing the same molecular lesion found in patients might alone provide information on the process of gliomagenesis; in other cases, having the actual patient's cells xenotransplanted in an animal could be necessary. In other occasions, a faithful recapitulation of histopathological features may be more important than achieving high accuracy in reproducing the molecular etiology.

The main subdivision in animal models is between those relying on the xenotransplant of patient-derived cells and those based on in vivo manipulations such as carcinogen administration or the aberrant expression of specific genes by germ-line transgenesis or by somatic gene transfer. Xenotransplants are particularly indicated in preclinical assays of new therapies, representing a fundamental extension of in vitro studies, allowing the accessibility of glioma cells to drugs or treatments to be addressed under conditions that are more realistic. Xenotransplants are also used to assess the relevance of molecular pathways for the maintenance of cell autonomous malignant traits, such as the ability to generate secondary tumors and to invade the brain tissues as well as for the analysis of the interactions between glioma cells and the local microenvironment. Especially in the latter cases, the possible differences between the human and the immunodeficient mouse host impose caveats in the interpretation of the results.

Syngeneic models are better suited to study gliomagenesis and tumor progression. The possibility to perform cell type-specific genetic alterations allows analyzing the role of candidate glioma-inducing insults while concomitantly addressing the potential for different target cells to act as the cell of origin of these tumors. Germ-line transgenic animals can also be used to study long-term predisposition to gliomagenesis. All these aspects are particularly important since the early phases of glioma development are poorly understood and are likely to depend on factors such as the initiating molecular alterations and the targeted cell type. While a better understanding of the early steps of human gliomagenesis would likely provide avenues for intervention against later-stage gliomas, access to such information is limited, because gliomas are commonly diagnosed when already progressed, thus presenting with a complex set of molecular alterations reflecting their history.

Syngeneic models can also be used to study the interplay between tumors and the immune system, which in xenotransplants is complicated by the need to reconstitute part of the immune system of an immunodeficient animal.

In the next sections, an overview of the different models of glioma will be provided followed by the detailed description of the popular and diverse family of proneural glioblastoma models based on aberrant platelet-derived growth factor (PDGF) signaling.

13.2 Xenotransplants

Xenotransplantation of human-derived glioma cells in rodents was first performed in the mid-1940s, using the Guinea pig anterior eye chamber as a transplantation site. Although these pioneer studies gave some interesting results, showing a correlation between grafting success and histopathological grade or patient's prognosis (Greene 1952; Krementz and Greene 1953), it became more commonly used only following the discovery of athymic immunodeficient mouse strains (Flanagan 1966) and the establishment of human glioma cell lines (Ponten and Macintyre 1968). Xenotransplants have been extensively used to test anticancer drugs, with mixed outcomes: until 1994, different drugs that proved somehow effective against xenotransplanted tumors such as BCNU and PCNU (Slagel et al. 1982; Houchens et al. 1983) were demonstrated to lack efficacy when tested in trials on human patients (Eyre et al. 1986; Dinapoli et al. 1993), showing the limits of those models. On the other side, xenotransplants have been fundamental in the development of temozolomide, the only compound showing a positive impact, however minimal, on glioblastoma patients' prognosis (Plowman et al. 1994).

One of the strongest limitations in xenotransplanting cell lines is their limited ability to recapitulate the pattern and extent of parenchymal infiltration observed in patients (Finkelstein et al. 1994; Mahesparan et al. 2003). Cells from established glioma lines tend to generate compact tumor masses and migrate as chains rather than individually as cells in high-grade gliomas. This might obviously impose bias in the responsiveness to drugs, since the accessibility of tumor cells infiltrating a healthy parenchyma may be very different from that of a cell belonging to a compact tumor mass.

More recently, methods to establish and maintain cultures of high-grade glioma-initiating cells (hGICs) from patients have been derived from protocols originally designed to culture neural stem cells (Ignatova et al. 2002; Singh et al. 2003). These protocols share the use of growth factors such as epidermal growth factor (EGF) and basic fibroblast growth factor (bFGF) and the lack of bovine serum. hGICs can be transplanted in immunodeficient mice, giving rise to tumors that strongly resemble the parental gliomas under many respects, including the highly infiltrative behavior (Galli et al. 2004).

One of the reasons for the popularity of hGICs was the observation that their gene expression profile does not change with culture passaging, unlike glioma cells maintained in serum-containing medium (Lee et al. 2006). After several passages hGIC were more similar to freshly derived glioma cells in comparison to those maintained in serum-containing medium, whose phenotype seemed to converge on those of established human glioma lines (Lee et al. 2006). The closer resemblance of primary tumors to serum-free cultured hGICs rather than to glioma cells cultured in serum and established glioma lines was further corroborated by analyses of genomic alterations induced by or selected for under different conditions. Single nucleotide polymorphism (SNP) array analyses showed that established glioma cell lines share many alterations with different non-glioma cell lines rather than with primary gliomas, suggesting they could represent culturing artifacts (Li et al. 2008). Along the same line, comparative genomic hybridization (CGH) analyses showed that the alterations found in primary tumors are poorly represented in glioma cell lines and serum-cultured glioma patient cells while they are maintained in serum-free hGIC conditions (Ernst et al. 2009).

13.3 Syngenic Models

Studies performed during the 1970s on the effects of mutagenesis induced in rats by nitroso-amide compounds (N-nitroso-N-ethylurea NEU and N-nitroso-N-methylurea NMU) led to the development of several models of endogenously generated CNS tumors. Unexpectedly, glioma-like lesions developed selectively because of the administration of the compound to pregnant rats or pups (Druckrey et al. 1966; Jones et al. 1973) and could be cloned and propagated in vitro. The extensive research performed on a handful of cell lines belonging to this class of models resulted in more than 5,000 publications (Table 13.1). These often pleiomorphic lines are all characterized by cells with glia-like phenotypes, capable of generating secondary gliomas either with sharp margins and little invasion, as is the case for C6 (Benda et al. 1968) and 9L (Benda et al. 1971), or with a highly infiltrative, glioblastoma-resembling, growth pattern, exemplified by BT4C (Laerum and Rajewsky 1975), CNS-1 (Kruse et al. 1994), F98, and RG2 (Ko et al. 1980).

Rat cell lines have been used to investigate a variety of issues. C6, the most popular, as well as 9L, has been employed to assay imaging techniques (Ikezaki et al. 1994; Yuan et al. 2006), strategies of drug delivery (Zhang et al. 2013a, b), the effects of the blood–brain barrier on drug distribution (Warnke et al. 1987), angiogenesis (Valable et al. 2008), and the efficacy of candidate therapies, spanning from traditional chemotherapy (Bencokova et al. 2008; Sheehan et al. 2008) to less conventional approaches like immunotherapy (Iwadate et al. 2005; Tanriover et al. 2008) and oncolytic virotherapy (Yang et al. 2004). It should however be considered that both C6 and 9L are strongly immunogenic (Parsa et al. 2000), and this imposes strong limitations to their use in assays of therapeutic efficacy. The outstanding results obtained in some studies, often including the cure of the animals, could later be attributed to the effects of the immunogenicity of the lines rather than to the efficacy of the therapy itself. This issue is less relevant in the case of BT4C, CNS-1, F98, and RG2, which are scarcely immunogenic and can be transplanted in the syngeneic inbred strain. It is known, however, that various treatments, such as exposure to interferon gamma, might increase the immunogenicity of some of them (RG2) reducing their lethality (Oshiro et al. 2001).

Due to the resemblance to human glioblastoma, F98, BT4C, and CNS-1 have been employed in a number of studies focusing on glioma invasion (Owens et al. 1998) and tumor angiogenesis (Zhang et al. 2009) and to assess the efficacy of combined chemo- and radiotherapy (Biston et al. 2004; Sandstrom et al. 2008), boron neutron capture therapy (Barth et al. 2003), and anti-angiogenetic therapy (Huszthy et al. 2006). For a detailed review about these aspects, see (Barth and Kaur 2009).

The availability of reliable techniques to manipulate genetic loci in the mouse, combined with the CRE/lox site-specific recombination techniques and somatic gene transfer abilities, led to the generation of a wealth of new glioma models in the mouse that will be discussed in more detail in the next paragraphs and summarized in Table 13.2.

Table 13.1 Summary of the most popular rat glioma lines

Name	Year	Mutagen	Exposition to mutagen	Description	Rat strain	Published articles at 2013	Reference (PMID)
C6	1968	MNU	Adult ip	Glioma, immunogenic	Wistar	4,201	Benda et al. (1968)
9L	1971	MNU	Adult iv	Gliosarcoma, noninvasive, immunogenic	Fisher344	815	Benda et al. (1971)
F98	1971	ENU	Fetal	Undifferentiated glioma, invasive, non-immunogenic	Fisher344	246	Ko et al. (1980)
RG2	1971	ENU	Fetal	GBM-like, malignant, invasive, non-immunogenic	Fisher344	52	Ko et al. (1980)
BT4C	1975	ENU	Fetal	High-grade glioma, angiogenic, infiltrative, non-immunogenic	BD IX	67	Laerum and Rajewsky (1975)
CNS1	1994	MNU	Vein adult	GBM-like, invasive, non-immunogenic	Lewis	28	Kruse et al. (1994)

These models have been very useful to assess the role of molecular alterations commonly found in human gliomas, thus representing models for the induction and progression of the pathology. Many of the alterations tested, like the enhancement of EGFR, PDGFR, and NF1/Ras pathways, have been later recognized to be characteristic lesions of the human glioma subtypes (classical, proneural, and mesenchymal) recently emerged from microarray data analyses (Verhaak et al. 2010). Other modifications, like the loss of tumor suppressors PTEN, Trp53, and p16^{Ink4A}/p19^{ARF}, are frequent co-mutations of the previous alterations. Indeed, in samples derived from patients, molecular alterations are usually found in combination, providing little information about the order in which they appeared.

The development of models where single molecular lesions can be introduced in a controlled manner has been important to appreciate the role of each alteration in the process of gliomagenesis. This is, for example, the case of the loss of function of Trp53 and NF1, which, as it will be detailed below, have dramatically different effects depending on the order in which they occur (Zhu et al. 2005).

Although the introduction of single molecular alterations is generally thought not to be sufficient to generate gliomas, there are exceptions, as in the case of the overexpression of PDGF-B in neonatal or embryonic neural progenitors (Uhrbom et al. 1998; Calzolari et al. 2008), the over-activation of the Ras pathway (Ding et al. 2001; Bajenaru et al. 2003), or the tissue-specific disruption of Trp53 (Zheng et al. 2008; Wang et al. 2009). In all these cases, however, cells reach full malignancy only by undergoing further alterations. This is indirectly shown by the slow acquisition of the ability to generate new tumors when transplanted (Calzolari et al. 2008), by the consistent finding of additional alterations in the

Table 13.2 Summary of the main syngeneic mouse models

Involved pathways	Type	Grade	Symptom onset (weeks)	Penetrance (%)	Genotype	Treatment	Time of treatment	Paper endnote
RTK	Oligo	3	14–29	40	wt	MMLV-PDGF B	Neonatal	Uhrbom et al. (1998)
RTK	Oligo	2–3	12[a]	60	*Nestin prom*-tv-A	RCASa-PDGF B	Neonatal	Dai et al. (2001)
RTK	Oligo/Mix	2–3	12[a]	40	*hGFAP prom*-tv-A	RCASa-PDGF B	Neonatal	Dai et al. (2001)
RTK	Oligo	2	8–32	60	*S100 prom*-vErbB	None		Weiss et al. (2003)
RTK	Ast	3–4	30–55	40	FigROS flxSTOP	Adenovirus-CRE	Adult	Charest et al. (2006)
RTK	Oligo	1–4	4–28	100	wt	MMLV-PDGF B	Embryonic	Calzolari et al. (2008)
RTK INK/ARF	Oligo	1–2	8–10	52	*Nestin prom*-tv-A Ink/Arf(wt/−)	RCASa-EGFRvIII	Neonatal	Holland et al. (1998)
RTK INK/ARF	Oligo	1–2	8–10	42	*Nestin prom*-tv-A Ink/Arf(−/−)	RCASa-EGFRvIII	Neonatal	Holland et al. (1998)
RTK INK/ARF	Oligo	3–4	12[a]	57	*Nestin prom*-tv-A Ink/Arf(−/−)	RCASa-PDGF B	Neonatal	Dai et al. (2001)
RTK INK/ARF	Oligo/Mix	3–4	12[a]	70	*hGFAP prom*-tv-A Ink/Arf(−/−)	RCASa-PDGF B	Neonatal	Dai et al. (2001)
RTK INK/ARF	Ast	3–4	4–8	100	Ink/Arf(−/−)	Transplant of ex vivo MMLV-EGFRvIII transduced astrocytes	Subadult	Bachoo et al. (2002)
RTK INK/ARF	Ast	3–4	4–8	100	Ink/Arf(−/−)	Transplant of ex vivo MMLV-EGFRvIII transduced NSC	Subadult	Bachoo et al. (2002)
RTK INK/ARF	Oligo	3	4–24	90	*S100 prom*-vErbB Ink/Arf(−/−)	None	–	Weiss et al. (2003)
RTK INK/ARF	Ast	3–4	8–40	100	FigROS flxSTOP Ink/Arf(wt/−)	Adenovirus-CRE	Adult	Charest et al. (2006)

(continued)

Table 13.2 (continued)

Involved pathways	Type	Grade	Symptom onset (weeks)	Penetrance (%)	Genotype	Treatment	Time of treatment	Paper endnote
RTK INK/ARF	Ast	3–4	8–40	100	FigROS flxSTOP Ink/Arf(−/−)	Adenovirus-CRE	Adult	Charest et al. (2006)
RTK INK/ARF	Gbm	4	5–22	100	*Col1a prom*-EGFR1 flxSTOP Ink/Arf(−/−)	Lentiv-Tgf-α-(IRES)-CRE	Adult	Acquaviva et al. (2011)
RTK AKT–PTEN INK/ARF	Gbm	4	5–10	100	*Col1a prom*-EGFRvIII flxSTOP PTEN(flx/flx) Ink/Arf(−/−)	Adenovirus-CRE	Adult	Zhu et al. (2009)
RTK AKT–PTEN INK/ARF	Gbm	4	7–13	100	*Col1a prom*-EGFR1 flxSTOP *Col1a prom*-EGFRvIII flxSTOP PTEN(flx/flx) Ink/Arf(−/−)	Adenovirus-CRE	Adult	Zhu et al. (2009)
RTK Trp53	Oligo	3	4–24	90	*S100 prom*-vErbB Trp53(wt/−)	None	–	Weiss et al. (2003)
NF1–RAS	Ast	2–3	8–24	ND	*hGFAP prom*-HrasV12	None	–	Ding et al. (2001)
NF1–RAS	Ast	1–2	32–52	100	Nf1(flx/−) *hGFAP prom*-CRE	None	–	Bajenaru et al. (2003)
RTK NF1–RAS	Oligo/Mix	3	2–13	100	*hGFAP prom*-EGFRvIII *hGFAP prom*-HrasV12	None	–	Ding et al. (2003)

RTK NF1–RAS	Ast/Mix	2–4	12	95	*hGFAP prom-* HrasV12	Adenovirus-EGFRvIII	Subadult	Wei et al. (2006)
NF1–RAS Trp53	Ast	2–4	15–55	30–75	Nf1(wt/–) Trp53(wt/–)	None	–	Reilly et al. (2000)
NF1–RAS Trp53	Ast	2–4	25–45	100	Nf1(wt/flx) Trp53(wt/–) *hGFAP prom-* CRE	None	–	Zhu et al. (2005)
NF1–RAS Trp53	Ast	2–4	10–20	100	Nf1(flx/flx) Trp53(–/–) *hGFAP prom-* CRE	None	–	Zhu et al. (2005)
NF1–RAS Trp53	Ast	2–4	25–45	100	Nf1(flx/wt) Trp53(flx/–) *hGFAP prom-* CRE	None	–	Kwon et al. (2008)
NF1–RAS Trp53	Ast/Oligo	3–4	18–26	97	Nf1(wt/–) Trp53(flx/flx) *hGFAP prom-* CRE	None	–	Wang et al. (2009)
NF1–RAS AKT–PTEN	Ast	4	9	26	*Nestin prom*-tv-A	RCASa-Kras(G12D) RCASa-AKT	Neonatal	Holland et al. (2000)
NF1–RAS AKT PTEN	Gbm	4	12[a]	60	*Nestin prom*-tv-A PTEN(flx/flx)	RCASa-Kras(G12D) RCASa-CRE	Neonatal	Hu et al. (2005)
NF1–RAS AKT–PTEN	Ast/Mix	3	6	100	*hGFAP prom-* HrasV12 PTEN(flx/wt)	Adenovirus-CRE	Subadult	Wei et al. (2006)
NF1–RAS AKT–PTEN	Ast	3–4	16–20	42	*hGFAP prom-* CRE	Lentiv-HrasV12 flxSTOP Lentiv-AKT flxSTOP	Adult	Marumoto et al. (2009)

(continued)

Table 13.2 (continued)

Involved pathways	Type	Grade	Symptom onset (weeks)	Penetrance (%)	Genotype	Treatment	Time of treatment	Paper endnote
NF1–RAS AKT–PTEN Trp53	Ast	3–4	12–24	100	Nf1(flx/wt) PTEN(flx/wt) Trp53(wt/–) *hGFAP prom*-CRE	None	–	Kwon et al. (2008)
NF1–RAS AKT–PTEN Trp53	Ast	3–4	11–24	100	Nf1(flx/wt) PTEN(flx/wt) Trp53(flx/–) *hGFAP prom*-CRE	None	–	Kwon et al. (2008)
NF1–RAS AKT–PTEN Trp53	Ast	4	22	100	*hGFAP prom*-CRE Trp53(wt/–)	Lentiv-HrasV12 flxSTOP Lentiv-AKT flxSTOP	Adult	Marumoto et al. (2009)
NF1–RAS AKT–PTEN Trp53	Ast	3–4	11–24	100	Nf1(flx/wt) PTEN(flx/wt) Trp53(flx/flx) *Nestin prom*-CRE-ERT	Tamoxifen injection	Embryonic	Alcantara Llaguno et al. (2009)
NF1–RAS AKT–PTEN Trp53	Ast	3–4	28–56	100	Nf1(flx/wt) PTEN(flx/wt) Trp53(flx/flx) *Nestin prom*-CRE-ERT	Tamoxifen injection	Subadult	Alcantara Llaguno et al. (2009)
NF1–RAS AKT–PTEN Trp53	Ast	3–4	11–24	100	Nf1(flx/wt) PTEN(flx/wt) Trp53(flx/flx)	Adenovirus-CRE in SVZ	Neonatal, subadult and adult	Alcantara Llaguno et al. (2009)

NF1–RAS AKT–PTEN INK/ARF	Ast	4	12	49	*Nestin prom* tv-A Ink/Arf(–/–)	RCASa-Kras(G12D) RCASa-AKT	Neonatal	Uhrbom et al. (2002)
NF1–RAS AKT–PTEN INK/ARF	Ast	4	12	42	*hGFAP prom*-tv-A Ink/Arf(–/–)	RCASa-Kras(G12D) RCASa-AKT	Neonatal	Uhrbom et al. (2002)
AKT–PTEN Trp53	Ast/Gbm	3–4	15–50	88	PTEN(flx/wt) Trp53(flx/flx) *hGFAP prom*-CRE	None	–	Zheng et al. (2008)
AKT–PTEN Trp53	Ast	3–4	12–54	100	PTEN(flx/flx) Trp53(flx/flx) *hGFAP prom*-CRE-ERT	Tamoxifen injection	Adult	Chow et al. (2011)
Rb	Ast	2	3–4	100	*mouseGFAP prom*-SV40 Tlarge	None	–	Danks et al. (1995)
Rb	Ast/Oligo	2–3	4–50	100	*hGFAP prom*-T121	None	–	Xiao et al. (2002)
RTK Rb Trp53	Oligo	1–2	8–10	42	*Nestin prom*-tv-A Ink/Arf(–/–)	RCASa-EGFRvIII RCASa-Cdk4	Neonatal	Holland et al. (1998)
RTK Rb Trp53 INK/ARF	Oligo	2–3	4–5	40	*Nestin prom* tv-A Trp53(wt/–) Ink/Arf(–/–)	RCASa-EGFRvIII RCASa-Cdk4	Neonatal	Holland et al. (1998)
AKT–PTEN Rb	Ast/Oligo	2–3	4–26	100	*hGFAP prom*-T121 PTEN(wt/–)	None	–	Xiao et al. (2002)

(continued)

Table 13.2 (continued)

Involved pathways	Type	Grade	Symptom onset (weeks)	Penetrance (%)	Genotype	Treatment	Time of treatment	Paper endnote
AKT–PTEN Rb	Ast	3	ND	100	*hGFAP prom*-T121 PTEN(flx/flx)	MSCV-CRE	Subadult	Xiao et al. (2005)
AKT–PTEN Rb Trp53	Ast	3–4	12–31	100	RB(flx/flx) PTEN(flx/flx) Trp53(flx/flx) *hGFAP prom*-CRE-ERT	Tamoxifen injection	Adult	Chow et al. (2011)
Trp53	Ast	3	30–65	17	*hGFAP prom*-CRE Trp53(flx/flx)	None	–	Zheng et al. (2008)
Trp53	Ast/Oligo/Medullo	3–4	22–52	85	*hGFAP prom*-CRE Trp53(flx/flx)	None	–	Wang et al. (2009)

[a]Animals were killed independently of neurological symptoms. "Subadult" indicates an age between 2 and 6 weeks. "flx" indicates alleles flanked by loxP sites. "flx-STOP" following the name of a gene indicates that gene is preceded by a transcriptional/translational stop cassette flanked by loxP sites

full-blown tumors (Ding et al. 2001; Wang et al. 2009), and by the increase of the degree of malignancy in models where the original mutation is accompanied by further molecular damages (Dai et al. 2001; Zheng et al. 2008; Wang et al. 2009).

13.3.1 Models Obtained by the Over-Activation of the NF1/RAS Pathway

Although Ras mutations are not found in glioma patients, over-activation of the Ras pathway is one of the most common features in human gliomas. The pathway is usually overstimulated either by misregulated receptor tyrosine kinase (RTK) signaling or by the loss of its inhibitor neurofibromin-1 (NF1). The Ras pathway impinges on three signaling branches regulating cell proliferation, survival, and invasion by acting on the phosphatidylinositide-3-kinase PI3K/AKT, Raf/MEK/ERK, and Rac1/Rho axes. To model the activation of the Ras pathway, researchers have either overexpressed constitutively active forms of Ras (Holland et al. 2000; Ding et al. 2001; Uhrbom et al. 2002; Hu et al. 2005; Wei et al. 2006; Marumoto et al. 2009) or knocked out NF1 (Reilly et al. 2000; Bajenaru et al. 2003; Zhu et al. 2005; Kwon et al. 2008; Alcantara Llaguno et al. 2009; Wang et al. 2009).

The expression of the constitutively active form HRasV12 under control of the human GFAP promoter (hGFAP) was able to induce gliomas of different grades of malignancy and very high penetrance (>95 %) in a dose-dependent manner (Ding et al. 2001). Multiple copies of the construct did not undergo germ-line transmission because the chimeric mice died within 2 weeks with glioblastoma-like lesions. Germ-line transmission was however possible with single copy insertion in both heterozygosis and homozygosis. In the first case, mice tended to develop mostly grade II astrocytomas with a median survival time of 3 months while the latter developed grade III astrocytomas with a median survival time of 4 weeks. These observations appear in contrast with those reported by other authors that failed to induce glioma by expressing HRasV12 alone in adult mice (Marumoto et al. 2009). It should be noted, however, that the timing of HRasV12 expression, and therefore the identity of targeted cells, differed between these experiments. When driven from the hGFAP (glial fibrillary acidic protein) promoter, HRasV12 was expressed starting from embryonic day 13 in radial glial cells. In the other study, HRasV12 was inserted at a later stage so that its activation, although controlled by the same promoter, was possible only in some astrocytes, including the stem cells of the subventricular zone (Nolte et al. 2001; Hack et al. 2004). Moreover, the difference could be partly explained by the difference in the number of targeted cells, which, in the somatic gene transfer paradigm, are few tens per brain, while they constitute the majority of cells in the germ-line transgenic model. It is also important to note that the transduction of the very same small number of cells resulted in the induction of gliomas when both HRasV12 and AKT were transduced and the penetrance was further increased to about 100 % when one Trp53 allele was disrupted (Marumoto et al. 2009).

Interestingly, simultaneous hGFAP promoter-driven expression of HRasV12 and a mutated form of EGF receptor (EGFR), EGFRvIII in a germ-line transgenic mouse line, drove faster development of high-grade gliomas, often possessing oligodendroglioma characteristics (Ding et al. 2003). When EGFRvIII was added to HRasV12 expressing cells at a later time, by adenoviral transduction, the shift toward the oligodendroglial lineage was less frequently observed, suggesting that not only the genetic alteration but also the identity of the targeted cell is fundamental to determine the tumor subtype (Wei et al. 2006). The majority of glioblastoma-like (grade IV) lesions originating after adenoviral transduction of EGFRvIII in cells expressing HRasV12 ended up lacking PTEN expression, and conversely, when PTEN was abrogated in HRasV12 expressing mice, EGFR1 was invariably upregulated (Wei et al. 2006), showing a strong correlation between these genetic alterations, similarly to the condition found in tumors from glioma patients.

When similar alterations (i.e., expression of KRasD12 and AKT) were operated by using an avian retrovirus (RCAS-A) in mouse lines expressing the RCAS receptor under control of either the Nestin (N-tva) or the hGFAP promoter (G-tva), gliomas were generated only by the Nestin-positive progenitors (Holland et al. 2000; Hu et al. 2005). The authors interpreted these observations as an indication that terminally differentiated astrocytes are less prone than immature neural progenitors to give rise to gliomas. This interpretation is however challenged by the finding that neonatal stem cells do express hGFAP and potentially represent an even less differentiated target than nestin-expressing progenitor cells. Most likely the lack of effects of somatic gene transfer in Gtv-a is due to the fact that the injections were performed far from the SVZ, where neural stem cells are located, and GFAP-expressing astrocytes are still relatively rare in newborn mouse parenchyma. This alternative interpretation is corroborated by recent observations suggesting that, in a Trp53 loss-of-function condition, differentiated astrocytes can dedifferentiate and give rise to gliomas following over-activation of the Ras pathway (Friedmann-Morvinski et al. 2012).

Although NF1 knockout acts on the same pathway as Ras-activating mutations, its effects appear more limited and cause gliomas only when additional mutations predispose cells to tumor development. As expected, these mutations are the same that have been demonstrated to cooperate with Ras mutations, such as PTEN loss or Trp53 disruption. Interestingly, the experiments demonstrating cooperation between NF1 and Trp53 loss also showed that the order in which these alterations occur significantly affects tumor development. By wisely exploiting the chromosomal localization of NF1 and Trp53 in the mouse genome, Zhu and collaborators showed that if NF1 loss occurs before the complete loss of Trp53, gliomas are not formed (Zhu et al. 2005). This observation is in contrast with what is seen in other tumors, including colon carcinoma, where Trp53 loss is a late event (Kinzler and Vogelstein 1996). The need for an early mutation of Trp53 may reflect the impact of processes such as oncogene-induced senescence (OIS) or apoptosis on early phases of tumorigenesis, whereby NF1-deficient cells could commit to apoptotic death or permanent cell cycle exit in the presence of unaltered Trp53. The observation that overexpression of HRas does not need to be anticipated by Trp53 loss to induce high-grade gliomas may be explained by NF1 loss and HRas activation triggering partially distinct signaling pathways or by subtler effects on the signaling dynamics of a common set of effectors, resulting in differential activation of Trp53-mediated cellular responses.

13.3.2 Models Obtained by Over-Activation of EGFR1

Among the first molecular defects modeled by somatic gene transfer in mouse there is the over-activation of EGFR1.

In human glioma, EGFR1 is over-activated either by amplification of the wild-type form or by mutations making it constitutively active. The most common of these mutated forms is the EGFRvIII consisting of a 267-amino acid deletion within the extracellular portion. Both these kinds of alterations have been reproduced in mice often by somatic gene transfer or, in alternative, by germ-line transgenesis followed by conditional activation driven by CRE-mediated recombination or via the activity of tissue-specific promoters.

With the exception of the overexpression of v-erbB, a viral EGFR1 homologue, alone capable of inducing the formation of low-grade oligodendroglioma-like lesions (Weiss et al. 2003), other means of forcing the over-activation of EGFR pathway were not sufficient to trigger the formation of gliomas. In contrast with PDGFRalpha-driven pathways, which will be extensively discussed below, EGFR1 over-activation induces gliomas only when occurring together with other alterations, such as the loss of a tumor suppressor like $p16^{Ink4A}/p19^{ARF}$ or Trp53. These additional alterations cooperate with the over-activation of EGFR1: in the case of the overexpression of v-erbB, they increase the grade of the resulting lesion and shorten the survival of mice (Weiss et al. 2003), while in the other cases they allow the formation of gliomas.

The first glioma models based on the over-activation of EGFR1-driven pathways date back to 1998 and were obtained by somatic transfer of EGFRvIII in $p16^{Ink4A}/p19^{ARF}$-deficient Ntv-a and Gtv-a mouse lines using the avian retrovirus RCAS-A (Holland and Varmus 1998). Tumors showed features resembling mainly low-grade (I/II) gliomas and expressed GFAP and Nestin. In the same study, Holland and colleagues failed to observe gliomagenic cooperation between EGFRvIII overexpression and the loss of Trp53, which was achieved only following Cdk4 overexpression. This result is not surprising, considering that Cdk4 is one of the main negative targets of p16 and that Trp53 and p19 act along the same molecular axis. The simultaneous inactivation of Trp53 and overexpression of Cdk4 can be predicted to reconstruct molecular conditions similar to those induced by the loss of function of $p16^{Ink4A}/p19^{ARF}$.

Similarly to what was described above for KRas, gliomas were induced more efficiently in Ntv-a than in Gtv-a mice. As already noted, however, the interpretation of this observation is not obvious, in the light of what we currently know about cell-lineage relationship between the cells activating the hGFAP and the nestin promoters. Moreover, other reports have later shown that $p16^{Ink4A}/p19^{ARF}$-defective neonatal astrocytes transduced in vitro with EGFRvIII efficiently induced high-grade gliomas when grafted in Nod/scid mice (Bachoo et al. 2002). This observation indicates that astrocytes are prone to EGFRvIII transformation, as are the nestin-positive cells of the Ntv-a mouse model. It is however possible that the Nod/scid background represents a more permissive environment for gliomagenesis, compared to the background used by Holland and colleagues.

An efficient induction of high-grade gliomas in immunocompetent mouse lines was obtained when the loss of $p16^{Ink4A}/p19^{ARF}$ was accompanied by that of PTEN (Zhu et al. 2009). These results were obtained by employing CRE recombinase-expressing adenoviruses in mouse lines defective for $p16^{Ink4A}/p19^{ARF}$, homozygous for a PTENflox/flox allele, and carrying EGFRvIII as an allele conditionally activated by the removal of a floxed transcriptional/translational STOP cassette (Zhu et al. 2009). In these conditions, 100 % of the treated animals developed high-grade gliomas in 5–10 weeks. It is important to note that, in the same conditions (loss of $p16^{Ink4A}/p19^{ARF}$ and PTEN), a small but significant proportion (6 %) of high-grade gliomas were generated also by the overexpression of the EGFR1 wild-type allele, although via the activation of a different downstream effector: mTorc1 rather than mTorc2 which is activated by EGFRvIII (Zhu et al. 2009).

Thus, although common in human gliomas, the overstimulation of EGFR1-activated pathways is not as efficient at driving gliomagenesis in mice as is the aberrant activation PDGF-B signaling, which will be further discussed below. These differences may be due to peculiarities of the murine nervous system, such as a different composition of the proliferative compartment or different dependencies on alternative growth-promoting stimuli.

13.3.3 Insights into the Biology of Proneural Gliomas: The PDGF-Driven Glioma Paradigm

While it is clear from the above discussion that a broad variety of animal models have been fruitfully exploited to gain insights into the process of gliomagenesis, few systems have been explored and characterized as deeply as the PDGF-driven family of proneural glioma models. While probably far from allowing a truly comprehensive description of proneural gliomagenesis, work done on these systems significantly advanced our understanding of the molecular and cellular events whereby this specific class of gliomas emerges, acquires, and maintains malignant properties. What follows is an attempt to provide a coherent picture of PDGF-driven glioma biology, together with a brief discussion of more translationally oriented approaches using this class of models.

13.3.3.1 Target Cells

Forced PDGF expression can drive gliomagenesis from a variety of cell types, as shown by in vivo experiments performed on embryonic, early postnatal, and adult neural progenitor cells. Radial glia (RG), the neural stem cell type found in the embryonic brain, can be forced to generate gliomas by transduction with a PDGF-overexpressing retrovirus (Appolloni et al. 2009). Similarly, early postnatal neural progenitors, most likely glial-restricted or oligodendrocyte progenitors, can be forced to generate gliomas by similar means (Dai et al. 2001). Most importantly, though, adult white matter oligodendrocyte progenitor cells (OPCs) are efficiently transformed by locally administered PDGF-encoding retroviruses (Assanah et al. 2006).

RG cells are a heterogeneous progenitor cell population (Hartfuss et al. 2001) and are directly or indirectly responsible for most or all neural cell types generated from early brain development (Malatesta et al. 2008). While initially neurogenic, these cells become predominantly gliogenic at late embryonic stages and generate the postnatal and adult neural stem cell (NSC) compartments, even if details of the transition are lacking (Merkle et al. 2004). In utero transduction of PDGF-B in RGCs eventually results in glioma formation, and in vitro data suggest that this may result from the ability of PDGF signaling to affect fate choices in immature RGCs and force them to acquire an oligodendroglial progenitor identity, as opposed to selectively stimulating already committed glial progenitors (Hu et al. 2008a; Appolloni et al. 2009). Such an effect may be exerted indirectly, though, since RGCs do not express PDGFR. It is anyway clear that an early step during RGC-derived PDGF-induced gliomagenesis is the acquisition of oligodendroglial lineage features. When symptoms arise, thus, tumors are very obviously and invariably oligodendroglial in nature (Appolloni et al. 2009). Analogously, models based on the retroviral targeting of early postnatal glial progenitors using PDGF-overexpressing vectors also efficiently result in the generation of gliomas, mostly displaying oligodendroglial features (Fomchenko et al. 2011; Dougherty et al. 2012). The cellular mechanism whereby a glial progenitor population with a mixed oligo/astrocytic potential is forced to generate purely oligodendroglial tumors is not clear but may entail biases in alternative fate choices or selective responses by lineage-restricted progenitors. While the former possibility is indirectly supported by the above-mentioned data (Hu et al. 2008b; Appolloni et al. 2009), some support exists for the latter scenario, as shown by the fact that differences in proliferative responses to PDGF overexpression are evident even between OPCs residing in distinct brain regions (Hill et al. 2013). In the adult, OPCs residing in the white matter are clearly capable of generating gliomas when forced to overexpress PDGF (Assanah et al. 2006), while data are lacking for gray matter OPCs, which differ in their in vivo behavior from their white matter counterparts (Dimou et al. 2008; Simon et al. 2011; Vigano et al. 2013). NSCs, the only other adult neural population harboring a significant fraction of actively proliferating cells under physiological conditions, were suggested to react to aberrant PDGF availability and form glioma-like hyperplastic lesions (Jackson et al. 2006); however later reports excluded that NSCs could be directly responsible for the reaction observed in the initial infusion experiments (Chojnacki et al. 2011). Still, it was clearly shown that an intraventricular infusion of PDGF could broadly inhibit adult neurogenesis, suggesting a link between PDGF signaling and NSC fate choices. Whether such a link would imply direct effects of PDGF signaling on NSC fate choices is now doubtful, based on the lack of PDGFR expression in NSCs and the seeming lack of mixed oligo/neuronal potential for individual NSCs in the adult sub-ependymal zone (Ortega et al. 2013). Far from being settled, though, this issue could be clarified, e.g., by clonal in vivo lineage tracing experiments. In general, in vivo evidence drawn from experiments performed by targeting different classes of neural progenitors supports the view of aberrant PDGF signaling as a powerful oncogenic stimulus.

13.3.3.2 Molecular and Cellular Mechanisms of Initiation, Progression, and Maintenance of Malignancy

Comparisons of murine PDGF-induced gliomas with normal neural cell types have allowed the definition of useful physiological references, with the ultimate goal of understanding gliomas in terms of their progressive deviation from the biology of the normal neural cell types they aberrantly resemble. Translating this approach to characterize available animal models has the potential to establish causative connections between the occurrence of specific molecular alterations and the development of specific cellular and tumoral phenotypes. It is of course essential to keep in mind, though, that the degree of similarity in molecular features has no necessary bearing on the identity of the cell of origin. Explicit comparisons between gene expression profiles of PDGF-induced gliomas and various normal neural cell types identified the former as most closely resembling OPCs (Appolloni et al. 2009; Dougherty et al. 2012). Such comparisons also offered a possible view into the set of molecular features that make OPC-like glioma cells so aggressively unlike normal OPCs. Recent work (Fomchenko et al. 2011; Lei et al. 2011; Dougherty et al. 2012; Helmy et al. 2012) has addressed this issue, by comparing the polysomal mRNA sets found in PDGF-induced proneural glioma cells and that of oligodendroglial cells. While these comparisons have not yet advanced our understanding of glioma biology much beyond the realization that murine proneural glioma cells behave as aberrantly proliferating OPCs, there is hope that a deeper understanding of OPC biology and a concomitant finer temporal and molecular dissection of the process of gliomagenesis will shed light on targetable aspects of brain tumor biology. From a cellular standpoint, aberrant proliferation and dispersal of OPCs represent early and sustained responses to increased PDGF signaling (Assanah et al. 2009; Lei et al. 2011), recapitulating physiological processes normally associated with elevated PDGF exposure, in vitro and in vivo (Calzolari and Malatesta 2010).

Retrovirally mediated PDGF-induced gliomagenesis from adult white matter cells has been shown to be facilitated by acute loss of PTEN, and the process is further accelerated by concomitant loss of Trp53 (Lei et al. 2011). Interestingly, while both PTEN conditional knockout (CKO) and PTEN;Trp53 CKO approaches yielded high-grade tumors closely resembling human proneural gliomas, gene expression signatures differed among the two and allowed subclassification of human proneural gliomas in distinct survival classes, suggesting persisting biological differences despite the acquisition of broadly overlapping cellular and molecular phenotypes. Using the RCAS/tv-a paradigm to force PDGF overexpression in early postnatal glial progenitors also showed facilitation of the tumorigenic process by deletion of p16^{Ink4A}, p19^{Arf}, or both (Tchougounova et al. 2007). It is however unclear to which extent a broadened target population and/or increased progression rates contribute to the observed increases in morbidity.

With this caveat in mind, it is however clear that this approach has proved extremely fruitful, revealing important roles for a variety of factors, among which various components of the cell cycle control machinery, such as CyclinD1 and Cdk4, whose absence limits PDGF-induced glioma progression or prevents it altogether, respectively (Ciznadija et al. 2011). Interestingly, activity of CycD1/Cdk4 was necessary not only within tumor cells, but also in stromal elements, as discussed further below; furthermore, and somehow surprisingly, this required an intact p21cip (Liu et al. 2007), due to its paradoxical positive effect on the nuclear localization and activity of CycD1/Cdk4 (Hukkelhoven et al. 2012; Liu et al. 2014). On the contrary, the cyclin-dependent kinase inhibitor (CDKI) p27 has been shown to play a more canonical oncosuppressor role, its loss promoting PDGF-driven tumorigenesis, likely due to its requirement for the proper execution of double-stranded DNA damage repair (See et al. 2010). Similarly, defects in DNA damage-induced apoptosis and G1/S checkpoint were shown to accelerate tumorigenesis and confer resistance to ionizing radiation (Squatrito et al. 2010). While these and other molecular alterations likely provide selective advantages to tumor cells and thus contribute to progression, they may provide opportunities for therapeutic purposes via exploitation of molecular vulnerabilities that may silently persist or emerge during tumor progression (Chan and Giaccia 2011; Muller et al. 2012; Szczurek et al. 2013). A possible example in this sense is represented by the amplification of the PTEN-targeting microRNA mir26, specifically observed in gliomas with monoallelic deletion of PTEN. Mir26 amplification promotes progression of PDGF-driven gliomas while simultaneously relieving selective pressure that favors inactivation or loss of the second copy of PTEN (Huse et al. 2009), suggesting that mir26 inhibition may acutely expose these gliomas to the oncosuppressive action of PTEN.

Human glioma progression may sometimes consist of the transition through states that are currently considered as distinct molecular/functional classes, e.g., proneural and mesenchymal subtypes (Brennan et al. 2013). Such transitions may be modeled in vivo in PDGF-driven gliomas, as suggested by the altered molecular phenotype and aggressiveness of tumors induced by the concomitant overexpression of PDGF and factors characterizing the mesenchymal subtype such as Stat3 (Doucette et al. 2012) or TAZ (Bhat et al. 2011). It is however noted that in these specific cases it is hard to assess whether the alternative phenotype is the result of a transition from a proneural to a mesenchymal condition or is due to a de novo generation of mesenchymal tumors. Along with the above observations, overexpressing PDGF may be a relevant way to model early steps of gliomagenesis, other than just a tool to generate high-grade gliomas to be used as systems to study the biology of the most aggressive forms. Corroborating the view that these models may provide insights into the mechanisms of progression of human gliomas is the observation that PDGF-driven murine gliomagenesis shows evidence for functional progression from low- to high-grade forms (Calzolari et al. 2008). Tumors induced by in utero targeting of radial glia belong to either an early or a late wave of tumors, the former characterized by low-grade histopathology and lack of tumor-propagating potential and the latter highly reminiscent of glioblastoma and capable of reinitiating tumor growth when orthotopically transplanted. Demonstrating progression, all tumor cells challenged at early time points are equally incapable of reinitiating tumor growth after transplantation. Comparing early- and late-onset tumors will prove valuable to identify genes that are relevant during the various phases of glioma progression, as exemplified by Btg2, a gene whose expression declines as PDGF-induced gliomas progress, and which decreases malignancy when reintroduced in high-grade gliomas (Appolloni et al. 2012b). Approaches such as the testing of candidates emerging from such comparative analyses or investigating the role of known genes have been complemented by unbiased screening approaches aimed at discovering oncosuppressors and oncogenes whose deregulated expression may cooperate with PDGF overexpression in driving proneural gliomagenesis. Most attempts have relied on the use of retroviral insertional mutagenesis, whereby proviral integration into the genome of target cells may cooperate with a driving oncogenic stimulus (Johansson et al. 2004, 2005). By evaluating the enrichment for integration-sites occurrences, various candidate oncosuppressors or oncogenes were identified. Interestingly, at least one of the candidate oncosuppressors found (Btg2) was independently observed to be depleted in fully progressed PDGF-induced gliomas, as opposed to low-grade ones, as mentioned above (Appolloni et al. 2012b). Common features of many of these attempts to address modifiers of tumorigenesis may however be somehow confounding. Specifically, most of the studies were conducted following early postnatal overexpression of PDGF, a phase in which the targeted neural precursor populations would still be heterogeneous and developmentally plastic. This could arguably result in a different set of selective pressures compared to tumorigenesis from already committed, fate-restricted, adult OPCs, the likely most important target population for the initiation of proneural gliomas. Specifically, lineage determinants may affect early steps of tumorigenesis from neural progenitors with multilineage potential, without affecting tumorigenesis from lineage-restricted cells. One example of a factor whose physiological role in promoting alternative cell fates makes it a potential oncosuppressor is the transcription factor Pax6. Involved in regulating neurogenesis at embryonic stages and in the adult NSCs, Pax6 antagonizes oligodendroglial development by opposing Olig2 (Hack et al. 2005; Jang and Goldman 2011). Although not directly proven to act as an oncosuppressor to restrain initiation of oncogenesis, Pax6 strongly limits the tumorigenic potential of high-grade PDGF-induced gliomas, an effect associated with loss of Olig2 expression (Appolloni et al. 2012a).

Based on current knowledge, explicit comparisons of the gene expression profiles of unchallenged OPCs and of low- and high-grade gliomas generated from this cell population are most likely to allow the identification of clinically relevant modifiers of glioma biology.

13.3.3.3 Mechanisms Underlying Malignancy

Furthering our understanding of various aspects of PDGF-driven glioma biology is likely to provide us with improved therapeutic tools to approach proneural gliomas. One crucial aspect, which has long been appreciated in tumors in general, is the phenotypic and, likely, functional heterogeneity of glioma cells. The cellular identity and characteristics of the various cellular subtypes and the possibly corresponding heterogeneous tumorigenic capabilities are, however, still somewhat enigmatic.

PDGF-induced gliomas have been shown to contain a variable minority of tumor cells with features sometimes associated with somatic stem cells, such as the exhibition of xenobiotic-extrusion capacity, and the ability to form floating spheres in culture (Bleau et al. 2009). Markers associated with this phenotype (ABCG2, eNOS) identify at least some tumors cells in close proximity to blood vessels, where a perivascular niche has been proposed to maintain glioma-propagating cells. Such "side population" character of glioma cells has been shown to be under the control of Akt and Notch signaling and to correlate with stronger tumorigenic potential following transplantation (Bleau et al. 2009; Charles et al. 2010). Whether this feature, per se, confers higher aggressiveness to glioma cells following transplantation is currently unknown.

In general, our understanding of the interrelationships among cellular subtypes within PDGF-induced gliomas is still very limited and mostly restricted to in vitro data using culture systems that are unlikely to recapitulate in vivo conditions faithfully. A recent report, highlighting the in vivo heterogeneity in Id1 expression within PDGF-driven gliomas, challenged the concept that in vivo tumorigenic abilities correlate with in vitro self-renewal (Barrett et al. 2012). It showed that while $Id1^{high}$ cells are poorly tumorigenic but able to efficiently self-renew in vitro, the $Id1^{low}$ fraction shows limited self-renewal in vitro and higher tumorigenicity. The possibility that the $Id1^{high}$ population is lineage-related to the $Id1^{low}$ bulk tumor cell population was proposed but not rigorously tested and in our opinion it is likely to represent a stromal component, such as trapped or perivascular reactive astrocytes, which in vitro and under proper conditions can behave as multipotent self-renewing NSCs (Buffo et al. 2008; Sirko et al. 2013).

The above-mentioned comparisons between PDGF-induced gliomas and OPCs suggest that at least part of the malignant biology of these tumors may be understood in terms of deviations from the otherwise tightly controlled proliferative behavior of OPCs. The Olig2 transcription factor is a master regulator of oligodendroglial lineage specification and maturation during development and adulthood (Ligon et al. 2006; Ono et al. 2009) and is further involved in regulating the response of oligodendroglial lineage cells to injury. Olig2 is also diffusely expressed by proneural glioma cells (Ligon et al. 2004), a condition recapitulated in PDGF-induced gliomas. In vitro, abrogation of Olig2 expression resulted in reduced proliferation and decreased abundance of cells expressing immature NPC markers, without effects on cell viability. More dramatically, in vivo, Olig2 depletion caused an almost absolute loss of tumor-propagating ability following orthotopic transplantation. Mechanistically, the process involved increased expression of ID4, a powerful negative modulator of oligodendrogenesis, whose forced expression mimicked the in vivo effects of Olig2 loss (Appolloni et al. 2012a). These observations, together with others performed in other models of proneural gliomagenesis (Ligon et al. 2007; Mehta et al. 2011; Sun et al. 2011), indicate that Olig2 is a crucial regulator of malignancy in this class of gliomas, possibly by modulating proliferative behavior and adaptation to stressful conditions such as DNA damage.

Self-sufficiency in growth-promoting signals is one of the typical hallmarks of cancer (Hanahan and Weinberg 2011) and can result from the aberrant activation of growth factor-mediated signaling. Conceivably, though, tumors driven by such stimuli may eventually strictly rely on them for their survival and growth, despite the accumulation of further genetic and epigenetic alterations during progression. Indeed, various lines of evidence suggest that activation of multiple alternative signaling (e.g., downstream of various RTKs) may be uncommon at the single cell level (Szerlip et al. 2012). This suggests a high degree of "centralization" of growth control by single activated pathways that may expose tumors to the effects of abrogating such signaling. Such reliance, which can manifest itself in abrupt remissive responses upon oncogene signaling interruption, has been termed "oncogene addiction." Evidence for such a phenomenon has indeed emerged in the context of PDGF-driven gliomagenesis. As mentioned above, PDGF-driven gliomas need to acquire further molecular alterations to exhibit full-fledged tumorigenic potential. Yet, despite such requirement, PDGF maintains an essential role in the biology of malignant PDGF-initiated tumors. Cells from established high-grade PDGF-induced gliomas normally exhibit potent tumor-propagating potential when transplanted orthotopically, generating aggressive tumors that resemble the parental gliomas. Such tumorigenic capacity, though, strongly relies on continued PDGF expression, since its loss results in the complete abrogation of tumor-propagating potential (Calzolari et al. 2008). Interestingly, the cell biological basis for such a dependency seems to reside, at least in part, in an unexpected role for PDGF signaling in regulating cell–cell interactions, somehow overcoming contact-mediated inhibition of proliferation (Calzolari et al. 2008). Other aspects of PDGF signaling likely contribute to make it essential to maintain malignancy. Some evidence suggests PDGF may impact on metabolic and immune-modulatory processes, thus favoring glioma cell biology. PDGF receptor signaling was shown to trigger and maintain the Warburg effect in glioma cells, in culture (Ran et al. 2013). While traditionally glycolytic tumor metabolism has been thought of as an adaptation to oxygen-poor conditions, it is now seen more as an anabolically oriented deviant, caused by the diversion of Kreb's cycle intermediates along biosynthetic paths. In this sense, aberrant signaling by PDGF or other growth factors will likely reshape cellular metabolism to facilitate growth and allow high proliferation rates. Furthermore, the metabolic reprogramming typically associated with tumors may remarkably result in the concurrent reshaping of immune-modulatory properties, endowing tumor cells with a complex suite of advantageous features (Kareva and Hahnfeldt 2013). In this sense tumors, which can be seen as "wounds that do not heal" (Dvorak 1986; Schafer and Werner 2008), may aberrantly resemble regenerating responses, normally characterized by abrupt increases in local progenitor proliferation, recruitment of repair-facilitating immune components, and promotion of local angiogenesis (all processes playing an important role during tumor progression). Further elucidation of the molecular and cellular mechanisms underlying glioma cell addiction to PDGF may open important therapeutic options and therefore represents an important area for further basic and translational research.

13.3.3.4 Tumor Composition Dynamics: Recruitment and Hijacking of Endogenous Resident Progenitors, Astrocytes and Immune Cells

Gliomas, like most tumors, represent heterogeneous entities comprising tumor cells along with various cellular elements of local and systemic origins. While cellular heterogeneity among glioma cells has already received a good deal of attention, especially due to the hope to isolate and characterize tumor-propagating subpopulations, less has been done to identify and characterize the non-tumoral components. It is however clear, especially thanks to work done in other systems, that non-transformed intratumoral components can play central roles in shaping tumor development and progression as well as in determining the outcome of therapeutic interventions (Quail and Joyce 2013). PDGF-driven gliomas harbor significant populations of proliferating recruited untransformed OPCs (Assanah et al. 2006; Calzolari et al. 2008; Fomchenko et al. 2011). Recruited OPCs likely contribute to tumor mass growth and may even actively drive tumor progression or gain independent tumorigenic potential, in cases in which predisposing mutations exist. Glioma-recruited OPCs from wild-type animals have been shown to be incapable of autonomous propagation after dissociation from their high-grade host tumors (Appolloni et al. 2009), while the lack of multiple oncosuppressor genes may endow them with some degree of independence from the PDGF-overexpressing cells when transplanted (Fomchenko et al. 2011). Interestingly, oncosuppressor-deficient OPCs recruited to PDGF-driven gliomas most closely resemble PDGF-expressing glioma cells from the same tumors rather than normal OPCs. Of note, cells lacking OPC features possessed some features associated with other glioma subtypes (e.g., mesenchymal). Coexistence of glioma cells with molecular features typical of various glioma subclasses has been reported within individual human glioma samples (Snuderl et al. 2011; Szerlip et al. 2012), and the possibility to model such degree of heterogeneity using a relatively simple in vivo murine model may be very useful to further our understanding of glioma biology. Interestingly, as mentioned above, the possibility that tumor or recruited cells from proneural gliomas may be driven toward non-proneural phenotypes has been experimentally assessed by forcing mesenchymal gene expression in PDGF-induced gliomas, suggesting that recruited OPCs may diverge and contribute to tumor progression and acquisition of alternative molecular signatures (Bhat et al. 2011; Doucette et al. 2012).

Astrocytes represent another set of cell types that are both abundant and remarkably plastic. Normally working to maintain ion and metabolite homeostasis and modulate neuronal network activity, astrocytes are also well known for their ability to respond to various kinds of acute and chronic stresses. Challenged astrocytes mount a reactive response, which can also result in remarkable morphological alterations and even cell cycle reentry and division. While the pervasive presence of reactive astrocytes within gliomas has long been appreciated (and sometimes mistakenly used to argue for the astrocytic nature of the analyzed tumors), little attention has been devoted to the in vivo elucidation of the roles that astrocytes may play when recruited or trapped into growing gliomas. Recent work has shown that proneural gliomas are clearly pervasively populated by reactive astrocytes in the less densely packed areas of tumors, where infiltration takes place or where growth has not yet resulted in displacement or elimination of normal brain cells (Calzolari et al. 2008; Appolloni et al. 2009). In areas more densely populated by glioma cells, astrocytes are mostly found associated with blood vessels. Whether such association results from the recruitment of astrocytes to newly generated blood vessels or if it represents remnants of astrocytic coverage existing before tumor growth is not clear. Recent analyses, though, have provided a preliminary molecular characterization of glioma-associated reactive astrocytes, comparing them to low-grade glioma-associated and normal astrocytes (Katz et al. 2012). The gene expression signature of tumor-associated astrocytes pointed to a specific role of such astrocytes in interacting with immune system components. Interestingly, focusing on high-grade glioma-associated astrocytes resulted in a signature whose expression correlates inversely with prognosis, highlighting the likely tumor-promoting nature of the reactive astrocyte population found in such advanced gliomas.

Apart from cells of neural origins, other cell lineages are strongly represented among the "stromal" component of gliomas. Immune cells likely represent a continuous presence during the dynamic life of a tumor, in the form of resident microglia (da Fonseca and Badie 2013) and perivascular macrophages, as well as of elements recruited from the systemic circulation. Compelling evidence has recently emerged for the in vivo relevance of specific macrophage subtypes in sustaining proneural glioma growth, and provided important details into the nature of the cross talk between tumor cells and recruited elements in shaping tumor malignancy. A clear example of this phenomenon is the observation that the impairment of gliomagenesis in CyclinD1 or Cdk4 knockout mice is due to a defective microglial/macrophagic activation rather than to a proliferative defect of tumor cells (Ciznadija et al. 2011). More recently it was shown that inhibition of colony-stimulating factor-1 receptor (CSF-1R), expressed by macrophages and normally activated by glioma cell-secreted CSF-1 (Coniglio et al. 2012), impairs PDGF-induced tumor growth and causes regression of established gliomas, in vivo (Pyonteck et al. 2013). Interestingly, CSF-1R inhibition did not affect macrophage survival but resulted only in the shift from a tumor-promoting M2-like phenotype to a M1-like state with tumor-restraining functions. This work clarifies the role of macrophage recruitment and modulation in the context of glioma formation and progression, but also shows how fine regulation of immune system function may be extremely promising for the eradication of gliomas, likely even more so than the overall elimination of such stromal components.

13.3.3.5 PDGF-Induced Gliomas as Translational Research Platforms

Since current therapies for high-grade gliomas are only slightly effective, appropriate models are essential as basic platforms to identify, characterize, and optimize potential therapeutic strategies. While, in general, the efficacy of promising cancer therapeutic candidates often fails to be replicated when tested in preclinical human settings, there is hope that recent advances in the molecular characterization of glioma subtypes will concur to define crucial molecular targets and identify potentially responsive patients (Brennan et al. 2013). Tumor accessibility is another main issue when dealing with brain tumors: the presence of the blood–brain barrier, the altered intratumoral hematic perfusion, and the possible existence of active xenobiotic-extruding mechanisms in at least some subsets of tumor-propagating glioma cells concur to protect gliomas from drugs. PDGF-induced gliomas have been employed to assess the in vitro and in vivo efficacy of various approaches, both diagnostic and therapeutic, some of which have yielded promising results. For example, in vivo studies using PDGF-induced gliomas assessed the correlation between MRI signals and tumor structural/functional features, such as distribution, architecture, and permeability of glioma-associated blood vessels and overall tumor grade (Koutcher et al. 2002), to be used as a starting point to test the efficacy of candidate antitumor compounds. MRI features can thus be employed to admit tumor grade-matched mice to trials to assess the effects of the alkylating agent Temozolomide treatment on high-grade tumors (McConville et al. 2007). Combined treatment of glioma-bearing mice with Temozolomide and Perifosine, an AKT inhibitor, building on an in vitro characterization of the effects of Perifosine on PDGF-transformed perinatal glial progenitors, revealed a partially additive effect of the two compounds on the in vivo proliferative index of PDGF-induced gliomas (Momota et al. 2005). The same or similar models have also been used to test alternative drug delivery approaches, such as in the case of convection-enhanced in vivo delivery of Topotecan, a DNA topoisomerase inhibitor, which was shown to strongly reduce PDGF-induced tumor growth, leading to temporary regression of established tumors, followed by recurrence after removal of the infusion system (Lopez et al. 2011; Yun et al. 2013).

PDGF-induced tumors have been used to assay unconventional therapeutic approaches such as oncolytic virotherapy. Viruses with wild-type or altered tropism have been used to target tumor cells. Promising attempts have been performed to target PDGF-driven proneural gliomas, in vivo, using completely retargeted oncolytic herpex simplex viruses (HSVs) directed against the HER2 receptor (Menotti et al. 2008). In these studies, PDGF-induced glioma cells engineered to express human HER2 were efficiently infected with retargeted HSVs and thus impaired in their in vivo tumorigenic potential (Gambini et al. 2012; Reisoli et al. 2012).

Conclusions

Animal models of gliomagenesis have been instrumental to improve our understanding of glioma biology. While early systems relied on the use of spontaneously arising or carcinogen-induced tumors, the last decade and a half has seen an ever growing use of transgenic and gene transfer models, despite the continued use of long-term established human cells lines. These have been repeatedly demonstrated to poorly recapitulate molecular and functional properties of primary human gliomas and should be progressively abandoned in favor of patient-derived cultures of tumor-propagating glioma cells. In vivo modeling using transgenic and gene transfer-based models has provided important insights into the molecular and cellular biology of glioma initiation, progression, and maintenance. However, in order to continue contributing to our understanding of glioma biology, the description of the features of existing and new murine models needs to be integrated with the evolving view of human glioma subtype distinctions. Understanding the extent to which currently recognized molecular subclasses of human gliomas are represented among the available mouse models will be necessary to focus on understudied groups. The molecular subtype most thoroughly analyzed, thanks to the existence of multiple models, is arguably the proneural one. Among the existing models, PDGF-driven ones have been the most heavily exploited. In spite of this, many areas remain poorly explored, such as the early molecular and cellular steps of gliomagenesis and the identity and behavior of tumor-maintaining cells.

References

Acquaviva J, Jun HJ, Lessard J, Ruiz R, Zhu H, Donovan M et al (2011) Chronic activation of wild-type epidermal growth factor receptor and loss of Cdkn2a cause mouse glioblastoma formation. Cancer Res 71:7198–7206

Alcantara Llaguno S, Chen J, Kwon CH, Jackson EL, Li Y, Burns DK et al (2009) Malignant astrocytomas originate from neural stem/progenitor cells in a somatic tumor suppressor mouse model. Cancer Cell 15:45–56

Appolloni I, Calzolari F, Tutucci E, Caviglia S, Terrile M, Corte G et al (2009) PDGF-B induces a homogeneous class of oligodendrogliomas from embryonic neural progenitors. Int J Cancer 124:2251–2259

Appolloni I, Calzolari F, Barilari M, Terrile M, Daga A, Malatesta P (2012a) Antagonistic modulation of gliomagenesis by Pax6 and Olig2 in PDGF-induced oligodendroglioma. Int J Cancer 131:E1078–E1087

Appolloni I, Curreli S, Caviglia S, Barilari M, Gambini E, Pagano A et al (2012b) Role of Btg2 in the progression of a PDGF-induced oligodendroglioma model. Int J Mol Sci 13:14667–14678

Assanah M, Lochhead R, Ogden A, Bruce J, Goldman J, Canoll P (2006) Glial progenitors in adult white matter are driven to form malignant gliomas by platelet-derived growth factor-expressing retroviruses. J Neurosci 26:6781–6790

Assanah MC, Bruce JN, Suzuki SO, Chen A, Goldman JE, Canoll P (2009) PDGF stimulates the massive expansion of glial progenitors in the neonatal forebrain. Glia 57:1835–1847

Bachoo RM, Maher EA, Ligon KL, Sharpless NE, Chan SS, You MJ et al (2002) Epidermal growth factor receptor and Ink4a/Arf: convergent mechanisms governing terminal differentiation and transformation along the neural stem cell to astrocyte axis. Cancer Cell 1:269–277

Bajenaru ML, Hernandez MR, Perry A, Zhu Y, Parada LF, Garbow JR et al (2003) Optic nerve glioma in mice requires astrocyte Nf1 gene inactivation and Nf1 brain heterozygosity. Cancer Res 63:8573–8577

Barrett LE, Granot Z, Coker C, Iavarone A, Hambardzumyan D, Holland EC et al (2012) Self-renewal does not predict tumor growth potential in mouse models of high-grade glioma. Cancer Cell 21:11–24

Barth RF, Kaur B (2009) Rat brain tumor models in experimental neuro-oncology: the C6, 9L, T9, RG2, F98, BT4C, RT-2 and CNS-1 gliomas. J Neurooncol 94:299–312

Barth RF, Yang W, Coderre JA (2003) Rat brain tumor models to assess the efficacy of boron neutron capture therapy: a critical evaluation. J Neurooncol 62:61–74

Bencokova Z, Pauron L, Devic C, Joubert A, Gastaldo J, Massart C et al (2008) Molecular and cellular response of the most extensively used rodent glioma models to radiation and/or cisplatin. J Neurooncol 86:13–21

Benda P, Lightbody J, Sato G, Levine L, Sweet W (1968) Differentiated rat glial cell strain in tissue culture. Science 161:370–371

Benda P, Someda K, Messer J, Sweet WH (1971) Morphological and immunochemical studies of rat glial tumors and clonal strains propagated in culture. J Neurosurg 34:310–323

Bhat KP, Salazar KL, Balasubramaniyan V, Wani K, Heathcock L, Hollingsworth F et al (2011) The transcriptional coactivator TAZ regulates mesenchymal differentiation in malignant glioma. Genes Dev 25:2594–2609

Biston MC, Joubert A, Adam JF, Elleaume H, Bohic S, Charvet AM et al (2004) Cure of Fisher rats bearing radioresistant F98 glioma treated with cis-platinum and irradiated with monochromatic synchrotron X-rays. Cancer Res 64:2317–2323

Bleau AM, Hambardzumyan D, Ozawa T, Fomchenko EI, Huse JT, Brennan CW et al (2009) PTEN/PI3K/Akt pathway regulates the side population phenotype and ABCG2 activity in glioma tumor stem-like cells. Cell Stem Cell 4:226–235

Brennan CW, Verhaak RG, McKenna A, Campos B, Noushmehr H, Salama SR et al (2013) The somatic genomic landscape of glioblastoma. Cell 155:462–477

Buffo A, Rite I, Tripathi P, Lepier A, Colak D, Horn AP et al (2008) Origin and progeny of reactive gliosis: a source of multipotent cells in the injured brain. Proc Natl Acad Sci U S A 105:3581–3586

Calzolari F, Malatesta P (2010) Recent insights into PDGF-induced gliomagenesis. Brain Pathol 20:527–538

Calzolari F, Appolloni I, Tutucci E, Caviglia S, Terrile M, Corte G et al (2008) Tumor progression and oncogene addiction in a PDGF-B-induced model of gliomagenesis. Neoplasia 10:1373–1382, following 82

Chan DA, Giaccia AJ (2011) Harnessing synthetic lethal interactions in anticancer drug discovery. Nat Rev Drug Discov 10:351–364

Charest A, Wilker EW, McLaughlin ME, Lane K, Gowda R, Coven S et al (2006) ROS fusion tyrosine kinase activates a SH2 domain-containing phosphatase-2/phosphatidylinositol 3-kinase/mammalian target of rapamycin signaling axis to form glioblastoma in mice. Cancer Res 66:7473–7481

Charles N, Ozawa T, Squatrito M, Bleau AM, Brennan CW, Hambardzumyan D et al (2010) Perivascular nitric oxide activates notch signaling and promotes stem-like character in PDGF-induced glioma cells. Cell Stem Cell 6:141–152

Chojnacki A, Mak G, Weiss S (2011) PDGFRalpha expression distinguishes GFAP-expressing neural stem cells from PDGF-responsive neural precursors in the adult periventricular area. J Neurosci 31:9503–9512

Chow LM, Endersby R, Zhu X, Rankin S, Qu C, Zhang J et al (2011) Cooperativity within and among Pten, p53, and Rb pathways induces high-grade astrocytoma in adult brain. Cancer Cell 19:305–316

Ciznadija D, Liu Y, Pyonteck SM, Holland EC, Koff A (2011) Cyclin D1 and cdk4 mediate development of neurologically destructive oligodendroglioma. Cancer Res 71:6174–6183

Coniglio SJ, Eugenin E, Dobrenis K, Stanley ER, West BL, Symons MH et al (2012) Microglial stimulation of glioblastoma invasion involves epidermal growth factor receptor (EGFR) and colony stimulating factor 1 receptor (CSF-1R) signaling. Mol Med 18:519–527 da Fonseca AC, Badie B (2013) Microglia and macrophages in malignant gliomas: recent discoveries and implications for promising therapies. Clin Dev Immunol 2013:264124

Dai C, Celestino JC, Okada Y, Louis DN, Fuller GN, Holland EC (2001) PDGF autocrine stimulation dedifferentiates cultured astrocytes and induces oligodendrogliomas and oligoastrocytomas from neural progenitors and astrocytes in vivo. Genes Dev 15:1913–1925

Danks RA, Orian JM, Gonzales MF, Tan SS, Alexander B, Mikoshiba K et al (1995) Transformation of astrocytes in transgenic mice expressing SV40 T antigen under the transcriptional control of the glial fibrillary acidic protein promoter. Cancer Res 55:4302–4310

Dimou L, Simon C, Kirchhoff F, Takebayashi H, Gotz M (2008) Progeny of Olig2-expressing progenitors in the gray and white matter of the adult mouse cerebral cortex. J Neurosci 28:10434–10442

Dinapoli RP, Brown LD, Arusell RM, Earle JD, O'Fallon JR, Buckner JC et al (1993) Phase III comparative evaluation of PCNU and carmustine combined with radiation therapy for high-grade glioma. J Clin Oncol 11:1316–1321

Ding H, Roncari L, Shannon P, Wu X, Lau N, Karaskova J et al (2001) Astrocyte-specific expression of activated p21-ras results in malignant astrocytoma formation in a transgenic mouse model of human gliomas. Cancer Res 61:3826–3836

Ding H, Shannon P, Lau N, Wu X, Roncari L, Baldwin RL et al (2003) Oligodendrogliomas result from the expression of an activated mutant epidermal growth factor receptor in a RAS transgenic mouse astrocytoma model. Cancer Res 63:1106–1113

Doucette TA, Kong LY, Yang Y, Ferguson SD, Yang J, Wei J et al (2012) Signal transducer and activator of transcription 3 promotes angiogenesis and drives malignant progression in glioma. Neuro Oncol 14:1136–1145

Dougherty JD, Fomchenko EI, Akuffo AA, Schmidt E, Helmy KY, Bazzoli E et al (2012) Candidate pathways for promoting differentiation or quiescence of oligodendrocyte progenitor-like cells in glioma. Cancer Res 72:4856–4868

Druckrey H, Ivankovic S, Preussmann R (1966) Teratogenic and carcinogenic effects in the offspring after single injection of ethylnitrosourea to pregnant rats. Nature 210:1378–1379

Dvorak HF (1986) Tumors: wounds that do not heal. Similarities between tumor stroma generation and wound healing. New Engl J Med 315:1650–1659

Ernst A, Hofmann S, Ahmadi R, Becker N, Korshunov A, Engel F et al (2009) Genomic and expression profiling of glioblastoma stem cell-like spheroid cultures identifies novel tumor-relevant genes associated with survival. Clin Cancer Res 15:6541–6550

Eyre HJ, Eltringham JR, Gehan EA, Vogel FS, Al-Sarraf M, Talley RW et al (1986) Randomized comparisons of radiotherapy and carmustine versus procarbazine versus dacarbazine for the treatment of malignant gliomas following surgery: a Southwest Oncology Group Study. Cancer Treat Rep 70:1085–1090

Finkelstein SD, Black P, Nowak TP, Hand CM, Christensen S, Finch PW (1994) Histological characteristics and expression of acidic and basic fibroblast growth factor genes in intracerebral xenogeneic transplants of human glioma cells. Neurosurgery 34:136–143

Flanagan SP (1966) 'Nude', a new hairless gene with pleiotropic effects in the mouse. Genet Res 8:295–309

Fomchenko EI, Dougherty JD, Helmy KY, Katz AM, Pietras A, Brennan C et al (2011) Recruited cells can become transformed and overtake PDGF-induced murine gliomas in vivo during tumor progression. PLoS One 6:e20605

Friedmann-Morvinski D, Bushong EA, Ke E, Soda Y, Marumoto T, Singer O et al (2012) Dedifferentiation of neurons and astrocytes by oncogenes can induce gliomas in mice. Science 338:1080–1084

Galli R, Binda E, Orfanelli U, Cipelletti B, Gritti A, De Vitis S et al (2004) Isolation and characterization of tumorigenic, stem-like neural precursors from human glioblastoma. Cancer Res 64:7011–7021

Gambini E, Reisoli E, Appolloni I, Gatta V, Campadelli-Fiume G, Menotti L et al (2012) Replication-competent herpes simplex virus retargeted to HER2 as therapy for high-grade glioma. Mol Ther 20:994–1001

Greene HS (1952) The significance of the heterologous transplantability of human cancer. Cancer 5:24–44

Hack MA, Sugimori M, Lundberg C, Nakafuku M, Gotz M (2004) Regionalization and fate specification in neurospheres: the role of Olig2 and Pax6. Mol Cell Neurosci 25:664–678

Hack MA, Saghatelyan A, de Chevigny A, Pfeifer A, Ashery-Padan R, Lledo PM et al (2005) Neuronal fate determinants of adult olfactory bulb neurogenesis. Nat Neurosci 8:865–872

Hanahan D, Weinberg RA (2011) Hallmarks of cancer: the next generation. Cell 144:646–674

Hartfuss E, Galli R, Heins N, Gotz M (2001) Characterization of CNS precursor subtypes and radial glia. Dev Biol 229:15–30

Helmy K, Halliday J, Fomchenko E, Setty M, Pitter K, Hafemeister C et al (2012) Identification of global alteration of translational regulation in glioma in vivo. PLoS One 7:e46965

Hill RA, Patel KD, Medved J, Reiss AM, Nishiyama A (2013) NG2 cells in white matter but not gray matter proliferate in response to PDGF. J Neurosci 33:14558–14566

Holland EC, Varmus HE (1998) Basic fibroblast growth factor induces cell migration and proliferation after glia-specific gene transfer in mice. Proc Natl Acad Sci U S A 95:1218–1223

Holland EC, Hively WP, DePinho RA, Varmus HE (1998) A constitutively active epidermal growth factor receptor cooperates with disruption of G1 cell-cycle arrest pathways to induce glioma-like lesions in mice. Genes Dev 12:3675–3685

Holland EC, Celestino J, Dai C, Schaefer L, Sawaya RE, Fuller GN (2000) Combined activation of Ras and Akt in neural progenitors induces glioblastoma formation in mice. Nat Genet 25:55–57

Houchens DP, Ovejera AA, Riblet SM, Slagel DE (1983) Human brain tumor xenografts in nude mice as a chemotherapy model. Eur J Cancer Clin Oncol 19:799–805

Hu X, Pandolfi PP, Li Y, Koutcher JA, Rosenblum M, Holland EC (2005) mTOR promotes survival and astrocytic characteristics induced by Pten/AKT signaling in glioblastoma. Neoplasia 7:356–368

Hu JG, Fu SL, Wang YX, Li Y, Jiang XY, Wang XF et al (2008) Platelet-derived growth factor-AA mediates oligodendrocyte lineage differentiation through activation of extracellular signal-regulated kinase signaling pathway. Neuroscience 151:138–147

Hukkelhoven E, Liu Y, Yeh N, Ciznadija D, Blain SW, Koff A (2012) Tyrosine phosphorylation of the p21 cyclin-dependent kinase inhibitor facilitates the development of proneural glioma. J Biol Chem 287:38523–38530

Huse JT, Brennan C, Hambardzumyan D, Wee B, Pena J, Rouhanifard SH et al (2009) The PTEN-regulating microRNA miR-26a is amplified in high-grade glioma and facilitates gliomagenesis in vivo. Genes Dev 23:1327–1337

Huszthy PC, Brekken C, Pedersen TB, Thorsen F, Sakariassen PO, Skaftnesmo KO et al (2006) Antitumor efficacy improved by local delivery of species-specific endostatin. J Neurosurg 104:118–128

Ignatova TN, Kukekov VG, Laywell ED, Suslov ON, Vrionis FD, Steindler DA (2002) Human cortical glial tumors contain neural stem-like cells expressing astroglial and neuronal markers in vitro. Glia 39:193–206

Ikezaki K, Nomura T, Takahashi M, Zieroth BF, Fukui M (1994) MRI contrast enhancement by Mn-TPPS in experimental rat brain tumor with peripheral benzodiazepine receptors. Acta Neurochir Suppl 60:353–355

Iwadate Y, Inoue M, Saegusa T, Tokusumi Y, Kinoh H, Hasegawa M et al (2005) Recombinant Sendai virus vector induces complete remission of established brain tumors through efficient interleukin-2 gene transfer in vaccinated rats. Clin Cancer Res 11:3821–3827

Jackson EL, Garcia-Verdugo JM, Gil-Perotin S, Roy M, Quinones-Hinojosa A, VandenBerg S et al (2006) PDGFR alpha-positive B cells are neural stem cells in the adult SVZ that form glioma-like growths in response to increased PDGF signaling. Neuron 51:187–199

Jang ES, Goldman JE (2011) Pax6 expression is sufficient to induce a neurogenic fate in glial progenitors of the neonatal subventricular zone. PLoS One 6:e20894

Johansson FK, Brodd J, Eklof C, Ferletta M, Hesselager G, Tiger CF et al (2004) Identification of candidate cancer-causing genes in mouse brain tumors by retroviral tagging. Proc Natl Acad Sci U S A 101:11334–11337

Johansson FK, Goransson H, Westermark B (2005) Expression analysis of genes involved in brain tumor progression driven by retroviral insertional mutagenesis in mice. Oncogene 24:3896–3905

Jones EL, Searle CE, Smith WT (1973) Tumours of the nervous system induced in rats by the neonatal administration of N-ethyl-N-nitrosourea. J Pathol 109:123–139

Kareva I, Hahnfeldt P (2013) The emerging "hallmarks" of metabolic reprogramming and immune evasion: distinct or linked? Cancer Res 73:2737–2742

Katz AM, Amankulor NM, Pitter K, Helmy K, Squatrito M, Holland EC (2012) Astrocyte-specific expression patterns associated with the PDGF-induced glioma microenvironment. PLoS One 7:e32453

Kinzler KW, Vogelstein B (1996) Lessons from hereditary colorectal cancer. Cell 87:159–170

Ko L, Koestner A, Wechsler W (1980) Morphological characterization of nitrosourea-induced glioma cell lines and clones. Acta Neuropathol 51:23–31

Koutcher JA, Hu X, Xu S, Gade TP, Leeds N, Zhou XJ et al (2002) MRI of mouse models for gliomas shows similarities to humans and can be used to identify mice for preclinical trials. Neoplasia 4:480–485

Krementz ET, Greene HS (1953) Heterologous transplantation of human neural tumors. Cancer 6:100–110

Kruse CA, Molleston MC, Parks EP, Schiltz PM, Kleinschmidt-DeMasters BK, Hickey WF (1994) A rat glioma model, CNS-1, with invasive characteristics similar to those of human gliomas: a comparison to 9L gliosarcoma. J Neurooncol 22:191–200

Kwon CH, Zhao D, Chen J, Alcantara S, Li Y, Burns DK et al (2008) Pten haploinsufficiency accelerates formation of high-grade astrocytomas. Cancer Res 68:3286–3294

Laerum OD, Rajewsky MF (1975) Neoplastic transformation of fetal rat brain cells in culture after exposure to ethylnitrosourea in vivo. J Natl Cancer Inst 55:1177–1187

Lee J, Kotliarova S, Kotliarov Y, Li A, Su Q, Donin NM et al (2006) Tumor stem cells derived from glioblastomas cultured in bFGF and EGF more closely mirror the phenotype and genotype of primary tumors than do serum-cultured cell lines. Cancer Cell 9:391–403

Lei L, Sonabend AM, Guarnieri P, Soderquist C, Ludwig T, Rosenfeld S et al (2011) Glioblastoma models reveal the connection between adult glial progenitors and the proneural phenotype. PLoS One 6:e20041

Li A, Walling J, Kotliarov Y, Center A, Steed ME, Ahn SJ et al (2008) Genomic changes and gene expression profiles reveal that established glioma cell lines are poorly representative of primary human gliomas. Mol Cancer Res 6:21–30

Ligon KL, Alberta JA, Kho AT, Weiss J, Kwaan MR, Nutt CL et al (2004) The oligodendroglial lineage marker OLIG2 is universally expressed in diffuse gliomas. J Neuropathol Exp Neurol 63:499–509

Ligon KL, Fancy SP, Franklin RJ, Rowitch DH (2006) Olig gene function in CNS development and disease. Glia 54:1–10

Ligon KL, Huillard E, Mehta S, Kesari S, Liu H, Alberta JA et al (2007) Olig2-regulated lineage-restricted pathway controls replication competence in neural stem cells and malignant glioma. Neuron 53:503–517

Liu Y, Yeh N, Zhu XH, Leversha M, Cordon-Cardo C, Ghossein R et al (2007) Somatic cell type specific gene transfer reveals a tumor-promoting function for p21(Waf1/Cip1). EMBO J 26:4683–4693

Liu Y, Raheja R, Yeh N, Ciznadija D, Pedraza AM, Ozawa T et al (2014) TRIM3, a tumor suppressor linked to regulation of p21(Waf1/Cip1). Oncogene 33(3):308–315

Lopez KA, Tannenbaum AM, Assanah MC, Linskey K, Yun J, Kangarlu A et al (2011) Convection-enhanced delivery of topotecan into a PDGF-driven model of glioblastoma prolongs survival and ablates both tumor-initiating cells and recruited glial progenitors. Cancer Res 71:3963–3971

Mahesparan R, Read TA, Lund-Johansen M, Skaftnesmo KO, Bjerkvig R, Engebraaten O (2003) Expression of extracellular matrix components in a highly infiltrative in vivo glioma model. Acta Neuropathol 105:49–57

Malatesta P, Appolloni I, Calzolari F (2008) Radial glia and neural stem cells. Cell Tissue Res 331:165–178

Marumoto T, Tashiro A, Friedmann-Morvinski D, Scadeng M, Soda Y, Gage FH et al (2009) Development of a novel mouse glioma model using lentiviral vectors. Nat Med 15:110–116

McConville P, Hambardzumyan D, Moody JB, Leopold WR, Kreger AR, Woolliscroft MJ et al (2007) Magnetic resonance imaging determination of tumor grade and early response to temozolomide in a genetically engineered mouse model of glioma. Clin Cancer Res 13:2897–2904

Mehta S, Huillard E, Kesari S, Maire CL, Golebiowski D, Harrington EP et al (2011) The central nervous system-restricted transcription factor Olig2 opposes p53 responses to genotoxic damage in neural progenitors and malignant glioma. Cancer Cell 19:359–371

Menotti L, Cerretani A, Hengel H, Campadelli-Fiume G (2008) Construction of a fully retargeted herpes simplex virus 1 recombinant capable of entering cells solely via human epidermal growth factor receptor 2. J Virol 82:10153–10161

Merkle FT, Tramontin AD, Garcia-Verdugo JM, Alvarez-Buylla A (2004) Radial glia give rise to adult neural stem cells in the subventricular zone. Proc Natl Acad Sci U S A 101:17528–17532

Momota H, Nerio E, Holland EC (2005) Perifosine inhibits multiple signaling pathways in glial progenitors and cooperates with temozolomide to arrest cell proliferation in gliomas in vivo. Cancer Res 65:7429–7435

Muller FL, Colla S, Aquilanti E, Manzo VE, Genovese G, Lee J et al (2012) Passenger deletions generate therapeutic vulnerabilities in cancer. Nature 488:337–342

Nolte C, Matyash M, Pivneva T, Schipke CG, Ohlemeyer C, Hanisch UK et al (2001) GFAP promoter-controlled EGFP-expressing transgenic mice: a tool to visualize astrocytes and astrogliosis in living brain tissue. Glia 33:72–86

Ono K, Takebayashi H, Ikenaka K (2009) Olig2 transcription factor in the developing and injured forebrain; cell lineage and glial development. Mol Cells 27:397–401

Ortega F, Gascon S, Masserdotti G, Deshpande A, Simon C, Fischer J et al (2013) Oligodendrogliogenic and neurogenic adult subependymal zone neural stem cells constitute distinct lineages and exhibit differential responsiveness to Wnt signalling. Nat Cell Biol 15:602–613

Oshiro S, Liu Y, Fukushima T, Asotra K, Black KL (2001) Modified immunoregulation associated with interferon-gamma treatment of rat glioma. Neurol Res 23:359–366

Owens GC, Orr EA, DeMasters BK, Muschel RJ, Berens ME, Kruse CA (1998) Overexpression of a transmembrane isoform of neural cell adhesion molecule alters the invasiveness of rat CNS-1 glioma. Cancer Res 58:2020–2028

Parsa AT, Chakrabarti I, Hurley PT, Chi JH, Hall JS, Kaiser MG et al (2000) Limitations of the C6/Wistar rat intracerebral glioma model: implications for evaluating immunotherapy. Neurosurgery 47:993–999, discussion 9-1000

Plowman J, Waud WR, Koutsoukos AD, Rubinstein LV, Moore TD, Grever MR (1994) Preclinical antitumor activity of temozolomide in mice: efficacy against human brain tumor xenografts and synergism with 1,3-bis(2-chloroethyl)-1-nitrosourea. Cancer Res 54:3793–3799

Ponten J, Macintyre EH (1968) Long term culture of normal and neoplastic human glia. Acta Pathol Microbiol Scand 74:465–486

Pyonteck SM, Akkari L, Schuhmacher AJ, Bowman RL, Sevenich L, Quail DF et al (2013) CSF-1R inhibition alters macrophage polarization and blocks glioma progression. Nat Med 19:1264–1272

Quail DF, Joyce JA (2013) Microenvironmental regulation of tumor progression and metastasis. Nat Med 19:1423–1437

Ran C, Liu H, Hitoshi Y, Israel MA (2013) Proliferation-independent control of tumor glycolysis by PDGFR-mediated AKT activation. Cancer Res 73:1831–1843

Reilly KM, Loisel DA, Bronson RT, McLaughlin ME, Jacks T (2000) Nf1;Trp53 mutant mice develop glioblastoma with evidence of strain-specific effects. Nat Genet 26:109–113

Reisoli E, Gambini E, Appolloni I, Gatta V, Barilari M, Menotti L et al (2012) Efficacy of HER2 retargeted herpes simplex virus as therapy for high-grade glioma in immunocompetent mice. Cancer Gene Ther 19:788–795

Sandstrom M, Johansson M, Bergstrom P, Bergenheim AT, Henriksson R (2008) Effects of the VEGFR inhibitor ZD6474 in combination with radiotherapy and temozolomide in an orthotopic glioma model. J Neurooncol 88:1–9

Schafer M, Werner S (2008) Cancer as an overhealing wound: an old hypothesis revisited. Nat Rev Mol Cell Biol 9:628–638

See WL, Heinberg AR, Holland EC, Resh MD (2010) p27 deficiency is associated with migration defects in PDGF-expressing gliomas in vivo. Cell Cycle 9:1562–1567

Sheehan J, Ionescu A, Pouratian N, Hamilton DK, Schlesinger D, Oskouian RJ Jr et al (2008) Use of trans sodium crocetinate for sensitizing glioblastoma multiforme to radiation: laboratory investigation. J Neurosurg 108:972–978

Simon C, Gotz M, Dimou L (2011) Progenitors in the adult cerebral cortex: cell cycle properties and regulation by physiological stimuli and injury. Glia 59:869–881

Singh SK, Clarke ID, Terasaki M, Bonn VE, Hawkins C, Squire J et al (2003) Identification of a cancer stem cell in human brain tumors. Cancer Res 63:5821–5828

Sirko S, Behrendt G, Johansson PA, Tripathi P, Costa M, Bek S et al (2013) Reactive glia in the injured brain acquire stem cell properties in response to sonic hedgehog. Cell Stem Cell 12:426–439

Slagel DE, Feola J, Houchens DP, Ovejera AA (1982) Combined modality treatment using radiation and/or chemotherapy in an athymic nude mouse-human medulloblastoma and glioblastoma xenograft model. Cancer Res 42:812–816

Snuderl M, Fazlollahi L, Le LP, Nitta M, Zhelyazkova BH, Davidson CJ et al (2011) Mosaic amplification of multiple receptor tyrosine kinase genes in glioblastoma. Cancer Cell 20:810–817

Squatrito M, Brennan CW, Helmy K, Huse JT, Petrini JH, Holland EC (2010) Loss of ATM/Chk2/p53 pathway components accelerates tumor development and contributes to radiation resistance in gliomas. Cancer Cell 18:619–629

Sun Y, Meijer DH, Alberta JA, Mehta S, Kane MF, Tien AC et al (2011) Phosphorylation state of Olig2 regulates proliferation of neural progenitors. Neuron 69:906–917

Szczurek E, Misra N, Vingron M (2013) Synthetic sickness or lethality points at candidate combination therapy targets in glioblastoma. Int J Cancer 133:2123–2132

Szerlip NJ, Pedraza A, Chakravarty D, Azim M, McGuire J, Fang Y et al (2012) Intratumoral heterogeneity of receptor tyrosine kinases EGFR and PDGFRA amplification in glioblastoma defines subpopulations with distinct growth factor response. Proc Natl Acad Sci U S A 109:3041–3046

Tanriover N, Ulu MO, Sanus GZ, Bilir A, Canbeyli R, Oz B et al (2008) The effects of systemic and intratumoral interleukin-12 treatment in C6 rat glioma model. Neurol Res 30:511–517

Tchougounova E, Kastemar M, Brasater D, Holland EC, Westermark B, Uhrbom L (2007) Loss of Arf causes tumor progression of PDGFB-induced oligodendroglioma. Oncogene 26:6289–6296

Uhrbom L, Hesselager G, Nister M, Westermark B (1998) Induction of brain tumors in mice using a recombinant platelet-derived growth factor B-chain retrovirus. Cancer Res 58:5275–5279

Uhrbom L, Dai C, Celestino JC, Rosenblum MK, Fuller GN, Holland EC (2002) Ink4a-Arf loss cooperates with KRas activation in astrocytes and neural progenitors to generate glioblastomas of various morphologies depending on activated Akt. Cancer Res 62:5551–5558

Valable S, Lemasson B, Farion R, Beaumont M, Segebarth C, Remy C et al (2008) Assessment of blood volume, vessel size, and the expression of angiogenic factors in two rat glioma models: a longitudinal in vivo and ex vivo study. NMR Biomed 21:1043–1056

Verhaak RG, Hoadley KA, Purdom E, Wang V, Qi Y, Wilkerson MD et al (2010) Integrated genomic analysis identifies clinically relevant subtypes of glioblastoma characterized by abnormalities in PDGFRA, IDH1, EGFR, and NF1. Cancer Cell 17:98–110

Vigano F, Mobius W, Gotz M, Dimou L (2013) Transplantation reveals regional differences in oligodendrocyte differentiation in the adult brain. Nat Neurosci 16:1370–1372

Wang Y, Yang J, Zheng H, Tomasek GJ, Zhang P, McKeever PE et al (2009) Expression of mutant p53 proteins implicates a lineage relationship between neural stem cells and malignant astrocytic glioma in a murine model. Cancer Cell 15:514–526

Warnke PC, Blasberg RG, Groothuis DR (1987) The effect of hyperosmotic blood-brain barrier disruption on blood-to-tissue transport in ENU-induced gliomas. Ann Neurol 22:300–305

Wei Q, Clarke L, Scheidenhelm DK, Qian B, Tong A, Sabha N et al (2006) High-grade glioma formation results from postnatal pten loss or mutant epidermal growth factor receptor expression in a transgenic mouse glioma model. Cancer Res 66:7429–7437

Weiss WA, Burns MJ, Hackett C, Aldape K, Hill JR, Kuriyama H et al (2003) Genetic determinants of malignancy in a mouse model for oligodendroglioma. Cancer Res 63:1589–1595

Xiao A, Wu H, Pandolfi PP, Louis DN, Van Dyke T (2002) Astrocyte inactivation of the pRb pathway predisposes mice to malignant astrocytoma development that is accelerated by PTEN mutation. Cancer Cell 1:157–168

Xiao A, Yin C, Yang C, Di Cristofano A, Pandolfi PP, Van Dyke T (2005) Somatic induction of Pten loss in a preclinical astrocytoma model reveals major roles in disease progression and avenues for target discovery and validation. Cancer Res 65:5172–5180

Yang WQ, Lun X, Palmer CA, Wilcox ME, Muzik H, Shi ZQ et al (2004) Efficacy and safety evaluation of human reovirus type 3 in immunocompetent animals: racine and nonhuman primates. Clin Cancer Res 10:8561–8576

Yuan H, Schroeder T, Bowsher JE, Hedlund LW, Wong T, Dewhirst MW (2006) Intertumoral differences in hypoxia selectivity of the PET imaging agent 64Cu(II)-diacetyl-bis (N4-methylthiosemicarbazone). J Nucl Med 47:989–998

Yun J, Rothrock RJ, Canoll P, Bruce JN (2013) Convection-enhanced delivery for targeted delivery of antiglioma agents: the translational experience. J Drug Deliv 2013:107573

Zhang D, Feng XY, Henning TD, Wen L, Lu WY, Pan H et al (2009) MR imaging of tumor angiogenesis using sterically stabilized Gd-DTPA liposomes targeted to CD105. Eur J Radiol 70:180–189

Zhang B, Sun X, Mei H, Wang Y, Liao Z, Chen J et al (2013a) LDLR-mediated peptide-22-conjugated nanoparticles for dual-targeting therapy of brain glioma. Biomaterials 34:9171–9182

Zhang J, Shin MC, David AE, Zhou J, Lee K, He H et al (2013b) Long-circulating heparin-functionalized magnetic nanoparticles for potential application as a protein drug delivery platform. Mol Pharm 10(10):3892–3902

Zheng H, Ying H, Yan H, Kimmelman AC, Hiller DJ, Chen AJ et al (2008) p53 and Pten control neural and glioma stem/progenitor cell renewal and differentiation. Nature 455:1129–1133

Zhu Y, Guignard F, Zhao D, Liu L, Burns DK, Mason RP et al (2005) Early inactivation of p53 tumor suppressor gene cooperating with NF1 loss induces malignant astrocytoma. Cancer Cell 8:119–130

Zhu H, Acquaviva J, Ramachandran P, Boskovitz A, Woolfenden S, Pfannl R et al (2009) Oncogenic EGFR signaling cooperates with loss of tumor suppressor gene functions in gliomagenesis. Proc Natl Acad Sci 106:2712–2716

Models to Study Glioma Cell Invasion

14

Imad Saeed Khan and Moneeb Ehtesham

Abstract

Gliomas represent the most common primary intracranial tumors. The aggressive clinical behavior of these tumors stems from the highly invasive properties of neoplastic glial cells. Fundamental to the development of an effective clinical treatment is the need to therapeutically target this invasive potential. Research into the mechanisms of invasion is imperative to better understand the various biological processes that could be targeted to treat the invading glioma cells. Such research is ideally conducted with reliable and relevant models. We describe some of the most commonly used models in glioma invasion research. Each model has its benefits and drawbacks and the nature of the experiments being conducted will help choose the type of model most suitable.

Keywords

Glioma • Invasion • In vitro models • Mathematical models • In vivo models

Abbreviations

ADAM	A disintegrin and metalloproteinase
ECM	Extracellular matrix
EGFP	Enhanced green fluorescent protein
EGFR	Epidermal growth factor receptor
ENU	N-Ethyl-N nitrosourea

I.S. Khan • M. Ehtesham (✉)
Department of Neurosurgery, Vanderbilt University Medical Center, T-4224 Medical Center North, 1161 21st Ave S, Nashville, TN 37201, USA
e-mail: moneeb.ehtesham@vanderbilt.edu

A. Sedo and R. Mentlein (eds.), *Glioma Cell Biology*,
DOI 10.1007/978-3-7091-1431-5_14, © Springer-Verlag Wien 2014

GBM	*Glioblastoma multiforme*
GFAP	Glial fibrillary acidic protein
MCS	Multicellular spheroid
MMP	Matrix metalloproteinase
MNU	Methynitrosourea
MRI	Magnetic resonance imaging
PDGF	Platelet-derived growth factor
PTEN	Phosphatase and tensin homolog deleted from chromosome 10
Rb	Retinoblastoma protein
RFP	Red fluorescent protein
VEGF	Vascular endothelial cell growth factor

14.1 Introduction

Gliomas represent the most common primary intracranial tumors. The most common and aggressive sub-type, *glioblastoma multiforme* (GBM), accounts for approximately 15,000 deaths every year in the USA alone (Demuth and Berens 2004). Despite the advent of modern diagnostic and treatment modalities, the prognosis for patients with these tumors remains dismal, with an approximate median survival of 15 months (Stupp et al. 2005). The aggressive clinical behavior of these tumors stems from the highly invasive properties of neoplastic glial cells. Fundamental to the development of an effective clinical treatment is the need to therapeutically target this invasive potential.

Standard treatment paradigms for GBM center on surgical resection followed by chemotherapy and radiation. Unfortunately, time has established that the utilization of these current modalities has been proven insufficient to effect durable cures. Decades ago Matsukado showed that more than 50 % of untreated brain tumors had already reached the contralateral hemisphere (Demuth and Berens 2004; Fukai et al. 2010). Supratentorial bilateral extension has been shown to occur because of rapid growth along myelinated structures across the corpus callosum and along the fornices toward the temporal lobes. Glioma cells have also been shown to extend around perivascular spaces.

Walter Dandy, perhaps understanding the significance and magnitude of invasiveness in these tumors, carried out hemispherectomies in patients presenting with preoperative hemiplegia (Demuth and Berens 2004). In spite of drastic surgical efforts like these by him and others, recurrences were detected in some of these patients as early as 3 months postoperatively (Bell and Karnosh 1949).

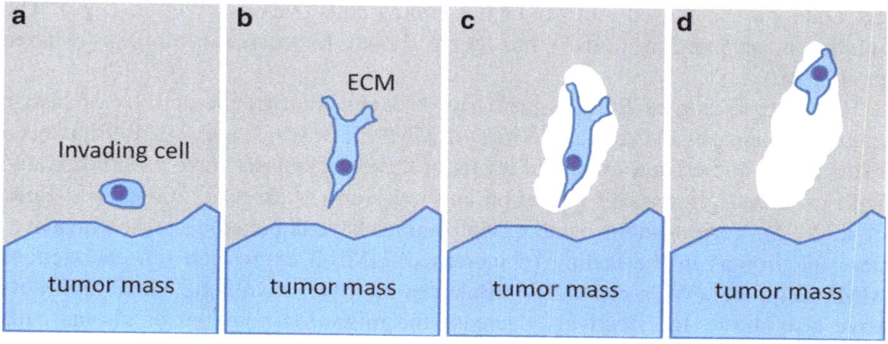

Fig. 14.1 Schematic representation of the mechanism of glioma invasion. Invasion requires four distinct steps: (**a**) detachment of invading cells from the primary tumor mass, (**b**) adhesion to ECM, (**c**) degradation of ECM, and (**d**) cell motility and contractility. (Figure from Onishi et al., 2011, Angiogenesis and invasion in glioma, Brain Tumor Pathol (2011) 28:13–24)

14.2 Molecular Mechanisms of Invasion

Glioma cell invasion is a multifactorial process (Onishi et al. 2011). The dynamic interplay between tumor cells, extracellular matrix, and adjacent cells leads to active movement of tumor cells. Physiologic cell motility is a tightly regulated process and the initiation of invasiveness in tumor cells likely stems from a loss of normal inhibitory control (Demuth and Berens 2004; Graham et al. 1994).

The initial step in invasion is the initiation of cell migration, which is achieved with a change in the shape and stiffness of the cell. To actively become motile, the cell has to become polarized, develop membrane protrusions, and form anchors via connections to the extracellular matrix. The contraction of the cytoskeleton will then advance the cell forward.

Tumor migration has been described in four steps (Onishi et al. 2011): (1) Detachment of invading cells from the primary tumor mass, (2) adhesion to the ECM, (3) degradation of the ECM, and (4) cell motility and contractility (Fig. 14.1).

Detachment of tumor cells requires destabilization and disorganization of the cadherin-mediated junctions that hold the primary tumor (Onishi et al. 2011). Concurrently there is decreased gap junction formation due to an inhibition of connexin 43 expression. There have been reports of increased malignancy in glioma specimens with decreased expression of connexin 43 and reduced in situ gap junction formation (Goodenough et al. 1996). These changes are associated with increased growth and invasion in gliomas (Demuth and Berens 2004; Soroceanu et al. 2001). This is followed by cleavage of cell surface CD44 (anchors the cell to the primary tumor mass) by the metalloproteinase ADAM (Nagano and Saya 2004).

Adherence of glial tumor cells to the ECM requires integrin expression. In particular, increased expression of integrin (alpha,v, beta,3) has been shown to correlate with increased motility of the glioma cells (Leavesley et al. 1992). The inhibition of integrin (avB3) has been shown to decrease motility (Platten et al. 2000).

The degradation of ECM is primarily brought upon by the activity of matrix metalloproteinases (MMPs). The role of MMP-2, MMP-3, and MMP-9 has been extensively studied and increased levels of these proteinases have been correlated with increased glioma cell migration and invasion (Wick et al. 2001; Wild-Bode et al. 2001). Various in vitro and in vivo studies have depicted decreased invasiveness of gliomas in the setting of increased TIMP-3 expression (an endogenous MMP inhibitor) (Wick et al. 2001; Baker et al. 1999). Synthetic MMP inhibitors have also shown to effectively decrease the migration potential of glioma cells (Tonn et al. 1999; Tonn and Goldbrunner 2003).

Glioma cell motility and contractility is brought about by cytoplasmic contraction. Myosin II is the major source of the contractile force and allows the cells to percolate through pores smaller than their nuclear diameter (Beadle et al. 2008). Recently, Ivkovic and colleagues were able to inhibit myosin II and block glioma invasion, lending weight to the hypothesis that myosin II represents a point of convergence for signal transduction pathways that drive glioma invasion and that its inhibition cannot be overcome by other motility mechanisms (Ivkovic et al. 2012).

In order to improve outcome in patients with glioma there is an urgent need to establish a fundamental understanding of the molecular mechanisms governing migration and invasion of tumor cells. The aim of this chapter is to review the various models used to study the invasion of glioma cells.

14.2.1 In Vitro Models

An experimental in vitro model needs to effectively mimic the basic characteristics of tumor cells as well as the interaction between the tumor cells and neighboring cells and ECM. In this section we will review some of the most commonly utilized in vitro assays to study migration and invasion in glioma cells.

14.2.1.1 Matrigel-Based Invasion Assays

Among various invasion assays, Matrigel-based invasion assays are the most reliable, reproducible, and representative of in vivo events (Kleinman and Jacob 2001). Matrigel is a solubilized basement membrane preparation extracted from the Engelbreth–Holm–Swarm mouse sarcoma. This tumor is rich in ECM proteins and contains all the major components of basement membranes—laminin, collagen IV, heparan sulfate proteoglycans, entactin, and nidogen, all components of basement membranes (M and Brooks 2001).

A thin layer of Matrigel is placed upon porous filters and placed in a Boyden migration chamber. This consists of multiple sets of paired chambers separated by a porous membrane. The tumor cells are placed in the upper well and a chemoattractant in the lower. The matrigel coating on the membrane acts as a natural barrier to the migration of the glioma cells. The ability of the tumor cells to attach to the ECM, invade through the ECM, and migrate through the pores replicates the crucial steps involved in in vivo cellular invasion.

The Boyden chamber is incubated from 3 to 10 h depending upon the type of tumor cells used. The filter is removed following the incubation period and the cells on the lower side of the filter are quantified. Apart from being quick, reliable, and quantitative, another advantage of the test is the ability to recover the cells that have migrated through the pores for subsequent study.

This test can be used to test various inhibitors of invasion in order to better understand the molecular mechanisms of invasion and migration. Before the Matrigel-based assay is used, it is important to optimize the setup first. The concentration of Matrigel solution needs to be high enough to impede weakly invasive cells but allow the more invasive cells through. The number of cells should also be optimized; too many cells may lead to clumping and erroneous results. Pretest optimization can be done using serial dilutions of the Matrigel solution in combination with differing numbers of tumor cells.

14.2.1.2 Spheroid Culture

Most cells in the human body interact with neighboring cells and ECM and are oriented in a 3-D fashion. These interactions with the surrounding structures are integral for normal cell physiology (Lin et al. 2008). A 2-D monolayer culture suffers from a loss of these interactions, which may lead to a less-applicable model to study invasion in tumor cells (Griffith and Swartz 2006; Pampaloni et al. 2007; Abbott 2003). Recently, there has been a push to develop 3-D cultures that mimic in vivo biological processes closely and will be able to yield more accurate invasion assays. The multicellular spheroid (MCS) model represents a valuable 3-D model for invasion studies and was first described by Holtfreter (1944) and Moscona and Moscona (1952).

The model is formed by a self-assembly process and yields spheroids that exhibit remarkable similarity to their parent tissue architecture. The relevance of 3-D culture methodology is reinforced by the significant change in RNA expression pattern seen when cells are grown in spheroids as opposed to 2-D cultures (Smalley et al. 2006). Cell architecture in MCS exhibits histological similarity to that seen in parent tissues establishing these as a good resource to study glioma cell invasiveness.

There are multiple methods of forming spheroids including liquid overlay, spinner flasks, NASA rotary system, hanging drop method, micromolding, and 3-D scaffolds, among others. The details of these methods are beyond the scope of this chapter and are reviewed elsewhere (Lin et al. 2008). It is also important to note that there are key differences between larger and smaller spheroids: larger spheroids are characterized by a central area of necrosis surrounded by a variable rim of hypoxic cells and a well-oxygenated and proliferating outer rim of cells, while smaller spheroids have a smaller hypoxic cell proportion (Lordo et al. 1987; Sutherland 1988; Sutherland et al. 1971).

Invasion is a process where the tumor cells interact with the host cells and matrix to migrate and invade through it. A confrontation model is where tumor spheroids are confronted with normal tissue spheroids and this experimentally mimics the in vivo conditions of the tumor/normal tissue interface (Lin et al. 2008). Interactions between tumor cells and dermal cells, fibroblasts (Djordjevic and Lange 2004; Seidl et al. 2002), endothelial cells (Oudar 2000; Timmins et al. 2004), and epithelial cells (Kim 2005) in a heterotypic spheroid have been used to study tumor cell invasion. Contact between cell populations is generally established on a semisolid nonadhesive agar at 37 °C overnight allowing mutual adherence. The culture period may last from 1 to 7 days. During this time the confronting spheroid pairs reorganize themselves into one spheroid. The composition of the heterotypic spheroid and degree of invasive tumor cell behavior can then be ascertained using either light or electron microscopy-based assessments (Fig. 14.2).

Investigators have described fluorescent labeling of cells prior to spheroid formation and then following the pattern of cell migration after confrontation of tumor cell spheroids with normal host tissue spheroids (Golembieski et al. 1999; Held-Feindt et al. 2010; Oxmann et al. 2008; Mohanam et al. 2002; Hattermann et al. 2011). The confrontation patterns observed may be described as invasive vs. noninvasive (De Ridder et al. 2000).

Glioma spheroids can also be embedded in collagen type I gels to study tumor cell invasion and proliferation. First described by Tamaki et al. using rat-derived C6 glioma spheroids, the collagen gel provides a suitable 3-D scaffold to test the interaction between the glioma cells and the ECM. The utility of this model lies in the fact that not only can interactions between adjoining tumor cells be altered (by using different sized spheroids) but furthermore the interaction between neoplastic cells and ECM can be modulated by modifying the components of the collagen matrix (Tamaki et al. 1997).

14.2.1.3 Organotypic Brain Slice Culture

Although the earlier described in vitro techniques to study invasion have various advantages, these models are still hampered by significant limitations as they do not account for the novel ECM configuration of the brain (Albini et al. 1987; Amar et al. 1994; Bernstein et al. 1990; Bjerkvig et al. 1986, 1990; Chicoine and Silbergeld 1995; Erkell and Schirrmacher 1988; Kramer et al. 1986; Pilkington et al. 1997). To specifically address this, organotypic coculture models have been developed (Bjerkvig et al. 1986, 1990; Pedersen et al. 1995; Engebraaten et al. 1990). Ohnishi et al. used rat brain slices obtained from the hippocampus or cortical regions of 2-day-old rats and maintained the brain slices in culture at the interface between air and the culture medium (Matsumura et al. 2000). Although this model lacks normal blood supply and immune responsiveness, it represents a valuable model to study invasion of glioma cells in situ.

Juvenile rat and mouse brain slice cultures can be kept viable for more than 2 months. Brain slice cultures can also be developed using human tissue. Jung et al. used brain tissue specimens obtained from patients undergoing temporal

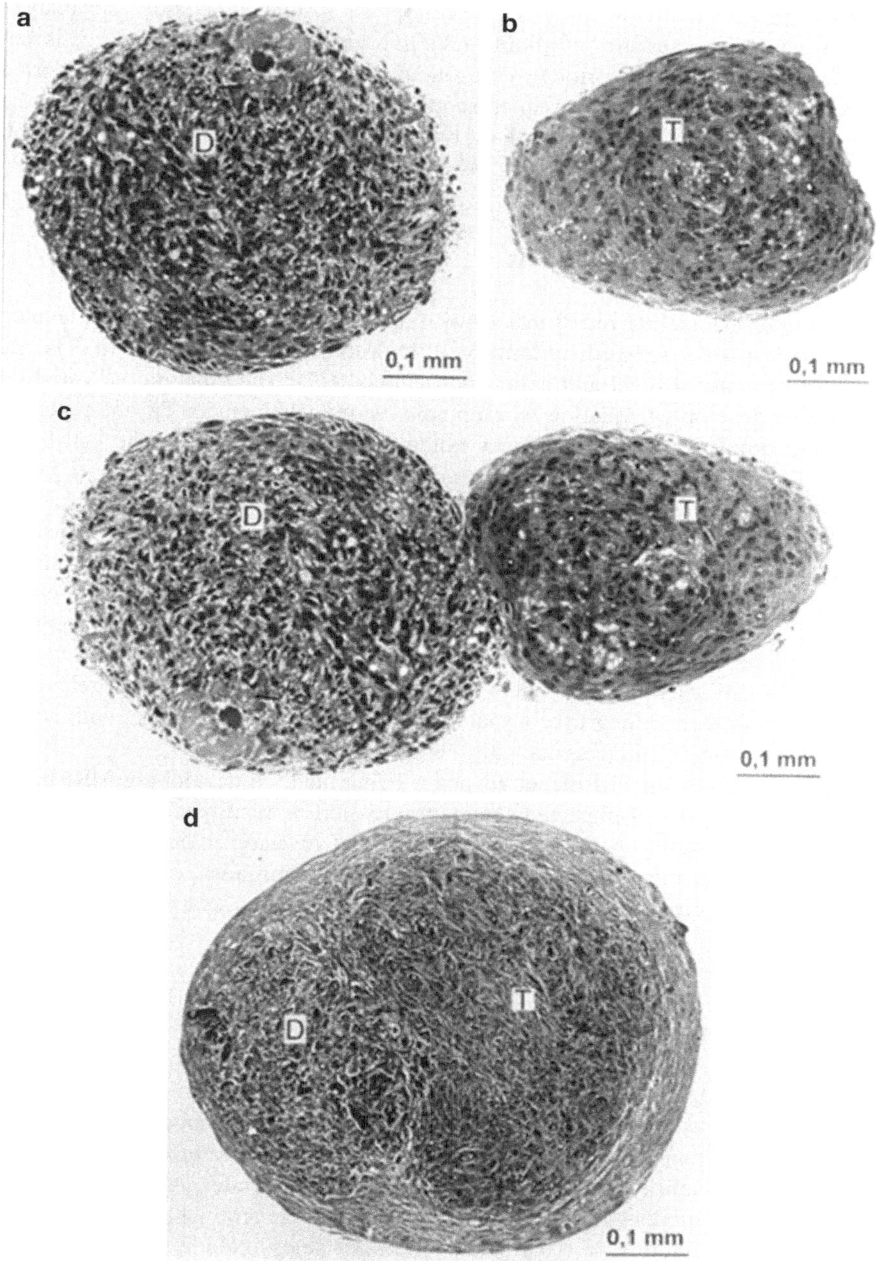

Fig. 14.2 Photomicrograph of 2-mm-thick sections from a confrontation. Autologous dermal spheroid confronted with tumor-derived spheroids originating from a meningioma. (**a**) Autologous dermal spheroid (D). (**b**) Meningioma-derived spheroid (T). (**c**) Confrontation between the two spheroids after 6 h on a semisolid medium. (**d**) Confrontation of the spheroids after 7 days in a gyratory shaker. Noninvasive pattern. (From de Ridder et al. Fig. 9, 2000, Autologous spheroid culture: a screening tool for human brain tumour invasion, Critical Reviews in Oncology: Hematology 36 (2000) 107–122)

lobectomies as a matrix to study glioma cell invasion (Jung et al. 2002). To analyze the invasion characteristics of glioma cells in brain slice matrices, tumor cells can be fluorescently labeled prior to confrontation. Matsumura et al. have reported using confocal laser scanning microscopy to spatially and serially analyze the extent of glioma cell invasion in these brain slice cultures over periods of several weeks (Matsumura et al. 2000) (Fig. 14.3).

14.2.2 Mathematical Models

Mathematical modeling represents a novel and exciting new modality to develop predictive models to study glioma cell invasion. Fundamental to this is the pioneering work of Steel and colleagues (Steel 1977). The quantitative variables utilized in developing equation-based predictive models include: Tp (the potential doubling time), Td (the actual gross volume doubling time), Tc (the cell cycle time), Ts (the time during which DNA is being synthesized within Tc), F (a factor defining the timing of Ts within Tc and the age distribution of the cells), LI (the labeling index), V (the cell loss factor, the proportion of newly formed cells that are lost), and GF (the growth fraction, the proportion of the total cells that are actually cycling) (Harpold et al. 2007). Alvord and Shaw have further added the amount of tumor left behind after surgical excision of the tumor as an additional input to further refine the mathematical assessment of tumor cell behavior (Alvord and Shaw 1991). Woodward and colleagues have demonstrated the clinical relevance of mathematical modeling by correlating their predictive assessments with actual clinical outcome data in patients with low grade gliomas.

Currently, there are significant advances being made in developing MRI-based predictive modeling of expected tumor progression or recurrence. These mathematical models define patient-specific differential invasion indices for grey and white matter and have been shown to predict tumor progression (as determined utilizing serial clinical imaging) with a high degree of accuracy (Woodward et al. 1996).

14.2.3 In Vivo Models

Animal models remain the most important experimental platform for investigating glioma cell invasion and testing the efficacy of treatment modalities. In vivo syngeneic or xenograft rodent glioma models have served as effective tools secondary to the reliability and reproducibility of their growth rates. Although multiple cell line-based models for these tumors exist, the general characteristics that constitute the suitability of a particular model for study include (Barth and Kaur 2009):

1. Should be derived from glial cells
2. It should be possible to grow and clone the glioma cells in vitro as continuous cell lines and propagate them in vivo by serial transplantation

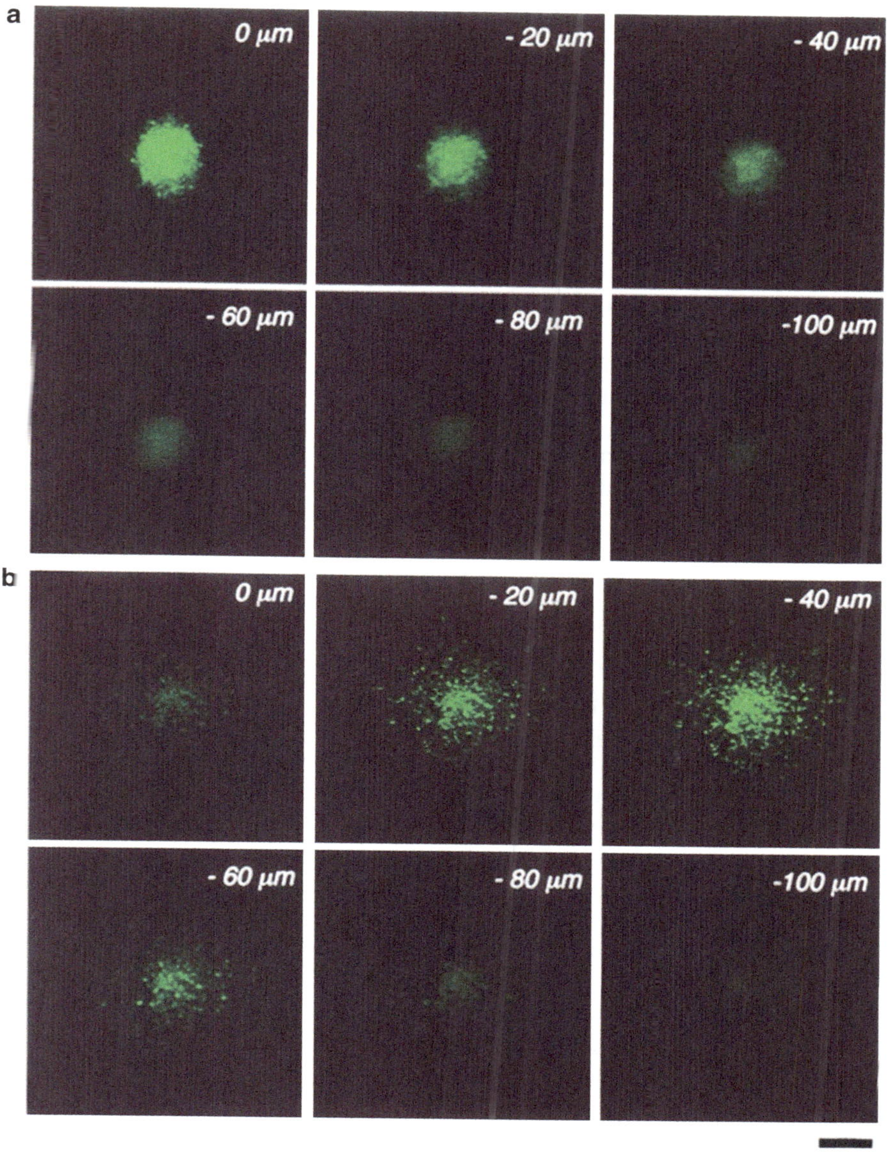

Fig. 14.3 Invasive patterns of T98G glioma cells at 24 h (**a**) and at 72 h (**b**) after the coculture of the glioma spheroid and brain slices. The migrating glioma cells, which were labeled with the PKH2 fluorescent dye, were detected under a confocal laser scanning (CLS) microscope. In each experiment, serial sections of brain slices were obtained every 20 mm downward from the basal plane (0 mm) to a deeper site (100 mm). Bar: 100 mm. [From Matsumura et al., Fig. 3, Quantitative Analysis of Glioma Cell Invasion by Confocal Laser Scanning Microscopy in a Novel Brain Slice Model, 2000, Biochemical and Biophysical Research Communications 269, 513–520 (2000)]

3. Tumor growth rates should be predictable and reproducible
4. The tumors should have glioma-like growth characteristics within the brain including neovascularization, alteration of the blood brain barrier (BBB), an invasive pattern of growth, and lack of encapsulation
5. Host survival time following tumor implantation should be of sufficient duration to permit therapy and determination of efficacy
6. The tumors should be either non or weakly immunogenic in syngeneic hosts for therapy studies
7. The tumor should not grow into the epidural space or extend beyond the brain
8. The response or lack thereof to conventional treatment should be predictive of the response in human brain tumors.

We will now present an overview of the different in vivo models used to study the different aspects of invasion of glioma cells.

14.2.3.1 Syngeneic Models

It was first reported in the 1970s that repetitive intravenous administration of nitrosourea compounds such as methynitrosourea (MNU) and N-ethyl-N-nitrosourea (ENU) produces glial-type neoplasms in immunocompetent rats. These studies lead to the development of various highly reproducible rodent tumor models without the need of the application of a toxic carcinogen to the rat brain (Lijun 2011).

Precise intracranial implantation of tumor cells is of utmost importance to get an accurate and reproducible model. Before the 1970s a freehand technique was used to implant glioma cells with imprecise results (Barth and Kaur 2009). The availability and applicability of stereotactic tumor cell implantation techniques has significantly decreased inter-animal variability thereby improving the reliability of intracranial implantation models (Barker et al. 1973). Further technique refinements such as concentrating cell samples in smaller volumes, use of 0.5–1.0 % low gelling agarose to prevent backflow of tumor cells, and rinsing of the brain surface with sterile water to destroy any extravasated cells (to eliminate spinal or extracranial dissemination) have all contributed toward improving the accuracy of in vivo intracranial glioma models (Landen et al. 2004; Kobayashi et al. 1980). Other developments include the permanent placement of metallic or plastic screws in the skull with an entry port during the initial implantation surgery. These ports then serve as avenues for future local therapeutic intervention without the need for separate stereotactic surgeries (Lal et al. 2000; Saini et al. 2004).

The C6 glioma line was produced by the repetitive administration of MNU to outbred Wistar rats over a period of approximately 8 months (Schmidek et al. 1971; Benda et al. 1968). The C6 glioma model is composed of pleomorphic population of cells with variably shaped nuclei and shows focal contiguous normal brain tissue. The cells have been reported to have a mutant p16/Cdkn2a/Ink4a locus with no expression of p16 and p19ARF mRNAs, and a wild-type p53 (Schlegel et al. 1999; Asai et al. 1994). More recent molecular characterization, which compared changes in gene expression between the C6 glioma and rat stem cell-derived astrocytes, revealed that the changes in gene expression observed in the C6 cell line were the most similar to those reported in human brain tumors (Barth and Kaur 2009; Sibenaller et al. 2005). Shen et al. using single-cell clonal analysis revealed that C6 cells also have cancer stem cell-like characteristics, including self-renewal, the potential for multi-lineage differentiation in vitro, and tumor formation in vivo (Shen et al. 2008).

One of the most widely used rat glioma models has been the 9L gliosarcoma. It was first produced in Fisher 344 rats by the intravenous injection of 5 mg/kg of MNU for 26 weeks (Schmidek et al. 1971; Benda et al. 1971). Microscopically, the tumor margins are sharply delineated with little obvious invasion into the contiguous normal brain. The 9L gliosarcoma has a mutant p53 gene but there is normal expression of p16 and p19ARF mRNAs, indicating that there is a wild-type p16/Cdkn2a/INK4a locus (Schlegel et al. 1999; Asai et al. 1994). Cancer stem cells (CSCs) have also been isolated from the 9L gliomsarcoma (Ghods et al. 2007). These cells form neurospheres in suitable chemical medium and express Nestin and Sox2. These 9L-derived CSCs form tumors in vivo that have been shown to be more aggressive compared to the parental tumor (Ghods et al. 2007). The 9L gliosarcoma model has been used in various glioma in vivo and in vitro studies (Cretu et al. 2005; Wang and Zhou 2012; Saito et al. 2004).

The RG2 glioma, classified as an anaplastic or undifferentiated glioma, was first produced by the intravenous administration of ENU (50 mg/kg body weight) to a pregnant Fischer 344 rat on the 20th day of gestation. The progeny of the rats subsequently developed the tumor (Ko et al. 1980). Due to its highly proliferative and invasive nature, it is a good representative model for GBM (Weizsäcker et al. 1982). They show increased gene expression of PDGFb, IGF-1, Ras, Erb3/HER3 precursor mRNA, and cyclin D2 (Sibenaller et al. 2005). They have also been shown to express a wild-type p53 and a concurrent loss in the expression of the p16/Cdkn2a/Ink4 gene locus (Schlegel et al. 1999). The growth and invasion of RG2 glioma has been described in rat (Krajewski et al. 1986) and chick embryo models (Cretu et al. 2005).

The F98 glioma is composed of a mixed population of spindle-shaped cells, the majority of which have fusiform nuclei, and a smaller number of polygonal cells with round to oval nuclei (Barth and Kaur 2009). This tumor shows characteristics similar to the human GBM-extensive invasion of contiguous brain tissue and islands of tumor cells at varying distances from the parent tumor, sometimes forming perivascular clusters (Barth and Kaur 2009). Like GBM these cells overexpress PDGFb, and Ras along with an increase in EGFR, cyclin D1, and cyclin D2 expression relative to rat astrocytes (Sibenaller et al. 2005). To carry out in vivo tracking of these glioma cells, Bryant et al. developed bioluminescence imaging. They transfected F98 glioma cells with the firefly luciferase gene and injected orthotopically into the brains of rats to carry out weekly imaging monitoring with high accuracy (Bryant et al. 2008). The F98 glioma cells can also be grown as xenografts in cats (Ernestus et al. 1992) but since the cells evoke an immune response, this model is of limited usefulness (Barth and Kaur 2009).

The CNS-1 glioma was derived from an inbred Lewis rat that had received weekly IV injections of MNU for 6 months (Kruse et al. 1994). It shows an infiltrative pattern of growth with leptomeningeal, perivascular, and periventricular spread and extension of the tumor into the choroid plexus (Kruse et al. 1994). Histologically, these tumors exhibited hypercellularity, nuclear atypia, and pleomorphism, and had necrotic foci. The glioma cells are arranged in a pseudopalisading pattern around the necrotic spaces (Candolfi et al. 2007). Owens et al. investigated the effect of increasing neural cell adhesion molecule (NCAM) 140 expression in a population CNS-1 glioma cells to their invasion potential (Owens et al. 1998). While there was decreased spread of NCAM 140 overexpressers, NCAM-140-overexpressing tumor exhibited a less cohesive pattern of growth near the site of tumor instillation and more individual cell infiltration of brain parenchyma with more pronounced perineuronal satellitosis (Owens et al. 1998). This suggested overexpression of NCAM 140 decreases long-range invasion but enhances local infiltration. Recently, this cell line was also used to carry out live imaging of the tumor dissemination in vivo by transfecting the glioma cells with the yellow fluorescent protein (YFP) variant Venus to enable standard one-photon and two-photon imaging (Madden et al. 2013).

14.2.3.2 Xenograft Models

Xenograft transplantation of human glioma cells in immunocompromised mice (SCID or athymic) has been an extensively used model to study glioma invasion in vivo (Lijun 2011). The benefit of such a model lies in the relative short latency period and the potential to monitor the invasive behavior of human cancer cells in brain tissue. However, the fact that most glioma cell lines are not as invasive when propagated in a murine model limits the usefulness in evaluating migration and invasion properties of glioma cells (Curtin et al. 2008). Another problem lies in the loss of some key genetic alterations (such as mutant EGFR) if the glioma cell lines are cultured in vitro (Tsurushima et al. 2007). Ozawa et al. reported that when the primary tumor cells were implanted in the murine brain without culturing them in vitro, the original genetic alterations found in the patient's tumor were maintained (Ozawa et al. 2005).

Zebrafish (*Danio rerio*) and their transparent embryos have recently emerged as promising xenograft tumor model system in glioma cell invasion studies (Yang et al. 2013). Being a vertebrate fish, Zebrafish are physiologically and genetically very similar to mammals (Jesuthasan 2002; Fishman 2001), and there is data that glioma cells interact with vessel tissues in Zebrafish embryos (Geiger et al. 2008). Other advantages include simplicity for genetic manipulation, inexpensive maintenance, easy visualization of internal structures, and rapid embryonic growth and development. Yang and colleagues described a glioma xenograft in Zebrafish embryos to study the invasion of glioma stem cells (Yang et al. 2013). To establish a stable red florescent protein (RFP)-expressing U87 cell line, they transfected the cells with a pcDNA3.1(+)-RFP vector (Fig. 14.4). The glioma cells were then injected into the middle of the embryonic yolk sac (Yang et al. 2013).

A few investigators have also used xenografts in chicken embryos to study tumor cell behavior (Chambers et al. 1990, 1991). Cretu and colleagues used established rat glioma cell lines (U-87, C6, and 9L) and injected them into the ventricles of the

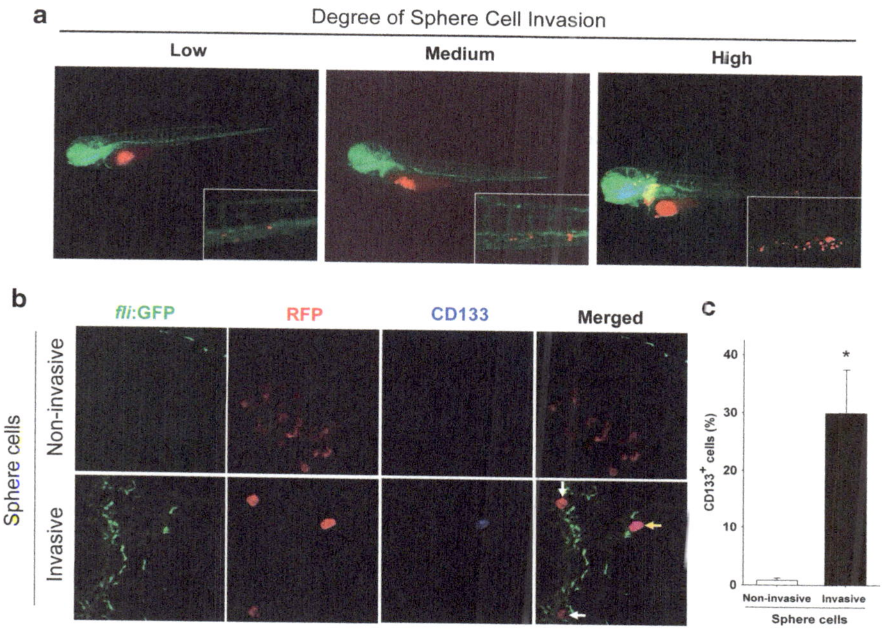

Fig. 14.4 Invasive U87 sphere cells express CD133. (**a**) U87 sphere cells with various invasion capabilities within zebrafish embryos. The extent of invasion was classified in three degrees: Low: less than 5 migrated cells; Medium: 5–20 migrated cells; High: more than 20 migrated cells. Representative images at higher magnification show the invasive RFP-labeled U87 sphere cell masses (*red*) in the tail region of the embryos via EGFP-labeled host vessels (*green*). (**b**) Detection of CD133 expression on noninvasive and invasive U87 sphere cells at 2 dpi by immunofluorescent staining. All of U87 sphere cells within injected embryos were stained with monoclonal anti-CD133 antibody (1:300) and examined by confocal microscopy. *Green*: Tg (fli1:EGFP)y1 microvessels; *red*: RFP-labeled U87 sphere cells; *blue*: CD133-positive U87 cells. (**c**) Quantitative analysis of CD133-expressing cells in noninvasive cell group ($n = 713$) and high-invasive cell group ($n = 175$) at 2 dpi. ($p = 0.001$). doi:10.1371/journal.pone.0061801.g002 (from Yang et al. 2013, Fig. 2, A Novel Zebrafish Xenotransplantation Model for Study of Glioma Stem Cell Invasion, PLoS ONE 8(4): e61801. doi:10.1371/journal.pone.0061801)

chicken embryos (Cretu et al. 2005). All the cell lines not only survived in vivo but formed well-vascularized tumors that migrated and invaded into the adjacent brain (Cretu et al. 2005).

14.2.3.3 Spontaneous Models

Genetic analyses of tissue from human brain tumors have revealed that various histopathological categories for brain tumors generally result from a limited number of mutation patterns (Sughrue et al. 2009). More recently, transgenic technology has allowed investigators to specifically target and alter the function of relevant glioma-related oncogenes and/or tumor suppressor genes within glial cells. This methodology has resulted in the development of several elegant spontaneous models of murine glioma which demonstrate a high degree of histopathologic fidelity that is not seen in the cell line-based transplantation models described earlier (Sughrue et al. 2009).

Most of the models created through this method generate a mixture of grade III and IV tumors (Rankin et al. 2012). These models utilize the Cre-Lox recombination system, which involves the targeting of specific sequences of DNA and splicing them with the help of an enzyme called Cre recombinase. The Cre recombinase system using an Adenovirus vector for efficient gene activation has been developed recently and is targeted to glioma-specific promoters: myelin basic protein (MBP), glial fibrillary acidic protein (GFAP), and nestin (Hardy et al. 1997; Maeda et al. 2006).

The combined deletion of *PTEN* and *p53* with or without Rb deletion, for example, generates gliomas with a long latency to tumor formation (Rankin et al. 2012). Zheng and colleagues showed that mice with dual, but not singular, inactivation of CNS *p53* and *PTEN* developed malignant gliomas with short latency that was remarkably similar to human disease both histopathologically and molecularly (Zheng et al. 2008) (Fig. 14.5).

Activated EGFR on the other hand has been shown to drive astrocytoma formation in some cases (Bachoo et al. 2002; Endersby et al. 2011; Wei et al. 2006), and oligodendroglioma in others (Ding et al. 2003; Weiss et al. 2003). The fact that the resulting tumors acquire somatic tumors similar to the spectrum of copy number alterations points toward similar selective pressures driving the development of gliomas in man and mouse (Rankin et al. 2012).

The specific method used to generate glioma models has also been shown to produce different results. The combined deletion of *PTEN* and *p53* consistently generates high-grade gliomas in multiple models (Zheng et al. 2008; Chow et al. 2011; Jacques et al. 2010). The additional deletion of *Rb*-mediated by *GFAP*-cre generated high-grade astrocytoma with shortened latency (Chow et al. 2011). Interestingly, the same combined deletion of *PTEN*, *p53*, and *Rb* induced via adeno-cre injection to the subventricular zone exclusively generated primitive neuroectodermal tumors (PNETs) instead of astrocytomas (Jacques et al. 2010).

Other models in this category include the spontaneous GBM in brachycephalic dog breeds, such as Boston terriers and Boxers (Heidner et al. 1991). These breeds are susceptible to develop spontaneous GBM (Stoica et al. 2004) and portray the same histopathological characteristics of the human disease, including necrosis with pseudopalizading, neovascularization, and endothelial proliferation (Rong et al. 2006). More importantly, the highly invasive GBM in these dogs exhibit the classical patterns of glioma invasion (Lijun 2011).

Canine GBM J3T and W&W cell lines allow in vitro experimentation as well. Stoica et al., have recently reported the characterization of cancer stem cells from a GBM in a Boxer (Stoica et al. 2009). These cells exhibit cancer stem markers, high proliferative rate, ability of self-renewal and differentiation. In vitro they form neurospheres and in vivo grow form GBMs that exhibit histopathological features of parent (dog) GBM (Stoica et al. 2009).

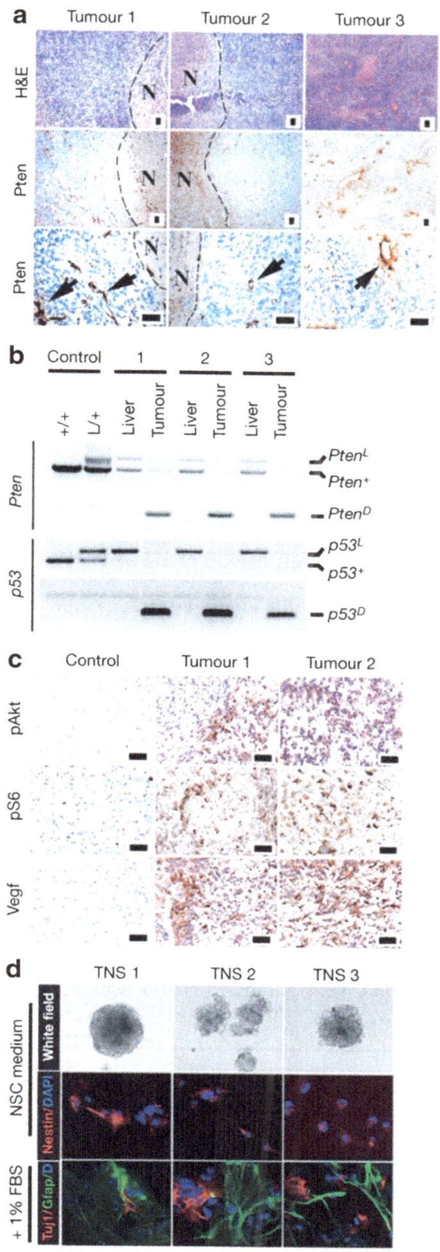

Fig. 14.5 hGFAP-Cre+; p53lox/lox; Ptenlox/+ gliomas mirror key features of human malignant gliomas. (**a**) PTEN expression is completely extinguished in tumor cells. Sections of three independent malignant gliomas were stained with hematoxylin and eosin (H&E) and an anti-PTEN antibody. 'N' indicates the adjacent normal regions of the tumor cells; the *arrows* point to

Conclusion

The insidious invasive nature of high-grade gliomas makes cure almost impossible in patients. Research into the mechanisms of invasion will provide further insight into the various biological processes that could be targeted to treat the invading glioma cells. It is imperative to conduct such research with reliable and relevant models. We have described some of the most commonly used models in glioma invasion research. Each model has its benefits and drawbacks and the nature of the experiments being conducted will help choose the type of model most suitable.

References

Abbott A (2003) Cell culture: biology's new dimension. Nature 424(6951):870–872

Albini A, Iwamoto Y, Kleinman HK, Martin GR, Aaronson SA, Kozlowski JM et al (1987) A rapid in vitro assay for quantitating the invasive potential of tumor cells. Cancer Res 47 (12):3239–3245

Alvord EC Jr, Shaw CM (1991) Neoplasms affecting the nervous system in the elderly. In: Duckett S (ed) The pathology of the aging human nervous system. Lea and Febiger, Philadelphia, PA

Amar AP, DeArmond SJ, Spencer DR, Coopersmith PF, Ramos DM, Rosenblum ML (1994) Development of an in vitro extracellular matrix assay for studies of brain tumor cell invasion. J Neurooncol 20(1):1–15

Asai A, Miyagi Y, Sugiyama A, Gamanuma M, Hong SH, Takamoto S et al (1994) Negative effects of wild-type p53 and s-Myc on cellular growth and tumorigenicity of glioma cells Implication of the tumor suppressor genes for gene therapy. J Neurooncol 19(3):259–268

Bachoo RM, Maher EA, Ligon KL, Sharpless NE, Chan SS, You MJ et al (2002) Epidermal growth factor receptor and Ink4a/Arf: convergent mechanisms governing terminal differentiation and transformation along the neural stem cell to astrocyte axis. Cancer Cell 1(3):269–277

Baker AH, George SJ, Zaltsman AB, Murphy G, Newby AC (1999) Inhibition of invasion and induction of apoptotic cell death of cancer cell lines by overexpression of TIMP-3. Br J Cancer 79(9–10):1347–1355

Barker M, Hoshino T, Gurcay O, Wilson CB, Nielsen SL, Downie R et al (1973) Development of an animal brain tumor model and its response to therapy with 1,3-bis(2-chloroethyl)-1-nitrosourea. Cancer Res 33(5):976–986

Barth RF, Kaur B (2009) Rat brain tumor models in experimental neuro-oncology: the C6, 9L, T9, RG2, F98, BT4C, RT-2 and CNS-1 gliomas. J Neurooncol 94(3):299–312

Fig. 14.5 (continued) PTEN-positive vascular cells embedded in the tumor. (**b**) The wild-type PTEN allele is lost in glioma cells. Genomic DNA isolated from liver tissues and brain tumor cells was subjected to PCR-based assays for genotyping *PTEN* and *p53* alleles. "+" designates the *PTEN* wild-type allele, "L" denotes the conditional allele, and "D" denotes the inactivated form of the conditional allele after Cre-mediated recombination. (**c**) Immunohistochemical staining of mouse normal brain or glioma sections with antibodies against activated phosphorylated Akt (pAkt), phosphorylated ribosomal protein S6 kinase (pS6), and VEGF. (**d**) TNS lines isolated from independent malignant gliomas were cultured in NSC medium or differentiation medium [1 % fetal bovine serum (FBS)] and immunostained for Nestin, GFAP, and Tuj1 as indicated. Scale bars, 50 μm (**a**, **c**); original magnification in (**d**) ×400. (Figure 2 from Zheng et al. 2008, p53 and Pten control neural and glioma stem/progenitor cell renewal and differentiation, Nature 455, 1129–1133(23 October 2008) doi:10.1038/nature07443)

Beadle C, Assanah MC, Monzo P, Vallee R, Rosenfeld SS, Canoll P (2008) The role of myosin II in glioma invasion of the brain. Mol Biol Cell 19(8):3357–3368

Bell E Jr, Karnosh LJ (1949) Cerebral hemispherectomy; report of a case 10 years after operation. J Neurosurg 6(4):285–293

Benda P, Lightbody J, Sato G, Levine L, Sweet W (1968) Differentiated rat glial cell strain in tissue culture. Science 161(3839):370–371

Benda P, Someda K, Messer J, Sweet WH (1971) Morphological and immunochemical studies of rat glial tumors and clonal strains propagated in culture. J Neurosurg 34(3):310–323

Bernstein JJ, Goldberg WJ, Laws ER Jr, Conger D, Morreale V, Wood LR (1990) C6 glioma cell invasion and migration of rat brain after neural homografting: ultrastructure. Neurosurgery 26 (4):622–628

Bjerkvig R, Laerum OD, Mella O (1986) Glioma cell interactions with fetal rat brain aggregates in vitro and with brain tissue in vivo. Cancer Res 46(8):4071–4079

Bjerkvig R, Tønnesen A, Laerum OD, Backlund EO (1990) Multicellular tumor spheroids from human gliomas maintained in organ culture. J Neurosurg 72(3):463–475

Bryant MJ, Chuah TL, Luff J, Lavin MF, Walker DG (2008) A novel rat model for glioblastoma multiforme using a bioluminescent F98 cell line. J Clin Neurosci 15(5):545–551

Candolfi M, Curtin JF, Nichols WS, Muhammad AG, King GD, Pluhar GE et al (2007) Intracranial glioblastoma models in preclinical neuro-oncology: neuropathological characterization and tumor progression. J Neurooncol 85(2):133–148

Chambers AF, Wilson SM, Tuck AB, Denhardt GH, Cairncross JG (1990) Comparison of metastatic properties of a variety of mouse, rat, and human cells in assays in nude mice and chick embryos. In Vivo 4(4):215–219

Chambers AF, Denhardt GH, Wilson SM, Cairncross JG (1991) Malignant properties of P635 glioma are independent of glial fibrillary acidic protein expression. J Neurooncol 11(1):43–48

Chicoine MR, Silbergeld DL (1995) Invading C6 glioma cells maintaining tumorigenicity. J Neurosurg 83(4):665–671

Chow LML, Endersby R, Zhu X, Rankin S, Qu C, Zhang J et al (2011) Cooperativity within and among Pten, p53, and Rb pathways induces high-grade astrocytoma in adult brain. Cancer Cell 19(3):305–316

Cretu A, Fotos JS, Little BW, Galileo DS (2005) Human and rat glioma growth, invasion, and vascularization in a novel chick embryo brain tumor model. Clin Exp Metastasis 22(3):225–236

Curtin JF, Candolfi M, Xiong W, Lowenstein PR, Castro MG (2008) Turning the gene tap off; implications of regulating gene expression for cancer therapeutics. Mol Cancer Ther 7(3):439–448

De Ridder L, Cornelissen M, de Ridder D (2000) Autologous spheroid culture: a screening tool for human brain tumour invasion. Crit Rev Oncol Hematol 36(2–3):107–122

Demuth T, Berens ME (2004) Molecular mechanisms of glioma cell migration and invasion. J Neurooncol 70(2):217–228

Ding H, Shannon P, Lau N, Wu X, Roncari L, Baldwin RL et al (2003) Oligodendrogliomas result from the expression of an activated mutant epidermal growth factor receptor in a RAS transgenic mouse astrocytoma model. Cancer Res 63(5):1106–1113

Djordjevic B, Lange CS (2004) Hybrid spheroids as a tool for prediction of radiosensitivity in tumor therapy. Indian J Exp Biol 42(5):443–447

Endersby R, Zhu X, Hay N, Ellison DW, Baker SJ (2011) Nonredundant functions for Akt isoforms in astrocyte growth and gliomagenesis in an orthotopic transplantation model. Cancer Res 71(12):4106–4116

Engebraaten O, Bjerkvig R, Lund-Johansen M, Wester K, Pedersen PH, Mørk S et al (1990) Interaction between human brain tumour biopsies and fetal rat brain tissue in vitro. Acta Neuropathol (Berl) 81(2):130–140

Erkell LJ, Schirrmacher V (1988) Quantitative in vitro assay for tumor cell invasion through extracellular matrix or into protein gels. Cancer Res 48(23):6933–6937

Ernestus RI, Wilmes LJ, Hoehn-Berlage M (1992) Identification of intracranial liqor metastases of experimental stereotactically implanted brain tumors by the tumor-selective MRI contrast agent MnTPPS. Clin Exp Metastasis 10(5):345–350

Fishman MC (2001) Genomics. Zebrafish–the canonical vertebrate. Science 294(5545):1290–1291

Fukai J, Nohgawa M, Uematsu Y, Itakura T, Kamei I (2010) Immunoglobulin D multiple myeloma involving the sella manifesting as oculomotor palsy: case report. Neurosurgery 67(2):E505–E506

Geiger GA, Fu W, Kao GD (2008) Temozolomide-mediated radiosensitization of human glioma cells in a zebrafish embryonic system. Cancer Res 68(9):3396–3404

Ghods AJ, Irvin D, Liu G, Yuan X, Abdulkadir IR, Tunici P et al (2007) Spheres isolated from 9L gliosarcoma rat cell line possess chemoresistant and aggressive cancer stem-like cells. Stem Cells 25(7):1645–1653

Golembieski WA, Ge S, Nelson K, Mikkelsen T, Rempel SA (1999) Increased SPARC expression promotes U87 glioblastoma invasion in vitro. Int J Dev Neurosci 17(5–6):463–472

Goodenough DA, Goliger JA, Paul DL (1996) Connexins, connexons, and intercellular communication. Annu Rev Biochem 65:475–502

Graham CH, Connelly I, MacDougall JR, Kerbel RS, Stetler-Stevenson WG, Lala PK (1994) Resistance of malignant trophoblast cells to both the anti-proliferative and anti-invasive effects of transforming growth factor-beta. Exp Cell Res 214(1):93–99

Griffith LG, Swartz MA (2006) Capturing complex 3D tissue physiology in vitro. Nat Rev Mol Cell Biol 7(3):211–224

Hardy S, Kitamura M, Harris-Stansil T, Dai Y, Phipps ML (1997) Construction of adenovirus vectors through Cre-lox recombination. J Virol 71(3):1842–1849

Harpold HLP, Alvord EC Jr, Swanson KR (2007) The evolution of mathematical modeling of glioma proliferation and invasion. J Neuropathol Exp Neurol 66(1):1–9

Hattermann K, Held-Feindt J, Mentlein R (2011) Spheroid confrontation assay: a simple method to monitor the three-dimensional migration of different cell types in vitro. Ann Anat 193(3):181–184

Heidner GL, Kornegay JN, Page RL, Dodge RK, Thrall DE (1991) Analysis of survival in a retrospective study of 86 dogs with brain tumors. J Vet Intern Med 5(4):219–226

Held-Feindt J, Hattermann K, Müerköster SS, Wedderkopp H, Knerlich-Lukoschus F, Ungefroren H et al (2010) CX3CR1 promotes recruitment of human glioma-infiltrating microglia/macrophages (GIMs). Exp Cell Res 316(9):1553–1566

Holtfreter J (1944) A study of the mechanics of gastrulation. J Exp Zool 95(2):171–212

Ivkovic S, Beadle C, Noticewala S, Massey SC, Swanson KR, Toro LN et al (2012) Direct inhibition of myosin II effectively blocks glioma invasion in the presence of multiple motogens. Mol Biol Cell 23(4):533–542

Jacques TS, Swales A, Brzozowski MJ, Henriquez NV, Linehan JM, Mirzadeh Z et al (2010) Combinations of genetic mutations in the adult neural stem cell compartment determine brain tumour phenotypes. EMBO J 29(1):222–235

Jesuthasan S (2002) Genetics and development. Zebrafish in the spotlight. Science 297 (5586):1484–1485

Jung S, Kim H-W, Lee J-H, Kang S-S, Rhu H-H, Jeong Y-I et al (2002) Brain tumor invasion model system using organotypic brain-slice culture as an alternative to in vivo model. J Cancer Res Clin Oncol 128(9):469–476

Kim JB (2005) Three-dimensional tissue culture models in cancer biology. Semin Cancer Biol 15 (5):365–377

Kleinman HK, Jacob K (2001) Invasion assays. Curr Protoc Cell Biol. Editorial Board Juan S Bonifacino et al. Chapter 12:Unit 12.2

Ko L, Koestner A, Wechsler W (1980) Morphological characterization of nitrosourea-induced glioma cell lines and clones. Acta Neuropathol (Berl) 51(1):23–31

Kobayashi N, Allen N, Clendenon NR, Ko LW (1980) An improved rat brain-tumor model. J Neurosurg 53(6):808–815

Krajewski S, Kiwit JC, Wechsler W (1986) RG2 glioma growth in rat cerebellum after subdural implantation. J Neurosurg 65(2):222–229

Kramer RH, Bensch KG, Wong J (1986) Invasion of reconstituted basement membrane matrix by metastatic human tumor cells. Cancer Res 46(4 Pt 2):1980–1989

Kruse CA, Molleston MC, Parks EP, Schiltz PM, Kleinschmidt-DeMasters BK, Hickey WF (1994) A rat glioma model, CNS-1, with invasive characteristics similar to those of human gliomas: a comparison to 9L gliosarcoma. J Neurooncol 22(3):191–200

Lal S, Lacroix M, Tofilon P, Fuller GN, Sawaya R, Lang FF (2000) An implantable guide-screw system for brain tumor studies in small animals. J Neurosurg 92(2):326–333

Landen JW, Hau V, Wang M, Davis T, Ciliax B, Wainer BH et al (2004) Noscapine crosses the blood-brain barrier and inhibits glioblastoma growth. Clin Cancer Res 10(15):5187–5201

Leavesley DI, Ferguson GD, Wayner EA, Cheresh DA (1992) Requirement of the integrin beta 3 subunit for carcinoma cell spreading or migration on vitronectin and fibrinogen. J Cell Biol 117(5):1101–1107

Lijun S (2011) Animal models of glioma. Glioma – Explor. Its Biol. Pr. Relev. [Internet]. Intech. Available from: http://www.intechopen.com/books/glioma-exploring-its-biology-and-practical-relevance/animal-models-of-glioma [cited 20 Jun 2013]

Lin R-Z, Lin R-Z, Chang H-Y (2008) Recent advances in three-dimensional multicellular spheroid culture for biomedical research. Biotechnol J 3(9–10):1172–1184

Lordo CD, Stroude EC, Del Maestro RF (1987) The effects of diphenylhydantoin on murine astrocytoma radiosensitivity. J Neurooncol 5(4):339–350

M D, Brooks SA (2001) In vitro invasion assay using matrigel®. Methods Mol Med 58:61–70

Madden KS, Zettel ML, Majewska AK, Brown EB (2013) Brain tumor imaging: live imaging of glioma by two-photon microscopy. Cold Spring Harb Protoc 2013(3):231–236

Maeda M, Namikawa K, Kobayashi I, Ohba N, Takahara Y, Kadono C et al (2006) Targeted gene therapy toward astrocytoma using a Cre/loxP-based adenovirus system. Brain Res 1081(1):34–43

Matsumura H, Ohnishi T, Kanemura Y, Maruno M, Yoshimine T (2000) Quantitative analysis of glioma cell invasion by confocal laser scanning microscopy in a novel brain slice model. Biochem Biophys Res Commun 269(2):513–520

Mohanam S, Chandrasekar N, Yanamandra N, Khawar S, Mirza F, Dinh DH et al (2002) Modulation of invasive properties of human glioblastoma cells stably expressing amino-terminal fragment of urokinase-type plasminogen activator. Oncogene 21(51):7824–7830

Moscona A, Moscona H (1952) The dissociation and aggregation of cells from organ rudiments of the early chick embryo. J Anat 86(3):287–301

Nagano O, Saya H (2004) Mechanism and biological significance of CD44 cleavage. Cancer Sci 95(12):930–935

Onishi M, Ichikawa T, Kurozumi K, Date I (2011) Angiogenesis and invasion in glioma. Brain Tumor Pathol 28(1):13–24

Oudar O (2000) Spheroids: relation between tumour and endothelial cells. Crit Rev Oncol Hematol 36(2–3):99–106

Owens GC, Orr EA, DeMasters BK, Muschel RJ, Berens ME, Kruse CA (1998) Overexpression of a transmembrane isoform of neural cell adhesion molecule alters the invasiveness of rat CNS-1 glioma. Cancer Res 58(9):2020–2028

Oxmann D, Held-Feindt J, Stark AM, Hattermann K, Yoneda T, Mentlein R (2008) Endoglin expression in metastatic breast cancer cells enhances their invasive phenotype. Oncogene 27 (25):3567–3575

Ozawa T, Hu JL, Hu LJ, Kong EL, Bollen AW, Lamborn KR et al (2005) Functionality of hypoxia-induced BAX expression in a human glioblastoma xenograft model. Cancer Gene Ther 12(5):449–455

Pampaloni F, Reynaud EG, Stelzer EHK (2007) The third dimension bridges the gap between cell culture and live tissue. Nat Rev Mol Cell Biol 8(10):839–845

Pedersen PH, Edvardsen K, Garcia-Cabrera I, Mahesparan R, Thorsen J, Mathisen B et al (1995) Migratory patterns of lac-z transfected human glioma cells in the rat brain. Int J Cancer 62 (6):767–771

Pilkington GJ, Bjerkvig R, De Ridder L, Kaaijk P (1997) In vitro and in vivo models for the study of brain tumour invasion. Anticancer Res 17(6B):4107–4109

Platten M, Wick W, Wild-Bode C, Aulwurm S, Dichgans J, Weller M (2000) Transforming growth factors beta(1) (TGF-beta(1)) and TGF-beta(2) promote glioma cell migration via Up-regulation of alpha(V)beta(3) integrin expression. Biochem Biophys Res Commun 268 (2):607–611

Rankin SL, Zhu G, Baker SJ (2012) Review: insights gained from modelling high-grade glioma in the mouse. Neuropathol Appl Neurobiol 38(3):254–270

Rong Y, Durden DL, Van Meir EG, Brat DJ (2006) "Pseudopalisading" necrosis in glioblastoma: a familiar morphologic feature that links vascular pathology, hypoxia, and angiogenesis. J Neuropathol Exp Neurol 65(6):529–539

Saini M, Roser F, Samii M, Bellinzona M (2004) A model for intratumoural chemotherapy in the rat brain. Acta Neurochir (Wien) 146(7):731–734

Saito R, Bringas J, Mirek H, Berger MS, Bankiewicz KS (2004) Invasive phenotype observed in 1,3-bis(2-chloroethyl)-1-nitrosourea-resistant sublines of 9L rat glioma cells: a tumor model mimicking a recurrent malignant glioma. J Neurosurg 101(5):826–831

Schlegel J, Piontek G, Kersting M, Schuermann M, Kappler R, Scherthan H et al (1999) The p16/Cdkn2a/Ink4a gene is frequently deleted in nitrosourea-induced rat glial tumors. Pathobiol J Immunopathol Mol Cell Biol 67(4):202–206

Schmidek HH, Nielsen SL, Schiller AL, Messer J (1971) Morphological studies of rat brain tumors induced by N-nitrosomethylurea. J Neurosurg 34(3):335–340

Seidl P, Huettinger R, Knuechel R, Kunz-Schughart LA (2002) Three-dimensional fibroblast-tumor cell interaction causes downregulation of RACK1 mRNA expression in breast cancer cells in vitro. Int J Cancer 102(2):129–136

Shen G, Shen F, Shi Z, Liu W, Hu W, Zheng X et al (2008) Identification of cancer stem-like cells in the C6 glioma cell line and the limitation of current identification methods. In Vitro Cell Dev Biol Anim 44(7):280–289

Sibenaller ZA, Etame AB, Ali MM, Barua M, Braun TA, Casavant TL et al (2005) Genetic characterization of commonly used glioma cell lines in the rat animal model system. Neurosurg Focus 19(4):E1

Smalley KSM, Lioni M, Herlyn M (2006) Life isn't flat: taking cancer biology to the next dimension. In Vitro Cell Dev Biol Anim 42(8–9):242–247

Soroceanu L, Manning TJ Jr, Sontheimer H (2001) Reduced expression of connexin-43 and functional gap junction coupling in human gliomas. Glia 33(2):107–117

Steel GG (1977) Growth kinetics of tumours: cell population kinetics in relation to the growth and treatment of cancer. Clarendon Press, Oxford

Stoica G, Kim H-T, Hall DG, Coates JR (2004) Morphology, immunohistochemistry, and genetic alterations in dog astrocytomas. Vet Pathol 41(1):10–19

Stoica G, Lungu G, Martini-Stoica H, Waghela S, Levine J, Smith R III (2009) Identification of cancer stem cells in dog glioblastoma. Vet Pathol 46(3):391–406

Stupp R, Mason WP, van den Bent MJ, Weller M, Fisher B, Taphoorn MJB et al (2005) Radiotherapy plus concomitant and adjuvant temozolomide for glioblastoma. N Engl J Med 352(10):987–996

Sughrue ME, Yang I, Kane AJ, Rutkowski MJ, Fang S, James CD et al (2009) Immunological considerations of modern animal models of malignant primary brain tumors. J Transl Med 7:84

Sutherland RM (1988) Cell and environment interactions in tumor microregions: the multicell spheroid model. Science 240(4849):177–184

Sutherland RM, McCredie JA, Inch WR (1971) Growth of multicell spheroids in tissue culture as a model of nodular carcinomas. J Natl Cancer Inst 46(1):113–120

Tamaki M, McDonald W, Amberger VR, Moore E, Del Maestro RF (1997) Implantation of C6 astrocytoma spheroid into collagen type I gels: invasive, proliferative, and enzymatic characterizations. J Neurosurg 87(4):602–609

Timmins NE, Dietmair S, Nielsen LK (2004) Hanging-drop multicellular spheroids as a model of tumour angiogenesis. Angiogenesis 7(2):97–103

Tonn JC, Goldbrunner R (2003) Mechanisms of glioma cell invasion. Acta Neurochir Suppl 88:163–167

Tonn JC, Kerkau S, Hanke A, Bouterfa H, Mueller JG, Wagner S et al (1999) Effect of synthetic matrix-metalloproteinase inhibitors on invasive capacity and proliferation of human malignant gliomas in vitro. Int J Cancer 80(5):764–772

Tsurushima H, Yuan X, Dillehay LE, Leong KW (2007) Radioresponsive tumor necrosis factor-related apoptosis-inducing ligand (TRAIL) gene therapy for malignant brain tumors. Cancer Gene Ther 14(8):706–716

Wang S, Zhou J (2012) Diffusion tensor magnetic resonance imaging of rat glioma models: a correlation study of MR imaging and histology. J Comput Assist Tomogr 36(6):739–744

Wei Q, Clarke L, Scheidenhelm DK, Qian B, Tong A, Sabha N et al (2006) High-grade glioma formation results from postnatal pten loss or mutant epidermal growth factor receptor expression in a transgenic mouse glioma model. Cancer Res 66(15):7429–7437

Weiss WA, Burns MJ, Hackett C, Aldape K, Hill JR, Kuriyama H et al (2003) Genetic determinants of malignancy in a mouse model for oligodendroglioma. Cancer Res 63 (7):1589–1595

Weizsäcker M, Nagamune A, Winkelströter R, Vieten H, Wechsler W (1982) Radiation and drug response of the rat glioma RG2. Eur J Cancer Clin Oncol 18(9):891–895

Wick W, Platten M, Weller M (2001) Glioma cell invasion: regulation of metalloproteinase activity by TGF-beta. J Neurooncol 53(2):177–185

Wild-Bode C, Weller M, Wick W (2001) Molecular determinants of glioma cell migration and invasion. J Neurosurg 94(6):978–984

Woodward DE, Cook J, Tracqui P, Cruywagen GC, Murray JD, Alvord EC Jr (1996) A mathematical model of glioma growth: the effect of extent of surgical resection. Cell Prolif 29 (6):269–288

Yang X-J, Cui W, Gu A, Xu C, Yu S-C, Li T-T et al (2013) A novel zebrafish xenotransplantation model for study of glioma stem cell invasion. PLoS One 8(4):e61801

Zheng H, Ying H, Yan H, Kimmelman AC, Hiller DJ, Chen A-J et al (2008) p53 and Pten control neural and glioma stem/progenitor cell renewal and differentiation. Nature 455(7216):1129–1133

Index

A. Sedo and R. Mentlein (eds.), *Glioma Cell Biology*,
DOI 10.1007/978-3-7091-1431-5, © Springer-Verlag Wien 2014

D

CPSIA information can be obtained at www.ICGtesting.com
Printed in the USA
LVOW01*1838171114

414132LV00001B/4/P